Multi-media Document Translation

ODA and the EXPRES Project

S
N
P

Jonathan Rosenberg
Bell Communications Research
Morristown, NJ 07962
USA

Mark Sherman
Ann Marks
Carnegie Mellon University
Information Technology Center
Pittsburgh, PA 15213
USA

Jaap Akkerhuis
Mt. Xinu, Inc.
Berkeley, CA 94710
USA

Printed on acid-free paper.

Camera-ready copy prepared by the authors using Scribe.
Printed and bound by Edwards Brothers, Inc., Ann Arbor, Michigan.
Printed in the United States of America.

9 8 7 6 5 4 3 2 1

ISBN 0-387-97397-4 Springer-Verlag New York Berlin Heidelberg
ISBN 3-540-97397-4 Springer-Verlag Berlin Heidelberg New York

Preface

This book discusses the work of our small research group at the Information Technology Center (ITC) at Carnegie Mellon University. We spent two years investigating the interchange of electronic multi-media documents among heterogeneous systems. This work was done as part of the the EXPRES project, which was funded by the National Science Foundation. The ITC EXPRES group wrote this book primarily to serve as a repository for all of the technical knowledge we gained during the project. The knowledge, and therefore this book, covers a wide range of experience, research and system building.

The book is composed of five parts. Part 1, which contains the majority of the prose, describes the bulk of our experiences. This part provides introductory material, such as descriptions of ODA and the general topic of multi-media document interchange. But, part 1 also includes technical details, such as the description of the conventions used by EXPRES participants to effect document interchange. Finally, part 1 ends with descriptions of our experiences with using ODA, in attempting multi-media document translation and in building several software systems.

As part of our investigations, we specified, designed and implemented two software tool kits: the CMU ODA Tool Kit, for manipulating ODA documents and the Raster Graphics Tool Kit, for handling raster images in a variety of popular formats. These tool kits are freely available, and parts 2 through 5 of the book are reference materials for these tool kits.

Because the scope and and the technical content of the book are so varied, we hope that it will appeal to a wide audience. We expect that it will be useful to people looking for a non-technical introduction to ODA, multi-media document architectures or document interchange, as well as to people looking for a technical introduction to these topics. In addition, the wealth of detailed technical information should make the book appealing to people involved in the design of document interchange standards or multi-media document systems.

Furthermore, many parts of the book are not deeply technical in nature and should be suitable for managers considering the choice of a multi-media document architecture or format for some application.

Finally, there is enough reference material so that this book will be valuable to programmers constructing ODA applications using our tool kits.

The end of the first chapter discusses in some length the organization of this book and provides several reading strategies to please virtually any reader.

Acknowledgements

Although we acknowledge many people and institutions in this section, the opinions expressed in this book are solely those of the authors and not of our funding agency, place of employment or other participants.

The EXPRES project would not have come about without the efforts of its NSF program director, Alvin Thaler, and his administrative assistant, Don Mitchell. They shepherded the project from its birth and ran cover for us so that we were free to pursue our work.

And speaking of air cover, we must thank Jim Morris, the principal investigator for the Carnegie Mellon University (CMU) EXPRES project, who had the original idea for our participation. In addition, Jim was the initial director of the ITC, and is, therefore, indirectly responsible for our ability to write this book.

We also extend a special thanks to Barry Silverman, who helped steer the EXPRES effort through some troubling times.

This is also the time to express our thanks to the entire ITC and to its funders, IBM. Both organizations have provided an environment in which it was possible to pursue this work and to take the time to write this book. In particular, we would like to thank Al Spector, the ITC director after Jim Morris, and John Howard, the current director of the ITC, who allowed us the time and resources necessary to produce this book.

Other people at CMU have been instrumental in our work. Marvin Sirbu, of the Department of Engineering and Public Policy, acted as a consultant during the early part of the project as we investigated document interchange formats. Marvin was an early advocate for ODA and greatly influenced our decision to use the standard as our intermediate representation. Paul Crumley and Maria Wadlow, of the ITC, participated in our efforts to understand ODA and in the first design of the CMU ODA Tool Kit. Ayami Ogura read much of the book in near-final form and provided many useful comments.

The support staff at the ITC performed above and beyond the call of duty. Carolyn Councill, Kris Hutchings, Susan Straub, Dave Mattis, Lynn Brown and Debbie Cavlovich put in many hours dealing with administrative issues and in setting up machines for demonstrations.

Many pieces of the tool kits were entirely written by students who worked at the ITC during the summer. Andy Werth specified, designed, documented and implemented the Raster Graphics Tool Kit (although much of the code was taken from various freely-available bitmap packages). The validation suite was implemented and documented by Jude George, and Jim Liu was responsible for the ISODE support for ODIF. Michael Zheng, who spent almost six months at the ITC, wrote the primary CMU ODA Tool Kit ODIF support. Finally David Zuhn began the implementation of a support package for CGM, although it never was robust enough for distribution.

The other NSF-funded EXPRES group was at the Center for Information Technology Integration at the University of Michigan (UM). Initially, Frank Giuffrida was our primary point of contact for the document interchange work, and he participated in much of our early planning. The Diamond translators were designed and written by Nigel Hinds, who put in many long days (at UM and CMU).

The collaboration with McDonnell Douglas Aerospace Information Systems was made possible by Clyde Miner, and was initiated by Greg Russell. Many of the initial planning sessions involved Karen Bishop and Bob Naiditch. The Interleaf translators were designed and implemented by Bob Schwarz and Ken Coppersmith, who also participated in the EXPRES demonstrations at Santa Fe.

The EXPRES advisory committee also deserves our thanks for their technical direction and advice. We would like especially to thank Vint Cerf, who was chairman of the committee during most of the project (and, who took the time to read and comment on part of this book). The comments of members Ken King, John Gage, Nick Shelness and Barry Leiner were also valuable. And, a special thanks to Steve Zilles who, not only participated actively on the committee, but volunteered to organize the ODA workshop held in 1989 at IBM Almaden Research Center.

Bob Morris was especially helpful in providing the EXPRES project with software and technical advice from Interleaf Corporation.

Jaap Akkerhuis
Ann Marks
Jonathan Rosenberg
Mark Sherman

Carnegie Mellon University
May 24, 1990

Contents

Part 1. ODA and Document Interchange

Part 2. Introduction to the Tool Kits

Figures

Tables

Examples

1 ODA and Document Interchange

1
Introduction

This book discusses the work of a small research group that spent two years, under the auspices of the National Science Foundation, investigating the interchange of electronic multi-media documents among heterogeneous systems. Ever since the advent of word-processing systems, the world has become familiar with the notion of an *electronic document*; that is, a document that exists in digital form within a computer system. The advantage of representing a document in this manner is that the processing power of a computer can greatly ease the tedious and time-consuming tasks of editing and formatting the document.

As users became more comfortable with word-processing systems and as the power of such systems increased, people began to explore richer electronic representations of documents, including other media common in paper documents: raster images, line drawings and photographs, for example. It turned out that systems could, in fact, handle several other media with relative ease. In particular, raster images and line drawings were amenable to inclusion in electronic documents.

As word-processing systems became more widely accepted, the market grew rapidly. A variety of systems became available, with a wide range of functionality and implementations on many different host computers. Such diversity is, in general, beneficial to consumers. But, as electronic documents grew to be the standard document form for many people, this diversity became a problem. The problem was that there was little, if any, compatibility among these systems. A document created on one system was unlikely to be usable on another system, even though the systems might be quite similar in capability.

The time period that encompassed the rise of word processing also saw the spread of personal workstations, networks and electronic mail in the scientific community. These developments encouraged collaboration in document preparation among scientists at geographically dispersed locations. Unfortunately, it became clear that this was next to impossible unless all authors were using the same document processing system, or one author was willing to be the editor and formatter of the document.

It was this problem—sharing an electronic multi-media document among diverse systems—that the team at the Information Technology Center tried to solve. Our work, performed at Carnegie Mellon University, was sponsored by the National Science Foundation (NSF) as part of the EXPRES project, whose purpose was to investigate the related problem of computerizing the NSF proposal process.

1.1 The EXPRES Project

The National Science Foundation (NSF) receives approximately 37,000 proposals for research funding annually [47]. Ten copies of each proposal are submitted, each consisting of an average of 50 pages that frequently contains not just text, but images and graphics. Over half of these proposals—those that pass an internal review—are evaluated by six to eight reviewers. Responses from the reviewers are mailed back to the NSF. Over 18.5 million pieces of paper every year are manipulated by the NSF through this complex process. The costs in time, material and personnel resources are significant.

In addition, the NSF was aware of the difficulty involved in preparing and submitting a proposal. It is common for a proposal to be the joint effort of a number of researchers who are not co-located. The lack of physical proximity makes collaboration difficult, especially if several individuals author the proposal. The problem is exacerbated in the (common) case that the researchers are using several different document processing systems. The typical scenario for putting together a proposal under these circumstances involves the cutting and pasting of paper documents. Not only is this process time consuming and prone to error, but it makes the production of intermediate versions of the proposal for review difficult.

The NSF decided to attack the compound document interchange problem in the context of the NSF proposal process. In June 1986, the National Science Foundation solicited proposals for the Experimental Research in Electronic Submission (EXPRES) project. EXPRES was to focus on the electronic submission and processing of proposals to NSF, as well as to improve the ability of the nation's research community to interchange multi-media documents. In addition, EXPRES was to investigate the feasibility of electronic proposal submission to the NSF.

The awardees would be expected to perform the following activities:

- install and demonstrate prototype proposal submission systems at several sites, including the NSF,
- demonstrate the exchange of multi-media documents among several diverse systems,

- participate in standards activities relating to multi-media document interchange,

- generate the specifications necessary for other systems to interoperate with the EXPRES systems.

In addition, the NSF hoped that the prototype efforts would evolve into a seamless system allowing the interchange of multi-media documents among heterogeneous environments.

In September of 1986, the NSF made equal three-year awards to the Information Technology Center (ITC) at Carnegie Mellon University (CMU), and to the Center for Information Technology Integration (CITI) at the University of Michigan (UM).

1.1.1 The EXPRES Participants

The Information Technology Center was established in 1982 as a joint effort between CMU and IBM. The original mission of the ITC was to establish a computing infrastructure for the CMU campus that would support sophisticated educational applications. These applications were expected to include computer aided instruction (CAI), laboratory simulations, document production environments and a wealth of other programs that could revolutionize the state of higher education.

The result of the first six years' of work by the ITC, an organization of 25 to 30 computer professionals, is the *Andrew* system. The Andrew system is variously taken to identify both the hardware and software environments fostered by the ITC's work. The hardware environment consists of about 600 high-function workstations[1] running Unix, several thousand lower-function workstations[2] and a campus-wide local area network.

The Andrew software environment consists of three major systems. The *Andrew File System* presents the high-function workstations with the appearance of a large monolithic Unix file system [24]. The *Andrew Toolkit (ATK)* is a subroutine library and collection of applications that supports multi-media documents [52]. The *Andrew Message System* is a multi-media, electronic mail and bulletin board system that supports thousands of users and thousands of public bulletin boards [11]. For the purposes of this book, the Andrew File System and the Andrew Message System are not relevant. But,

---

[1]High-function workstations are those with a local disk, a local area network connection, several megabytes of primary memory and at least a million pixel bit-mapped display. In the Andrew environment these workstations include IBM, DEC and Sun workstations.

[2]These include IBM PC-compatibles and Apple Macintosh systems.

ATK is central to some of our work and is discussed in greater detail in section 5.1.1.

The Center for Information Technology Integration (CITI) at the University of Michigan (UM) is a networking research center that undertakes a variety of projects, which are funded jointly by industrial partners and UM. For the University, CITI provides a focus for internal initiatives and coordination of efforts aimed at establishing an advanced information technology environment in support of higher education.

The EXPRES project at UM is based on the Diamond multi-media system [60], which was developed under a DARPA contract by Bolt, Beranek and Newman, Incorporated[3]. Diamond is a system that allows the creation and modification of multi-media documents. Like ATK, the Diamond system runs on high-function workstations and provides the capability for dealing with a number of multi-media objects, including multi-font text, raster images, hierarchical line drawings, spread sheets, equations and voice.

While Diamond and ATK are quite similar in their capabilities, they were developed independently and are strikingly different in their underlying implementations. This provided the EXPRES project an opportunity to investigate document interchange among dissimilar systems.

The EXPRES project was managed from the Directorate of Computer, Information Science and Engineering at the NSF. In order to aid in directing and evaluating the technical progress of the ITC and CITI EXPRES teams, the NSF formed an advisory committee of experts in areas related to the goals of the project. The committee, which consisted of members drawn primarily from industry, met regularly with the EXPRES participants and NSF program manager to establish technical milestones, discuss technical issues and evaluate progress.

The funded EXPRES participants were joined for part of the project by a group at McDonnell Douglas Aerospace Information Systems. This group was charged by McDonnell Douglas Corporation (MDC) with investigating the feasibility of multi-media document interchange within the company.

MDC produces enormous quantities of multi-media documentation in response to requests for proposals and as documentation for their aircraft. MDC faced many of the same problems that EXPRES was investigating: multiple authors, multi-media documents and a heterogeneous document production environment. The group from MDC agreed to become collaborators in the EXPRES document interchange experiment by writing a set of translators. Because MDC is a heavy user of the Interleaf document production system, they committed to build a set of translators for the Interleaf document format [46].

---

[3]Variants of the Diamond system are also known by the names UM EXPRES and Slate.

Several vendors provided goods and services for the EXPRES project. Interleaf Corporation provided a variety of software support tools and technical assistance. Sun Microsystems, DEC, IBM and Apple provided workstations and software.

1.1.2 Proposal Submission

In choosing the NSF proposal development, submission and review process as the vehicle for experimentation, the NSF was hoping to:

1 improve the ability of scientific researchers to interchange multi-media documents among diverse systems (and, in particular, with NSF), and

2 investigate the requirements of a system that would allow the proposal process to be performed electronically[4].

The NSF proposal process embodies many of the aspects of scientific document interchange: geographically dispersed community, heterogeneous environments and the need for multi-media documents. Thus, if the EXPRES participants could actually demonstrate an electronic proposal process, it would have addressed many of the important issues for scientific document interchange.

There are many issues that must be addressed to improve the ability of researchers to interchange multi-media electronic documents. The tasks include improved network infra-structure, improved mail transport protocols for large multi-media documents, better multi-media editors and effective methods for interchanging multi-media documents among diverse systems. The effective transmission of proposals to the NSF is even more problematical, because of the enormous volume of electronic information arriving at a single location. Solving this task would also require a system for administering the proposals. The administration system would be responsible for extracting bookkeeping information from proposals; tracking the status of a proposal from submittal, to reviewers, back to the NSF and final response; and preparing regular summary reports on the proposals received.

This is obviously an enormous agenda for a three year project. As the participants investigated the tasks, one issue emerged as the key technical issue on which EXPRES should focus: the problem of effective interchange of multi-media documents among diverse systems. Although the other issues were ob-

---

[4]And, in the best of all worlds, to actually begin setting up that system.

viously important, and far from solved, other capable efforts were underway attacking those problems; document interchange appeared to be relatively unexplored.

1.1.3 Scientific Document Interchange

The NSF proposal process comprises development, submission and review. In looking at the proposal process, the EXPRES participants determined a set of requirements for effective scientific document interchange. These requirements are that the interchange process must provide support for multi-media documents, must preserve the processability of a document and must allow formatting information to be interchanged.

Due to the nature of the information scientific documents (including funding proposals) convey, they often contain media other than text. It is common for such documents to contain graphic images of various forms (raster graphics or geometric graphics), tables or equations[5].

When authors collaborate on the creation of an electronic document, it is usually most efficient if all authors can participate in the actual editing of the document as it evolves. Such participation is usually impossible if the authors are using different document processing systems. In these cases, one person usually assumes the role of editor and takes the responsibility for printing and distributing copies of the document as it evolves. The author collects the written comments of the coauthors and merges these changes into the document.

This process has the advantage of centralizing changes to an evolving document and works just fine for minor revisions. But, when coauthors have a large number of comments on a version of a document, this imposes an enormous burden on the editor and can slow down the process.

For this reason, the EXPRES participants believed that it was crucial to allow documents to be interchanged in their *processable* form. That is, as the document is interchanged among systems, it must be possible for each author to edit the document. This requirement turned out to be a key issue for EXPRES, and we will have much more to say about preserving processability.

Authors are particular about the way their documents are formatted. This is especially true for proposals, since so much is riding on the appraisal of the document. In fact, it is quite common for authors to debug the format of a document from its inception, much as the content is revised as the document develops. We, thus, felt that it was important that information about the format

---

[5]This is not, of course, an exclusive list. Scientific documents may also contain photographs, for example.

of a document be interchanged along with the content. Furthermore, the formatting instructions also needed to be editable.

1.2 Multi-media Document Interchange

The proposal submission process, like the scientific collaboration process, involves many different hardware and software tools. Therefore, any attempt to automate these processes must account for broad heterogeneity. Devising an interchange scheme that accounts only for a fixed set of multi-media document formats (for example, the formats in use at CMU and UM) fails to address a critical requirement: the ability of a new document system to participate in free interchange among existing systems.

The obvious technique of performing direct translations between each pair of systems is impractical, because the addition of a new system would require the construction of new translators for all of the existing systems. In order to attack this problem efficiently, the EXPRES project used a standard representation that documents are translated to and from for each system. This technique requires no modifications to the existing set of translators for the entry of a new system.

1.2.1 Choice of an Intermediate Representation

The choice of intermediate representation was an important issue because the efficacy of our work would be partially dependent on the suitability of the representation. In addition, because we were planning for success, we wanted it to be as easy as possible for others to continue along the path we had started.

For these reasons, the EXPRES participants formulated a set of requirements for the intermediate representation. We determined that the representation must:

- support several media, including multi-font structured text and raster graphics,
- be extensible for additional media,
- specify the structure of a document as well as its formatting,
- be publicly available.

The requirement that the representation support multi-media documents is obvious given the goals of the EXPRES project. Furthermore, we felt it important that our work not restrict future support for additional media in any way. We, thus, demanded that the representation be extensible, allowing the addition of new media types without doing violence to existing parts of the representation.

In an earlier section of this book, we discussed how authors insist on the ability to edit both the organization of a document and its appearance. This requirement led to the necessity that the representation have support for both structure and formatting information.

Lastly, because EXPRES was designed specifically to improve the ability of researchers to interchange documents, we were concerned with the long-term effects of our decisions. This meant that the software we used and created had to be publicly available. Not only did we require that our intermediate representation be publicly available, but we favored representations that we felt had a good chance of becoming a standard (either de facto or de jure).

1.2.1.1 Candidates for an Intermediate Representation

The EXPRES participants spent several months investigating alternative intermediate representations. Because designing a new format was outlawed by our requirements, we considered only existing or evolving formats. The serious contenders emerged quickly: the Standard Generalized Markup Language (SGML) [33], Digital Document Interchange Format (DDIF) [18], Document Content Architecture (DCA) [25], Rich Text Format (RTF) [6] and the Office Document Architecture (ODA) [30].

SGML is an international standard intended for the flexible markup of documents. In other words, SGML is designed principally to allow humans to annotate documents. These annotations may serve many purposes, but in particular are frequently used to delineate the structure of a document.

The problem with SGML, from our point of view, was that it has no formatting or document organization semantics. Specifically, the SGML standard defines a syntax that may be used to perform document markup, but it attaches no meaning to any use of the syntax. The power of SGML is in its freedom from meaning—this allows the designer to define his own semantics.

Although we had ruled out defining our own semantics, there was work proceeding on defining an SGML semantics for documents by the American Association of Publishers [5]. Unfortunately, this standard did not include any provisions for formatting information and so was insufficient. There is now an effort underway to define a rich logical and layout semantics for SGML, known as the Document Style and Semantics Specification Language (DSSSL) [26]. Our latest information is that the final results of this work will not be available for several years yet.

DCA was intended as a common document format for IBM word processors. We were able to eliminate quickly DCA as a viable candidate for several reasons. It has no support for non-text media and it does not support the kind of structure we required. In addition, although there was a defining document for DCA, in practice it appeared that DCA was actually defined by particular implementations. This allowed for multiple, inconsistent uses of the standard and was unacceptable for our purposes.

The problem of different interpretations of a standard by different implementations was shared by another candidate for our intermediate format: Rich Text Format (RTF). RTF was developed by a group of companies led by Microsoft as a standard for encoding formatted text and graphics to allow the transfer of documents between DOS applications and Apple Macintosh applications. RTF provides a rather rich set of structuring and formatting capabilities and seemed ideal for EXPRES. Unfortunately, like DCA, RTF was problematic in that each implementation used its own subset of the standard. This would have limited severely the utility of our efforts and so we reluctantly abandoned RTF.

In designing DDIF, DEC began with an early version ODA. Because of this, DDIF bears a strong resemblance to ODA although DDIF has diverged significantly over the years. In many ways, DDIF appeared to be the ideal candidate for the EXPRES project. DDIF has the strengths of ODA but, in addition, has much richer facilities for supporting a wide range of media. Unfortunately, at the time we began this work, the details of DDIF were not publicly available. Furthermore, DEC could not provide us with an expected date for the availability of DDIF. It was not at all clear that DDIF would be made public before the project ended; this made the standard inappropriate for our purposes.

We determined that ODA met our needs to a large extent: it supported multimedia documents, it could be extended for additional media, it contained a rich set of structuring facilities for both organization and formatting and it was about to become an international standard. We defer a detailed discussion of ODA until chapter 3.

1.2.2 Document Interchange Goals

The primary technical goal of EXPRES was to demonstrate the feasibility of interchanging processable multi-media documents among diverse systems. We determined to do this by building translators for several document production systems. Each translator would translate from a particular document format to ODA or from ODA to the document format. These programs would then be used to interchange multi-media documents.

To keep the task manageable, the EXPRES participants decided to implement translations for just two media: multi-font structured text and raster images. ODA, and many document processing systems, also support structured graphics. Although this is a popular and useful medium, we felt that the implementation of two distinct media (text and rasters) would be sufficient to demonstrate multi-media capabilities.

When considering candidate document production systems and formats on which to base translators, we formulated several criteria. We were primarily concerned that the document format allow for the media types we had targeted and provide a rich set of structuring facilities. In addition, the format had to be well-defined and available to the EXPRES participants. Finally, the document system had to run in a hardware and software environment with which some of the EXPRES participants were familiar.

The ATK system at the ITC provided a natural candidate and the ITC EXPRES participants wrote translators between ODA and the ATK format. In addition, the group at the ITC acquired a troff [51] expert about halfway through the project and so we also built a translator from ODA to troff. The CITI EXPRES participants were working with the Diamond system and implemented a pair of translators between the Diamond format and ODA. McDonnell Douglas Corporation is a heavy user of the Interleaf document processing system [46] and so the group at MDC built translators between ODA and the Interleaf ASCII document format.

In order to implement a set of cooperating translators, we found it necessary to agree on a document model and to impose a set of conventions on the use of the intermediate representation (chapter 4). The determination of a viable document model and the specification of the conventions was an important goal of EXPRES. We hoped that these specifications would be a useful long-term result of the project.

It occurred to the EXPRES participants at an early stage that it would be useful for all implementers to share a common tool kit for manipulating documents represented in ODA. For this reason, the ITC group undertook the design, specification, implementation and distribution of a portable tool kit: CMU ODA Tool Kit. We expended a significant amount of effort in the construction of this tool kit in the hopes that it, like the document model and conventions, would prove to be a piece of work that had utility beyond the life of the project.

1.3 Notational Conventions for this Book

Whenever applicable, this book uses the notation specified in ISO IS 8613 for designating entities:

- An attribute is designated by enclosing its name within double quote characters, for example, "line layout table". Capitalization is not significant within an attribute name.

- A parameter or sub-parameter of an attribute is designated by enclosing its name within double quote characters. For example, "leading edge" is a parameter of the attribute "separation". Capitalization is not significant within a parameter name.

- A distinguished value is designated by enclosing its name within matching single quote characters. For example, the sub-parameter "centreing" of the parameter "positioning rule" of the attribute "position" may have one of the distinguished values 'none', 'width', 'height' or 'both'.

Most C program examples in this book are presented in a fixed-width font with reserved words underlined:

```
/* Check type & cef fields */
if (class != CONTEXT_SPECIFIC_class_tag) {
    TKError = ASN1_FORMAT_err;
    /* Format error message here */
    return -1;
}
```

Type definitions are not presented in the order that they would be within the source code. Instead, they are ordered in a top-down manner so that, whenever possible, a type is defined after its use in a superior type. We believe that this improves the reader's chances of understanding a complex type definition.

Frequently, this book makes references within expository text to identifiers that are defined by a tool kit. In such cases, the identifiers are represented in the same fixed-width font as used for the program examples. For example, the identifier used to designate the boolean value true is denoted as `BOOL_true`.

1.4 Organization of this Book

This is a large book, covering a wide range of material related to multi-media documents and ODA. We hope, therefore, that this book will be useful to many people, including:

- people looking for a non-technical introduction to ODA, multi-media document architectures or document interchange,
- people looking for a technical introduction to these topics,
- people involved in the design of document interchange standards or multi-media document systems,
- managers considering the choice of a multi-media document architecture or format for some application,
- programmers constructing ODA applications or tool kits.

After describing the basic organization of the book, we suggest reading strategies for each of these groups of readers.

1.4.1 Overall Organization

This book is composed of five major parts. The bulk of the explanatory prose of the book is contained in part 1; parts 2 through 5 are comprised almost entirely of reference material for the CMU ODA Tool Kit and the Raster Graphics Tool Kit.

The parts of the book and a brief description of each is as follows:

- **Part 1. ODA & Document Interchange** provides an introduction to our work, an overview of ODA and multi-media document translation, details of our translation strategy, discussions of our experiences with translation and ODA, and some overall conclusions.
- **Part 2. Introduction to the Tool Kits** contains introductory reference material that is common to both tool kits, including notational conventions and a description of site-dependent requirements. This part begins the technical reference material.
- **Part 3. The CMU ODA Tool Kit** is the application programmer's interface reference manual for the CMU ODA Tool Kit.

- **Part 4. The Raster Graphics Tool Kit** is the application programmer's interface reference manual for the Raster Graphics Tool Kit.
- **Part 5. References and Appendixes** contains additional tool kit reference material.

As should be obvious, part 1 is the only part of the book likely to be of interest to non-technical people and may be read independently of the other parts. Although parts 3 and 4 are independent, part 2 is common to both and should be read first. Part 5 is strictly reference material relating to the tool kits. Each of these parts is discussed in the following sections.

1.4.2 Part 1: Introduction to ODA and Document Interchange

Part 1, which should be of interest to all readers, comprises the following chapters:

- **1. Introduction** included brief descriptions of the EXPRES project and its participants, the problems we were investigating and our goals. This chapter is non-technical.
- **2. Document Models and Interchange Fidelity** introduces our document model, including high-level notions of document architecture and document format translation. This chapter is largely non-technical.
- **3. Introduction to the Office Document Architecture** is an introduction to the ODA standard and is intended for those people who do not already have a good grasp of the fundamentals of ODA and some of its details. This chapter is somewhat technical, but requires no previous exposure to ODA.
- **4. Document Translation Using ODA** is a discussion of the EXPRES document translation strategy and is divided effectively into two parts. The first part, from the beginning through section 4.4, is largely non-technical and provides an overview of the EXPRES document model. The second part of this chapter, section 4.4.1 until the end, presents the details of our interchange conventions. This part is highly technical and demands a thorough understanding of ODA, even beyond that provided in chapter 3.
- **5. Experience Using ODA** contains our experiences using ODA as an intermediate representation for document translation. It includes brief descriptions of the document production systems for which translators

were written, the document features that we translated and our demonstrations of document translation. This is followed by a discussion of some problems we faced in performing high-fidelity document translation and a discussion of the use of ODA in this context.

- **6. Experience Implementing ODA** discusses our experience implementing the CMU ODA Tool Kit.
- **7. Conclusions** presents our overall conclusions in five areas: the EXPRES interchange requirements, document interchange, the Office Document Architecture, the two tool kits we constructed and our implementation strategies for the tool kits and the translators.

As mentioned earlier, part 1 of the book should appeal to all readers, although they will probably make use of different chapters. We recommend the following strategies:

- *For readers interested in a a non-technical introduction to ODA* read chapters 1, 3 and section 7.3 of the conclusions.
- *For readers interested in a non-technical overview of multi-media document architectures* read chapters 1, 2 and 4 (through section 4.4).
- *For readers interested in a non-technical overview of multi-media document interchange* read chapters 1, 2 and sections 7.1 and 7.2.
- *For programmers constructing ODA applications or tool kits* read chapters 1, 3 (if necessary), 6 and sections 7.3 through 7.5.
- *For readers interested in a complete overview without the implementation details* read chapters 1 through 5 and 7 (only read chapter 4 through section 4.4).

1.4.3 Part 2: Introduction to the Tool Kits

Part 2 of this book consists of reference information that is common to both tool kits:

- **8. Introduction to the Tool Kits** is a brief introduction to the CMU ODA Tool Kit and the Raster Graphics Tool Kit and discusses their relationship to each other.
- **9. Notational Conventions** discusses notational conventions used throughout the remainder of the book.

- **10. Site Dependencies** is a description of the site-dependent parts of the tool kits and of the definitions and code to be written locally.

The reader interested in either of the tool kits should read chapters 8 and 9. Chapter 10 is necessary only if the reader is actually going to be using either of the tool kits.

Installation instructions and ordering information for both tool kits can be found in appendixes B and K.

1.4.4 Part 3: The CMU ODA Tool Kit

This part consists of the application programmer's interface manual for the CMU ODA Tool Kit, and consists of the following chapters:

- **11. Introduction to the CMU ODA Tool Kit** provides a general introduction to the tool kit and its use.
- **12. Using the Tool Kit** discusses details about using the tool kit in an application: include files, global variables and linking instructions.
- **13. Types** provides details about common user-visible types in the tool kit.
- **14. Initialization and Finalization** describes the means for initializing and finalizing the tool kit.
- **15. Constituents** describes the manner in which ODA constituents are manipulated in the tool kit.
- **16. Attribute Routines** contains descriptions of the routines used for manipulating attributes.
- **17. Specific Attribute Descriptions** describes in detail the tool kit support for each ODA attribute.
- **18. Expressions** lists the data types and routines for manipulating ODA expressions: string expressions, numeric expressions, object identifier expressions and construction expressions.
- **19. Traversing an ODA Document** describes the tool kit support for iterating over parts of an ODA document.
- **20. External Representations** is a description of the tool kit's support for reading and writing ODA datastreams.

- **21. Affecting Run-time Behavior** describes routines intended to affect the run-time behavior of the tool kit.

- **22. Debugging Support** discusses the application debugging support provided by the tool kit.

Before reading any of these chapters, the reader should read part 2 of the book. In addition, programmers will find useful information in the appendixes (see section 1.4.6), especially appendix A, which contains examples of application programs that use the tool kit.

1.4.5 Part 4: The Raster Graphics Tool Kit

Part 4 contains the programmer's reference manual for the Raster Graphics Tool Kit, and comprises the following chapters:

- **23. Introduction to the Raster Graphics Tool Kit** is a general introduction to the purpose and use of the tool kit.

- **24. Tool Kit Data Structures** describes the user-visible data types in the tool kit.

- **25. Raster Graphics Tool Kit Routines** describes the user-visible routines available as part of the tool kit.

Before reading any of these chapters, the reader should read part 2 of the book. In addition, programmers will find useful information in the appendixes (see section 1.4.6), especially appendix A, which contains examples of applications programs that use the tool kit.

1.4.6 Part 5: References and Appendixes

Part 5 of the book consists of a list of references and a set of appendixes that contain reference material. Most of this reference material will be useful only to programmers who are using one of the tool kits.

The appendixes, and a brief description of each, are as follows:

- **A. Example Programs** contains several annotated example programs that use the CMU ODA Tool Kit and the Raster Graphics Tool Kit.

- **B. Building and Installing the Tool Kits** provides complete installation instructions for both tool kits.

- **C. Error Codes** lists the error codes returned by the tool kits, along with the generic description of each error code.
- **D. Current Restrictions and Bugs** describes the current bugs and known limitations of the tool kits.
- **E. C Portability Guidelines** is a listing of the restrictions we adhered to in our use of C to make the tool kits portable.
- **F. Alphabetical List of Routines** contains two alphabetized listings of all user-visible routines from both tool kits. One listing is organized by real name and the other by routine alias.
- **G. Alphabetical List of Attributes** provides alphabetical listings of all ODA attributes, specifying the section that describes each. There is a listing alphabetized by ODA name and one by CMU ODA Tool Kit identifier.
- **H. Alphabetical List of Types** is an alphabetized listing of all user-visible types from both tool kits, with defining section references.
- **I. List of Real and Alias Names** is meant for programmers involved in debugging an application (or for programmers modifying the tool kits). This appendix contains two alphabetized listings of all real name and alias name pairs in the tool kits: one list is organized by real name, the other by alias name.
- **K. Ordering Information** contains information on how to obtain the tool kits.

2
Document Models and Interchange Fidelity

This chapter provides a discussion of the document models and features we considered during the EXPRES project. We describe the structures of a document, the kinds of content a document contains, and the way in which formatting information is provided. We then examine how users edit a document, especially through the use of a style system. After examining each of these aspects of a document model, we consider briefly their impacts on document translation.

2.1 Parts of a Document Model

For our purposes, a document is composed of an organization, a collection of content and some formatting instructions. Each is considered in the following sections.

2.1.1 Document Organization

A document's *organization* consists of the way that an author structures the information without regard to format. Most parts of a document's organization are specified and provided by the author, such as title pages, chapters and sections. For example, we specified the organization of this chapter into sections and provided the text that is contained in each section.

Typically, the parts of a document are large, such as a chapter. However, much smaller parts are possible. For example, a footnote, a reference to another part of the document and a literature citation are all parts of documents.

The parts of a document are not isolated, but are related by their structure. For example, a chapter is made up of a sequence of sections, a figure is composed of a drawing and a caption, and a section heading is composed of a section number and a title.

Parts of a document can be specified by the author but generated automatically by the document system. For example, systems such as Scribe [61] will generate automatically tables of contents, indices and bibliographies. As part of automatic document generation, a document system will guarantee some relationship among the document parts. For example, the system may automatically number sections, guaranteeing that they are consecutive and use the same representation. Automatic numbering or labeling can also be provided for citations, footnotes, equations, tables and figures. Document systems may also replace a specification of a cross reference with an appropriate reference, either by section number or by page number.

2.1.2 Document Content

The parts of a document specify its organization, but the information in each part is called the document's *content.* The overwhelming majority of most documents' content consists of text. However, others kinds of content are becoming increasingly popular. Most widely-used systems, such as Microsoft Word [44] and MacWrite [35], support at least raster graphics and some kind of structured graphics. Some systems, such as Diamond, Interleaf, and the Andrew Toolkit, also provide support for tables, equations, animations and audio.

Although content itself may have structure, content usually resides within the document structure and does not contribute to it. For example, a graphic image defined with PostScript is structured by the rules specified by the PostScript language [2, 3]. However, the image does not specify any more document structure to the book in which the image is contained. Naturally, there are exceptions, especially with content whose structure can act as a substrate or composer of other content. One frequent example is a sidebar, whose purpose is to provide a *minidocument* within a document. Systems such as FullWrite Professional [7] and Interleaf provide this feature. Another example is a table as provided in the Andrew Toolkit, where a cell of the table could be a complete document part, such as a labeled figure, or a cross reference to another part of the document.

2.1.3 Document Presentation

A document's organization reflects how the author structured the information. How the document appears on a page is called its *presentation*[6]. The process of

[6] A presentation can be made on any imageable medium, such as a screen, window or microfilm. In this chapter, the word *page* is used to denote any imageable medium.

creating a presentation is called *formatting*. The process is performed within a piece of content, on a page and between pages. In this section, we give a brief discussion of each aspect of formatting.

2.1.3.1 Content Formatting

The most localized formatting is performed on content. Typically, a formatting system is given a specific area and a piece of content, and creates the necessary marks in the area to represent the content. For example, given a stream of text, a formatter would determine where line breaks and hyphenations should be inserted, how much space to insert between words, and where each character should be placed.

2.1.3.2 Page Layout

Areas within a page are allocated to different pieces of content. This allocation process is called *page layout* and can range from simple to complex. Simple systems restrict document organization and provide the ability to place only one type of content in some specific location on the whole page. For example, early batch systems allowed only paragraphs of text to be formatted sequentially on a fixed part of a page. In more complex systems, different parts of a document can be directed into different parts of several pages. An example is a newspaper layout system that permits advertising to be placed at one point, and has stories flowing into selected columns of several pages. Full-function systems also provide ways for collections of document parts to be formatted as a unit. For example, the formatting process may require that a figure's caption appear on the same page as the figure, or that a footnote must appear on the same page as its reference.

2.1.3.3 Relationships among Document Pieces and Formatting

Page layout and content formatting are related in two ways. First, *generated* content usually contains information about page layout. A trivial example is automatically generated page numbers, which are generated content that is based on the pages where the content is formatted. A similar example is footnote numbering. Page references between parts of a document are also generated based on the formatting of the document. Generated content is determined only by applying the formatting process to a document.

A second interaction between content and page layout occurs when some content will not fit in an allocated space. For example, a figure may not fit in the amount of space available on a page. In this case, the figure may be floated from one page to another as the page layout process attempts to find space for the content. In a second example, the content formatting process for the text in a paragraph may find that the space provided by the page layout process is insufficient. Resolution of the incompatibility usually involves splitting the paragraph into multiple pieces; however, this can be complex, depending on the formatting criteria used for widow and orphan elimination, the amount of acceptable character and word spacing, hyphenation rules and river elimination.

Some contents, such as graphics, can be stretched or shrunk. Thus, if a page layout process discovers extra room on a page, a graphic could be expanded; if the page layout process finds too little room, the graphic could be shrunk.

The way that formatting is done provides a wealth of differences between document systems. However, the ways that a document can be edited by a user are also important and provide further differentiation. A discussion of the editing process, and its relationship to our document model, is presented in the next section.

2.2 Document Editing

Initially, we are less concerned with the actual gestures[7] that a user makes when manipulating a document then the consequences of the gesture. However, some systems closely bind the gestures of a user to the representation of a document. Thus, we also briefly examine some relationships between editing gestures and document formats.

2.2.1 Manipulating Document Organization

Editing the document's organization involves naming the parts of an organization, constraining the possible organizations and describing the ways that the constraint can be changed. These are considered in turn.

2.2.1.1 Naming Structures

There is a wide variety among systems as to how they name the document's organizing structures. At one extreme are text editors like Emacs [21] and

---

[7] i.e., keystrokes, mouse clicks, menu selections

generic systems like thc Andrew Toolkit that impose no structure beyond a sequence of characters. A little more organization is provided by systems that give a specific list of possible structures, such as paragraphs, sections and titles. For example, Diamond contains the concept of a paragraph and Microsoft Word contains a collection of heading structures.

Newer editing systems provide an abstract structure, which does not name the parts of a document with the familiar nomenclature of *paragraph* or *section*. Instead, they provide primitives out of which these common features can be built. For example, ODA defines the parts of a document to include *composite logical objects* and SGML defines parts of a document to be *elements*. The concepts are implemented by the Multos system [10, 16], which uses the ODA definitions as a way to name document structure, and by Author-Editor [41], which uses the SGML naming system. In both systems, the concept of a *section* emerges as a special case of the more general document part.

Giving a name to a structure does not impart any meaning to it. One collection of semantics associated with a named document part is the way it can be edited. This is considered in the next section.

2.2.1.2 Constraining Structure

One of the motivations for naming (and thus distinguishing) different parts of a document is so that the relationships among pieces of structure can be defined. The usual constraint on a document is that a certain organization must be followed. For example, the document part *subsection* may only appear within the document part *section*. Another common example is that the document part *figure* is composed of a sequence of the document parts *graphic*, *number* and *caption*. Another common constraint on a document is that certain parts must always be present. For example, a business letter would always have to contain an *inside address* part and a *salutation* part.

Positional constraints can apply to generated parts of a document as well as authored parts. For example, the table of contents may always be constrained to be behind the title page while the index is constrained to be at the end of a book.

2.2.1.3 Constraining the Constraint

Providing only the abstract document parts, instead of predefined document parts, allows the user to define a class of documents that follow a certain structure. For example, one can define a class of business letters that contain a return address, a salutation, a body and a closing. Once such a structure is defined, the editing system can automatically enforce that structure. For example, ODA defines *generic structures* that constrain how a document may be constructed[8]. An ODA editor would use the generic structure to guide the editing process and ensure that a document's structure meets the imposed constraints. Similarly, SGML defines a *document type definition*, that serves the same purpose. An SGML editor, such as Author-Editor, would use this constraint to limit the ways that a user could edit a document, thereby guaranteeing acceptably structured documents.

The existence of a machine-processed, externally-specified description of a document class leads to a meta-issue: how to manipulate the description. The trivial answer is that the specification can be edited with an appropriate editor. A deeper question arises when organizational goals are embedded within the structure. For example, a company may require that its logo and phone number appear on every letter that is sent. The abstract structure specification for letters may require that both logo and phone-number document parts appear. But if an employee may change the structure specification, a letter could be produced without the required information. Similarly, the military may impose certain restrictions on the formats of sensitive information, such as the inclusion of periodically printed warnings. The editing systems used to manipulate sensitive documents should not be able to change the warning requirements. To the extent that the specification of a document's possible structure is part of a document, one may want to include the requirement that the constraints on structure may not be edited.

Few systems support constraints on editing a document's abstract structure. Typically, a particular structure is enforced by eliminating the general nature of the editor and embedding the abstract structure in the editor. For example, there are editors that will edit exactly those SGML documents that conform to the abstract document definition required by the CALS project as specified in MIL Standard 28001 [45] for technical documentation. Thus, documents manipulated by those systems will always obey the structure constraints. Another example occurs in programming-language structure editors, such as the Cornell Synthesizer [54] and Gandalf [23]. These editors will edit only the abstract structure as specified for a programming language. Thus, they will only accept and produce, say, legal Pascal programs. There is no way within the editor to change the definition of a legal Pascal program, and so there is no way for the system to produce an illegal Pascal program. Of course, there exist meta-systems for both systems that are capable of building new editors from new specifications.

The way a document can be manipulated gets more complicated when considering content. The next section discusses this topic.

---

[8]ODA's generic structures are discussed in section 3.1.3.

2.2.2 Manipulating Document Content

At the heart of most editing systems is the creation and editing of content, for that is where most of the document's information resides. Typically, there are few restrictions on the content that may appear within a document. Two restrictions that appear in practice are national character sets and media representations.

National character sets are collections of graphics that are needed for particular languages. For example, French requires characters with grave accents and German uses umlauts. Editors developed for a particular country or language normally provide the ability to express content using a national character set. However, few systems permit the user to customize a program for an arbitrary national character set. Thus, a user is constrained to use the character set provided by the editing system.

Media representations are also limited. Although some systems, like Interleaf, permit the user to create a wide variety of media, the user is usually constrained to exactly those media provided with the editor. For example, there is no way to extend the available content in Interleaf to provide animations. Some systems try to enlarge the available media. For example, programs on the Macintosh appear to accept media from any other Macintosh program. In fact, there is a minimum interchange format that programs adhere to. In the case of the Macintosh, programs must be prepared to exchange text and PICT data [9]; any other media are not, in general, interchangeable.

Users of a system may also wish to constrain the contents in a document. A simple example is that an entire document may be read-only, which does not permit any changes to the content. Another common example is a document that can only be appended, such as one used for auditing purposes.

One can envision more selective content restrictions. For example, a form or template for a business transaction may always contain the name of the company doing business. Further, the restricted content need not be invariant; it could be generated by the document system as necessary. For example, the document could specify that the date be generated automatically in a particular place in the document whenever it is edited.

Besides controlling the content, the user can also control the formatting. Some aspects of formatting control are considered in the next section.

2.2.3 Manipulating Formatting Information

It is possible to separate the presentation of a document from its organization. At one extreme, SGML defines only the organization of information without specifying its presentation. At the other extreme, the presentation and organization of a document could be related intimately. For example, early word processing systems, such as runoff [17], did not differentiate substantively between the concepts of presentation and organization.

Systems employ a variety of techniques for representing formatting information. In this section, several approaches to representing formatting information are presented. Later, we will investigate some implications of the formatting method on the fidelity of interchange.

2.2.3.1 Embedded Commands

Early systems, such as runoff, required user-visible codes within the content to designate certain formatting actions. Frequently, these codes were defined with reference to a specific formatting model that was described in terms of line breaks, paragraph breaks and page breaks[9].

Embedded-code systems originated in the days when document systems ran in batch mode on one machine and produced output intended for another machine or device (either display or printer). With the proliferation of inexpensive machines with bit-mapped displays, the concept of directly manipulating the formatting information became common.

2.2.3.2 Direct Manipulation

The embedded-code systems had the property that the editor of a document had to see the formatting commands in the content. With the ability to see the results of the formatting in the editing system, a style of editing called WYSIWYG[10] became popular. Initial versions of the technique were similar to the embedded-code systems: a user designated some piece of text to be formatted along with some formatting instructions. The system then displayed the properly formatted information.

Although the user was spared the sight of formatting codes in the content, the document system performed the same embedding in the document. Thus, initial versions of WYSIWYG systems only hid the visibility of the formatting commands; the formatting information was still embedded in the content. Some contemporary systems essentially use the same technique.

---

[9] One can argue that a paragraph is a logical entity, but early systems defined such concepts only to allow paragraph specific formatting, such as indentation, to be performed. There was little else provided for document organization.

[10] *W*hat *Y*ou *S*ee *I*s *W*hat *Y*ou *G*et

2.2.3.3 Named Manipulation

Early embedded-code systems and early WYSIWYG systems had the property that many formatting commands were low-level. For example, both systems provided separate commands to center a line, make a font bold face and increase the size of a font. Users who want to make a heading for a report used all three commands to get a large, bold, centered title.

To relieve the user from having to specify three commands repetitively, systems with macro facilities, like troff, became popular. In such systems, users could name a new kind of formatting command, such as *MakeTitle*, and associate a list of other commands to execute when the new command was selected, such as *large, bold, center*.

However, these macro expansion systems lost the structure of the command application. The interpretation of the macro resulted in the low-level commands being applied to the document, not a macro invocation being associated with the text. There is no difference in the document between applying the macro and executing the low-level commands individually. If the macro were to change, such as, a title were to be made italic instead of bold, each application of the macro would have to be found by the user and changed manually.

2.2.3.4 Style Systems

Style systems evolved to allow users to change the definitions of formatting commands and have the new definitions applied when formatting a document. This allowed users to denote abstractly that a formatting operation should be performed on some part of a document, while allowing changes in the definition of the formatting operation. Style systems can be specified with codes in batch oriented systems, such as Scribe [61] or in interactive WYSIWYG systems, such as Interleaf, Diamond and Andrew.

Style systems come in a variety of forms. To discuss the differences between style systems, we need first to define some terms for a general style system. We can then describe the common restrictions placed on the general system.

Parts of Style Systems

Style systems control the definition and use of *formatting state vectors*, *styles*, and *environments*. Each of these is defined and discussed in turn.

There is a host of formatting properties available in most document systems. These include such concepts as margins, font information, line spacing, raster densities and page orientation. This information is collected into a data structure called a *formatting state vector*. The value of this vector changes as the document is formatted. For example, as different parts of the document are formatted, the value for the left margin may be changed.

A *style* is a function that maps a formatting state vector into another formatting state vector. Most systems define styles (sometimes called *style sheets, property sheets, font deltas, rulers* or *attributes*) as a set of rules that describe how to change individual components of the formatting state vector. A trivial example of a style rule is "set the left margin to one inch."

Many systems allow the composition of styles, that is, one style may be defined in terms of another. For example, the *second-level-heading* style can be defined as "apply *first-level-heading* style and then make the font bold." Styles that are defined in terms of another are called *inherited styles*. Some systems allow more complicated calculations in their styles, such as references to current values in the formatting state vector. One such example is "set the left margin so that it is 4 inches to the left of the right margin." Styles that refer to the current value of the formatting state vector in setting new values are called *relative styles*. Styles that do not refer to the current value of the formatting state vector in setting new values are called *absolute styles*.

A style may be associated with several regions of a document, for example, with several different paragraphs. Each such region is called an *environment*, and has an associated value for a formatting state vector. The value for the formatting state vector is calculated as explained below.

The initial value of the formatting state vector is call its *root value*. As a document is formatted, the formatting system encounters environments with associated styles. The formatting system applies the style to the current value of the formatting state vector in order to produce the new value of the formatting state vector. The result is the formatting state vector associated with the environment. The part of a document associated with the environment is called the *scope* of the environment, or less precisely, the scope of the style. For completeness, a root environment is associated with the root value of the formatting state vector and has the scope of the entire document.

As specified either by the style or by the individual document system, changes to a formatting state vector made by a style have either local or global effects.

A *local* effect takes place only within the scope of the environment. The value of the formatting state vector after the scope of an environment reverts to its value before the style was applied. For example, a style called *indent* may specify a left and right margin of two inches. Let us assume that before the *indent* style was encountered, the left and right margins are one inch, and that *indent* is applied to a coming paragraph. When the environment for the paragraph is entered, the values for the left and right margins are changed to two inches, as specified by the *indent* style. When the paragraph is finished, the values of the left and right margin revert back to their values before the *indent* style was applied, that is, one inch. Previous and subsequent paragraphs are beyond the scope of the application of *indent* and will use whatever margin values are specified in the formatting state vector before *indent* was applied—only the specified paragraph will be indented.

If a *global* change is specified, then all formatting state vectors in subsequent environments of the document will be changed. For example, changes to the *page header* component of the formatting state vector are usually global—the specified page header remains in effect until a new *page header* state is specified. If our previous example had specified that the change to the left margin were global, then the value of the left margin component of the formatting state vector beyond the end of the paragraph would have remained at two inches instead of reverting back to one inch.

Restrictions on Style Systems

Our discussion has described a general, unrestricted style system. The style systems provided on contemporary document systems contain a variety of restrictions. Some of these restrictions are considered here.

The most common restriction is that environments be associated with particular parts of a document's organization. For example, only paragraphs or titles of a document may have a style applied. In the model presented before, an environment (or style application) could be applied to any contiguous region of the document. Most systems do not let a user start an environment in the middle of one paragraph and end it in the middle of another.

Another common practice is to require an environment for every part of a document's organization. For example, the system may require that every paragraph have an associated style—the system will generate an anonymous style for a paragraph if the user does not provide one. This contrasts with the general model that assumes that current values of the formatting state vector are used if no environment boundaries are specified.

Many document systems partition the features of a style system to enforce a particular kind of functionality for a class of styles. One example is partitioning the formatting state vector components so that different classes of styles manipulate different components. Systems typically provide different names for the different partitions. In some systems, for example, styles that may affect the margin and indentation components of the formatting state vector are called *property sheets* while styles that affect font characteristics are called *deltas.* Frequently, the partitioning of the formatting state vector components is used in conjunction with another partitioning: limiting the style's association with document parts. For example, property sheet styles may be applied only to document structures that are larger than a paragraph while deltas may be applied only to document structures that are smaller than a paragraph.

Style systems have really provided mini-programming languages for users to describe the formatting of their documents. Thus, style systems have provided a great deal of flexibility and functionality to document systems, but have also made the problem of interchange more difficult. In the next section, we start looking more closely at the issue of interchanging multimedia documents.

2.3 Translating Documents

When translating a document created on one system to a document that will be viewed and edited on another system, it is necessary to decide the kind of *fidelity* desired. The simplest kind of fidelity is *hardcopy fidelity*: the document should look the same when printed or imaged on both systems. The next level of fidelity is *content fidelity*. If the same abstract text, raster images and other content can be translated from one system to the other, then we have achieved content fidelity. Beyond content fidelity is *structural fidelity*, in which we are interested in retaining the organization of the document in addition to its contents. The highest level of fidelity that we define is *editing fidelity*, which allows the recipients to perform the same kinds of document manipulation that the sender could. Each of these levels of fidelity is considered in the next sections.

2.3.1 Imaging Fidelity

If the purpose of an interchange is to provide the recipient with an electronic document that can be imaged in exactly the same way as the sender intended, then one should use an image-oriented interchange medium. This is rather easy to accomplish by interchanging a *page description language* representation of the document, such as PostScript or Interpress< [63]. Assuming the recipient has an appropriate imaging engine, for example, a PostScript printer, then reproducing an image of the transmitted document is simple.

However, providing a high imaging-fidelity interchange of a document between two *document systems* is quite difficult. The use of a page description language may be inadequate, since the receiving system probably does not provide the same level of imaging functionality as the page description language. For example, it would be exceedingly difficult to translate a PostScript document into Microsoft Word such that the document displayed in Microsoft Word looked the same as the document printed by a LaserWriter interpreting the PostScript.

Fortunately, parties who wish to interchange images can usually find a common imaging representation. For example, facsimile images are in daily use and provide adequate imaging fidelity between sender and receiver for a wide variety of applications.

2.3.2 Content Fidelity

Content fidelity is provided by most document systems through their document importation and exportation facilities. This amounts to making sure that the content of the document, without regard to formatting or organizational considerations, can be moved from one system to another. Within the United States, content fidelity among word processing systems usually means ASCII text interchange. To facilitate this level of interchange, nearly every system provides the ability both to produce and accept "plain ASCII text".

Attempts to move beyond simple text raise several problems. One problem is interchange of national language text. As noted before, many countries use different character encodings to represent graphics specific to their native tongue. Interchange assuming plain ASCII text, or even the more precise ISO IS 646 format [29], will lose many national language features.

Another problem with content interchange comes from media with implied formatting semantics. Text only has to convey sequences of codes without much concern for appearance of individual characters. However, a variety of structured graphics and raster graphics forms are being incorporated into document systems. The semantics involved for graphics media are more complex than for text since by their nature, graphics have a great deal of implied imaging semantics. There is even less agreement on a common interchangeable representation for graphics (or even the general imaging model) than for text, hence it is harder to interchange a graphical content.

A third problem with content interchange comes from new media. The kinds of media that are being incorporated into systems is expanding rapidly. There are systems with equations, spread sheets, charts, clocks, voice, sound and video. The semantics of these contents, that is, what information is to be interchanged, is not yet well defined, so a representation for that information is not available. In such circumstances, trying to interchange with a high level of content fidelity is impossible.

So far, the discussion of content fidelity has been limited to user-provided content. Generated content should also be preserved in a translation. For example, a document may specify that *today's date* should be the content at a particular place in the document. Replacing the abstract *today's date* with the translation date loses part of the content and reduces the fidelity of the translation.

2.3.3 Structural Fidelity

Structural fidelity refers to the amount of a document's organization that is preserved during a transformation from one format to another. Maintaining the organization of a document involves describing the kinds of structures that can be interchanged and keeping those structures intact. For example, if a document were organized into chapters and sections, a transformation of the document into another format should preserve the chapters and sections, rather than changing the document into a sequence of paragraphs.

Moderate levels of structural fidelity are possible with today's systems. The key feature is the level of sophistication of the interoperating systems. For example, most document systems today distinguish paragraphs from lines, and many distinguish special collections of content, like footnotes and tables of contents, from the main text of a document. One observes the attempt to maintain structural fidelity in the translation process of many word processors when users are asked for coding of new line characters, whether to add line breaks where they appear on the screen, and how to merge footnotes, sidebars, tables of contents and indices during translation.

An example of preserving structural fidelity is shown in example 2-1. If one preserves some abstract notion of a paragraph, for example, a receiving system could change the window size of a display and force new wrapping of the words among displayed lines in the paragraph. In example 2-2, the conversion of the document has preserved the abstract paragraphs. Although the line wrapping is different, the words are still grouped together. Example 2-3 illustrates how structural fidelity can be lost, even with low functionality systems. The receiving system has a shorter line length than the originating system, and the distinction between paragraphs of text and formatted lines has been lost.

Unfortunately, most document systems embed knowledge about document structure. They are unable to define abstractly a new kind of structure. Therefore, they can accept only those structures that are part of their system. For example, very few systems allow a user to specify a *chapter* and *section* as parts

Input 1A should be clocked at no more than
5 Mhz in order to prevent data loss. If a higher
speed is desired, then model 40K must be
used.

Input 2A should be clocked at no more than
half the speed of 1A. This allows the latches
to store the inputs for further processing.

Example 2-1: Original Paragraphs in Document

Input 1A should be clocked at no more than 5 Mhz in order to
prevent data loss. If a higher speed is desired, then model 40K
must be used.

Input 2A should be clocked at no more than half the speed of 1A.
This allows the latches to store the inputs for further processing.

Example 2-2: Paragraphs Preserving Structure

Input 1A should be clocked at no more
than
5 Mhz in order to prevent data loss. If a
higher
speed is desired, then model 40K must be
used.

Input 2A should be clocked at no more
than
half the speed of 1A. This allows the
latches
to store the inputs for further
processing.

Example 2-3: Lost Paragraph Structure

of a document[11], though they do provide for footnotes. Therefore a document produced on a system that can define a logical chapter will lose some of its structure when translated to a target system that lacks logical chapters.

2.3.4 Editing Fidelity

The hardest kind of fidelity to obtain is *editing fidelity*, which requires that the document must be editable in a manner similar to that on the originating system. This is particularly important for EXPRES, because we are concerned with allowing collaboration on the construction of a multi-media document among people using dissimilar systems.

---

[11]Do not confuse document organization with formatting commands. Many systems do allow pieces of text to be marked as *chapter*, but they usually mean "apply the style named *chapter* to format the marked text as a chapter header" and not the organization of the document into chapters and sections.

Our most compelling example of an editing feature to be retained across translations is *style* information. However, as described in section 2.2.3.4, style systems vary greatly among document systems. The structure of styles, the components available, the equations available to define the style and the possible ranges of environments differ substantially. However style systems are represented in a document, they are the most general way of specifying formatting and editing that we are considering. Therefore, if we can preserve the information present in a style system, we will preserve the lesser amount of information present in the simpler formatting definition schemes.

Other aspects of editing fidelity that could be retained across translations are the restrictions on content, structure and editing. For example, special contents denoting a document as confidential should not be removable by a recipient system if the content can not be removed by an editor on the originating system. If the originating system restricts the organization in some way, such as insisting that a title page must be present in the document, a high-fidelity translation would also guarantee that the document retains its title page. Similarly, if the originating system does not permit meta-editing, that is, editing of the organization description or of the styles, then the receiving system should also be so constrained.

Very few systems provide the ability to limit organization, content or meta-editing. Hence, the ability to interchange such information is extremely limited. Fortunately, the needs of EXPRES do not require this fidelity of interchange. Although this chapter has provided a survey of some features of document systems and some general issues about translating those features among systems, we will be focusing on the model provided by ODA and how to use it. We start in the next chapter, where we consider, in some detail, the subset of document model features provided by ODA.

3
Introduction to the Office Document Architecture

The Office Document Architecture (ODA) [30] is designed to aid the representation and interchange of office documents such as memoranda, letters and reports. It provides comprehensive facilities for describing the structure and content of complex multi-media documents.

This chapter provides an informal introduction to the main features of ODA. There are five sections. The first gently introduces the reader to ODA and the four main ODA document structures. It uses a simple example to show how the structures are related and how documents are laid out. The second section gives a more formal overview of the complete ODA document architecture and processing model. The third section covers the different types of content allowed, and the fourth section explains sanctioned subsets of ODA called *document application profiles*. The last section explains the relationship between ODA and other standards for document description and interchange.

3.1 Informal Introduction to the ODA Document Structures

The preparation of international standards is a slow process. Conflicting requirements and vested interests add to the technical difficulties of producing a workable, coherent and clearly-defined standard. *The Office Document Architecture (ODA) and Interchange Format* [30] spent several years making its way through the various stages of the standards processes of the European Computer Manufacturers' Association (ECMA), the International Telegraph and Telephone Consultative Committee (CCITT) and the International Standards Organization (ISO) before emerging as an official ISO standard in 1988.

This chapter was written by Heather Brown at the University of Kent.

During this process, the scope and size of the standard have grown enormously. From a relatively short document in 1983, it grew to 250 pages in 1985, to 600 pages in 1987 and eventually to 750 pages for the final text in 1988. (As an additional 200 pages already exist in draft form, it is safe to predict that the 1,000-page mark will soon be reached.) The descriptions and examples given in this chapter are necessarily brief and in some cases are simplified to the extent that they are not strictly accurate. The intention is to introduce the basic principles of ODA in a gentle fashion, and then to expand a little on the more interesting features without getting lost in too many details.

ODA effectively provides a static description of a document that can be exchanged using an interchange format (ODIF or ODL). However, it specifically caters to the interchange of *editable* documents and provides an underlying architectural and processing model for documents. It is this model that is the most interesting feature of ODA and the principal subject of this chapter.

3.1.1 Introduction to the ODA Document Structures

The ODA model of a document is hierarchical and object-oriented. It is simplest to think of the model as a tree, where the shape of the tree gives the *structure* of the document and the leaves of the tree hold all the *content* of the document. This separation of structure and content is crucial as it provides an important part of the framework necessary for multi-media documents.

Attributes provide information about the objects in the document. These generally give information about the objects themselves, but they may also define relationships between objects that are not implicit in the tree structure. Attributes are introduced in this chapter wherever they are needed to illustrate important aspects of the document model, but no attempt is made here to give a full coverage.

One attribute is worth mentioning at this stage. This is the "content architecture class" attribute that is used to define the type of content associated with each leaf object. The use of this attribute, together with the separation of structure and content mentioned above, allows different types of content to co-exist within the document. At present, ODA defines three types of content: character, raster graphics and geometric graphics. Further types, such as sound, video and tables, will be added.

3.1.2 Logical and Layout Structures

In the ODA model, a document is described by two structures: the *logical* structure and the *layout* structure. The logical structure divides and subdivides the content of the document into items that mean something to the human reader or writer. These may be general items such as sections, paragraphs, diagrams and footnotes, or they may be items specific to a particular document type such as logos, prices and dictionary or catalog entries.

The layout structure, on the other hand, is concerned with the visible appearance of the document. It divides and subdivides the content into page sets, pages and rectangular areas within pages. These rectangular areas can have nested rectangular areas defined within them. Areas with nested areas defined within them are known as *frames*, while the lowest-level areas are known as *blocks*. By definition, only blocks have contents. A frame might be used to represent a column of text, for example, with nested blocks representing the areas used for individual paragraphs. Similarly, a frame could represent the area set aside for a diagram, with two nested blocks representing the diagram itself and its caption. Frames and blocks are always positioned so that their edges are parallel to the edge of the containing page.

To illustrate the two structures and to show how they relate to one another, we shall use this chapter as an example document. The main body of the chapter is made up of major divisions called *sections*, each of which has a title followed by a number of *subsections*. Figure 3-1 shows a possible logical structure for a fragment taken from the start of a new section. It contains the title of the section, the title of the first subsection and three paragraphs that comprise the first subsection.

When the document has been formatted, a corresponding layout structure exists. This associates the logical content with pages, frames and blocks. The two structures are clearly dependent on one another and come together at the level of the content. Figure 3-2 shows how the two structures might be related if we

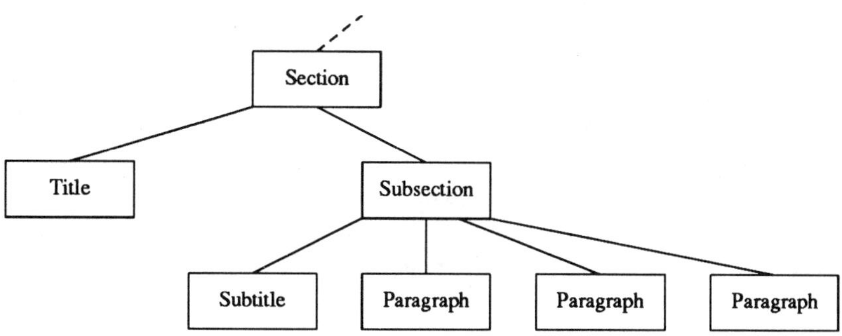

Figure 3-1. Logical Structure

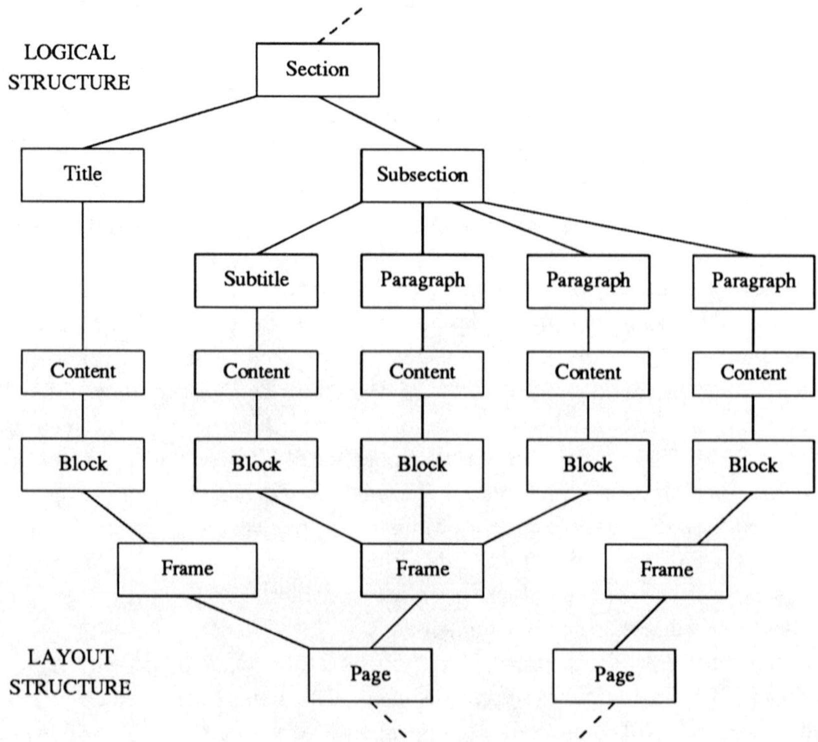

Figure 3-2. Logical and Layout Structures

assume that the section starts on a new page and that there is only room for the first two paragraphs on the page. Figure 3-3 shows how the document might look on paper.

The content of a leaf logical object will frequently correspond to the content of a block thus giving the neat one-to-one correspondence between the leaves of the logical and layout structures shown in figure 3-2. But this is not always so. A paragraph may need to be split over two pages, for example, and in this case the content for the paragraph is divided into two portions and associated with two separate blocks belonging to two different pages. Figure 3-4 shows how this might be reflected in the combined structures.

An object can have more than one content portion, but a content portion can only be associated with one logical and one layout object. As we shall see later, the content portions of two or more logical objects may be run together into a single paragraph so that they belong to a single block.

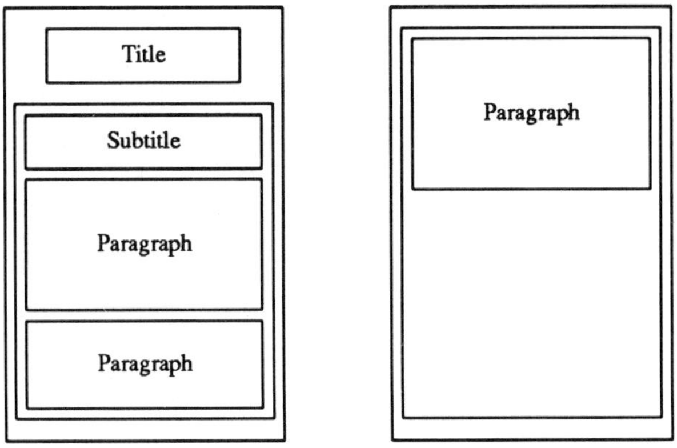

Figure 3-3. Actual Layout on Pages

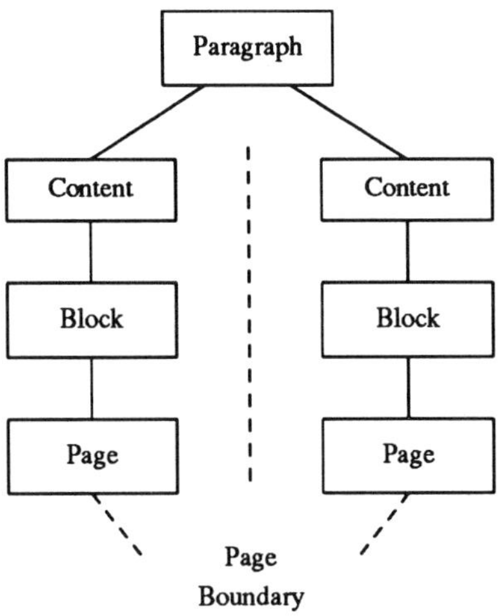

Figure 3-4. Splitting Content Over a Page Boundary

3.1.3 Specific and Generic Structures

Each document has its own *specific* logical and layout structures as described in the previous section. However, the creation of these specific document structures is guided and controlled by *generic* structures for that particular style or class of document. The generic structures are sets of object definitions (one set for logical objects and one for layout objects) that specify the types and combinations of objects allowed for that class. We introduce generic structures in this section.

The definition for each non-leaf generic object has an attribute called "generator for subordinates" that describes how the object can be made up from subordinate objects. These attributes, which may be thought of as providing a simple grammar for the document class, indicate that the occurrence of subordinates may be one of the following:

- optional (OPT): 0 or 1 occurrence,
- required (REQ): exactly 1 occurrence,
- repetitive (REP): 1 or more occurrences,
- optional and repetitive (OPT REP): 0, 1 or more occurrences.

In addition, a collection of possible subordinates may be expressed as

- a sequence (SEQ): occurring in the order specified,
- an aggregate (AGG): occurring in any order,
- a choice (CHO): only one occurs.

To show how this works in practice, we shall return to the example used in the previous section. The generic logical structure for a section of this chapter might be defined as shown in figure 3-5. This indicates that a section is made up of a mandatory title, followed by an optional author's name, followed by one or more subsections. Each subsection begins with an optional subtitle and then

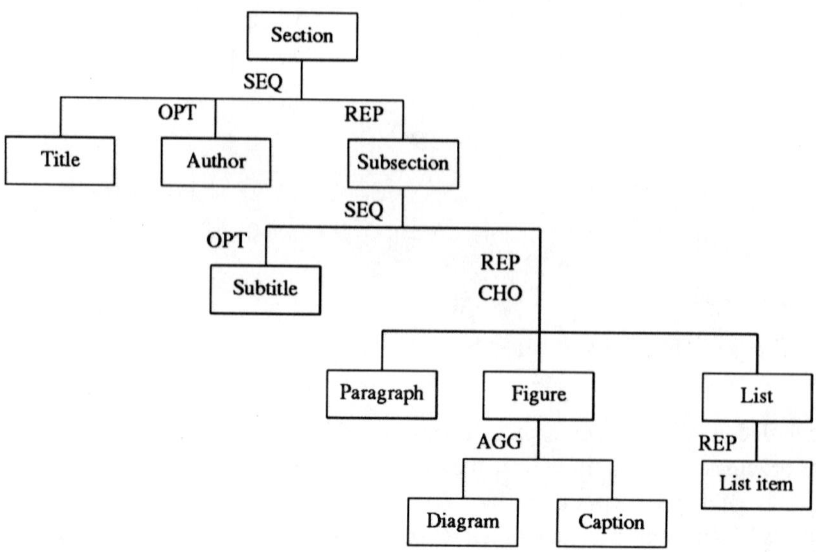

Figure 3-5. Generic Logical Structure

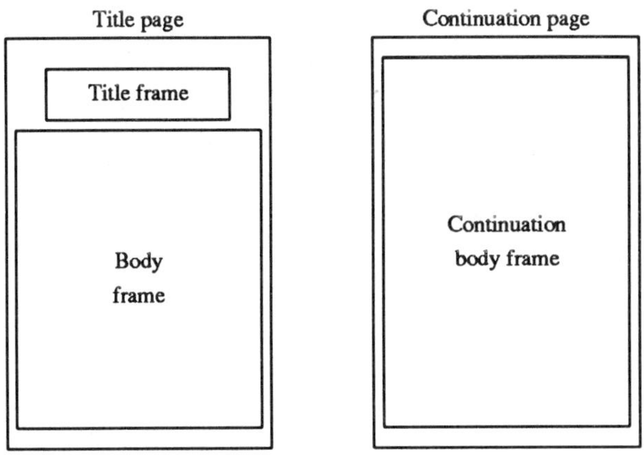

Figure 3-6. Page Styles

consists of a mixture of paragraphs, figures and lists. REP CHO indicates a series of one or more of these items occurring in any order. Figures in turn consist of a diagram and a caption (in either order), and lists consist of one or more list items.

Figure 3-1 showed a very simple example of a specific instance of this structure with no author's name and only a single subsection of three paragraphs. More complex instances could contain an author's name and several subsections containing different mixtures of paragraphs, figures and lists.

An appropriate generic layout structure corresponding to this generic logical structure is more difficult to describe without going into details of the document layout process. However, at the top level it would probably define two different page styles: a *Title page* for the first page of the section and a *Continuation page* for all subsequent pages. Figure 3-6 shows the two page styles, and figure 3-7 shows the top levels of the generic layout structure that might be used to define them.

In practice, further top level frames would be needed for running headers, and more complex chapters might need further nested frames within the body frames to provide separate areas for marginal notes or to allow for text flowing around figures.

Blocks are not included in the generic layout structure. Instead they are assigned to pages and frames during the *layout process*.

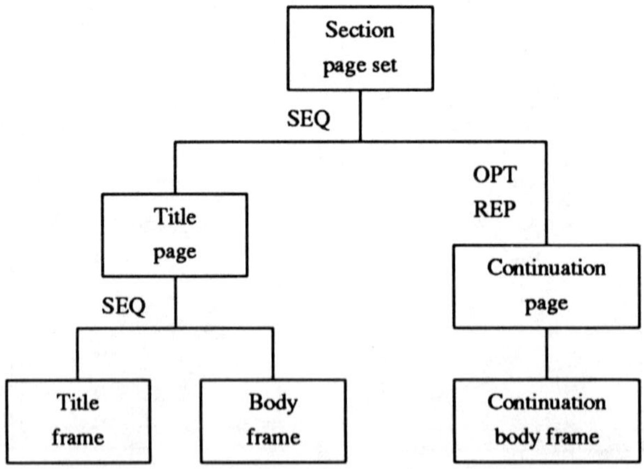

Figure 3-7. Generic Layout Structure

3.1.4 The Layout Process

The previous sections introduced the four main ODA structures: generic logical structure, specific logical structure, generic layout structure and specific layout structure. To understand how they all fit together, it is helpful to look briefly at ODA's layout process. This is what decides exactly where each item of the document is to be placed and on which page. It uses the specific logical structure, both generic structures and the content architectures to create the specific layout structure of the document. It works at two levels:

- *Content layout* takes portions of content and lays them out into blocks. This stage is dependent on the content architectures involved and on sets of attributes known as *presentation styles.*
- *Document layout* places these blocks into appropriate pages and frames. This stage is dependent on sets of attributes known as *layout styles.*

The content layout process is thus responsible for dealing with fine details like the use of particular character sets and the positioning of characters, lines, or pixels within blocks, while the document layout process decides how to place the blocks within pages and frames.

3.1.4.1 An Example of Document Layout

A crucial aspect of the document layout process is deciding what type of layout object can be used for any given logical object. Three important attributes used to make this decision are "layout object class", "layout category" and "permitted categories". Two of these, "layout object class" and "layout category", apply to logical objects and are part of the sets of attributes known as the layout styles. The third, "permitted categories", applies to frames.

"Layout object class" is used to indicate that an entire logical object (with all its subordinates) must be laid out in a single instance of a layout object derived from a given layout definition. No other part of the document may share the same layout object. "Layout object class" is typically used:

- to direct a major logical division of a document into its own page or page set,
- to ensure that a particular type of frame is generated to hold a logical object.

In our example, a logical section could have its "layout object class" defined as *Section page set.* Each section would then be laid out in a separate instance of the page set, with a single title page and as many continuation pages as necessary. The start of a new section would automatically signal the need for a new page set.

The two attributes "layout category" and "permitted categories" are provided to associate leaf logical objects with frames. "Layout category" associates a name with a leaf logical object. The contents of the object can then only be laid out in a frame that has the same name specified as one of its "permitted categories". A frame may have several different permitted categories, so objects from several different layout categories may appear within a single frame.

In our example, we shall keep a simple relationship between leaf logical objects and frames, and use just three layout categories called *Head*, *Body* and *Picture*. Figure 3-8 shows the generic logical structure with appropriate "layout object class" and "layout category" names added.

Head is used twice as the "layout category" of the section title and the author's name. It would also be given as the only permitted category of the *Title frame* of figure 3-7, thus ensuring that only the section title and author's name (if present) would be laid out in the title frame.

Either *Body* or *Picture* is used as the "layout category" name of all the remaining leaf objects. Both names would also be given as the "permitted categories" of the *Body frame* of the title page and of the *Continuation body frame* of the continuation page, thus ensuring that the subtitles and any following paragraphs, figures and lists are laid out in order in these frames. For this example, we could just as well discard the category name *Picture* and use only *Head* and

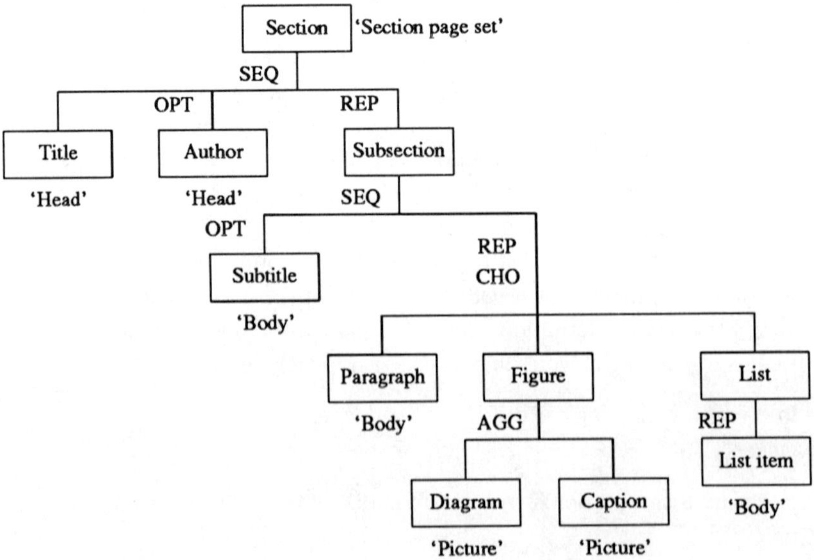

Figure 3-8. Generic Logical Structure with Layout Attributes

Body. As we shall see in section 3.1.4.3, however, there are advantages in giving different category names to items of essentially different types.

3.1.4.2 Positioning of Blocks within Frames

The previous section outlined the method of fitting contents into pages and frames. This section illustrates the final stage, fitting blocks into frames, by introducing a few more attributes and then looking at a variation of the example.

The actual size of a block is influenced by its contents. In the case of a paragraph of text, for example, the block would have a given width but would grow downwards just far enough to accommodate the lines of text it contained. Graphics blocks may be of a fixed size or they may, under certain circumstances, be scaled to fit into the area available in a frame. This sizing of blocks is the one area where the content architecture interacts with the higher-level document layout process.

The process of fitting blocks into frames is influenced by five separate attributes. Four of these ("fill order", "concatenation", "offset" and "separation") are part of the layout styles and apply to leaf logical objects, while the remaining one ("layout path") applies to pages or frames.

The "layout path" of a page or frame gives the direction of filling. If the layout path of a frame is downwards, blocks will normally be positioned with the first block at the top and subsequent blocks coming underneath. The "fill order" of the logical objects can be used to alter this order. If "fill order" has the value 'normal order', then the situation is as described above, but if it is 'reverse order', the blocks will be positioned with the first block at the bottom. In the 'reverse order' case, subsequent blocks will not be placed above the first one, but will push it up to make room for themselves underneath. The two cases are illustrated in figure 3-9.

The exact positioning of the blocks is influenced by "offset" and "separation" attributes, which define minimum distances between adjacent blocks or between blocks and their containing frame.

To illustrate the use of the "concatenation" attribute and to show a more complex version of filling of blocks into frames, we shall use a new version of the *paragraph* in our example. In this variation, a paragraph can be subdivided into strings of text and footnotes. The new generic logical definition is shown in figure 3-10 with the values of the "fill order" and "concatenation" attributes added. The intention is to allow the paragraph to contain footnote text that belongs logically within the paragraph but is to be laid out in the traditional manner at the bottom of the page. All the strings of text, however, are to be treated as one and to be laid out in a single block (unless a page break happens to intervene).

The desired effect can be achieved by giving the strings of text normal fill order but giving the footnote text reverse fill order. In addition, the first string of text and the footnotes are specified as 'non-concatenated' while the subsequent strings of text are specified as 'concatenated'. 'Concatenated' objects share the same layout block as previous objects with the same content architecture, layout category and fill order. So in this case, the subsequent strings are concatenated with the initial string rather than the intervening footnotes. If we assume there are three paragraphs and that the middle one contains two embedded footnotes, the final layout will be as shown in figure 3-11.

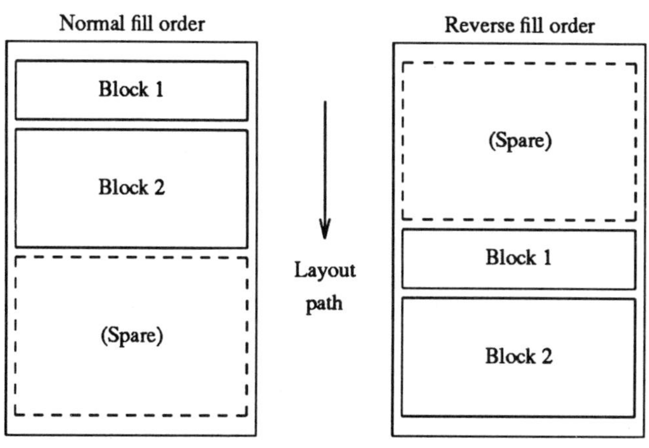

Figure 3-9. Layout in Normal and Reverse Order

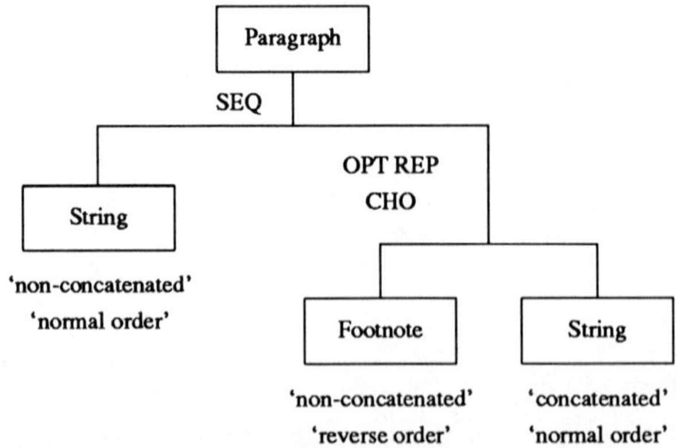

Figure 3-10. Generic Logical Structure for New Version of Paragraph

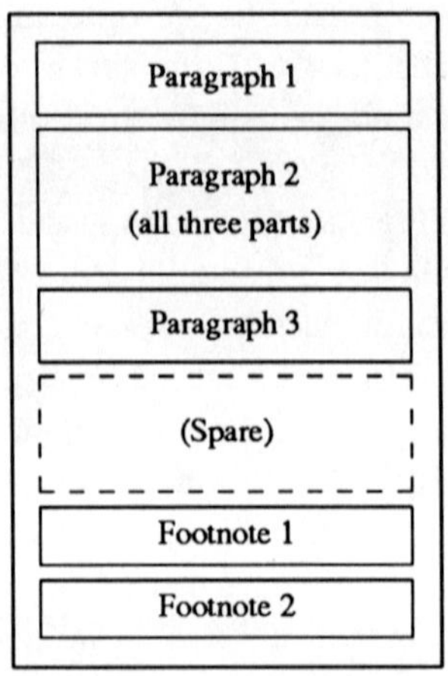

Figure 3-11. Layout of Paragraphs and Footnotes

3.1.4.3 Changing the Layout

The brief sketch of the ODA layout process given above should be sufficient to show that the appearance of a specific logical document can be altered significantly by making judicious changes to its generic layout structure. As a simple example, deleting the *Body frame* from the *Title page* of figure 3-7 would cause each section to be laid out with only the section title and author's name (if present) on the title page. There would be no frame on the first page with *Body* as a permitted category, so the first subsection would have to start on a new page in a *Continuation body frame*.

More radical layout changes can be achieved by altering the attributes that make up the layout and presentation styles. Although these attributes apply to logical objects, the specific objects contain only the identifier of the appropriate style. The styles themselves are held separately. This provides a more concise document representation and allows the styles to be altered without altering the logical structures.

The layout styles include the important "layout object class" and "layout category" attributes described earlier, as well as attributes governing the selection of frames and the positioning of blocks within frames. The "same layout object" attribute, for example, specifies that the block containing the contents of the logical object must share the same frame as the block containing the contents of another specified object, while "new layout object" specifies that the block containing the contents of the object must start a new frame. "Synchronization" specifies that the content of the logical object must be laid out at the same height as a specified object on a page (strictly speaking, along a line orthogonal to the direction of the layout path).

The presentation styles guide the lower-level content layout process and thus affect the appearance of the content within individual blocks. They contain different attributes for different content architectures. For character content, for example, they include attributes affecting the indentation of the first line and the distance between lines.

Changing both the generic layout structure and the styles can lead to significantly different views of the same logical document. Page and margin sizes can vary, single or double column layout can be used and paragraph spacing and font size can change. In particular, different house styles can be imposed on the same document in this way.

As a final example, we shall consider a couple of variations to the layout of our example chapter. The first variation is to turn footnotes into sidenotes (margin notes) placed in a narrow frame down one side of the page. This can be done by adding a new side frame beside each body frame in the generic layout, and using a new category name (such as *Notes*) for the "layout category" of the footnotes and the "permitted categories" of the new side frames. Instead of using reverse fill order for the footnotes, we would give them a "synchronization" attribute to ensure that they were positioned beside the appropriate paragraph.

The second variation is to place figures on the left side of the page and allow text to flow round them. This variation demands nested frames of variable size to be defined within the body frames of figure 3-7. Without going into details, it should be sufficient to say that it is possible to define a series of nested frames occupying the full width of the enclosing body frame but only taking just enough height to hold the item currently being laid out. Two sorts of these

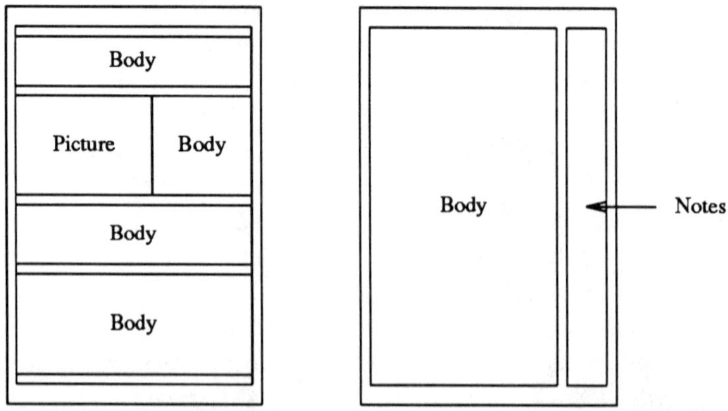

Figure 3-12. Two Variations of Layout

nested frames would be allowed:

- one without any further nested frames and with *Body* as its only permitted category,
- one with two further nested frames, the left-hand one called a *Figure frame* with *Picture* as its only permitted category and the right-hand one with *Body* as its only permitted category.

The use of different category names effectively divides the contents into separate streams and directs it into different frames. It does not necessarily ensure that exactly one figure is put into one figure frame. To make sure of this we also need to give *Figure frame* as the "layout object class" of the generic *Figure* object.

Figure 3-12 shows two possible page layouts that might result if we used these variations (text flowing round figures on the left, side notes on the right). The frames are labeled with their "permitted categories".

3.2 ODA Document Architecture and Processing Model

The previous section introduced the four main ODA document structures and used a simple example to show how the different structures were related. This section fills in some important gaps concerning document constituents and covers some of the wider issues of the ODA document architecture and processing model.

The main new topics are

- the document profile,
- the three stages of document processing—editing, layout and imaging—together with a brief description of the imaging process,
- the three document architecture classes—formatted document, processable document and formatted processable document,
- the use of expressions to specify the values of attributes and the methods used to derive the values of attributes.

3.2.1 The Document Profile

Figure 3-13 shows all the constituents of an ODA document, and indicates how the constituents are grouped together. The descriptions in the previous section introduced all the main constituents of the document body, but did not cover the *document profile*. This provides information about the document as a whole, and consists of three sets of attributes described in the next three sections.

3.2.1.1 Presence of Document Constituents

Not all parts of the document need to be present (as will be explained in section 3.2.4). The attributes given in this set indicate the presence of constituents in the document body, and typically have the value 'present'. Thus, if the document contains a specific logical structure, then the profile will contain an attribute called "specific logical structure" with the value 'present'.

Sometimes a document body may refer to the styles or generic structures of another document class. The other document class definition is known as an *external document* or *resource document* (depending on how it is referenced). In this case, the profile will contain one or both of the attributes "external document class" and "resource document" to give the appropriate document class name.

3.2.1.2 Document Characteristics

The attributes in this set cover a range of overall information about the document, including a specification of any facilities it uses (beyond a certain basic set). A few examples are given below to give a flavor of the information provided:

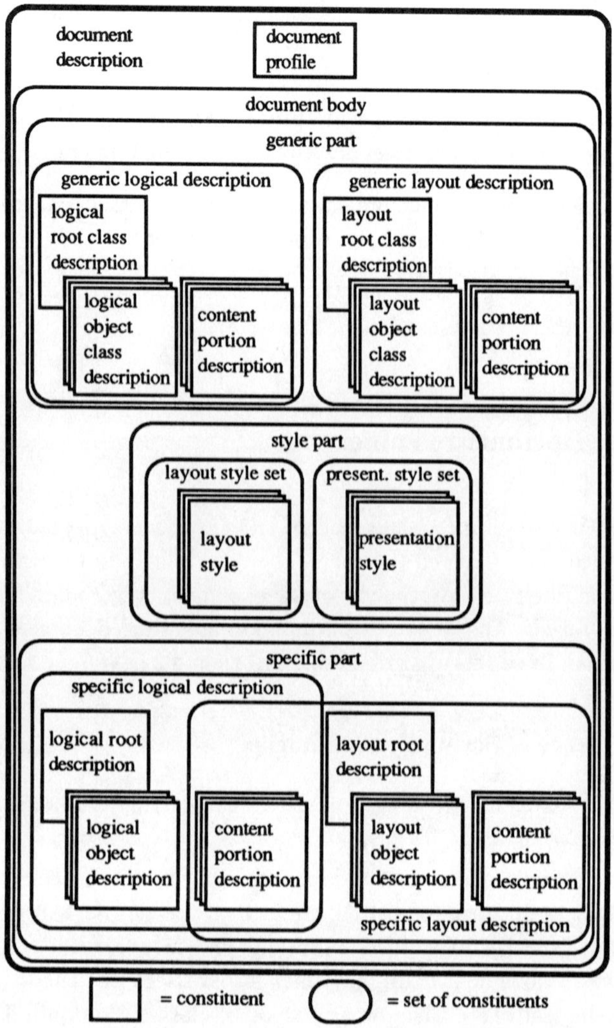

Figure 3-13. Document Components

- "ODA version": specifies the version of ODA to which this document conforms.
- "Unit scaling": defines a scaling factor to be applied to all attribute values that specify absolute or relative positions or dimensions.
- "Fill orders": specifies any values of "fill order" (other than 'normal order') used in the document.
- "Fonts list": lists the character fonts used in the document.

Also in this set is the "document application profile defaults" attribute, which gives default values for a number of other attributes. The full model for deriving default values (section 3.2.6) uses these values from the document profile if they are not specified elsewhere.

3.2.1.3 Document Management Attributes

This set of attributes covers items like the document's title, type, abstract, keywords, creation date and time, revision history, authors, distribution list, number of pages, security classification, the language(s) it is written in, and so on.

Most of this management information is informal. Thus, the originators of the document provide attribute values that are strings of characters for human consumption rather than machine processing. But a few attributes, such as "revision history" and those giving names or dates, must be in a form that conforms to other ISO standards or to CCITT recommendations.

The document profile contains vital information for the storage and interchange of documents, and though it is not often mentioned, it provides information needed by the three stages of document processing described in the next section.

3.2.2 ODA Document Processing Model

The ODA document processing model consists of three essentially separate processes—editing, layout and imaging—that make use of different document constituents. Figure 3-14 shows how all the different constituents and processes are related.

The *editing process* includes both document creation and document revision. Like the layout process, it divides into two related parts. *Content editing* creates or modifies content portions, and is dependent on the content architectures. *Logical structure editing* creates or modifies specific logical objects.

The generic logical structure is used to guide the editing process. The layout structures and the styles are not used to control editing, but the styles (and, of course, the specific layout structure that will eventually be produced) may be changed during editing.

The layout process has already been described in some detail. Figure 3-14 shows how this is central to the processing model with many document constituents feeding in to it. One point, which was not made clear earlier, is the distinction between layout and imaging. The result of the layout process is the specific layout structure and specially processed versions of the content portions in what is known as *formatted form*. Character content portions, for example, have control codes added to represent hard line terminators, the size of spaces and similar formatting details.

The *imaging process* is the final stage. It takes the specific layout structure, including the formatted content, and displays a visible version of the document on a suitable presentation medium. This may be paper or a computer terminal screen. As shown in figure 3-14, the imaging process uses information from the generic layout structure and the presentation styles.

The distinction between layout and imaging is useful and familiar to anyone who has made use of device-independent document preparation systems[12]. Although there are certain areas of overlap between the two, the layout process concentrates on deciding where items are to be placed and the imaging process concentrates on providing a visible representation of the items. Thus, both may need to know details about character widths and kerning pairs, but only the layout process knows about hyphenation and line breaking, while only the imaging process knows about the shape of characters. Similarly, both need to know about frames and blocks, but only the layout process knows about positioning them on pages, while only the imaging process knows about displaying overlapping blocks. As ODA is primarily a standard for document storage and interchange, it is clearly vital that it should maintain a distinction between the two processes if it is to provide a means of storing or transmitting documents in a device-independent formatted form.

3.2.3 The Imaging Process

Although ODA defines the form of the input to the imaging process, the process itself is not formally defined. This section presents an outline of the method used to combine the images of overlapping layout objects. The method is guided by three key attributes: "transparency", "colour" and "imaging order". Before discussing the use of the attributes, however, it may be helpful to give a few more details of the hierarchy of specific layout objects that can exist.

A specific layout structure defines a number of page sets and pages. Each page is considered to be an independent imaging surface, so there is no question of defining overlapping pages or having any priority of imaging of pages. Within each page, a hierarchy of frames and blocks is defined, all positioned so that their edges are parallel to the edges of the containing page. Frames and blocks may partially or fully intersect, but are always completely contained within their immediately superior frame or page.

---

[12] TeX [38] and typesetter-independent troff are well-known examples of formatters that lay out a document but leave the imaging to some further process.

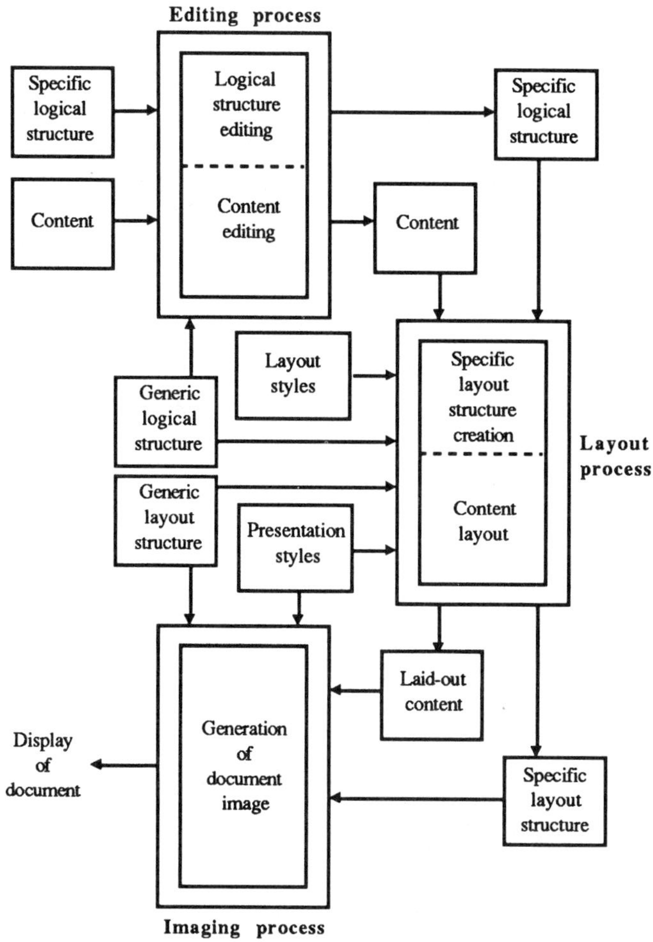

Figure 3-14. Document Processing Model

Two attributes from the presentation styles, "transparency" and "colour", define the *texture* of pages, frames, and blocks, and a further attribute, "imaging order", defines the order in which the subordinates of a particular page or frame are to be imaged. The values that these attributes may take and their effect is described briefly below.

"Transparency" may take the value 'transparent' to indicate that any underlying content is visible, or 'opaque' to indicate that any underlying content is not imaged. For pages, 'transparent' and 'opaque' are identical.

"Colour" may take the value 'colourless' to indicate that the background color of the area is the color of the imaging medium, or 'white' to indicate that the

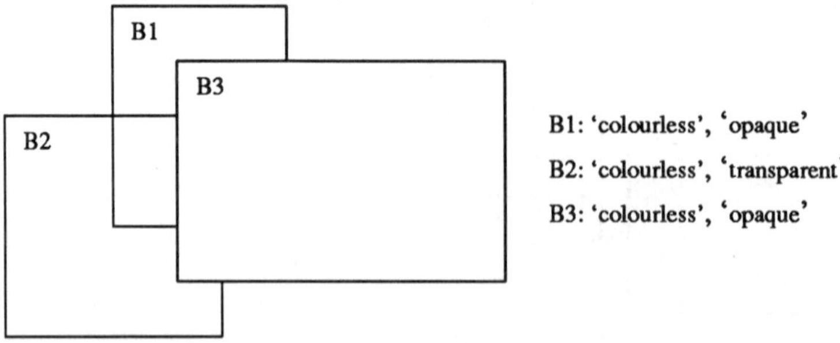

Figure 3-15. Imaging of Overlapping Blocks

background color is white. 'White' is only allowed with 'opaque'[13].

"Imaging order" specifies the precedence for imaging the immediate subordinates of a page or frame. It does this by giving a list of integers corresponding to the immediate subordinates (so that each subordinate occurs exactly once in the list). If "imaging order" is not specified, it is taken to correspond to the left-to-right sequential order of the subordinates. Objects later in the imaging order overlay objects earlier in the order.

Figure 3-15 shows a simple example of three blocks, B1, B2 and B3, with imaging order such that B1 is imaged first, then B2 and finally B3. The values of the "transparency" and "colour" attributes are shown in the figure.

B1 is imaged first. It is visible and combined with the content of the transparent B2 where they overlap, but not visible where it is overlaid by the opaque B3. B2 is imaged second. Because B2 is transparent, it is combined with B1 where they overlap, but is not visible where it is overlaid by the opaque B3. B3 is imaged last and is all visible.

Frames and blocks may have borders consisting of a border area and a border line. The border of a frame lies *within* the frame and reduces the area available for the layout of subordinate objects. The border of a block lies *outside* the block and takes up space in the containing frame. The layout process must take account of the size of borders. The imaging process is responsible for producing a visible border line.

The attribute "border" is used to specify the details. It does this by specifying up to four parameters (one for each of the four edges) that specify a "border line width", a "border line type" and a "border freespace width". Figures 3-16 and 3-17 show the effects of these for blocks and frames. The line type may be 'solid', 'dashed', 'dot', 'dash-dot', 'dash-dot-dot' or 'invisible'.

---

[13]A color addendum to ODA already exists in draft form, so changes and additions to this aspect of the imaging model are likely.

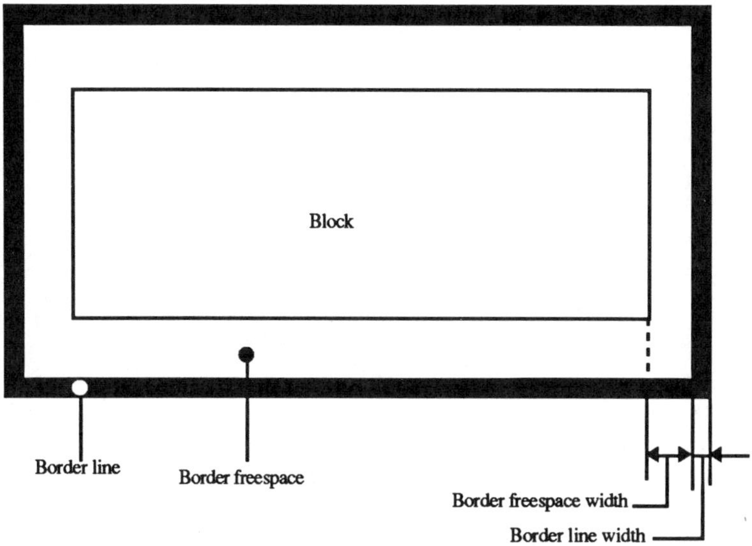

Figure 3-16. Border for a Block

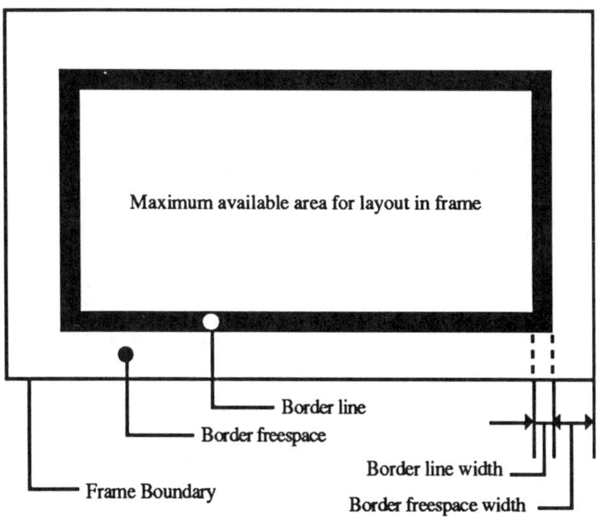

Figure 3-17. Border for a Frame

3.2.4 Document Architecture Classes

One of the basic principles of ODA is that documents may be stored or interchanged to allow presentation as intended by the originator and to allow processing, such as editing and reformatting.

It should be clear from previous sections that the layout structures are needed to allow presentation as intended by the originator, and the logical structures are needed for editing and reformatting. In fact, ODA allows for documents containing either or both of these and, thus, allows for three classes of documents: formatted form, processable form and formatted processable form.

Formatted form allows presentation of the document. A document of this class contains a document profile and a specific layout structure. It may also contain a generic layout structure, layout styles and presentation styles.

Processable form allows processing of the document. A document of this class contains a document profile and a specific logical structure. It may also contain a generic logical structure, a generic layout structure, layout styles and presentation styles.

Formatted processable form allows both presentation and processing. A document of this class contains a document profile, a specific logical structure, a specific layout structure and a generic layout structure. (The generic layout structure may be omitted if reference is made to the generic layout structure in an external document.) It may also contain a generic logical structure, layout styles and presentation styles.

Document architecture classes are simply sets of rules for defining the structure and representation of documents in these forms. The document profile contains an attribute, "document architecture class", that takes one of the values 'formatted', 'processable' or 'formatted processable' to specify the architecture class of the document.

The content portions (which must, of course, always be present for a specific document) must be in the appropriate form for the architecture class. Thus, character content portions must contain the appropriate control codes for the document architecture class. Some details on content portions and architecture classes are given in section 3.3.

3.2.5 Attribute Values

Most of the attributes described so far have values given by one of the following:

- One or more numeric values, representing positions or sizes.
- A string of characters, representing names (such as category names like *Body*) or information (such as the "title" given in the document profile).
- A value from a set of elements defined for that attribute, representing a fixed number of alternatives (such as "fill order" being either 'normal order' or 'reverse order').
- A reference to another document constituent—representing some connection with another element of the document (such as "layout style" referring to the layout style to be used for this object).

These are the common ways of specifying attribute values, and for most attributes it is the only way they can be expressed. A more complex specification can be given by means of an *attribute expression.* Attributes that use attribute expressions are:

- "generator for subordinates",
- "content generator",
- "bindings",
- "same layout object", and
- "synchronization".

"Generator for subordinates" has its own special form of attribute expression as described in section 3.1.3. "Same layout object" and "synchronization" specify layout constraints with respect to another object. The use of expressions for these attributes is relatively straightforward and should be clear from the description of the attribute expressions given below. The use of "content generator" and "bindings" is more complex, however, and needs some preliminary explanation.

These two attributes, together with attribute expressions, form the basis of ODA's mechanism for automatic numbering of document constituents. "Bindings" is an attribute consisting of pairs of parameters, each pair in turn consisting of a "binding name" and a "binding value". The binding value may be specified by a fixed value or an attribute expression. Binding values themselves may be referenced from attribute expressions. Indeed, the whole purpose of bindings is to allow their values to be referenced from the attribute expressions in other binding values and in "content generator" attributes.

Section 3.1.2 indicated that each specific leaf object must have an associated content portion. In fact, the objects must have *either* a content portion *or* a content generator. If a content generator is used, it consists of an attribute expression that can be evaluated to yield a string of characters. The string is then used in place of a character content portion. Figure 3-18 shows an outline of the use of bindings and content generators for automatic numbering of chapters. A binding, *Nr*, is used to generate a number one greater than the value of the same binding in a previous object, and a content generator is used to turn this number into a string of content characters. Further details of this example are given later.

The discussion above gives an outline of the main intended use of attribute expressions, and should provide a suitable background for the following brief explanation of the expressions themselves. Three types of expression are allowed: string, numeric and object identifier. String and numeric expressions may refer to binding values. The three types are described separately below.

3.2.5.1 String Expressions

A string expression is made from string literals, references to binding values and string functions. The string functions are designed to take an integer parameter and turn it into an appropriate string form. The functions provided are:

- MAKE-STRING: to give a decimal representation (negative, zero or positive).

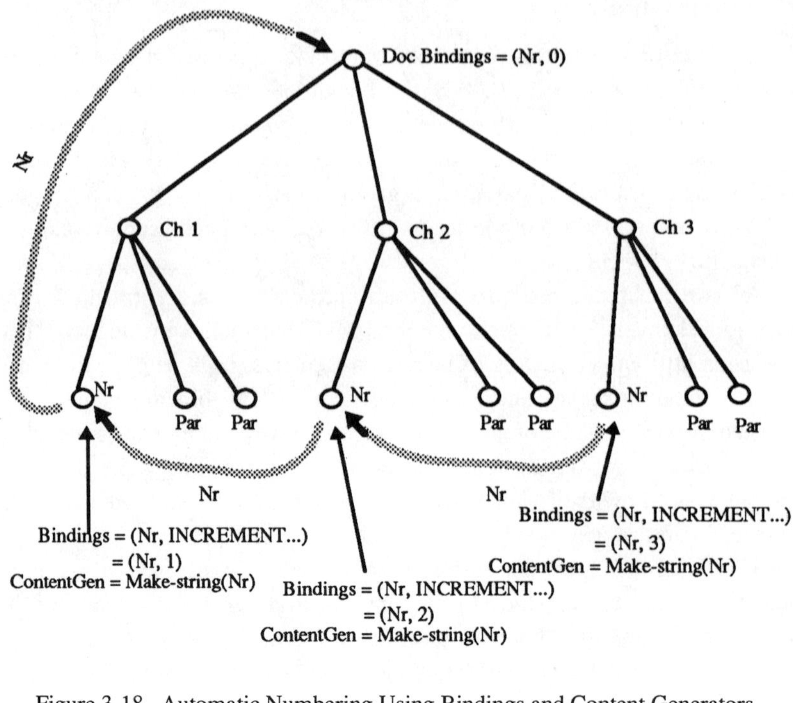

Figure 3-18. Automatic Numbering Using Bindings and Content Generators

- UPPER-ALPHA: to give an upper case letter A to Z (corresponding to the values 1 to 26).
- LOWER-ALPHA: to give a lower case letter a to z (corresponding to the values 1 to 26).
- UPPER-ROMAN: to give a Roman numeral representation using upper case letters (C, D, I, L, M, V and X).
- LOWER-ROMAN: to give a Roman numeral representation using lower case letters (c, d, i, l, m, v and x).

If several items occur in the expression, each is evaluated independently and their results are concatenated.

3.2.5.2 Numeric Expressions

A numeric expression is made from numeric literals, references to binding values and numeric functions. The two main numeric functions are INCREMENT and DECREMENT, which add or subtract one from the value of their numeric parameter.

3.2.5.3 Object Identifier Expressions

These expressions provide a flexible method of identifying objects. They consist of object selection functions that, with one exception, take an object as a parameter. The exact interpretation of the functions is complex. The following list is a simplified version:

- CURRENT-OBJECT: a function without a parameter that returns the identifier of the current object (for use in one of the following functions).
- CURRENT-INSTANCE(object): selects the current object if it is of the same type as the given object, otherwise it selects the nearest superior of the current object that is the same type as the given object.
- SUPERIOR-OBJECT(object): selects the nearest superior of the given object.
- PRECEDING-OBJECT(object): selects the nearest preceding object (in sequential order) of the given object.

The same functions may be used in references to binding values (but with two of the names slightly abbreviated to SUPERIOR and PRECEDING). In this context, the functions may be applied repeatedly until an object with a binding

```
INCREMENT(
    BINDING-REFERENCE(
        PRECEDING(CURRENT-OBJECT, Nr)))
```

Example 3-1: Expression for Binding Value of *Nr*

of the specified name is found. Thus PRECEDING, for example, may be used to search back through the document until the previous occurrence of a particular binding is found.

Only the form of a binding reference is now needed in order to present the full details of the example shown in figure 3-18. This uses the function BINDING-REFERENCE with two parameters: the first giving the target object and the second giving the binding name. The function returns the corresponding binding value. We can now explain that the full attribute expression specified for the binding value of *Nr* is as shown in example 3-1. The BINDING-REFERENCE starts at the current object and searches backwards until it identifies an object containing the binding called *Nr*. The whole expression therefore picks up the previous value of *Nr* and adds one to it.

The full attribute expression for the content generators is as shown in example 3-2.

3.2.6 Default Attribute Values

The final subject of section 3.2 is the method used for deriving the default values of attributes. An informal description is given first and then the full ODA mechanism is outlined.

ODA attributes are classified as mandatory, defaultable and non-mandatory. Mandatory attributes must always be specified for each relevant object, defaultable attributes have their values derived by the method given below, and non-mandatory attributes do not apply unless they are given explicitly for each relevant object. A document description will clearly be more concise if the maximum use is made of default attribute values.

```
MAKE-STRING(
    BINDING-REFERENCE(
        CURRENT-OBJECT, Nr))
```

Example 3-2: Content Generators

Many of the ODA attributes are defaultable and, in the majority of cases, the default values of these attributes are held in the object class definitions in the generic structures. There is then no need to include the value in a specific instance of that object, unless the default value is to be overridden.

Holding the default values in the object class descriptions is suitable if every instance of the object is to use the same value. Sometimes this is not sufficient, and there is a need for specific instances to use different values depending on their position in the document. It might be desirable, for example, to use a single *normal-paragraph* object class throughout the document, but use a different font for normal paragraphs occurring in the abstract than the font used for normal paragraphs occurring in the main body of the document. The ODA answer to this problem is to provide an attribute "default value lists", which provides default values for given attributes. Subordinate objects may then inherit these default values rather than using a default value from their object class description.

ODA specifies the following sequence of places to look for the value of a defaultable attribute. The first value found is the one used.

1 A value given in the specific object itself.

2 A value given in a style referenced by the specific object.

3 A value given in the generic object class description.

4 A value given in a style referenced by the generic object class description.

5 A value given in an object class description in an outside resource document used by the generic object class description.

6 A value given in a style referenced by an object class description in an outside resource document used by the generic object class description.

7 A value given in a default value list at a higher level of the hierarchical structure. (If several default value lists are applicable, the value from the one at the lowest hierarchical level is used.)

8 A value given in the document profile.

9 A value specified by ODA.

Put in simpler terms, this says first try the object itself, then the object definition, then any outside object definition, then default value lists, then the document profile. If all these fail, ODA provides a standard default value as a last resort.

3.3 ODA Content Architectures

The emphasis of the earlier sections has been on the ODA document structures and overall principles. Document content has been mentioned in several contexts, but only in very general terms. The two main points made so far are that ODA caters to three different types of content (character, raster graphics and geometric graphics) and that this content may be in a slightly different form depending on whether the document is in processable or formatted form.

This section focuses on the details of the internal structure of content portions and tries to give a feel for the level of content allowed, especially for character content.

A few points need to be made before discussing the three separate content architectures. The first is a reminder of the relationship between the content architectures and the rest of the document architecture. Most of the higher-level document structures are independent of the content. The content dependencies come in the attributes for the leaf objects. There are content-related attributes in the presentation styles in addition to the attributes that identify the actual content portion and its type.

A second point is that each of the three content architectures defines one or more *content architecture classes*. These are similar to the 'formatted', 'processable' and 'formatted processable' document architecture classes introduced previously, but may contain further classes catering to different levels of facilities. The character content architecture, for example, includes a very simple level for normal left-to-right placing of characters in a single font, and a complex level that allows different character orientations, right-to-left character placing, a variety of fonts and extras like blinking and crossed-out characters.

A content architecture class definition consists of specifications for the following:

- A set of presentation attributes and other content-dependent attributes.
- A set of content elements.
- A set of control functions.
- The type(s) of coding used.

The specification of attributes and control functions includes their permissible values and a recommended default value. Permissible values are classified as *basic* or *non-basic*, and any non-basic values used must be defined in the document profile. If several levels of content architecture classes are defined, the level defines any restrictions. Thus, the overall character content architecture defines all possible character orientations and fonts, and the definition for the simple level mentioned above specifies that many of the items are either not permitted or must take one particular value.

A third point is that new content architectures can be added by the use of *private content architectures.* The structure of a private content architecture is like the three defined architectures: it defines presentation and content-dependent attributes, content elements, control functions and types of coding. Because content architecture is separated from document architecture, one can easily fit the new kind of content into an existing ODA system and use that content only among systems that understand the private content architecture. While not strictly conforming, the use of private content architectures allows groups of ODA users to expand the capabilities of ODA to meet their needs.

3.3.1 Character Content

A character content portion consists of a string of characters made up of graphic characters and control characters. Graphic characters represent the basic content elements, and control characters represent the control functions. Not all control characters are allowed in all the different architecture classes. In particular, some are only meaningful in formatted form and some are only meaningful in processable form.

The positioning of graphic characters within a block is governed by several different attributes. The default positioning is for upright characters, placed left-to-right in lines that start at the top of the block and progress downwards, but several other possibilities are allowed. A representative selection of the attributes that control the positioning of characters is given below. Although the list is far from exhaustive, it is chosen to give a good impression of the scope of the content architecture.

- "Character path": the direction of progression of successive character images along a line (may take four values: 0, 90, 180 and 270 degrees).
- "Line progression": the direction of progression of successive lines within a block (may take two values: 90 and 270 degrees relative to the character path).
- "Character orientation": the orientation of the character relative to the character path (may take four values: 0, 90, 180 and 270 degrees).
- "Alignment": the positioning of characters within a line (may take the values 'start-aligned', 'end-aligned', 'centred' or 'justified').
- "Line layout table": sets up the positions and types of a series of tabulation stops (text may be 'start-aligned', 'end-aligned', 'centred' or 'justified' with respect to tabulation stops).

- "First line offset": allows for starting the first line at a position offset from subsequent lines (positive and negative offsets are allowed).

Further attributes define the position of the start and end of lines within the block, and the overall *positioning area* that is available for use within the block.

All the above attributes reflect positioning requirements that apply to complete content portions (though some may be overridden temporarily by means of control codes). Other requirements may refer to the positioning or imaging of single graphic characters or short strings of characters within a content portion. Such requirements are generally given by control codes embedded within the content. Some examples of the control codes available are as follows:

- Partial line down and partial line up (PLD/PLU): indicate the start and end of subscripts and superscripts.

- Graphic character composition (GCC): indicates that two or more following characters are to be combined into a single graphic symbol.

- Select graphic rendition (SGR): indicates how the following characters are to be imaged. Parameters to SGR allow up to ten different fonts to be selected, provide for changes in the weight, posture and spacing of characters and allow a number of variations such as blinking, underlined or crossed-out characters. The initial graphic rendition can be given by a presentation attribute called "graphic rendition".

- Start reverse string (SRS): marks the start and end of a string of characters that is to be imaged in the opposite direction to that of the preceding text.

- Parallel Texts (PTX): marks passages of text which appear one after the other but are intended to be presented in parallel with one another.

It should be clear from the above lists that the ODA character content architecture contains facilities for reasonably high-quality formatting and for dealing with non-European languages. Although it does not provide general PostScript-like facilities allowing text to appear at any orientation or to be stretched or distorted in a variety of ways, it does allow a significant amount of flexibility in the positioning and appearance of characters. The SRS and PTX control functions have been provided especially to cope with the problems of including left-to-right numeric text in right-to-left Arabic or Hebrew text, and of representing Japanese Ruby information where Kana characters are used to provide pronunciation and interpretation information for Kanji characters.

Figure 3-19 illustrates some basic concepts of character positioning, figure 3-20 shows the use of tabulation stops, and figure 3-21 shows how left-to-right and right-to-left text can be interspersed.

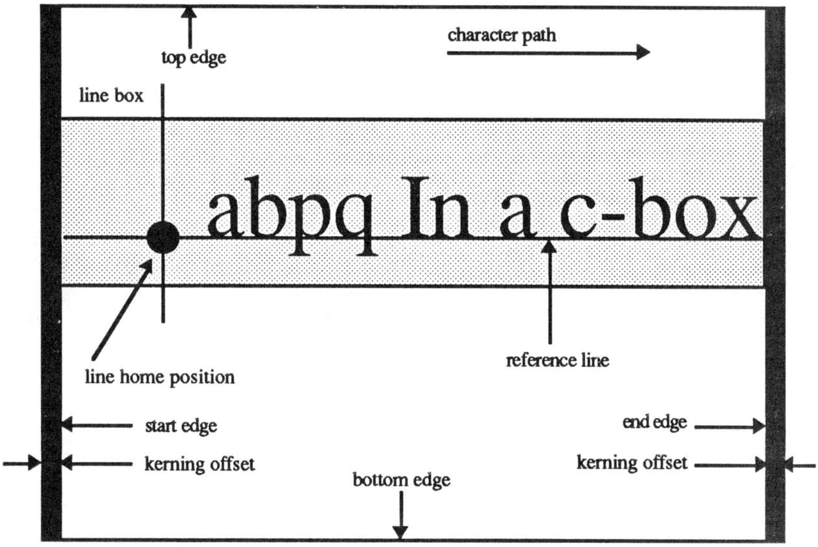

Figure 3-19. Some Character Positioning Concepts

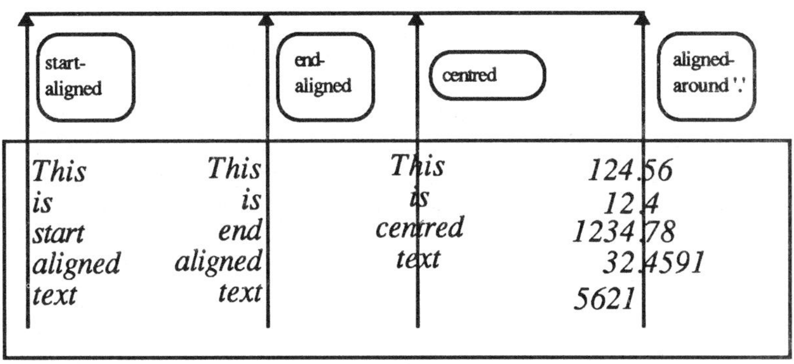

Figure 3-20. Tabulation

3.3.2 Raster Graphics Content

A raster graphics content portion represents a two-dimensional pictorial image in the form of a rectangular two-dimensional array of picture elements (pels). The array is represented in one of three ways:

1 *Bitmap encoding scheme:* Each element in the array has one of two distinct states: *set* (1) or *unset* (0). The array is encoded row-by-row

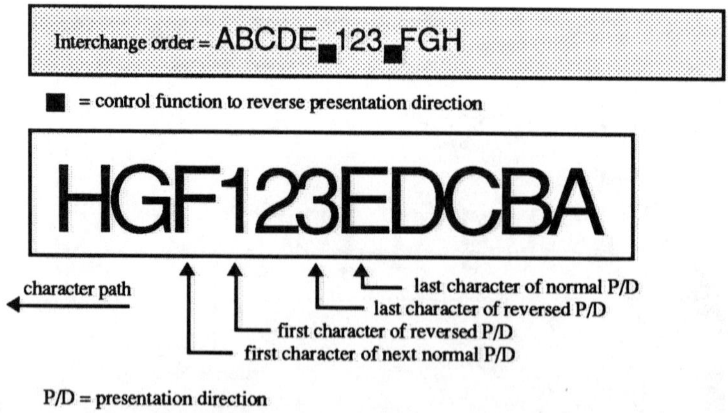

Figure 3-21. Reverse Character Ordering

as a string of octets. If the number of bits in the row is not a multiple of eight, then it is extended by the minimum number of 0 bits to ensure that its encoding ends on an octet boundary.

2 *Group 4 facsimile encoding scheme:* The array is encoded according to CCITT Recommendation T.6 (where *black* and *white* are considered to be equivalent to *set* and *unset*, respectively).

3 *Group 3 facsimile encoding scheme:* The array is encoded according to the one- or two-dimensional encoding schemes defined in CCITT Recommendation T.4 (where *black* and *white* are considered to be equivalent to *set* and *unset*, respectively).

In all cases, the picture elements in the array have a defined order. The array is an ordered sequence of rows, and each row is an ordered sequence of picture elements representing a line of the image.

As for character content, a large set of attributes is used to control the presentation of the content. A representative selection of these is given below:

- "Pel path": the direction of progression of successive pels along a line (may take four values 0, 90, 180, and 270 degrees).
- "Line progression": the direction of progression of successive lines within a block (may take two values: 90 and 270 degrees relative to the pel path).
- "Initial offset": the point relative to which all pels are positioned within a block.
- "Clipping": gives two coordinate pairs that define the subregion of the pel array to be imaged (the first coordinate pair gives the first pel that is

part of the selected subregion, the second pair gives the last pel that is part of the subregion).

- "Image dimensions": gives the intended dimensions of the block that is to contain the clipped pel array. Several parameters allow "preferred" and "minimum" dimensions to be given, and may allow the image to be scaled and the "aspect ratio" to be changed to fit the block area.

It can be seen from the above list that images may be presented at various orientations, and that they may be both clipped and scaled to fit the available area. Figure 3-22 illustrates the basic concepts of positioning of pels in the image.

3.3.3 Geometric Graphics Content

This content architecture is perhaps the easiest of the three to describe as it is based entirely on the Computer Graphics Metafile (CGM) standard [8]. Thus, the primitive graphics objects are *markers*, *lines*, *filled areas* and *text*. These are positioned in *virtual device coordinates* (VDC), and have a large number of attributes that control their rendition. A line, for example, has attributes

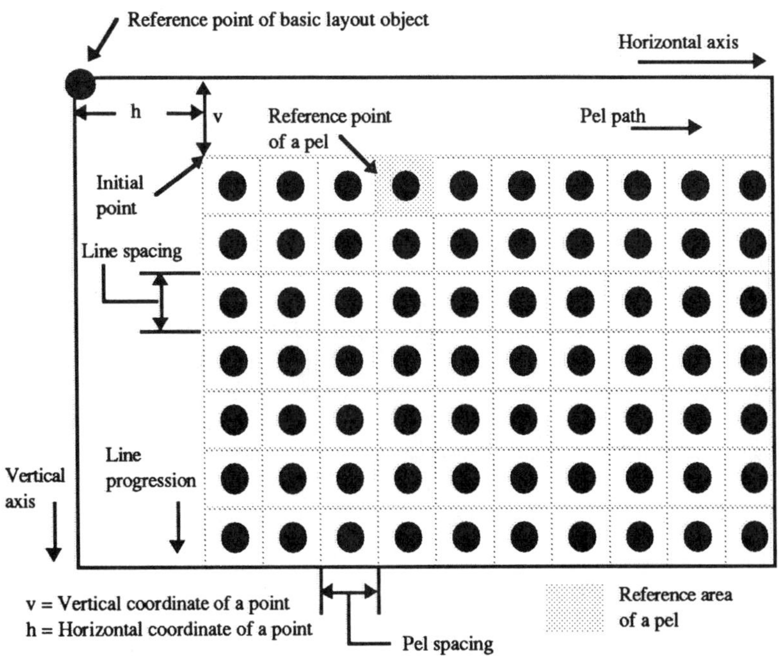

Figure 3-22. Positioning of Pels

describing its *width*, *type* and *colour*, while a filled area has attributes describing details such as its *interior style*, *fill colour* and *type of hatching* used. A few of the details of the objects and attributes (such as defaulting rules) are modified to fit the overall ODA model, but the functionality provided is entirely that of CGM. Each geometric graphics content portion represents a single pictorial image.

Instead of describing the CGM model in any detail, this section gives a representative list of the presentation attributes for geometric graphics content. These define the default values of the CGM attributes to be used during the imaging of the content.

CGM uses the concept of *bundles* of attributes for describing several characteristics of a particular type of graphics primitive. Thus a *line bundle* contains three attributes giving the line's type, size and colour. The line can then have all three attributes specified by giving the number of the appropriate attribute bundle. Several of the presentation attributes listed below specify a *bundle index* instead of (or as well as) values for the attributes that make up the bundle. A few also allow one or more bundles of attributes to be defined.

- "Line rendition": sets the default values for the presentation of line primitives, including the line bundle index, the line type, the line width, the line colour and zero or more line bundle specifications.

- "Text rendition": sets the default values for the presentation of text. This includes lists of font and character sets that may be used, a *character coding announcer* to define the 7- or 8-bit coding used, the text bundle index, precision, character spacing, colour, height, orientation and path.

- "Filled area rendition": sets the default values for the presentation of the interior of filled area primitives. This includes the fill bundle index, the interior style, type of hatching used, pattern index and zero or more pattern specifications.

- "Colour representations": sets the default value for background colour and defines the initial colour representations to be used for imaging objects.

- "Region of interest specification": sets a rectangular area of the VDC Extent to be used in imaging. The area is given by two sets of VDC coordinates or is specified as 'automatic' (in which case the entire VDC Extent is used).

- "Picture orientation": specifies which corner of the basic layout object is to coincide with the first corner of the region of interest (allows four values, 0, 90, 180 and 270 degrees for bottom left, bottom right, top right, and top left corners).

- "Picture dimensions": allows the width and height of the final image to be scaled according to preferred and minimum sizes specified.

Figure 3-23 shows the use of the "region of interest" and the "picture orientation" attributes.

It is worth noting some differences between character strings in geometric graphics content portions and character strings in character content portions. The essential difference is that each graphics character string is conceptually independent of any other string in the content portion, whereas each character content portion is a single string. Further important differences are listed below.

- Graphics character strings may be positioned anywhere within the VDC space, but character content is sequential in nature and positioned according to strict line and path conventions.

- The size of a graphics block is not affected by the amount of text in the content.

- Graphics text may be set at any angle and subjected to arbitrary scaling and rotation. Character content only allows text at four angles (0, 90, 180 and 270 degrees).

- Embedded control functions have a carefully defined meaning for character content, but are not defined for graphics text.

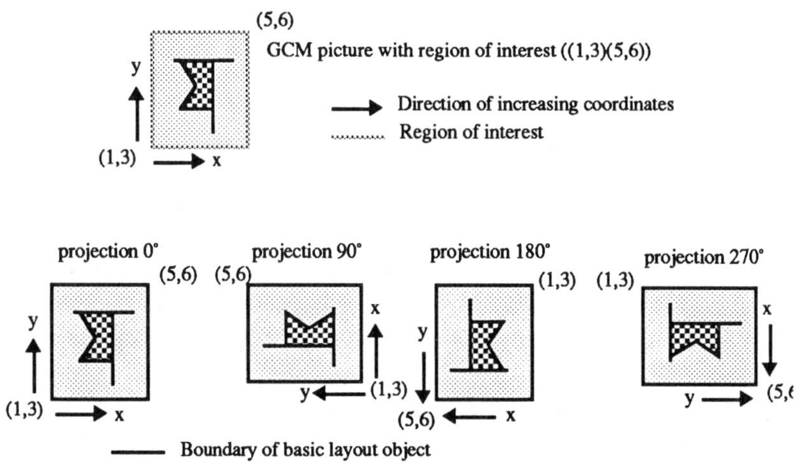

Figure 3-23. Use of Region of Interest and Picture Orientation

3.4 Document Application Profiles

The examples in section 3.1.2 assumed that we could represent this chapter in ODA as a collection of sections, subsections, paragraphs, titles and subtitles. ODA, however, defines none of these concepts. Instead, only the abstractions provided by composite logical objects and basic logical objects are defined. Composite logical objects are non-leaf objects like sections and subsections, while basic logical objects are leaf objects, like titles and paragraphs. Fortunately, one can define a collection of concrete objects for a particular class of documents, such as business letters or book chapters, through the use of a *document application profile*. In this section, we discuss document application profiles and list some profiles that are in use.

A *document application profile*, defined by the ODA standard, specifies a subset of ODA. Each part of the ODA standard—logical structures, layout structures and content architectures—can be restricted in a document application profile. For example, the profile used by the EXPRES participants defines concepts such as passages, paragraphs and phrases. By defining restrictions on attribute values, a document application profile allows the interpretation of constituents as document features. The attribute "application comments" is often used for this. For example, a profile may specify that the value for this attribute be "Paragraph" for constituents that represent document paragraphs. All senders and receivers can identify a document paragraph by seeing if the value of the "application comments" attribute is "Paragraph". Further restrictions help guide the construction of structures specified in a document application profile. For example, the profile used by EXPRES participants requires that composite logical objects that represent paragraphs must be subordinate to objects that represent passages.

Layout concepts are also provided by a document application profile. The profile used by the EXPRES participants defines header, footer and footnote areas within a page.

Content architectures may be limited in a number of ways. One typical way is to limit the encodings used by the content. For example, the profile may define that character content may include only certain character sets and control functions, or that raster content may use only group 3 encoding. A profile may even eliminate entire content types.

Various groups in Europe, North America and the Pacific Rim are working on three levels of document application profiles. Level 1 is intended for

> Basic documents consisting of character content only. This level of document contains the functionality found in most word processor products [48].

A level 2 profile encompasses

> Extended documents consisting of character, raster graphics and geometric graphics content. This level of document contains enhanced functionality to support multi-media document processors evolved from word processor products [48].

The third level is intended to be used for

> Advanced documents consisting of character, raster graphics and geometric graphics content. This level of document contains the functionality traditionally found in most "desktop publishing" or personal publishing products [48].

The most common profile for each of these levels is designated by the European Workshop on Open Systems (EWOS) as *Q111*, *Q112* and *Q113*. *Q112* and *Q113* are being aligned with similar level 2 and level 3 profiles defined by the OSI Implementor's Workshop, a group sponsored by the U.S. National Institute of Standards and Technology, which recommends implementation agreements for the U.S. government [48]. Other profiles are being defined, including a special profile for interchanging documents consisting of tiled rasters (for use in the CALS project [58]) and, so called, H-series profiles for participants in the CeBit ODA demonstrations held in Hannover each year [22].

3.5 ODA and Other Standards

It should already be clear from the previous sections that ODA has close connections with other ISO standards and with CCITT Recommendations. This section introduces three new topics:

- the relationship with CCITT Recommendations,
- the use of ASN.1 as an interchange format,
- the relationship with SGML.

3.5.1 Relationship with CCITT Recommendations

ODA has been developed in parallel with the T.410 series of CCITT Recommendations (1988): *Open Document Architecture (ODA) and Interchange Format.* The text of Parts 1-8 of the ODA standard is identical to the text of the correspondingly numbered Recommendations T.411 to T.418 "except for mandated stylistic differences and provisions of ISO 8613 that are outside the scope of these Recommendations" [14].

Provision has been made in ODA for compatibility with some older CCITT Recommendations. This has been done primarily by defining content architecture levels to correspond to the CCITT Recommendations and by allowing raster graphics portions to use facsimile encoding.

3.5.2 Use of ASN.1

An ODA document is interchanged by representing it in a form known as ODIF. This is an abstract data syntax (described in Part 5 of the ODA standard) in which the document constituents and attributes are represented by a hierarchy of data structures and data items appearing in a prescribed order. These structures and items are specified using the Abstract Syntax Notation, ASN.1 [31, 32], which represents each item by a type field, a length field and a value field. ASN.1 provides a semi-readable notation and an encoded form in 8-bit bytes. To give a flavor of the representations, example 3-3 shows an example of a generic logical object for a paragraph of text in ODIF/ASN.1 form (the semi-readable form is shown on the left and the encoded form on the right).

3.5.3 Relationship with SGML

The Standard Generalized Markup Language (SGML) is an ISO standard [33] that has a lot in common with certain aspects of ODA. There has, indeed, been a good deal of argument and rivalry between supporters of ODA and supporters of SGML. (Each side naturally likes their own way of doing things and pours

```
logical-object-class{                   A52E
  object-type basic-logical,            020102
  descriptor-body{                      3129
    object-class-identifier "2 1 1",    41053220312031
    user-visible-name "Body-paragraph",
                                        8E0E426F64792D70617261677270
    layout-style "4 9",                 9303342039
    presentation-style "5 3",           9103352033
    presentation-attributes{            A606
      content-architecture-class        060458020601
      {2 8 2 6 1} } } }
```

Example 3-3: ODIF/ASN.1 Form of a Generic Logical Object

scorn on the other way.) The purpose of this section, however, is to indicate some of the main relationships between ODA and SGML, not to attempt any critical evaluation.

SGML is a generic markup system. An SGML document is a file containing the text of the document interspersed with markup tags to define the beginning and end of each logical item in the document. An SGML *Document Type Definition* (DTD) defines the permissible relationships between the logical items. The marked-up document is thus the equivalent of the ODA specific logical structure, and the DTD is the equivalent of the ODA generic logical structure. Although the concepts involved overlap considerably, ODA adopts an object-oriented approach while SGML adopts a markup approach.

Perhaps the biggest difference between the two is that SGML makes no direct provisions for describing the layout of documents or for defining any form of content other than text. Layout characteristics can be described by attributes given in the markup tags, but these are not in any way part of the standard. Similarly, SGML provides a mechanism for including different types of content, but they also are not part of the standard.

Just as ODA has an interchange format called ODIF, SGML has an interchange format called SDIF. In addition, there is a standardized SGML representation of ODA documents known as *Office Document Language* (ODL). This allows ODA documents to be interchanged using ODL and SDIF. Annexes to several parts of the ODA standard define ODL and show how the various constituents of an ODA document are expressed in ODL.

An ODL document can come in processable and/or formatted form. To give a flavor of the form of ODL, examples 3-4 and 3-5 show possible ODL representations of two *Paragraph* objects (from our example document in section 3.1.2). The contents of the paragraphs are represented by a /* ... */ comment. Example 3-4 is for a processable document where straightforward markup tags signal the start and end of objects. Example 3-5 is for a formatted document. Here the tags represent layout objects. In particular, the *cf* tags represent blocks containing formatted character content. Attributes within the *cf* tags are used to supply information about the size and position of the block and about the layout and spacing of the content.

This chapter has explored the ODA document model and its description in terms of generic and specific structures, logical and layout structures, content architectures and attributes. In chapter 2, we described document models considered for the EXPRES. The next chapter brings the two together, showing how to represent an EXPRES document in ODA.

```
<Paragraph>
/* unformatted contents of first paragraph */
<Paragraph>
/* unformatted contents of second paragraph */
```

Example 3-4: ODL Representation of Two Unformatted Paragraphs

```
<Continuation page>
<Continuation body frame>
<cf pos="360 360" dim="7800 1550" calign=j
        cfloff=1200 clinespc=400>
/* formatted contents of first paragraph */
<cf pos="360 3200" dim="7800 2625" calign=j
        cfloff=1200 clinespc=400>
/* formatted contents of second paragraph */
```

Example 3-5: ODL Representation of Two Formatted Paragraphs

4
Document Translation Using ODA

In chapter 2, we discussed many features that could be incorporated into a document model and examined the fidelity of translation that is possible given the plethora of features in contemporary systems. We narrowed the scope of document models in the previous chapter, where we described the subset of features found in the ODA model. In this chapter, we discuss the general document model used by the EXPRES participants and how that model is represented using ODA and the document application profile published by the National Institute of Standards and Technology (NIST).

4.1 Motivating Another Document Model for EXPRES

One might ask why we needed to define a document model beyond that provided by ODA. The ODA model certainly provides a general document model with a large numbers of features and options. We found, however, that the ODA model had some deficiencies for EXPRES purposes:

- Many word processing systems have much more editing structure in a document than ODA provides, especially in their style systems. Researchers and reviewers edit the documents as they move among the authors, then to the NSF program manager, to the reviewers, back to the NSF and back to the proposer. Therefore, it is important that the editing structure of the document be preserved.
- ODA does not provide the same set of primitive descriptions of data as many other systems. Research proposals contain more than text, structured graphics and raster graphics. They include, but are not limited to, tables of numbers (which include spreadsheets of budget data), chemical structures, equations and circuit designs.

- ODA is highly redundant in that several features can be used for representing similar features. Our goal is to exchange proposals effectively, not investigate nuances of ODA implementation or application. Therefore, we had little need for multiple ways of representing the same feature.

The natural consequence of first two deficiencies is that many translators from native formats to ODA format will attempt to save their extra-ODA information using particular structures and encodings of ODA objects. We wanted to minimize any extra-ODA information by careful selection of our document model.

The third deficiency is asymmetric. When generating ODA format, a translator could minimize the third deficiency by picking one of the possible representations in ODA. However, if a translator from ODA format to a native format has no a priori knowledge about the generator of the ODA file, the translator could not assume that any particular conventions were followed. Thus, the translator would have to deduce the particular way that ODA was used to encode features that have multiple representations.

To assist document interchange in EXPRES, we created a document model, such as used by multi-media document editors like Andrew and Diamond. We then suggested a way to use ODA, along with the NIST document application profile, to represent this model. Our intention was to permit easy recovery of information. Before we can discuss the details of our model and its representation, we need to introduce some notation for documents.

4.2 Document Model Notation

When discussing translation techniques, it is useful to have a precise description of the document. A visual image is imprecise, because various relationships among styles, environments and the logical structure of the document may not be evident. In this section, we introduce some notation for document organization, styles, content and environments. Most of the notation is introduced through the use of examples.

All styles and parts of a document's organization are named. A style consists of an optional base style (if inherited), followed by a set of equations for adjusting components of the state vector. Equations have the form

```
Formatting state vector component←Expression
```

where `Formatting state vector component` is any piece of the formatting state vector allowed in the document product system and `Expression` is any expression consisting of formatting state vector components, constants, arithmetic operators, logical operators and other operators as appropriate for that component. (For now, the goal is to illustrate the document notation and not focus on the semantics of any particular style system.) The symbol $\leftarrow$ denotes local assignment, while the large arrow $\Leftarrow$ denotes global assignment. When interpreting a style, it is assumed that any reference to values of the formatting state vector on the right side of a style equation refer to the value of the state vector before any part of the equation has been applied.

To simplify the initial discussion, only character content and its structure will be described. Parts of the logical structure of a document, along with their environments (and associated styles), are shown as names interspersed in text, preceded by an at sign (@) and with the content enclosed by braces ({}). If a style is associated with the document piece, it will follow the name of the document piece, separated by a hyphen (-). Finally, we assume that all examples follow whatever nesting restrictions apply to document pieces; such nesting specifications will not be represented explicitly.

Six examples of this notation are provided to illustrate different document organization, style and environment configurations. Together, they show how to denote various kinds of calculations on formatting state vectors. For the first five examples, we assume that the formatting state vector has two local components: *FontFace*, which is the face code of a font, and *FontSize*, which is the size of a font in points. The root environment specifies initial values of empty for the *FontFace* and 10 for *FontSize*. The names of the styles have been chosen to be familiar and realistic, but there is no intention that the names of the styles, such as *Emphasis*, denote identical concepts with the same name in ODA or the NIST Document Application Profile. Similarly, the names of different parts of a document's organization are unimportant.

A simple document with two *text* phrases and no style definitions or environments is shown in example 4-1. The example shows a very simple document that consists of the string "abcdef". The structure of the document is denoted by use of braces. The organization of the document is a *Root* part that contains two *Text* parts. Although indentation and new lines are not part of the notation, we will sometimes use them to help show the structure of the document. In this example, none the document's parts has an associated style.

Example 4-2 shows the definition of styles and their use in environments. The first style, called *Emphasis*, sets the value of *FontFace* to *Italic*. Text formatted

```
@root{
   @Text{abc}
   @Text{def}
}
```

Example 4-1: Document with Two Phrases

with a *FontFace* of *Italic* will have an italicized font. Similarly, the *Heading* style sets the component *FontFace* to *Bold.*

The text in example 4-2 should be formatted to look like "**abc***def***ghi***jkl*". The *Heading* style associated with "abc" causes the formatting state vector component *FontFace* associated with the environment enclosing "abc" to be assigned *Bold.* When the next environment is reached (*Emphasis* style associated with "def"), the formatting state vector component *FontFace* is changed to *Italic.* The formatting state vector as specified by *Emphasis* lasts only through "def", at which point the formatting state vector is again the value that resulted from applying the *Heading* style. Thus, "ghi" is also displayed in a bold font. Once the *Heading* environment is finished, the formatting state vector has its initial values of the *root* environment. The second environment with the *Emphasis* style again changes the *FontFace* component of the formatting state vector to *Italic*, and the content "jkl" is displayed in an italic font.

Example 4-3 shows the effects of styles that interact in an environment. The text in example 4-3 looks like "**abc*def*ghi***jkl*". The *Heading* style causes the formatting state vector component *FontFace* to include *Bold.* Hence the text "abc" uses a bold font. When the next environment is reached (*Emphasis* associated with "def"), the formatting state vector component *FontFace* is changed to include *Italic.* Note the difference between example 4-2 and example 4-3 in the style definition for *Emphasis.* Example 4-3 defines the new value of *FontFace* to be the old value of *FontFace* with *Italic* added, while in example 4-2, the definition assigns *FontFace* to be exactly *Italic.* Hence the new value for *FontFace* in the environment for "def" in example 4-3 is both *Bold* and *Italic.* As before, the environment associated with the *Emphasis* style lasts only until the end of "def", at which point the formatting state vector is

Style Definitions:

```
Emphasis =  FontFace←Italic;
Heading =   FontFace←Bold;
```

Example text:

```
@root{
   @Phrase-Heading{
       @Text{abc}
       @Text-Emphasis{def}
       @Text{ghi}
   }
   @Text-Emphasis{jkl}
}
```

Example 4-2: Simple Style Definitions and Environments

Style Definitions:

```
Emphasis = FontFace←FontFace + Italic;
Heading =  FontFace←FontFace + Bold;
```

Example text:

```
@root{
   @Phrase-Heading{
      @Text{abc}
      @Text-Emphasis{def}
      @Text{ghi}
   }
   @Text-Emphasis{jkl}
}
```

Example 4-3: Use of Relative Styles

again the value that resulted from applying the *Heading* style. Thus "ghi" is also displayed in a bold font. Once the *Heading* environment is finished, the formatting state vector has its root values. A second environment for *Emphasis* is associated with "jkl". Using the *Emphasis* style, the *FontFace* component is changed to include *Italic*. The formatting state vector being changed is the root value, which originally has the value empty for *FontFace*. Thus, the new value is only *Italic*, not *Italic* and *Bold* as in the environment for "def". As a consequence, the content "jkl" is displayed in an italic font.

Example 4-4 shows how an inherited style is denoted and the semantics of associating it with an environment. The formatted text that results from example 4-4 looks like "**abcdef**". Like example 4-2, the association of *Heading* with "abc" causes the formatting state vector component *FontFace* for content

Style Definitions:

```
Heading = FontFace←Bold;
MajorHeading = Heading; FontSize←16;
```

Example text:

```
@root{
   @Text-Heading{abc}
   @Text-MajorHeading{def}
}
```

Example 4-4: Simple Use of Inherited Styles

starting with "a" to be changed to be *Bold*. When the environment associated with the *Heading* style is finished, the formatting state vector has its initial values. When the next environment is reached (association of *MajorHeading* with "def"), the formatting state vector is changed according to the definition for *MajorHeading*. The definition for *MajorHeading* specifies that it inherits equations from *Heading*, and therefore, the changes specified in *Heading* should be applied, followed by the assignment specifying a new *FontSize* component. To apply the *MajorHeading* style, one first changes the *FontFace* to the value *Bold* as specified by the *Heading* style and then changes the *FontSize* to the value of 16 points. The result is that the text "def" is displayed using a bold font that has a design size of 16 points.

The next example, 4-5, illustrates a combination of inherited and relative styles. The formatted text in example 4-5 should look like "**abc***def*". Like example 4-4, the association of *Heading* with "abc" causes the formatting state vector component *FontFace* associated with the content starting with "a" to be changed to be *Bold*. When the environment associated with *Heading* is finished, the formatting state vector has its initial values. When the next environment is reached (associated with *MajorHeading* and "def"), the formatting state vector is changed according to the definition for *MajorHeading*. The definition for MajorHeading specifies that it inherits from *Heading*, and therefore the changes specified in *Heading* should be applied, followed by the rule changing the *FontFace* component. By our definitions, all uses of a component on the right side of an equation refer to the value of the formatting state vector before the environment was applied. Thus the application of *Heading* to the state vector first changes the *FontFace* value from empty to *Bold*. The additional equation in *MajorHeading* also assigns the value of *FontFace* by referring to the value before the environment, hence it adds *Italic* to the empty set yielding just *Italic*. The effect of *Heading* on the *FontFace* has been lost.

Style Definitions:

```
Heading = FontFace←Bold;
MajorHeading = Heading;
               FontFace←FontFace + Italic;
```

Example text:

```
@root{
  @Text-Heading{abc}
  @Text-MajorHeading{def}
}
```

Example 4-5: Use of Inherited and Relative Styles

Other systems use a slightly different definition: the value of a reference to the formatting state vector on the right side of an equation is the formatting state vector returned from the latest parent style (if any) rather than always from the immediately enclosing environment. If these semantics were followed, then the result of applying the *FontFace* equation in *MajorHeading* wouid be to add *Italic* to the current value of *Bold*, and one would see "**abc*def***".

A global change is shown in example 4-6. Example 4-6 differs from the previous examples in that it uses different components of the formatting state vector, and makes global changes to all of the following formatting state vectors of enclosing environments. The details of this example are explained below.

We assume that the formatting state vector for the document system has a *pageheader* component, which corresponds to some string value to be used in running headers. Whenever a new page is started, the value of the *pageheader* component will be used for the page header. Initially, the value of the *pageheader* is empty, that is, no header is produced. Whenever the *UseHeader* style is applied, the text in the associated environment becomes the value of the *pageheader* component.

The example document consists of five *paragraph*s and two *UseHeader* applications. Several of the *paragraph*s are collected into a *section* (*paragraph*s "ccc" through "fff"). The text, or main body, of the document is the sequence of paragraphs: "aaa", "ccc", "ddd", "fff", "ggg"—"bbb" and "eee" are two possible

Style Definitions:

```
UseHeader = pageheader ⇐ set the page header
                         to a new, global
                         string value
                         equal to text in
                         document part
```

Example text:

```
@root{
   @paragraph{aaa}
   @text-UseHeader{bbb}
   @section{
      @paragraph{ccc} ...
      @paragraph{ddd}
      @text-UseHeader{eee}
      @paragraph{fff}
      } ...
   @paragraph{ggg}
}
```

Example 4-6: Simple Use of Global Component

page headers. This content may fit entirely on one page or on several pages. Each page that is produced contains a header. Since the root value of *pageheader* is defined to be empty, the pages on which the content "aaa" is formatted will have an empty header. If the *pageheader* component were local, then use of the *UseHeader* style in the environment starting with "bbb" would effectively cause no change in the formatted document, since it would affect the value of the *pageheader* component only in the content nested in the associated *text* part of the document. When "ccc" is formatted, it uses a formatting state vector associated with its environment. This formatting state vector is inherited, without any modification by a style, from its containing environment. The containing environment is associated with the *section* part of the document, which in turn gets the value of the formatting state vector of *root*. Hence the value of *pageheader* normally associated with "ccc" is normally the root value, or empty.

However, *UseHeader* specifies that a global change be made. Thus, using *UseHeader* will cause the string "bbb" to be imaged for page headers from the place where the style is used until the *pageheader* component is changed explicitly by another style in a subsequent environment. Thus, all new pages started for the contents of *text* "ccc" and "ddd" will have "bbb" for page headers. When the *UseHeader* environment associated with "eee" is reached, the same semantics will be applied and the new value of *pageheader* will be "eee" until changed again. Thus, the pages started by the content "fff" and "ggg" will have page headers of "eee".

Through these examples, we have introduced enough notation for describing the EXPRES document model. As the translation process is defined throughout this chapter, we will use this document notation to give examples of the document and its representation in ODA.

4.3 Available Content

We used two criteria for selecting the content types to be supported by EXPRES. First, we wanted a fair representation of the needs for proposal interchange. Second, we wanted to limit the scope of the project to ensure that we could complete it with available resources. To meet these criteria, we selected two collections of content types: one set from the ODA standard, and one set for the internal use of EXPRES. These are described in the next sections.

4.3.1 Predefined Content Architectures

We decided to use the processable versions of character and raster content architectures. We did not include either the formatted or formatted-processable versions since we were not interested in exchanging formatted information. In our community, most researchers have PostScript-compatible printers, and the desire to exchange formatted data is easily served by exchanging PostScript files. However, we did expect documents exchanged within EXPRES to be edited, so we needed to exchange the processable forms.

We elected to omit geometric graphics content for two reasons. First, we felt that limiting ourselves to two predefined types was sufficient to test the feasibility of using ODA as a multi-media interchange format. Second, there was no uniform graphics representation used by the word processing systems we were considering. As a result, the translation from some native graphics format to CGM would have been a substantial project on its own. This view is shared by other researchers in the field [42].

4.3.2 Additional Content Architectures

Our systems offered some content types that are included in proposals, but not supported by ODA, such as tables and equations. To see if ODA could be extended, we pursued implementations of these two additional contents. Rather than define a complete private content architecture, we chose to use some external (non-ODA) conventions. First, we created two new kinds of character content, tagged through the use of the "application comments" attribute to be content portions for tables and equations. Second, we chose a representation that matches the formats used by our system. Early in the project, we considered devising an interchange representation for these contents, but did not have time to complete the design.

4.4 Document Organization

The logical structure provided by ODA is abstract. As discussed in section 3.1.2, composite logical objects specified by the standard can be used to represent structures such as sections and diagrams. However, there is no a priori assignment of meaning to any logical object in ODA. As mentioned in section 3.4, one way to provide more structure in ODA is to define a *document*

application profile, which can define interpretations of some logical and layout structures. Superficially, it would appear that our needs would be best served by a level 2 profile, since the systems we were considering were essentially "multi-media document processors". However, two facts swayed our choice of a level 3 profile. First, at the time we started EXPRES (and even at the time of this writing), the NIST had defined only a level 3 profile. We wanted to use approved standards, and therefore, had no alternative to the level 3 profile. Second, the level 2 profiles, such as Q112, did not contain some ODA features we required. In particular, the level 2 profiles did not allow the use of multiple fonts or default value lists. As will be seen in later sections, we use both of these features heavily. Therefore, our document organization is based on the model provided in the document application profile published by NIST [48], technically part of the *Stable Implementation Agreements for Open Systems Interconnection Protocols.*

4.4.1 Subsetting the NIST Document Application Profile

The NIST document application profile defines a collection of document elements that denote various document structures. The set of document elements supported by the NIST document application profile is richer than those needed for EXPRES. Therefore, we decided to limit ourselves to a subset of their elements. Our list of logical components from the profile are shown in table 4-1.

We limited ourselves to the elements in table 4-1 in the hope that we would be able to interchange those features among our systems. As already noted, we did not process geometric graphics contents. In addition, we explicitly excluded end notes, multiple columns, numbered sections, page references, cross references and multiple page-number formats. But we expanded the list of available contents—the NIST profile does not define *Equation* or *Table*, but we included them as part of our explorations.

In the same way that we limited the possible document logical elements, we also limited the ways that they could be composed. The way we specified the ways that logical objects could be composed is through the use of a logical structure diagram. The notation in a logical structure diagram for describing document application profile constraints is essentially identical to the notation for generic structures. The interpretation of the diagram is that all legal specific logical structures must be derived by the diagram as if the diagram were a generic structure. Figure 4-1 shows a modified logical structure diagram of possible logical structures. This diagram is derived from NIST document ap-

Logical Object (Element)	Use
LogDoc	The entire document, composed of passages, paragraphs and figures
Passage	Collection of paragraphs, figures and other passages
Number	An integer (a leaf)
Paragraph	Collections of text, figures, phrases and footnotes
Phrase	Collection of text and other phrases
FNote	Collection of two numbers and footnote body
FNBody	Content of foot note
Figure	A raster, equation or table
Text	Collection of characters (a leaf)
Raster	Picture composed of pixels (a leaf)
Equation	Description of a mathematical equation (a leaf)
Table	A collection of columns of information (a leaf)

Table 4-1. Logical Objects From NIST Document Application Profile

plication profile diagrams[14] to show the universe of acceptable document structures.

The diagram for logical structures greatly reduces the kinds of logical objects that are permitted, and reduces the ways that the remaining kinds can be combined. Specifically, note that a document (*LogDoc*) is now a series of *passages*, *paragraphs* and *figures*, where a *passage* may contain the same elements as a document. A *paragraph* is now a sequence of *text*, footnotes (*fnote*), *figures* and *phrases*. A *phrase* may be nested and contains the same kinds of logical objects as a *paragraph*. The two additional kinds of basic logical objects, *table* and *equation*, are defined to be identical to the restrictions for the *text* basic

[14]See figures 16.4, 16.5, 16.6, and 16.7 of the NIST document application profile. The original diagrams do not fit on a single page, so the circled letters are used to denote connections between diagrams.

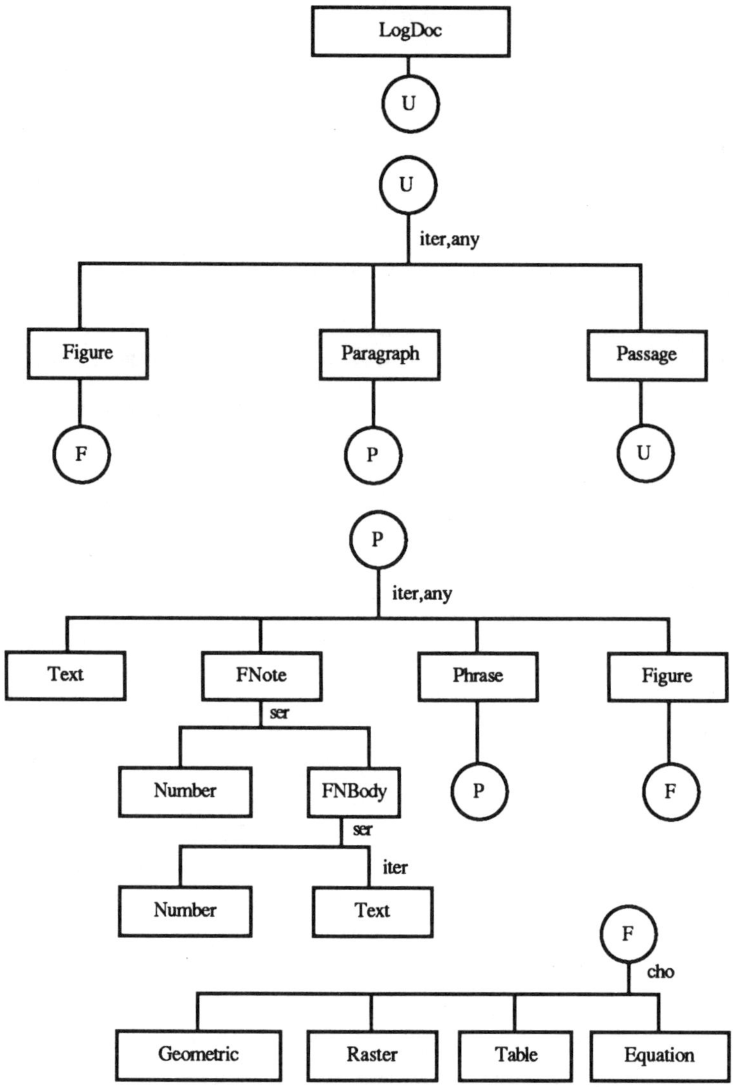

Figure 4-1. Use of the NIST Document Application Profile

logical object, except that the "application comments" attribute has the string "Table" or "Equation". These objects are used to hold octet strings that represent table or equation information understandable only to EXPRES participants, and are short term expedients until ODA has established standards for tables and equations.

4.4.2 Mapping Document Organization to ODA

The organization of a particular document provides its structure. We assume that each piece of the document's organization will be represented by an appropriate NIST logical element, though we do not attempt to define the mapping. The result should be a tree structure, where interior nodes of the tree are specific composite logical objects and the leaves are content portions (with associated basic logical objects). The edges in the tree are defined by the "subordinates" attribute.

A simple example will illustrate the ODA representation. We start with the document shown in example 4-7.

Example 4-7 shows two pieces of *text* that make up a *paragraph.* The *paragraph* is the only element under the root, *LogDoc.* The resulting complete specific logical structure for the document is illustrated in figure 4-2. In the figure, specific logical objects are boxes drawn with black lines, and the "subordinates" attributes are shown as black lines from one box to another. Each box is labeled with the name of the document element from the NIST document application profile that it represents. For example, the box labeled *paragraph* is a specific composite logical object representing a *paragraph* element. Technically, each piece of content should generate two boxes, a basic logical object (*text*) and a content portion that contains the characters. ODA provides several

```
@LogDoc{
   @Paragraph{
      @Text{aaa}
      @Text{bbb}
   }
}
```

Example 4-7: Document with a Paragraph

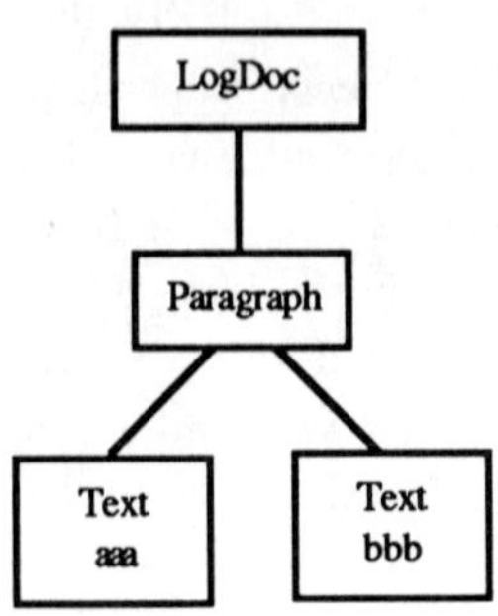

Figure 4-2. Simple Logical Structure for a Document

ways to associate content with a basic logical object, but this is the only one we will use, and so we will only draw one box to represent both a basic logical object and its associated content portion.

The logical elements defined by the NIST document application profile should be thought of as prototypes or metaclasses. For example, the NIST *paragraph* element can correspond to many kinds of paragraphs in a document: one for quotations, one for definitions and one for main-body text, for example. A native document system would manipulate three entities—say, quotations, definitions and body text—but all three would be NIST *paragraphs*. Fortunately, there is a way to distinguish between the different kinds of *paragraphs*: use of ODA's generic logical structure. In the next section, we explain how we use the generic logical structure for representing the native document's elements.

4.4.3 Interpreting Generic Logical Structure

We use the generic logical structure to distinguish parts of a document's organization and to describe style information. Like other parts of ODA, the facilities provided in the standard for a generic logical structure are quite general, and therefore, hard to use. Instead, we start with the specification of a generic logical structure provided by the NIST document application profile. Like those for logical elements, the provisions in the NIST document application profile for generic logical structure are still too rich for EXPRES purposes. For example, one of the features of generic structure in ODA is to restrict the ways that a document may be organized. Using this feature, one can require, for example, that an abstract must always be present. Although we did want to exchange information about available document structures, we decided that we did not want to exchange restrictions on document structure. None of the document systems we were using, or planned to use, could enforce such restrictions. Therefore, we circumvented the restriction specifications by ensuring that the generic structure generated by the EXPRES translators allowed arbitrary document organizations. This meets NIST document application profile requirements while not limiting the ways that a document could be manipulated by translators that enforced the restrictions in the generic structure.

We create a generic structure by building a structure parallel to the specific structure. In general, we need to define a separate generic logical object for each combination of document element and applied style, but a full discussion of the role of styles will be deferred until section 4.5.2. For now, we will omit styles and just assume that there is some collection of document elements that can be mapped to the NIST document elements, such as quotation to *paragraph*.

If the generic logical structure were an exact duplicate of the specific structure, then it too would form a tree. The interior nodes of the tree would be *composite logical object class descriptions* and the leaves would be *basic logical object class descriptions.* The nodes would be linked together using "generator for subordinates" attributes rather than "subordinates" attributes. This last difference represents a small problem.

In the specific logical structure, the value of the "subordinates" attribute lists the children of a parent. In the generic logical structure, the value of the "generator for subordinates" attribute describes possible subordinates that could be realized in the specific structure. Hence the value of the "generator for subordinates" attribute could define some restrictions on the editing of subordinates. As mentioned in section 3.1.3, the exact way that the restrictions are specified is through the use of a construction expression. In EXPRES, we permit only one form of construction expression in the generic logical structure, as shown below:

```
opt-rep(cho(req(id),req(id),...,req(id)))
```

In this expression, *id* refers to the identification of a child node. There is one *req(id)* in the expression for each child. The meaning of the expression is that the subordinates permitted in the specific structure may have any number of any of the children (including none) in any order. For example, this expression could be used to express the requirement that a *passage* contains any number of *figures* and *paragraphs*, in any order. A full graphical representation of this expression is shown in figure 4-3.

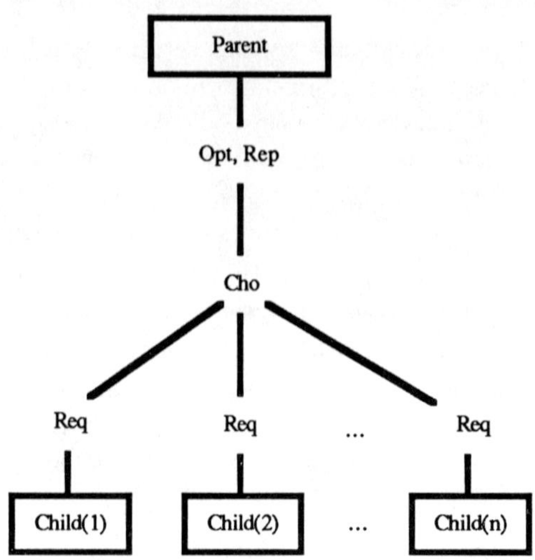

Figure 4-3. Expanded Diagram for "Generator for Subordinates"

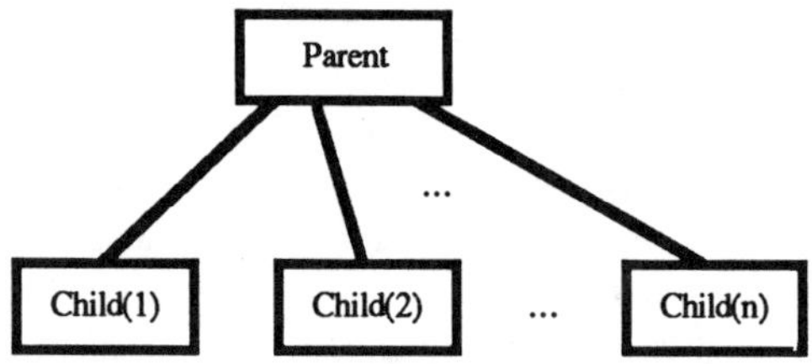

Figure 4-4. Shorthand for "Generator for Subordinates"

Drawing the entire construction expression clutters diagrams. Instead, we will use a conventional diagram showing the children of a parent, as illustrated in figure 4-4. However, the meaning is that each collection of edges from a parent represents the appropriate construction expression.

We have assumed, so far, that the generic structure exactly matches the specific structure. In fact, the generic structure could be a directed acyclic graph if it shares children, or could be an arbitrary directed graph with cycles. This situation would arise if one mapped nested parts, such as sections and subsections, of a document model into *passages*. The way that one specifies that each kind of section can contain a subsection (and subsubsection, ad infinitum) is by placing a loop in the generic structure: a *passage* would contain itself. Thus, any corresponding document structure could always be nested an additional level. The graphical representation of such a situation would show an edge from a node back to itself or to an ancestor node.

4.4.3.1 Building the Generic Logical Structure

In this section, we provide some rules that describe how to build a generic logical structure given a specific structure. For purposes of discussion, we assume that the specific structure is being built incrementally—as a translator reads part of the native format document, it builds the corresponding ODA structure.

When a new specific logical object is created for a particular part of the document, a generic logical object for the same kind of document part must exist and be in the right place of the generic logical structure. Specifically, the generic logical object must be present in the "generator for subordinates" for the generic logical object referred to by the specific logical object's parent[15]. Thus there are two steps: finding an appropriate generic object and ensuring that it is located properly in the generic structure. We will describe the steps and then work out some examples.

[15]See section 4.5.2, where this rule has to be modified to take into account styles.

The first step is relatively easy: if a generic object already exists for the document part, then the new specific logical object will refer to the generic object using the specific object's "object class" attribute. If the generic object does not already exist, then it must be created before being referenced by the "object class" attribute.

Once an appropriate generic logical object is found or created, it must be correctly positioned in the generic logical structure. To check for correct position, one searches from the new specific logical object to its parent specific logical object, and then to the generic logical object of the parent. If the desired generic logical object is already in the "generator for subordinates" attribute of the parent's generic logical object, then the generic structure is already complete. If the generic object for the document part already exists, but is not already in the "generator for subordinates" attribute of the parent's generic logical object, then the generic object must be added to the "generator for subordinates" attribute of the generic object logical referred to by the parent's "object class" attribute.

These rules can be illustrated by continuing the example shown in example 4-7. Our diagrams show generic logical objects as boxes with heavy black lines, while specific logical objects are boxes with thinner lines. The "object class" attribute is shown by a thick dashed line from the specific logical object to its generic logical object. Each box is labeled with the document part it represents. We start with a generic logical object for the entire document, called *LogDoc*. As the root of the tree, it is not subordinate to any other object. The resulting diagram is shown in figure 4-5.

Next, the *paragraph* of the document is added to the specific side. There needs to be a corresponding generic object. Since there does not yet exist a generic object for *paragraph* parts of a document, we need to create a generic *paragraph*. Because the specific *paragraph* is a subordinate of the specific *LogDoc* object in the specific logical structure, the generic *paragraph* must be a child of the the generic *LogDoc* in the generic logical structure. The result is shown in figure 4-6.

One follows the same procedure for adding the first specific *text* object and its corresponding generic object. The result is shown in figure 4-7.

Adding the second specific *text* object illustrates a difference in constructing the generic and specific logical structures. As before, we add a new specific logical object for the *text* object and add it to the subordinates of the containing *paragraph*. However, we have already created a generic logical object for a *text* document piece, so we do not need to create another one. Therefore, we can set

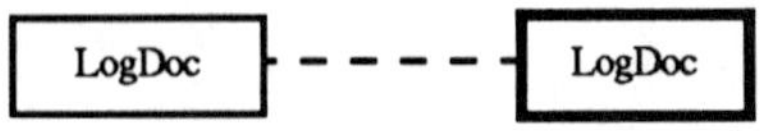

Figure 4-5. Starting the Generic Logical Structure for a Simple Document

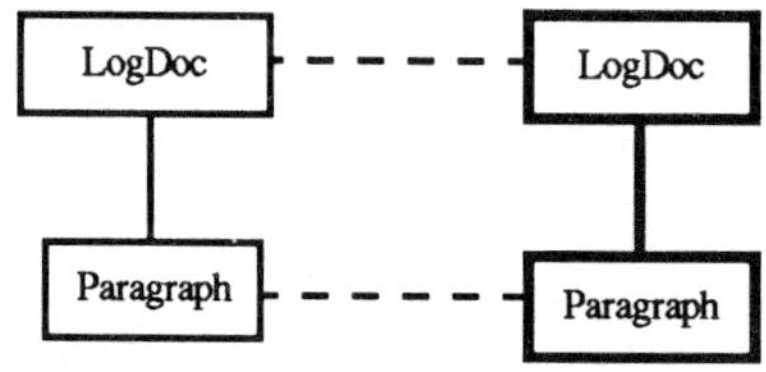

Figure 4-6. Partially Built Generic Logical Structure for a Simple Document

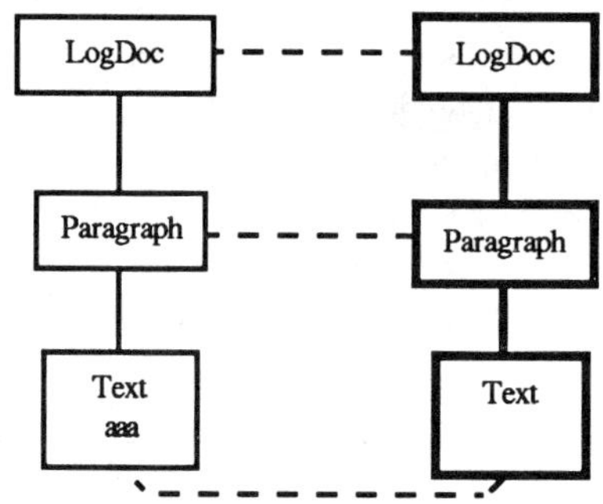

Figure 4-7. Adding Text to Partially Built Generic Logical Structure

the attribute "object class" of the newly created specific *text* object to refer to the previously created generic *text* object. However, we are not done yet. We must make sure that the edges in the generic logical structure are complete. This involves checking to see if the specific parent of the newly created specific object refers to a generic parent of the previously created generic object. Here, this means checking to see if the generic *paragraph* referred to by the specific *paragraph* is a parent of the generic *text* referred to by the newly created specific *text*. In the early examples we will be considering, it will be trivially true, but in later examples, we will need to change the value of the "generator for subordinates" attribute to ensure that the generic logical structure is *complete*[16]. The result of translating our simple document is shown in figure 4-8.

Another example shows how the same NIST document element can be used for different kinds of document parts that are provided by a native system. The example document is shown in example 4-8. Here, we distinguish the different kinds of *text* parts in the document by adding a parenthesized number where necessary. Example 4-8 is similar to example 4-7, except that the two *text* parts

---

[16]The NIST document application profile requires that the generic logical structure be *complete*, which roughly means that the actual document must be generable from the generic structure.

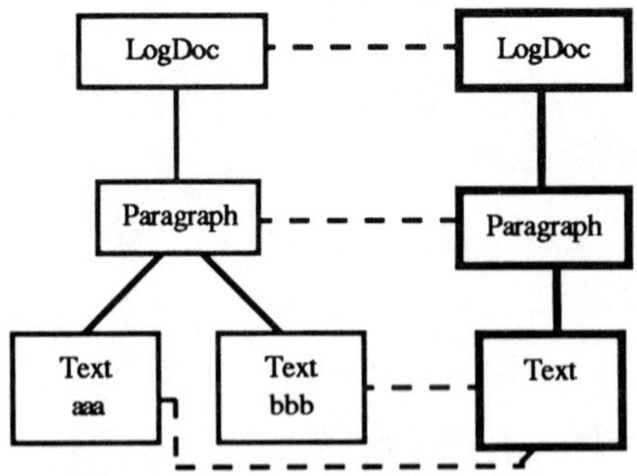

Figure 4-8. Generic Logical Structure for a Simple Document

```
@LogDoc{
   @Paragraph{
      @Text{aaa}
      @Text(2){bbb}
   }
}
```

Example 4-8: Reusing NIST Document Elements

within the paragraph are two separate kinds of document parts, though they both meet the specifications imposed by the NIST document application profile for *text* objects. For example, a native system might define one kind of text to be used for abstracts and another to be used for descriptive prose. These two kinds of *text* parts would both meet the *text* specifications, but to denote them as distinct parts of a document model, one would use two different generic logical objects. Thus, the generic structure for this example would have two generic *text* document parts in the generic structure, as shown in figure 4-9.

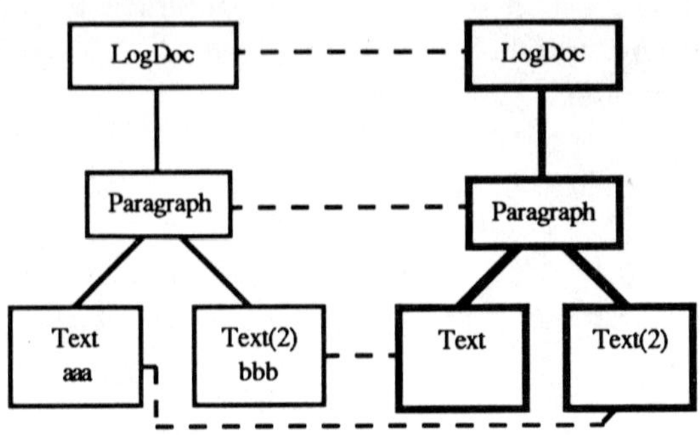

Figure 4-9. Differentiating Between Document Pieces

```
@LogDoc{
   @Paragraph{
      @Text{aaa}
   }
   @Paragraph(2){
      @Text{bbb}
   }
}
```

Example 4-9: Reusing NIST Document Elements for Interior Nodes

The example document in example 4-9 illustrates how the generic structure can be a directed acyclic graph instead of a tree. Like the document in example 4-8, the document in example 4-9 uses two different instances of an NIST document element. The two *paragraph* objects could be, say, a quotation and a definition, which are two distinct document parts in the native document model. Therefore, each of these document parts needs its own generic object. However, each paragraph may use the same kind of *text* object—the native system does not distinguish between different kinds of character content that may be part of either a quotation or a definition. Therefore, only one generic logical object for *text* is needed.

The interesting feature in this example is that each *text* document part is in a different kind of *paragraph* document part. If this document were processed in order, then a generic object for the first *text* document part already exists when the specific logical object for *text* document part containing "bbb" is created. Thus the "object class" attribute for the second *text* document part would refer to the same generic logical object as the first *text* document part. However, at the time that the specific logical object for the second *text* document piece is created, the generic *text* document part is a child of the generic logical object for the document part *paragraph.* But the generic *text* document part must also be a child of the generic logical object for the document part *paragraph(2).* Thus, the generic *text* object must be added to the "generator for subordinates" attribute of the generic logical object for *paragraph(2).* This results in the structure shown in figure 4-10.

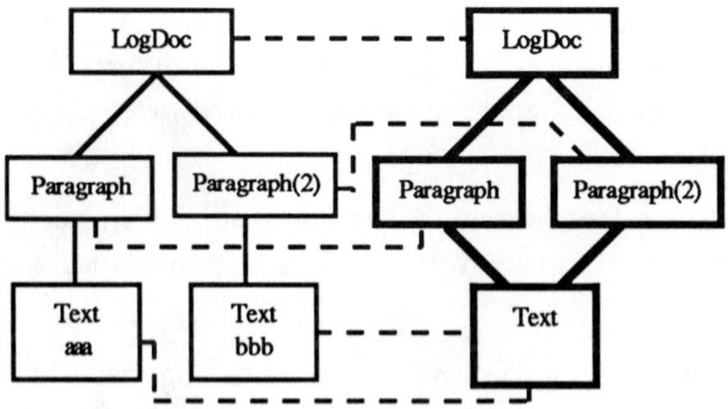

Figure 4-10. Sharing Generic Logical Structures

So far, our discussion has been limited to the logical structure in a document. We are also interested in exchanging formatting information. In the following sections, we give a description of our formatting model and then show how to translate that model into ODA.

4.5 ODA Structures for Styles and Environments

Styles in our document model contain both structure and information. The structure in styles results from the use of inherited styles: there is a structural relationship between the parent style and the inherited style. The information in a style is the set of instructions for how a formatting state vector should be changed. We do not define how the structure of styles should be represented in ODA[17]. Therefore, we limit the discussion in this section to how the information in styles should be represented.

In general, styles and environments are represented as pairs of layout styles and presentation styles. To distinguish ODA's styles from the styles used in the abstract document model by EXPRES, the latter will be referred to as *document styles*.

Most of the information of the formatting state vector can be represented as an attribute in a presentation or layout style. For purposes of exposition, we will assume that every document style and every environment will use a presentation and layout style. In practice, it is expected that only those presentation and layout styles needed to hold particular attributes will be defined and used.

---

[17]There is a proposal for adding a new attribute to layout and presentation styles ("variant of") that would define a structural relationship between styles, and if adopted, the new attribute would be used for this purpose.

Although both document styles and environments are represented using presentation and layout styles, the details of their use differ. The differences in use are related to the differences in meaning between a document style and an environment:

- A document style specifies *changes* to a formatting state vector while an environment is associated with the *value* of a formatting state vector;
- A document style may act on a *subset* of a formatting state vector while an environment describes an *entire* formatting state vector;
- A document style's interpretation is independent of where it is used while two environments that result from the association of the same document style can differ.

Because the values of nearly all layout directive attributes and presentation attributes in ODA are constants (rather than expressions), only absolute document styles can be represented perfectly in ODA. Later, we will discuss how relative document styles should be represented.

4.5.1 Use of Default Value Lists

Many of the document style representations use layout directives that apply only to basic logical objects, not to composite logical objects. Other document style representations use presentation attributes that also can be specified only for basic objects. However, the document part that the style is associated with may be represented as a composite logical object. For example, one may specify that an entire *paragraph* should be indented, which is represented by the layout directive "offset". Because a *paragraph* can be composed of several *phrases*, a *paragraph* is a composite logical object. Therefore, changes to an "offset" attribute would not apply to it. The effect one desires is that all descendent basic logical objects of the *paragraph* would have their "offset" attributes adjusted.

One can specify the values for attributes in subordinate basic objects through the use of the "default value lists" attribute on the ancestor composite object. For example, one can specify a value for the "default value lists" attribute for the *paragraph* such that the "offset" attribute for descendent basic logical objects reflects the desired indentation. Thus, when the information in a document style should be associated with a composite logical object, the appropriate attributes will be placed in presentation and layout styles on the composite object's default value list for descendant basic logical objects, not directly on the composite object. Some examples illustrating this rule are given in the next section.

4.5.2 Adding Constituents for Representing Document Styles

Before going into the details of how the values are calculated for individual attributes, we present some examples showing how styles and environments affect the logical structure of an ODA document. In particular, these examples illustrate how layout and presentation styles are used in the default value lists of specific and generic logical structures.

The document in example 4-10 shows a very simple document with some styles applied. Example 4-11 shows how the document styles and environments are represented. Associated with each logical object (both specific and generic) is a layout style and presentation style. Styles associated with the specific logical objects contain the relevant pieces of information describing *current values* of a formatting state vector associated with an environment. Styles associated with the generic logical objects contain the relevant pieces of information describing *changes* to be made to a formatting state vector by a document style. As mentioned before, some attributes apply to composite logical objects, and some to basic logical objects. We will ignore the distinction for now. Even so, note that three different document styles are used, in three separate environments. This results in 12 style objects[18].

Figure 4-11 also illustrates some new diagram notations. Styles are represented using boxes with dashed lines, labeled with the name of the associated document style. The dotted line from a logical object to a style represents a "layout style" or "presentation style" attribute in the logical object. Normally, the attribute would be in a default value list for basic logical objects, but as will be explained in section 4.9.11, there are circumstances when the attribute must be placed on a composite logical object directly.

For now, we use the convention that both a layout style and a presentation style will be specified. Therefore, we will cease making the distinction between layout and presentation styles, and just refer to a single style from each logical object. Similarly, we also make no distinction between a style referenced in a

```
@LogDoc-root{
   @Paragraph-style1{
      @Text-style2{aaa}
   }
}
```

Example 4-10: Document without Shared Styles

---

[18](3 document styles + 3 document environments)*(2 ODA style objects)

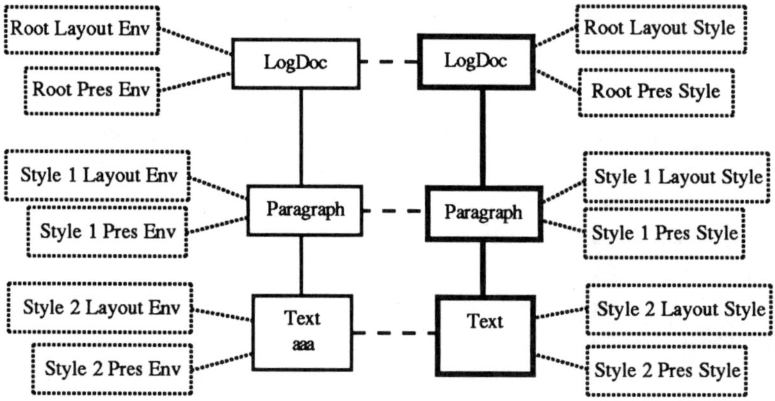

Figure 4-11. Styles on Logical Components

logical object and being referenced through a default value list on the logical object. Thus, we can reduce some of the clutter in the diagram, to give us figure 4-12. Our next document is specified in example 4-11. It shows multiple document parts using the same application of styles. The diagram in figure 4-13 shows how the structure of the document would be represented in ODA. In the figure, similar specific logical objects refer to the same generic logical objects. For example, both specific *paragraphs* refer to the same generic *paragraph*. The figure also shows how, in this particular instance, layout and presentation styles for the environments can be shared. In this example, it is assumed that the environments for both paragraphs are the same and thus, the ODA styles can be the same.

Our next example shows another common situation. In example 4-12, we show a document that contains the same kind of document parts, but with different document styles associated with them. As mentioned in section 4.4.3, one must take into account the use of document styles when creating the generic

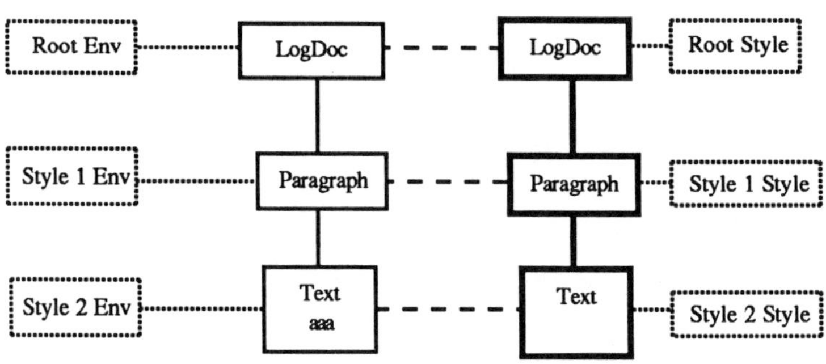

Figure 4-12. Collapsing Layout and Presentation Styles

```
@LogDoc-root{
   @Paragraph-style1{
      @Text-style2{aaa}
   }
   @Paragraph-style1{
      @Text-style2{bbb}
   }
}
```

Example 4-11: Multiple Document Parts with Repeated Document Styles

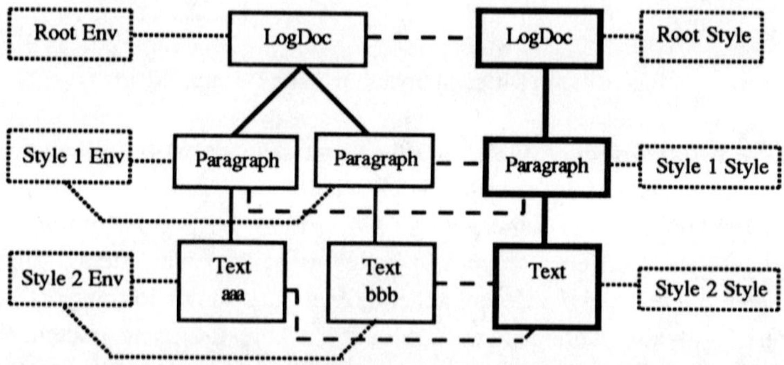

Figure 4-13. Reusing Document Styles

```
@LogDoc-root{
   @Paragraph-style1{
      @Text-style2{aaa}
   }
   @Paragraph-style3{
      @Text-style2{bbb}
   }
}
```

Example 4-12: Same Document Parts with Different Document Styles

logical structure. In particular, a new generic logical object is needed for every pair of document part and document style rather than just for every document part. The result of applying this rule is shown in figure 4-14.

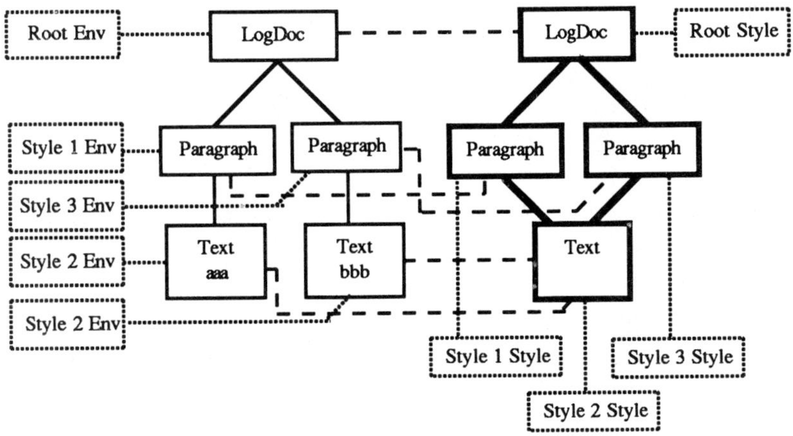

Figure 4-14. Different Styles on Same Document Parts

We have two *paragraphs*, which are the same abstract entity in the document. Unlike the document in example 4-11 where one could have both specific *paragraphs* refer to the same generic *paragraph*, the two *paragraphs* in example 4-12 have different document styles associated with them. Therefore, they need to refer to two different generic objects, as shown in figure 4-14.

Figure 4-14 also demonstrates a situation where the *same* document style could result in *different* environments. Although document style *style2* is applied twice, the first time it is within an application of document style *style1* while elsewhere it is within an application of document style *style3*. Thus the resulting environments for the different association of document style *style2* could result in two different environments. The diagram assumes that different environments resulted.

The next document, shown in example 4-13, contains two different parts of a document that are associated with the same document style.

```
@LogDoc-root{
   @Paragraph-style1{
      @Text-style2{aaa}
   }
   @Paragraph-style2{
      @Text-style3{bbb}
   }
}
```

Example 4-13: Different Document Parts with Same Document Style

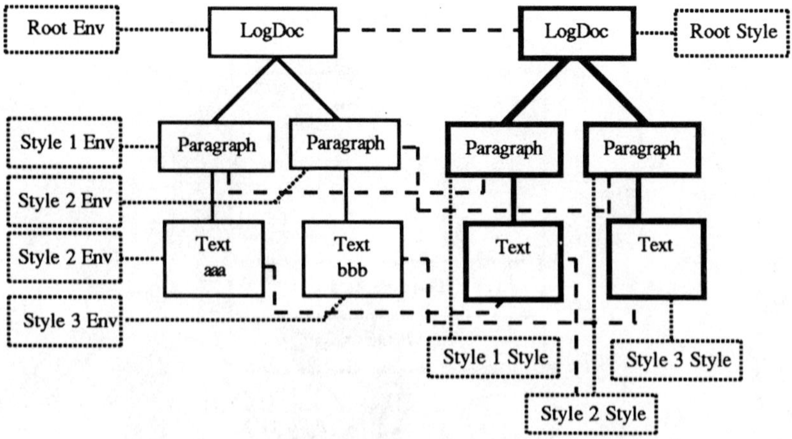

Figure 4-15. Sharing Document Styles Among Different Document Parts

The resulting structure for this document is shown in figure 4-15, which illustrates how sharing is accomplished when the same document style is associated with different document parts. Here, document style *style2* is used by both a *paragraph* document piece and a *text* document piece. This results in two different style *style2* environments, because they are applied in different contexts. Something not shown explicitly in this diagram is that the presentation and layout styles are used slightly differently for *paragraphs* and *text*. Since *paragraphs* are composite objects, the references to the styles are specified in the default value list while the styles are specified directly in the *text* document piece, which is a basic logical object.

So far, we have been concentrating on how document styles affect the logical structure of a generated ODA document. Document styles also affect page layout, which is described by the layout structure. The way that styles are represented in the layout structure is the topic of the next two sections.

4.6 General Page Layout

There is a great variety of possible formatting state vector components and page formats available in contemporary document systems. In the EXPRES project, we chose to restrict ourselves to a simple page layout, as illustrated in figure 4-16. Parts of a document that are designated to go into a header or footer area are placed into the top or bottom part of a page respectively, and are repeated for each new page of the document. Footnotes are placed into the *Footnote* area on the same page that they are referenced. Content that is not in a footnote,

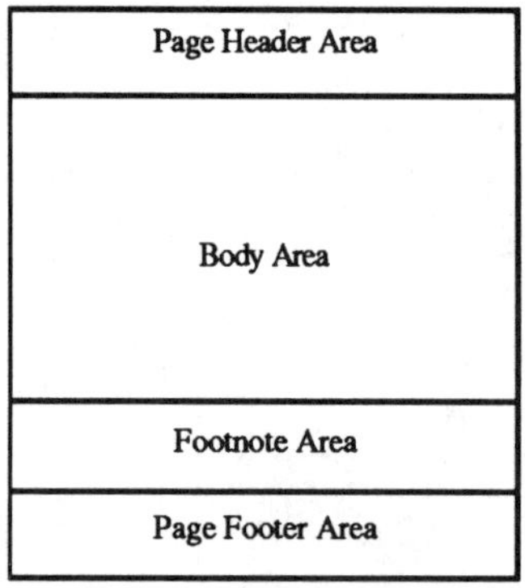

Figure 4-16. EXPRES Page Layout

header or footer is placed in the *Body Area*. The *Page Header Area* and *Page Footer Area* are fixed. The combined area for the *Body Area* and the *Footnote Area* is also fixed, but the partition of space between the body and footnotes areas may change as a result of the amount of body content and footnote content to be placed on the page.

As explained in the next section, it is easy to represent our page layout in ODA.

4.7 Generic Layout Structures for General Page Layout

ODA uses a generic layout structure to represent page layout. Just as the generic logical structure limits the available document organizations, the generic layout structure limits the available page organizations. Translation becomes easier when the range of possible page layouts is restricted.

The NIST document application profile limits the possible generic layout structures that can be used. The EXPRES project further restricts the available generic layout structures to match the page description shown in figure 4-16. The restrictions are summarized in the diagram in figure 4-17[19].

Most of the boxes describing layout elements are straightforward. The generic layout document root, *LayDoc*, must have a single page set, *PageSet*. The generic page set may have any number of *pages*, where each *page* represents an "example" or "template" page that meets our requirements for a header, footer,

[19]This figure is derived from figures 16.8, 16.9, 16.10 and 16.11 in the NIST document application profile.

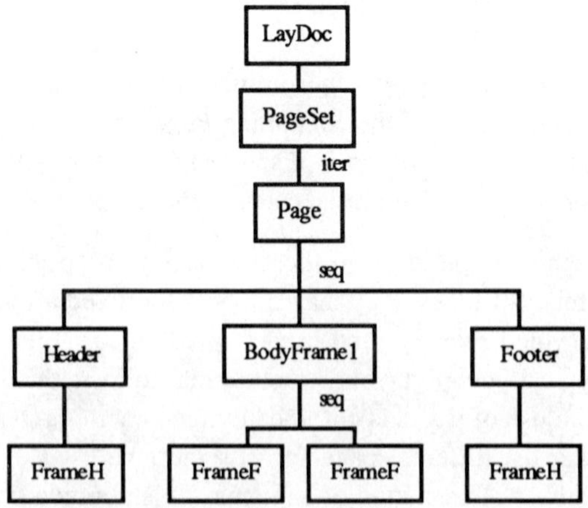

Figure 4-17. Detailed Generic Layout Structure for EXPRES

body and footnote areas. These requirements are met through the use of a frame for a header (*Header*), a body (*BodyFrame1*) and a footer (*Footer*). Further, each body must provide a frame for the body blocks (*FrameC*) and a frame for the footnote blocks (*FrameF*). The essential restrictions on this structure specified by the NIST document application profile are:

1 *Header* and *Footer* frames are at a fixed position. However, content within these frames can be positioned by attributes placed on the contained *FrameGs*.

2 Generic objects constrained by *FrameC* requirements have a fixed horizontal size but may contain any number of blocks, expanding their vertical size as needed ('RuleB'). This allows body content to expand to fill the body area until footnotes are reached.

3 Generic objects constrained by *FrameF* requirements have a fixed horizontal size but may contain any number of blocks, expanding their vertical size as needed ('RuleB'). This allows footnote content to expand to fill the footnote area until the body area is reached.

To ensure that the generic layout structure meets the EXPRES document model, we add further restrictions to those imposed by the NIST document application profile:

1 The "fill order" parameter of the "position" attribute for *FrameC* has a value of 'normal order'. This fills all body content from the top of the page while footnotes fill from the bottom of the page.

2 The "permitted categories" attributes in *FrameC* and *FrameF* are assigned values of the form "Text-N"[20] and "Footnotes-N", where N is an integer. The value of N is the ordinal position of the containing *Page* in the subordinates of the containing *PageSet*. The exact use the "permitted categories" attribute is discussed in section 4.9.12.1. All other frames have a null "permitted categories" attribute.

3 Although each example page in the generic layout structure may be used any number of times, they may not be intermixed arbitrarily, and the order in which they may be used is predefined. Figure 4-18 illustrates a typical generic layout structure that follows this restriction. The sequential use of pages is enforced by the generic layout structure specifying that the different page formats must be used in sequence. Thus, any number of each kind of page may exist, but we fix the order in which each kind of page may be used. Once a certain kind of page is finished being used in formatting, it may not be used again.

4 The *FrameH* used for headers and footers is restricted to have a single *Text* or *Number* basic logical object referenced by its "logical source" attribute.

The result of these restrictions is a simple page layout with a fixed header and footer. The detailed formatting of contents is controlled by values in the formatting state vector. In the next section, we discuss the components of the formatting state vector and how they are represented in ODA.

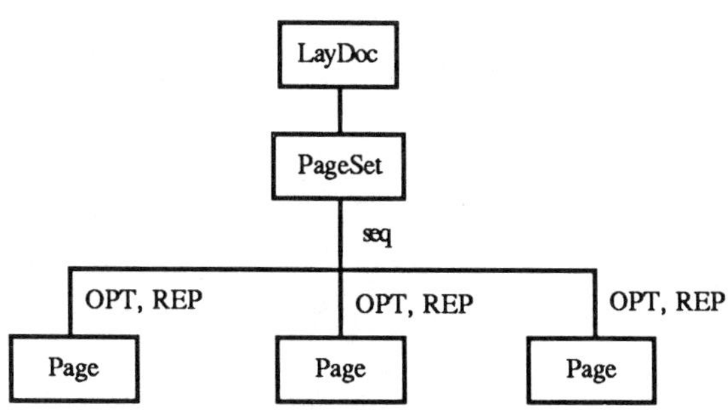

Figure 4-18. Using a Sequence of Pages for Different Formats

[20]Although we use the word "text", any kind of content may be formatted in the frame. In retrospect, "Body-N" would have been a better designation.

4.8 Formatting State Vector

For the EXPRES project, we selected a formatting state vector that contains important features found in the systems we were using. We tried to define a model that met our needs rather than one that was the intersection or union of the demonstration systems. The components we defined are displayed in tables 4-2 and 4-3.

FSV Component	Meaning
Left Margin	The amount of space between left edge of paper and left edge of content in the body area
Right Margin	The amount of space between left edge of paper and right edge of content in the body area
Top Margin	The amount of space from top edge of paper to top-most edge of text area (this is the same as available Page Header Area)
Bottom Margin	The amount of space from top edge of paper to bottom-most edge of footnote area
Line Spacing	The amount of space from one baseline of text to the next
Baseline Offset	The amount of space from baseline to vertical placement of content, such as subscripting
Paragraph Spread Above	The minimum amount of space between current paragraph and preceding
Paragraph Spread Below	The minimum amount of space between current paragraph and following
Paragraph Indentation	The amount of space the that first line of the paragraph should be indented
Justification	Alignment of content relative to margins: left, right, centered or both
Font Family	The family of typeface to use for characters, such as Times or Helvetica

Table 4-2. Formatting State Vector Components for EXPRES

FSV Component	Meaning
Font Face	Modifications of the typeface, which is a subset of {bold, italic, shadow, outline}
Font Presentation	Modifications of the characters appearance, which is a subset of {underline, strike, overbar, fixed spacing, condensed spacing}
Font Size	Design size of typeface in points
Tab Stops	A list of tabs stops with an interpretation of each stop as left adjusted, right adjusted, centered or aligned around a specific character (along with the character)
Line Break	Whether the content should start a new line
Paragraph Break	Whether the content should start a new paragraph
Page Break	Whether the content should start a new page
Header	Current header value
Footer	Current footer value
Footnote	Content targeted for the footnote area
Text	Content targeted for the main-body area

Table 4-3. Formatting State Vector Components for EXPRES (continued)

Obviously, individual systems can have more or less information in their corresponding formatting state vectors, but we found this collection of features to be adequate for the EXPRES demonstration.

4.9 ODA Attribute Values for Styles and Environments

We can now discuss the details of translating formatting state vectors, document styles and environments into ODA attribute values. The section starts with a brief overview of how formatting state vector components are represented in ODA and then explains the diagrams that we will be using in further illustrations. The rest of this section is organized by component of the formatting state vector. Within each section, there is a discussion of how a particular component of an environment's formatting state vector can be represented in the styles referred to by the specific logical structure, followed by a discussion of how to represent, in the generic structure, a change in that component specified by a document style.

4.9.1 General Mapping of Formatting State Vector to ODA

Some components of the formatting state vector are represented in the logical structure of a document, others in the layout structure. Therefore, we divide the discussion of representation into two parts. The next section discusses the parts of the formatting state vector that are represented in the logical structure while the following section discusses how some of the information is represented in the layout structure.

4.9.1.1 Attribute Representations

Most components of the formatting state vector are presented as layout directive attributes or presentation attributes in the ODA logical structure. As noted in section 4.1, ODA provides a variety of ways to represent certain pieces of information[21]. Table 4-4 shows the representation that the EXPRES participants decided to use for components in the formatting state vector.

The mapping specified in table 4-4 does not give the details of how to calculate the values for the attributes and parameters, nor where to put the attributes. The details depend a great deal on the way that styles and environments are represented and so will be considered later on a case-by-case basis.

FSV Component	ODA Representation
Left Margin	"Offset", "right-hand offset" and "indentation"
Right Margin	"Offset" and "left-hand offset"
Line Spacing	"Line spacing" and "proportional line spacing"
Baseline Offset	"Character fonts", "fonts list" and SGR control function
Paragraph Above Spread	"Separation" and "trailing edge"
Paragraph Below Spread	"Separation" and "leading edge"
Paragraph Indentation	"First line offset"
Justification	"Alignment" and "block alignment"
Font Family	"Character fonts", "fonts list" and SGR control function
Font Face	"Character fonts", "fonts list" and SGR control function
Font Presentation	"Character fonts", "fonts list" and SGR control function
Font Size	"Character fonts", "fonts list" and SGR control function
Tab Stops	"Line layout table" (in conjunction with STAB control function)
Line Break	CR, LF control functions
Paragraph Break	"Concatenation"

Table 4-4. Attributes for Representing Formatting Features

4.9.1.2 Generic Layout Representations

Some components of the formatting state vector are presented as attributes within the generic layout structure, or require the creation of new generic layout structures. The generic layout strategy is intended to ease the translation of components of the formatting state vector that affect page layout. The EXPRES participants decided to use the representations given in table 4-5 for components in the formatting state vector that affect page layout.

FSV Component	ODA Representation
Top Margin	"Dimension", "position" and new generic *page*
Bottom Margin	"Dimension", "position" and new generic *page*
Page Break	"New layout object"
Header	"Logical source" and new generic *page*
Footer	"Logical source" and new generic *page*
Footnote	"Permitted categories" and "layout category"
Text	"Permitted categories" and "layout category"

Table 4-5. Attributes for Representing Page Layout Features

The mapping specified in table 4-5 does not give the details of how to calculate the values the attributes and parameters, nor where to put the attributes. The details depend a great deal on the way that styles and environments are represented, and so will be considered later on a case-by-case basis.

4.9.2 Diagrams of Attributes

So far, our diagrams have concentrated on illustrating structural attributes, that is, attributes that describe how various components are related. The values of these attributes have been lines between components. However, the representation of formatting information will involve primarily layout directive attributes and presentation attributes that are enumerated values or numbers. Thus, it makes more sense to represent these attributes textually instead of graphically.

Technically, layout directive and presentation attributes should be placed in an appropriate style. For example, if a document has a root document piece that has a style that sets the "indentation" attribute to 1, the diagram should appear as in figure 4-19. The value for "indentation" is denoted by the *IN* value. Unlike the details of previous diagrams, the structural details illustrating the style information is pedantic, clutters the diagram and makes it more difficult to find the

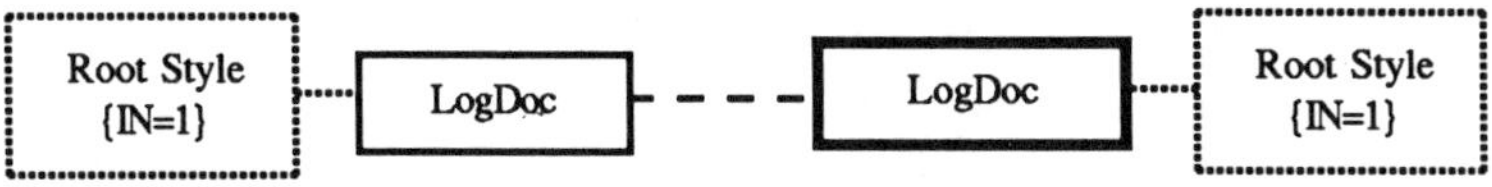

Figure 4-19. Complete Diagram Showing Layout Style Attributes

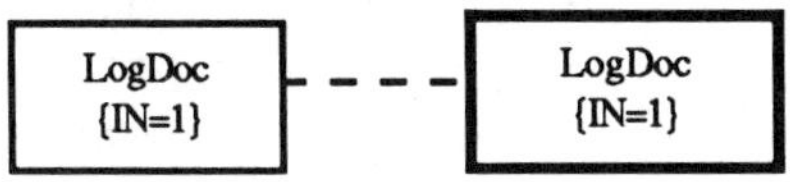

Figure 4-20. Collapsing Style Information into Other Components

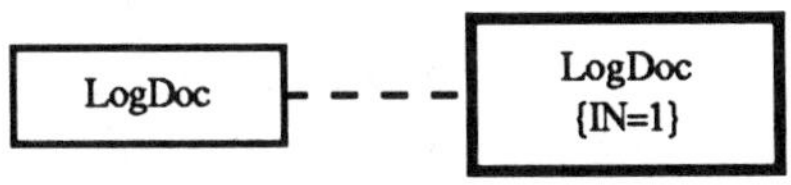

Figure 4-21. Eliminating Redundant Styles from Specific Logical Structure

necessary information. Therefore, unless required, the attributes will be denoted as being placed directly on the logical object, such as shown in figure 4-20. Further, it will be the case that most values specified for attributes will be identical in the styles specified by both generic and specific logical objects. When this is the case, the styles for the specific logical object are unnecessary and would normally be left off (in the ODA document as well as the diagram). This would result in a diagram as shown in figure 4-21. As appropriate, we will use the least cluttered version of a diagram.

When a part of a document is formatted, the formatting state vector is changed as specified by the associated document style. It is anticipated that most document styles will make few changes to the formatting state vector. Thus, the styles referred to by the generic logical objects, which represent document style information, will contain only those attributes that represent changes to be made to a formatting state vector. Similarly, styles will be needed for the specific logical object only when an attribute for an environment differs from the corresponding attribute for a document style. Thus, the styles specified by specific logical objects will specify only the necessary changes, and not the entire contents of the formatting state vector. The inheritance mechanism of ODA will retrieve all unspecified attributes from enclosing logical objects.

4.9.3 Distinguishing Style and Environment Information

ODA's attributes naturally reflect the values of a formatting state vector in a document piece or environment. For example, at any given point in the document, there is a particular right margin, and thus, the ODA attribute (or parameter) that represents a right margin should have the same value. However, document styles are not strictly values—document styles describe how to change values. In come cases, the change is simple to represent. For example, absolute document styles contain rules of the form "set the value to X". An obvious way to represent absolute document styles is by specifying an appropriate attribute that has the value X. Unfortunately, such a simple rule will not work for relative document styles, such as "add 1 inch to the current right margin." Using the value 1 could be misleading, since there is no way to distinguish between the addition of 1 inch and setting an absolute value of 1 inch. Further, there is no way to discern what base value is being referenced. For example, a new right margin might be measured from the old right margin, from the line width, from the edge of the paper or from some other fixed point. Thus, even if we knew that the information represented a relative document style, the "1" in the attribute does not convey enough information about the form of the relative calculation. None of this information is naturally represented in ODA.

The approach used for representing document styles in ODA is that they will be represented as environments generated by applying the document style to the root environment of a document. For example, if a root environment specifies that a right margin is 1 inch, and the document style "add 1 inch to the current right margin" is applied, the new right margin will be 2 inches. The value of 2 will be used in the appropriate ODA attribute to represent the document style information. The problem of deciding how a style should be translated back from ODA into a native document style system is system dependent.

Earlier, tables 4-4 and 4-5 listed the ODA attributes to be used for various pieces of the formatting state vector. The rest of this chapter explains in detail how each component of the formatting state vector is represented for environments and document styles.

4.9.4 Left Margin, Right Margin and Paragraph Indentation

This section discusses the representations used for values of and changes to the left margin, right margin and paragraph indentation. Paragraph indentation is included since there are interactions between the left margin calculations and indentation calculations. After discussing how the margins should be represented, several examples are worked out.

4.9.4.1 Left Margin and Paragraph Indentation

According to our document model, the left margin is where the body of the paragraph lines up and paragraph indentation is a signed number that describes where the first line should be started relative to the left margin. This is illustrated in figure 4-22.

The semantics for left margin in our model differ from ODA semantics. ODA defines a left margin of a block (relative to an enclosing frame) as the "right-hand offset" parameter of the "offset" attribute[22]. However, using the "offset" attribute as the value of the left margin would prohibit the use of outdenting (also called overhang). This is because no content may extend beyond a block boundary and the outdented part of the first line would extend beyond the left edge of the block. To compensate for this effect, the left margin and paragraph indentation must be used together to specify the values for the "right-hand offset" parameter of the "offset" attribute, the "first line offset" presentation attribute and the "indentation" presentation attribute. The relationships among

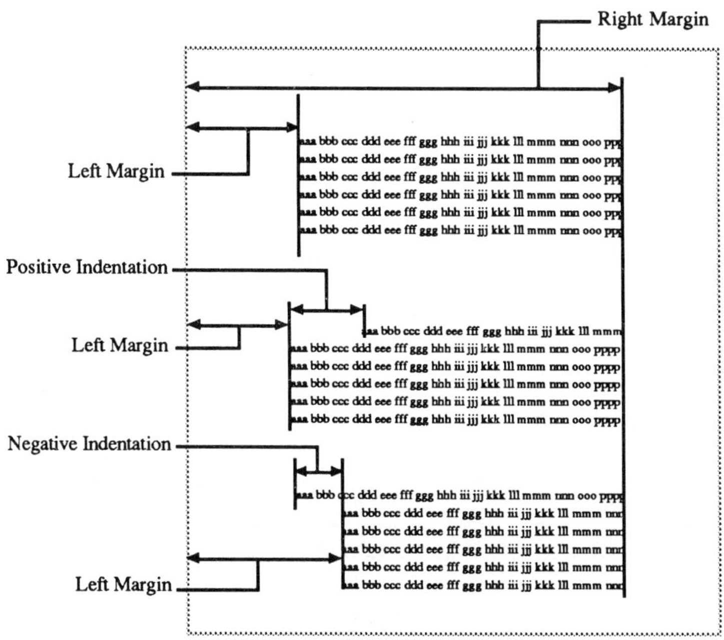

Figure 4-22. Typical Paragraph Layouts

[22]To simulate the page orientation defined by ODA, pretend you are standing at the top of the page and walking towards the bottom. The left margin then appears to be on the right. Similarly, the right margin is on the left, "leading" is towards the bottom of the page and "trailing" is towards the top of the page.

these attributes, and the left margin and paragraph indentation, are summarized in examples 4-14 and 4-15.

For these and subsequent calculations, we make no assumptions about units or scaling. We assume that a translator implementing our strategy will scale appropriately between the native system's units and ODA's scaled measurement units.

In general, when a change to the left margin is specified by a document style, a layout style for a generic logical object must be created to hold the attributes that represent the changes, and a layout style must be created for a specific logical object to represent the environment's value of the left margin after the change has been carried out. As mentioned before, if the resulting value is the same for both the document style and the environment, then the layout style for the specific object is unnecessary. The values for margins will differ only when relative styles are encountered.

We must take care to implement properly the definition of left margins in the presence of non-text content. When an outdented paragraph is specified, we change the "offset" attribute to match the appearance of our model. For a raster, however, the left margin is not outdented, and so we must not compensate the

```
oda compensation =
  if paragraph indentation < 0
     then paragraph indentation
     else 0

"right-hand offset" parameter =
  left margin + oda compensation

"indentation" presentation attribute =
  - (oda compensation)

"first line offset" attribute =
  paragraph indentation
```

Example 4-14: Converting Left Margin Value from EXPRES Model to ODA

```
left margin =
  "right-hand offset" parameter
  + "indentation" presentation attribute

paragraph indentation =
  "first line offset" attribute
```

Example 4-15: Converting Left Margin Value from ODA to EXPRES Model

value for the "offset" attribute. Thus, when a document style specifies an outdented paragraph, and a basic logical object contains raster, table or equation content, then the compensation to allow overhanging must be eliminated. In this case, the specific logical object must contain a layout style that will adjust the "right-hand offset" so that no compensation is added, and the "right-hand offset" will be identical to the left margin.

4.9.4.2 Right Margin

The value of the right margin is represented as the "left-hand offset" parameter of the "offset" attribute. In general, changes to the right margin will result in two layout styles being generated: one for the document style that specified the change to the right margin, the other for the environment that has the new value of the right margin. Like the styles for left margins, the style for the environment should be used only if the "left-hand offset" parameter would have a different value than the style referred to by the generic logical object.

4.9.4.3 Examples of Changed Margins

For purposes of discussion, we assume that changing margins will not force any kind of line or paragraph break. Our document model factors out this information into other components of the formatting state vector. Systems that combine these components would always represent multiple changes at the same time. For purposes of the examples, we also assume that the root environment starts with a left margin of 1, a paragraph indentation of 0 and a right margin of 7.5. Further, we assume that the scaled measurement units used by various ODA attributes and parameters are inches. (Technically, scaled measurement units must be integers, but for purposes of the examples, they can be assigned real values. An actual implementation would describe the scale for scaled measurement units in the document profile and the numbers from the formatting state vector would have to be scaled and rounded appropriately.)

For the following examples, assume that the following document styles have been defined (*LM* denotes left margin, *RM* denotes right margin, *PI* denotes paragraph indentation):

```
LeftIndent = LM←2;

RightIndent = RM←6.5;

BothIndent = LM←2; RM←6.5;

MoveIn = LM←LM + 1;

PIndent = PI←1;

Outdent = PI← -0.5;
```

We assume that no line breaks occur, so it is possible that no visible effect may be seen (since the effect of the environment would never be visible at a line boundary and, therefore, never get an opportunity to affect line formatting decisions). Recall that the abstract document model expresses right margins as measured from the left edge of the paper. This is unimportant, except for the calculations present in the examples. In the examples, we will dispense with extraneous definitions of document pieces and styles; many document pieces will have no explicit style. Finally, the examples show only the part of the logical structure under discussion, rather than filling in all of the details.

Example 4-16 shows a document that has a changed left margin. A diagram of the translated document is shown in figure 4-23.

Figure 4-23 shows what happens when a left margin is set to a new value. The "right-hand offset" is assigned a new value and the layout style is associated

```
@LogDoc-root{
   @Paragraph-LeftIndent{ ... }
}
```

Example 4-16: Document with Changed Left Margin

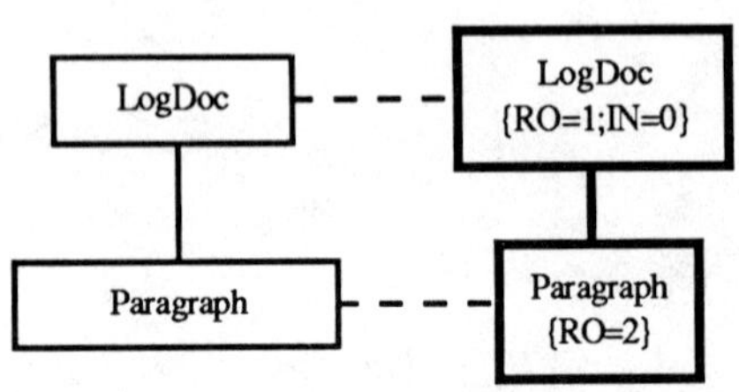

Figure 4-23. Representing a Change to the Left Margin

```
@LogDoc-root{
   @Paragraph-RightIndent{ ... }
}
```

Example 4-17: Simple Right Margin Changing

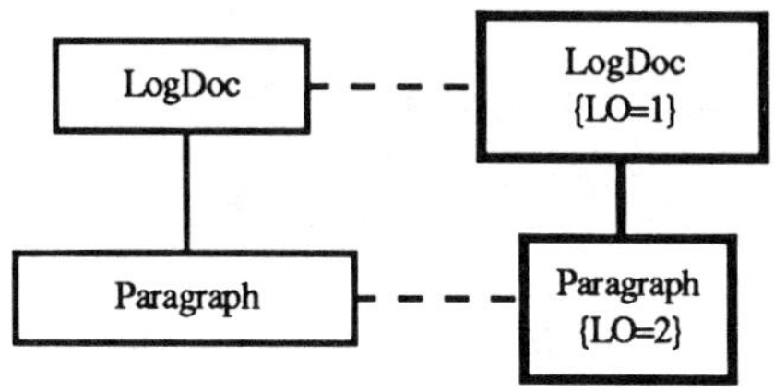

Figure 4-24. Representing a Change to the Right Margin

with the generic logical object. Because resulting attribute values for the environment and the document style are the same, there is no need to have a style specified by the specific logical object for *paragraph.*

The next example shows how to represent a change to the right margin. The document is specified in example 4-17.

The translated ODA document, as shown in figure 4-24, illustrates how the "left-hand offset" parameter of the "offset" attribute is set when changing the right margin to a new value. Like the previous example, the value calculated for the document style and for the environment is the same, so only the generic logical object needs to refer to a style.

The next example, 4-18, illustrates two simultaneous changes to both the left and right margin. The result of translating the document in example 4-18 is shown in figure 4-25.

In the translated document, the use of *LeftIndent* in the first *paragraph* changes only the left margin and is similar to example 4-16. The use of *BothIndent* in the second *paragraph* changes both margins via one document style, and there-

```
@LogDoc-root{
   @Paragraph-LeftIndent{
      @Text-RightIndent{ ... }
   }
   @Paragraph-BothIndent{ ... }
}
```

Example 4-18: Document with Both Margins Changed

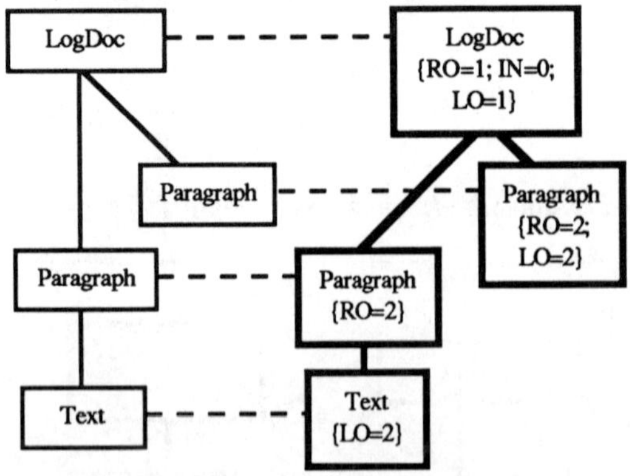

Figure 4-25. Representing Multiple Margin Changes

fore must record changes to two components of the formatting state vector. Thus, new values for both "right-hand offset" and "left-hand offset" are specified in the style. The use of *RightIndent* in the *text* document part occurs in a different environment than the root environment (which is where all other styles have been applied). Because the document style changes only the right margin, only the "left-hand offset" parameter is specified. Although the environment for the *text* document piece has an indented left margin as well, the value for the changed left margin is specified in the containing *paragraph*. Like the previous two examples, the values calculated for both the document style and the environment are the same, so no styles are needed for the specific logical objects.

The document in example 4-19 performs a paragraph indentation. We see the representation of example 4-19 in figure 4-26.

Figure 4-26 illustrates how an indented paragraph would be represented. The *PIndent* style specifies a new indentation value, which is represented by the "first line offset" attribute. Because the indentation is positive, only the "first line offset" attribute needs to be adjusted; the "left-hand offset" parameter need not change. Like in previous examples, the same value for "first line offset" attribute is calculated for the environment and document style, so only the generic logical object needs a style.

```
@LogDoc-root{
  @Paragraph-PIndent{...}
}
```

Example 4-19: Document with Indented Paragraph

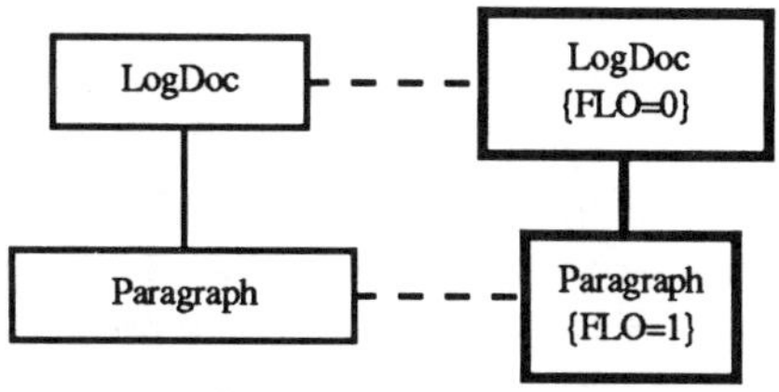

Figure 4-26. Representing Paragraph Indentation

Our next example shows how a single change in the formatting state vector can affect more than one ODA attribute. The document in example 4-20 contains an overhang. In figure 4-27, we see the representation of this document.

Figure 4-27 illustrates a complicated kind of paragraph indentation: a negative one. Note that the negative paragraph indentation requires that the "right-hand

```
@LogDoc-root{
   @Paragraph-Outdent{
      @Text{aaa}
      @Figure{
         @Raster{...}
      }
   }
}
```

Example 4-20: Document with Overhang

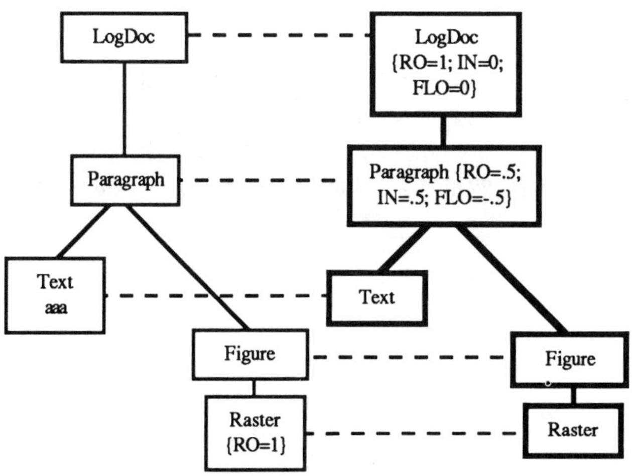

Figure 4-27. Representing Overhangs (Outdenting)

offset" parameter, the "indentation" attribute and the "first line offset" attribute also need to be set. This ensures that enough space for the first line of the paragraph is provided in the block used to hold the character content. Because the values for the document style are the same as for the environment, no style is created for the specific logical object for the paragraph.

However, this example also contains a *raster* in a *figure*. In this case, the left margin should not be compensated—we do not need to provide space for an outdented paragraph. The "first line offset" and "indentation" presentation attributes do not apply to the *rasters*, and so the *raster* would be positioned too far to the left. Thus, a style must be referenced by the specific logical object for the *raster* to remove the compensation added to the "right-hand offset" parameter for *text*. Thus a "right-hand offset" is calculated that ignores the paragraph indentation requirements, and the resulting value is used for the *raster*. It may not be clear whether to place the new style on the generic *raster* object or the specific *raster* object. Because the styles referenced from the generic logical structure are intended to correspond with document styles, placing the style on the generic *raster* object would misleadingly imply that a new kind of document style that set a left margin was being applied to the *raster* document part. Therefore, the style should be placed in the specific logical structure to represent an environment value.

Our next example considers the representation of relative attributes. The document in example 4-21 specifies the use of a relative margin change. We can see the representation of the document in figure 4-28.

The *MoveIn* document style specifies that the left margin should be moved in one inch from where it is. The document style *MoveIn* has a single representation, illustrated by the layout style labeled *MoveIn Style*. Because two different document parts refer to this document style, the ODA style is shown explicitly, as well as the references from the first *paragraph* and the *text* document part. For the first *paragraph*, the environment value for the left margin is the same as for the document style, and therefore no ODA style is needed for the specific logical object representing the first *paragraph*. However, the *text* document part is in an environment that has already altered the left margin (in the containing *paragraph* that has the associated *LeftIndent* document style).

```
@LogDoc-root{
   @Paragraph-MoveIn{ ... }
   @Paragraph-LeftIndent{
      @Text-MoveIn{bbb}
   }
}
```

Example 4-21: Document with Relative Margin Changes

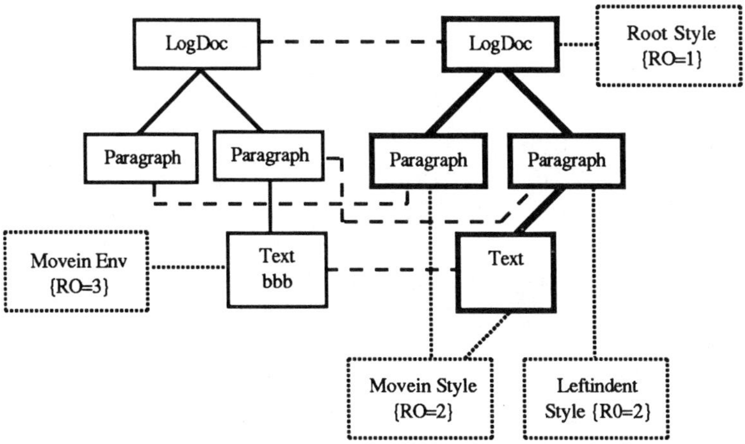

Figure 4-28. Representing Relative Margin Changes

Therefore, applying the *MoveIn* document style to the environment associated with the second *paragraph* results in a left margin that is 3 inches instead of 2 inches. Therefore, the value in the ODA style for *MoveIn Style* is not applicable and a new ODA style, referenced from the specific logical object for the *text* document part, is created to hold an appropriate value for the "right-hand offset" of 3.

4.9.5 Paragraph Spread

The amount of space between paragraphs is controlled by two components of the formatting state vector: paragraph spread above and paragraph spread below. Informally, the paragraph spread describes the minimum amount of space that should be left between two paragraphs. The above and below components further refine this specification to the minimum amount of space that should be present before a paragraph and after a paragraph. If we assume that paragraph P1 is on top of paragraph P2, then the amount of space between them must be at least:

```
max(P1.Paragraph Spread Below,
    P2.Paragraph Spread Above)
```

A natural way to specify the distance between paragraphs is to specify the distance between the blocks in which the logical objects are formatted. This specification is made with the "separation" attribute. Specifically, paragraph spread below can be represented using the "leading edge" parameter of the "separation" attribute and paragraph spread above can be represented using the "trailing edge" parameter of the "separation" attribute. Because of the conventions in this chapter, no two blocks with different layout paths will be specified in a frame, so there is no need to specify the value of the "centre separation" parameter of "separation".

Spread values may be specified independently. The only values needed are those represented by parameters in the "separation" attribute. Other parameters are left unspecified and will be obtained by the ODA defaulting rules.

As with the representation for margins, two layout styles may be necessary for paragraph spread: one for the environment and one for the document style. The layout style for the environment's formatting state vector (referenced by a specific logical object) holds the actual value of the "separation" attribute while the layout style for the document style (referenced by a generic logical object) holds the value used to represent changes to the formatting state vector. As for margins, the method for determining the value to be used in the layout style for the generic object is obtained by applying the document style to the root environment and using the resulting spread values.

Some examples illustrate how various spread calculations are carried out. The root environment is assumed to have an above and below spread of 1. In the document styles below, *PSA* is an abbreviation for *paragraph spread above* and *PSB* is an abbreviation for *paragraph spread below*.

```
BigAbove = PSA←3;

BigBelow = PSB←4;

BigBoth= PSA←3; PSB←4;

IncAbove = PSA←PSA + 1;
```

The first example of a document with paragraph spreads, specified in example 4-22, shows an application of two document styles that each adjusts one component of the formatting state vector. Applying the rules for representing paragraph spreads, we get the representation shown in figure 4-29.

The document specified in example 4-22 is analogous to the margin example documents in examples 4-16 and 4-17, where each margin is changed

```
@LogDoc-root{
   @Paragraph-BigAbove{...}
   @Paragraph-BigBelow{ ... }
}
```

Example 4-22: Document with Independently Adjusted Spreads

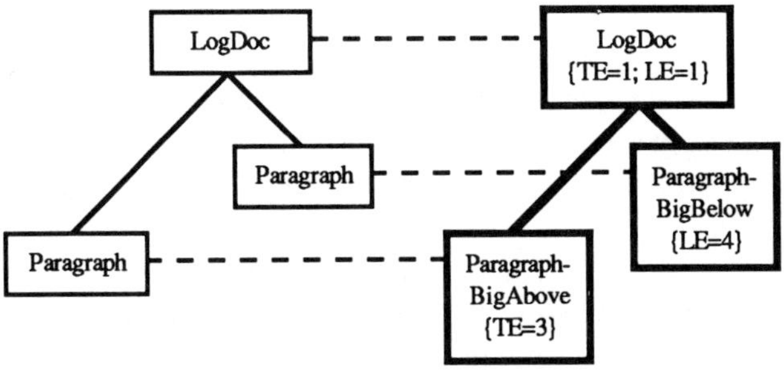

Figure 4-29. Representing Simple Interparagraph Spreads

separately. The difference is that the appropriate parameters for the "separation" attribute are being set instead of the parameters for the "offset" attribute.

The document in example 4-23 illustrates some interactions among the document styles as they adjust paragraph spreads. The representation of that document is shown in figure 4-30.

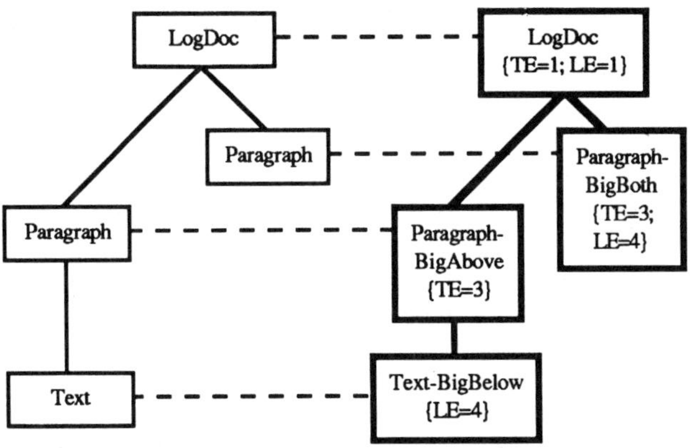

Figure 4-30. Representing Different Kinds of Paragraph Spreads

```
@LogDoc-root{
  @Paragraph-BigAbove{
     @Text-BigBelow{...}
  }
  @Paragraph-BigBoth{...}
}
```

Example 4-23: Document with Multiple Paragraph Spreads

The document is analogous to the document in example 4-18, where the margins were first changed one at a time and then together, except that the "separation" attribute is being specified instead of the "offset" attribute.
As with margins, the spread between paragraphs can be represented by relative styles. One such document is shown in example 4-24. Its representation is shown in figure 4-31.

Like the document in example 4-21 that used a relative left margin, this example uses a relative spread for space above the *paragraph*. Note that the generic objects can be reused, since both *paragraphs* have the same document style associated with them. The specific logical object for the *text* document piece needs its own style since the value for the "trailing edge" parameter differs from the value used in the style referred to by the generic object.

4.9.6 Paragraph Breaks

Paragraph breaks occur both explicitly and implicitly on most systems. To maintain a processable document, we are interested primarily in preserving the explicit paragraph breaks.

```
@LogDoc-root{
   @Paragraph-IncAbove{...}
   @Paragraph-IncAbove{
      @Text-IncAbove{...}
   }
}
```

Example 4-24: Document with Relative Paragraph Spreads

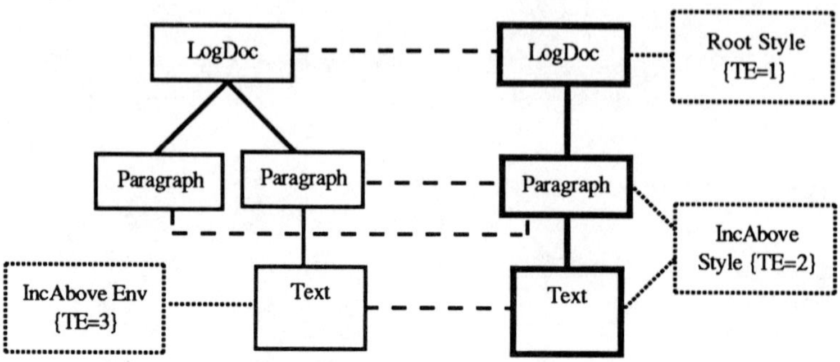

Figure 4-31. Representing Relative Paragraph Spreads

Paragraph breaks are technically part of a document's organization. However, many systems treat paragraph breaks only implicitly, and, thus, they appear to be formatting information. For example, Andrew only has implicit paragraph breaks within the content itself—two carriage-return characters in a row. In the EXPRES document model, paragraph breaks can be part of a style, where we assume that a style specifying a paragraph break will cause any associated environments to start a new paragraph.

Because we are using the NIST document application profile, paragraph breaks occur only with the creation of new *paragraph* document elements. All character contents descended from a *paragraph* composite logical object, with the exception of footnotes, are part of the same paragraph. The way that all the text content under a *paragraph* logical object is gathered together into a formatted paragraph is through the use of a "concatenation" attribute. Specifically, the first *text* logical object will have the value of 'non-concatenated' for the "concatenation" attribute. Normally, the attribute defaults to this value through the default value mechanism defined in ODA. All other *text* logical objects descended from the same *paragraph* should have the value 'concatenated' for their "concatenation" attribute. This value is specified typically in a layout style referred to by the *text* specific basic logical object, although it is possible that the layout style referred to by a *text* generic basic logical object would have the appropriate value for the "concatenation" attribute, or that a default value list associated with an enclosing phrase would provide the value.

Some examples illustrate the use of the "concatenation" attribute. In the diagrams that follow, basic logical objects that have a black box in the upper left corner refer to a style that has a "concatenation" attribute with the value 'non-concatenated' while basic logical objects that do not have a black box in the upper left corner refer to a style that has a "concatenation" attribute with the value 'concatenated'.

The example document in example 4-25 contains two paragraphs. The ODA representation for the document is shown in figure 4-32.

```
@LogDoc{
   @Paragraph{
      @Text{...}
   }
   @Paragraph{
      @Text{...}
   }
}
```

Example 4-25: Document with Two Paragraphs

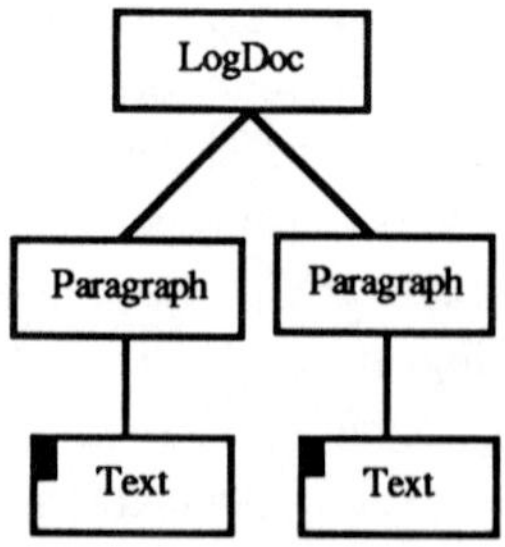

Figure 4-32. Representing Paragraphs

Figure 4-32 shows two *paragraphs* in a document. The first *text* document part in each *paragraph* document part starts a new paragraph, and, therefore, is not concatenated with the previous *text* document part. Hence, both *text* document parts have black squares indicating that the value for their "concatenation" attribute should be 'non-concatenated' (the default ODA value).

The next example, 4-26, shows a paragraph made from sequences of text. A representation for the document in example 4-26 is provided in figure 4-33.

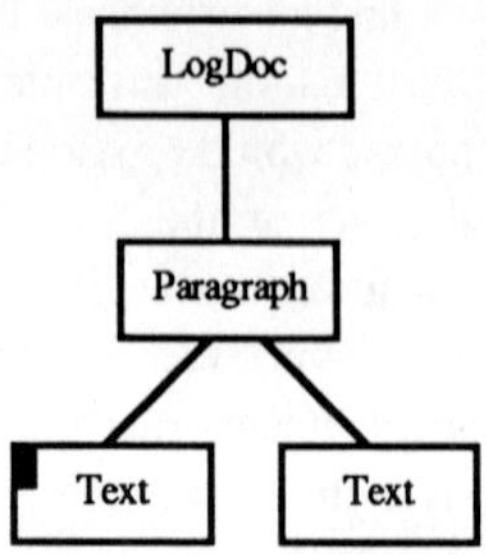

Figure 4-33. Representing Sequences of Text

Figure 4-33 illustrates the "concatenation" attribute when two *text* document parts are in the same *paragraph*. Although not exploited here, one reason why the two *text* document parts might be specified rather than a single *text* docu-

```
@LogDoc{
  @Paragraph{
      @Text{...}
      @Text{...}
  }
}
```

Example 4-26: Document with Sequences of Text

ment part is that each *text* document part has its own document style. The second *text* document part logically follows the first and is contained in the same paragraph, thus, it should be concatenated by setting its "concatenation" attribute to 'concatenated'.

Through the use of *phrases*, text can be gathered together from a whole subtree rather than just a sequence. A document illustrating this behavior is specified in example 4-27.

Figure 4-34 shows the ODA representation of the document. The figure shows two *text* document parts (possibly with different document styles) that are in the same *paragraph* but are not siblings. Because both *text* document parts are descents of the same *paragraph* document part, they are logically concatenated together, so the first *text* document part must have a "concatenation" attribute with the value 'non-concatenated' while all other text document parts in the *paragraph*, even if they are higher in the tree, have the "concatenation" attribute of 'concatenated'.

As permitted by the NIST document application profile, *phrases* can be nested arbitrarily. As a result, all of the text in nested phrases must be gathered up. Example 4-28 specifies a document with nested *phrases*. Its representation in ODA is given in figure 4-35.

```
@LogDoc{
  @Paragraph{
     @Phrase{
        @Text{...}
     }
     @Text{...}
  }
}
```

Example 4-27: Document with Nested Text

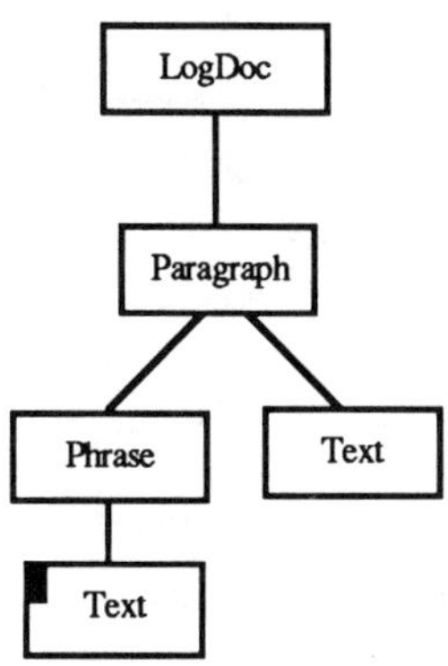

Figure 4-34. Representing Nested Text

```
@LogDoc{
  @Paragraph{
     @Phrase{
        @Phrase{
           @Text{...}
        }
        @Text{...}
     }
     @Text{...}
  }
}
```

Example 4-28: Document with Nested Phrases

The document in example 4-28 is an expanded version of the document in example 4-27, showing more possible intermediate layers of phrases. The same rule applies: within a paragraph, only the first text document part has the value of 'non-concatenated' for its "concatenation" attribute. All other text document parts in the same paragraph have the value 'concatenated' for their "concatenation" attributes.

Although paragraph breaks and the use of the "concatenation" attribute are quite simple, they play an important role in the representation of font information. This is covered in the next section.

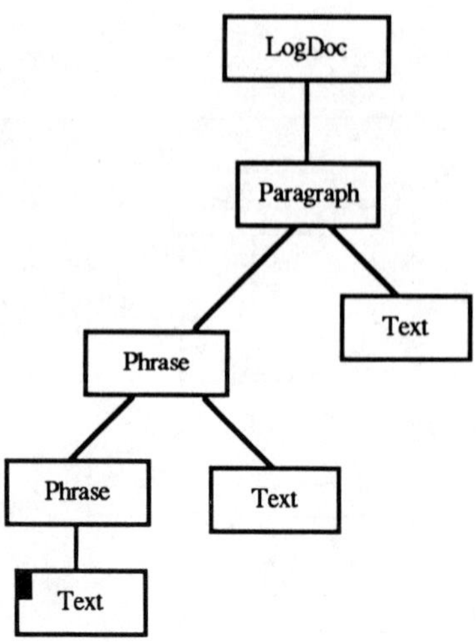

Figure 4-35. Representing Nested Phrases

4.9.7 Font Manipulations

In our model, document styles can make four orthogonal changes to font components of a formatting state vector: family, face, presentation and size. Each of these components can be represented in a font reference in ODA. In addition, the font reference facility of ODA allows the specification of subscripts and superscripts, which corresponds to the baseline offset component of the formatting state vector or document style. Therefore, the same strategy for using ODA font references will be applied for representing all five formatting state vector components. The general strategy is:

1 Calculate an appropriate font reference that represents the font information desired.

2 Include that font reference in the "fonts list" attribute of the document profile.

3 Include a reference to that font reference in the "character fonts" attribute of a basic logical object.

4 Invoke the font as appropriate with a *select graphic rendition* (SGR) control function in the content.

Each of these steps will be described below.

Different components of the formatting state vector component are represented by different parts of a font reference. Table 4-6 lists the representation of components in the formatting state vector as attributes and subattributes of the font reference as specified by the *Font and Character Information Interchange Standard* [34] used by ODA.

The values that are assigned to each of the attributes and subattributes are shown in table 4-7. Naturally, the values given in table 4-7 are for the features in our document model. Should a system support additional functionality, for example, more than one kind of bold face, then other values for these attributes may be required. In this particular example, other ISO/WEIGHT values can be used. How to interpret the values is up to each receiving system.

FSV[23] Value	FSV Component	ISO DIS 9541 Attribute
Font Family	Font Family	ISO/FAMILY, ISO/DSCLASS
Font Size	Font Size	ISO/DSNSIZE
Bold Face	Font Face	ISO/WEIGHT
Italic Face	Font Face	ISO/POSTURE
Outline Face	Font Face	ISO/STRUCTUR
Shadow Face	Font Face	ISO/STRUCTUR
Underline	Font Presentation	ISO/MODES, ISO/SCORES, ISO/RSCORE
Overbar	Font Presentation	ISO/MODES, ISO/SCORES, ISO/LSCORE
Strike through	Font Presentation	ISO/MODES, ISO/SCORES, ISO/TSCORE
Fixed Face	Font Presentation	ISO/MODES, ISO/ESCLASS
Condensed	Font Presentation	ISO/PROPWDTH
Subscript	Baseline Offset	ISO/MODES, ISO/VARSCRPT, ISO/LSCRPT, ISO/VYOFFSET
Superscript	Baseline Offset	ISO/MODES, ISO/VARSCRPT, ISO/RSCRPT, ISO/VYOFFSET

Table 4-6. Representing Font Information Using ISO DIS 9541

The representation of the font information in a formatting state vector is straightforward. The information about a font's face, size, presentation, style and baseline offset are described using the appropriate font attributes from table 4-7. If a font reference with those same attributes is already in the "fonts list" attribute of the document profile, then a translator would reuse the associated

[23]FSV is the formatting state vector.

ISO DIS 9541 Attribute	Value
ISO/FAMILY	name of family
ISO/DSCLASS	index into design table
ISO/DSNSIZE	real-valued size of font in mm
ISO/WEIGHT	5= no bold, 7= bold
ISO/POSTURE	1= no italic, 4=italic
ISO/STRUCTUR	1= solid, 2=outline, 4=shadow
ISO/RSCORE	thickness of underline, placement of underline
ISO/LSCORE	thickness of overbar, placement of overbar
ISO/TSCORE	thickness of strikeout, placement of strikeout
ISO/ESCLASS	1=fixed, 2= variable
ISO/PROPWDTH	medium=5, condensed =3, expanded =7
ISO/LSCRPT	subscript selection
ISO/VYOFFSET	size of baseline offset
ISO/RSCRPT	superscript selection

Table 4-7. Font Reference Values

font identifier. If no existing font reference were present in the document profile, a translator would add the font reference to the "fonts list" attribute with a newly created font identifier. In either case, the presentation style associated with a document part should have a "character fonts" attribute that includes the font identifier (along with the size defined by that font reference), and, if necessary, a "graphic rendition" attribute that invokes the identified font. (The "graphic rendition" attribute is necessary only if the font reference is not the first font in the "character fonts" attribute, since the semantics of ODA will automatically invoke the first font listed in the "character fonts" attribute).

One complication of this approach comes from the ODA's specification that some attributes be ignored by the layout process under certain circumstances. In particular, layout using concatenated basic logical (text) objects will ignore any defined "character fonts" attributes of all but the first object in the sequence of concatenated basic objects. Thus, the first specific basic logical object in the sequence of concatenated objects must have all font references in a paragraph added to its "character fonts" attribute.

The representation of font information for a style is straightforward. Like all other style information, the technique is to apply the style to the root environment and then represent changes in the resulting formatting state vector in the presentation and layout styles referred to by the generic logical object. Thus, the font reference must be present in the document profile's "fonts list" attribute so that it can be referenced in the "character fonts" attribute of the presentation style referred to by the generic logical object.

Some examples will illustrate the process of representing font information. These examples illustrate most manipulations that can be carried out on font components. By necessity, the diagrams representing the translated documents show great detail. We use the following conventions in the diagrams:

- When representing font components, we use D<string> to be the descriptor of a font in the document profile.
- We use F<number> to mean select the <number>th font in a "character fonts" list.
- We use SGR xxx to mean either include the select graphic rendition control function for font xxx in a content portion or use the value xxx for the "graphic rendition" attribute.

The examples use points as the unit of measurement. Because the representation requires the "fonts list" attribute of the document profile, the "fonts list" attribute is shown in a table following each diagram. The first basic logical object of each paragraph is marked with a black square in the upper left hand corner for ease of location.

The examples assume the following style definitions:

```
Head = FontFace←Bold;

Emphasis = FontFace←FontFace + Italic;

Art = FontFace←Shadow;

Define = FontPresentation←Underline;

Shout = FontSize←24;

Incr = FontSize←FontSize + 2;

Typewriter = FontFamily←Courier;

FakeType = FontPresentation←Fixed
```

```
Sub = BaselineOffset← -2;

Super = BaselineOffset←2;
```

The font for the root environment is a 12 point *Helvetica* font with no faces or presentations.

The first example makes some changes to the faces and presentations of some fonts. Example 4-29 specifies the document to be translated. The ODA document that corresponds to example 4-29 is shown in figure 4-36 and table 4-8.

```
@LogDoc-root{
  @Paragraph{
     @Text{aaa}
     @Text-Head{bbb}
     @Text-Emphasis{ccc}
     @Text-Art{ddd}
     @Text-Define{eee}
  }
}
```

Example 4-29: Document with Face and Presentation Changes

	D0	D1	D1d	D2	D2d	D3	D3d	D4	D4d
Family	Helv		Helv		Helv		Helv		Helv
DsClass	Sans		Sans		Sans		Sans		Sans
DsnSize	12		12		12		12		12
Weight	5	7	7		5	5	5		5
Pos	1	1	1		1	4	4		1
Struc	1	1	1		1	4	4		1
Scores								RSco	RSco
Esclass	2		2		2		2		2
Scrpt									
Prop	5		5		5		5		5

Table 4-8. Font References for Font Faces and Presentations

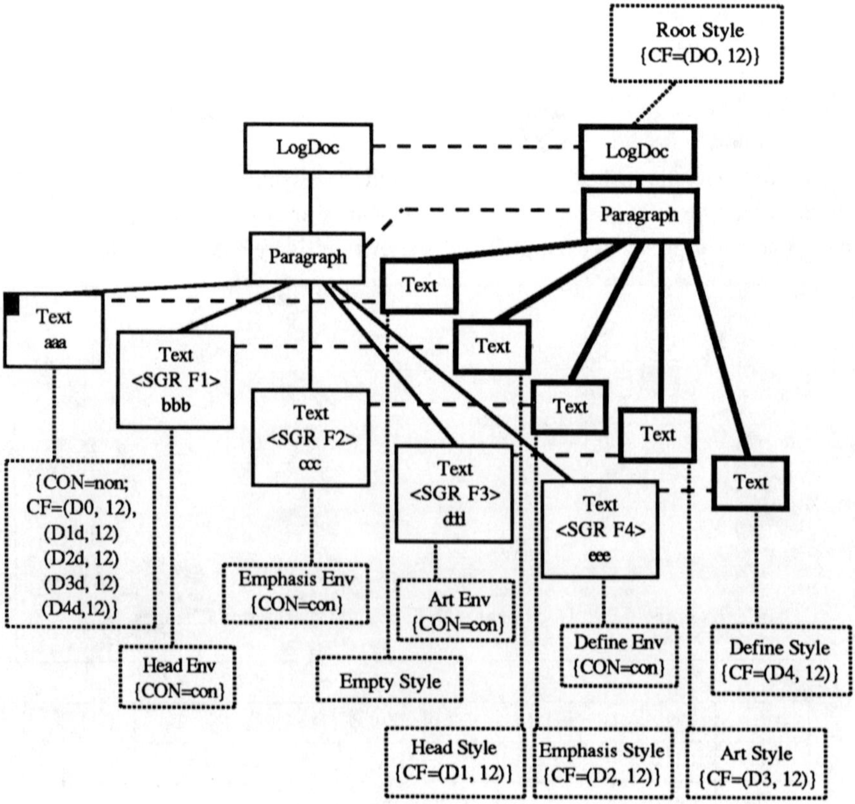

Figure 4-36. Using "Character Fonts" Attribute for Font Faces and Presentations

Example 4-29 illustrates a variety of changes. The *Head* style changes the face to bold, the *Emphasis* style adds italic to the face, the *Art* style uses a *shadow* face, and the *Define* style changes the presentation to underlining. In each case, a font reference is defined by giving the changes specified by that style. Because the *Head* style uses a bold face, the *D1* font reference defines a weight of 7, and sets the posture and structure to 1, forcing the face to be bold alone. (The selection of *D1* as the identification for the font reference has no special meaning; any unique symbol would do.) The *Emphasis* style is slightly different, in that the *Emphasis* style adds italic to the face, rather than setting the face to italic. Its font reference, shown as *D2*, sets the posture to 4 for italic, but does not affect the other attributes that control font faces, such as weight and structure. The third style, *Art*, has the representation shown for *D3*. Like the *Head* style, the *Art* style forces the face to be a set value (shadow), which requires setting the weight to 5 (to remove any possible bold), the posture to 1 (to remove any possible italic) and the structure to 4 (to define shadow). All font references that are representations of document styles, are associated with the generic object for a document part through the use of the "character fonts" attribute in a presentation style.

Note that each ODA style referenced by the generic objects has a single font in its "character fonts" attribute, so there is no need to provide a "graphic rendition" attribute specifying which font should be used first by that document part. However, none of these font references are sufficient for use in the specific structure, since they only describe parts of the font information from the formatting state vector. Thus, a font reference must be defined for the formatting state vector in each environment as well. When the *Head* style is applied to the root environment, the resulting font is the same as the root, except that the face is now bold. The resulting font reference that shows how the text ought to be displayed is designated *D1d*. Careful examination shows that *D1d* is the font that results from applying the style *Head* to the default font reference (designated as *D0*). The same explanation holds for the calculation of font references *D2d*, *D3d* and *D4d*, which result from applying the *Emphasis*, *Art* and *Define* styles to the default font value.

Even with the correct font descriptions for the specific logical structure, the "character fonts" attribute of each basic logical object cannot be used. This is because only the "character fonts" attribute of the first non-concatenated basic logical object is examined by the layout process. Thus, the text document parts for "bbb", "ccc", "ddd" and "eee" need to have their font references listed in the "character fonts" attribute of the first specific logical object, here the basic logical text object holding the "aaa" content. Further, each basic logical object needs to invoke explicitly the desired font, hence the SGR control function must be present in the "content information" attribute of the content portions for the latter four text objects[24]. Finally, note that each of the four latter text objects needs to specify that it is concatenated with the preceding basic logical object, so that each refers to a layout style to hold the "concatenation" attribute. In this case, there is no need to create separate layout styles to hold the "concatenation" attribute, so in fact, each of the four latter text objects could refer to the same layout style. Further, there is no need for the first generic text object, which has no style associated with it, to refer to an empty layout style. The diagram with these changes is shown in figure 4-37.

The next document example, specified in example 4-30, shows changes being made to the sizes and families of fonts. Figure 4-38 and table 4-9 show the representation of this document in ODA.

---

[24]For a font that is invoked at the beginning of a "content information" attribute, one can specify the select graphic rendition code either by using the "graphic rendition" attribute or by placing the characters in the "content information" byte sequence. We will always use the latter in our discussions.

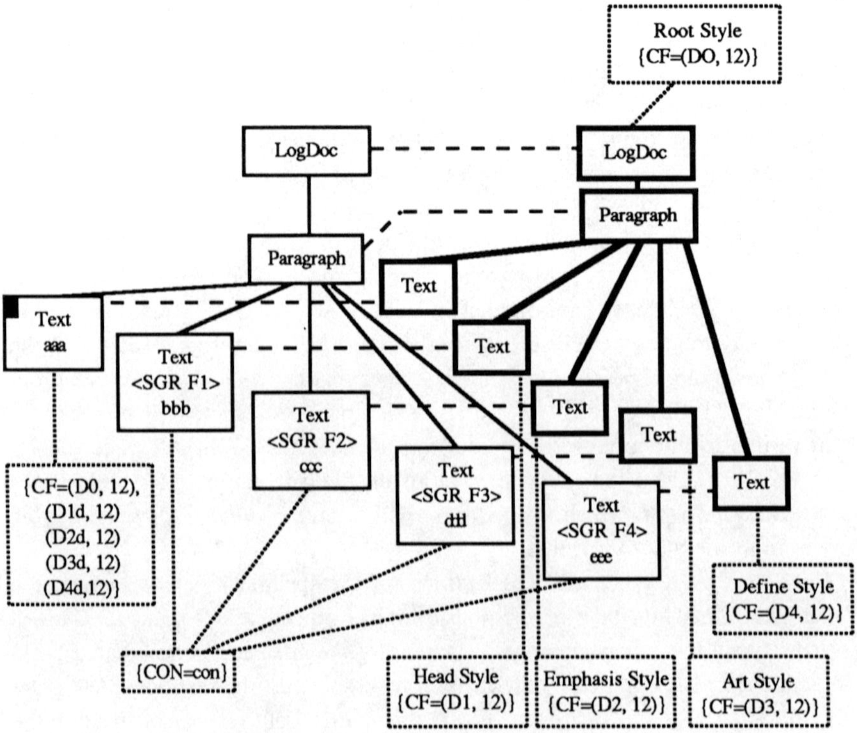

Figure 4-37. Optimizing Style Usage

```
@LogDoc-root{
  @Paragraph{
     @Text{aaa}
     @Text-Shout{bbb}
     @Text-Incr{ccc}
     @Text-Typewriter{ddd}
     @Text-FakeType{eee}
  }
}
```

Example 4-30: Document Changing Font Sizes and Families

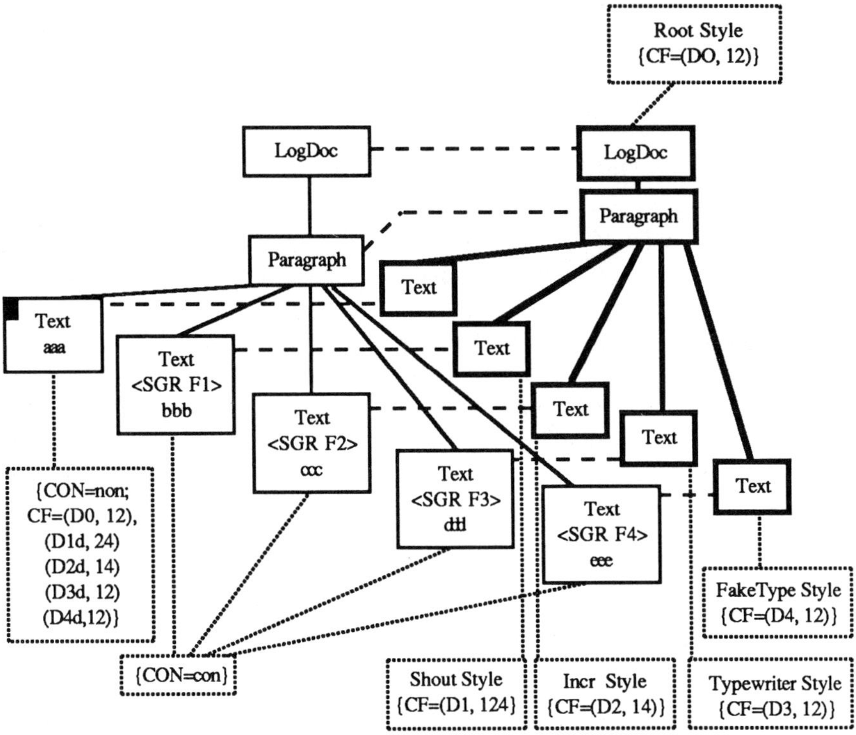

Figure 4-38. Representing Changes to Font Sizes and Families

	D0	D1	D1d	D2	D2d	D3	D3d	D4	D4d
Family	Helv		Helv		Helv	Cour	Cour		Helv
DsClass	Sans		Sans		Sans	Type	Type		Sans
DsnSize	12	24	24	14	14		12		12
Weight	5	7	7		5	5	5	5	5
Pos	1		1		1		1	1	1
Struc	1		1		1		1	1	1
Scores									
Esclass	2		2		2		2	1	1
Scrpt									
Prop	5		5		5		5		5

Table 4-9. Font References for Font Sizes and Families

The document specified in example 4-30 is essentially the same as the document specified in example 4-29, except that some changes are being made to the size (by the *Shout* and *Incr* styles) and to the font family (to *Courier* by the *Typewriter* style). The font reference for the *Shout* document style is given by *D1*, and shows that the value of 24 points was selected. The *Incr* style is a relative style, and when applied to the root formatting state vector (which specifies a font size of 12) results in a font size of 14. This is indicated in the font reference for *D2*. The selection of a new family name by the *Typewriter* style causes a new font family to be specified (*Courier*) as well as a new design class (*Typewriter*). In fact, the value of the design class is an index into the table of design classifications specified by ISO DIS 9541, where the *SansSerif* specified for *Helvetica* would be 5 (or 5.1.2 depending on how much information was to be conveyed), while the design class value for *Typewriter* specified by the *Typewriter* style would be 4.6.5. The fourth style, *FakeType*, is unusual in that it shows a way that a fixed width (fixed escapement) font might be specified on a system that cannot guarantee the existence of fixed width fonts. The *FakeType* style specifies that a fixed width layout (escapement) of a font should be used, even if the font has characters of different size. The font reference for *D4* specifies this requirement with a value of 1 for *esclass*.

As always, a font reference that contains all of the information in the formatting state vector is needed by the specific logical objects. Hence, there are font references generated by applying the document style to the root formatting state vector, yielding font references *D1d*, *D2d*, *D3d* and *D4d*. Like the previous example, these font references are listed in the "character fonts" attribute of the first specific logical object for the paragraph, and invoked explicitly in the "content information" attribute of the concatenated specific basic logical objects.

The descriptors for fonts can specify more than just the face and size of fonts. They can be used to describe subscripts and superscripts as well. Consider the document specified in example 4-31. The representation of the document in ODA is shown in figure 4-39 and table 4-10.

```
@LogDoc-root{
   @Paragraph{
      @Text{aaa}
      @Text-sub{bbb}
      @Text-super{ccc}
   }
}
```

Example 4-31: Document with Subscripts and Superscripts

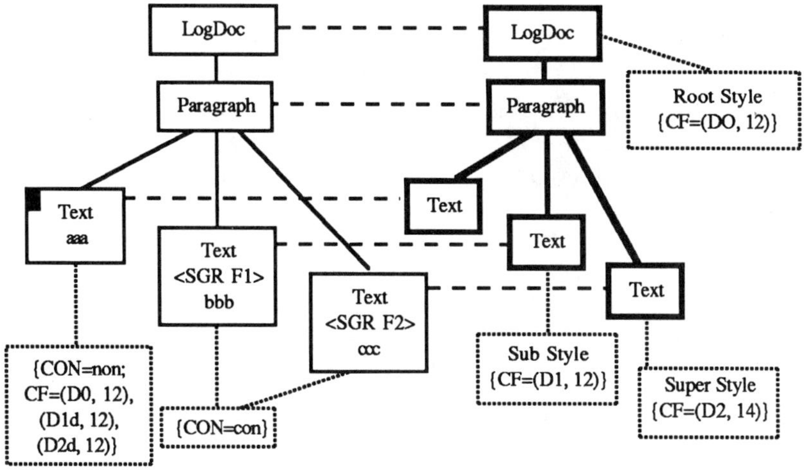

Figure 4-39. Representing Subscripts and Superscripts

	D0	D1	D1d	D2	D2d
Family	Helv		Helv		Helv
DsClass	Sans		Sans		Sans
DsnSize	12		12		12
Weight	5		5		5
Pos	1		1		1
Struc	1		1		1
Scores					
Esclass	2		2		2
Scrpt		L(1/6)	L(1/6)	R(1/6)	R(1/6)
Prop	5		5		5

Table 4-10. Font References for Subscripts and Superscripts

Figure 4-39, along with table 4-10, illustrates how subscripts and superscripts can be represented. As before, there are font references for both the document styles (*D1* and *D2*) and the environments' formatting state vectors (*D1d* and *D2d*). The selection of the left script or right script controls the selection of superscripts or subscripts respectively. The font standard requires that the values given for the offsets in script attributes be made in character scales, which means relative to the size of the character. Because the font references are being calculated relative to the default formatting state vector (which specifies a 12 point font), a movement of 2 points represents a movement of 1/6 of the character height, hence the value of 1/6 in each font reference. Should the font size change, the offset would have to change as well.

Styles that nest give rise to a larger number of font references than expected. For example, the document specified in example 4-32 contains nested document style applications that result in several new font references.

```
@LogDoc-root{
  @Paragraph-Emphasis{
     @Text-Head{aaa}
     @Text-Define{bbb}
  }
}
```

Example 4-32: Document with Nested Font Specifications

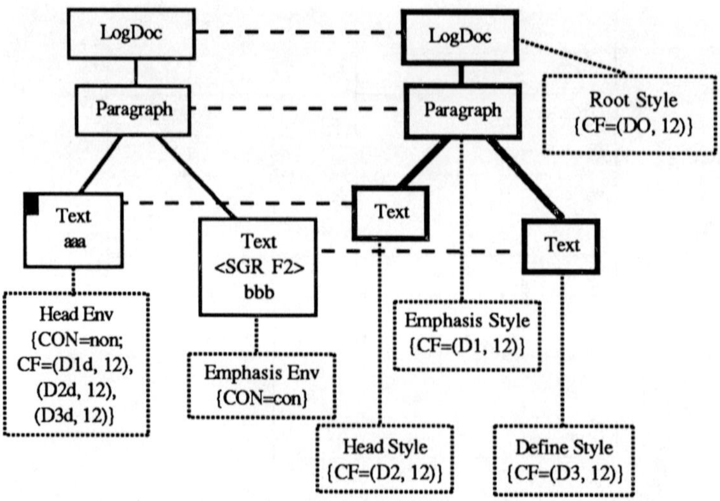

Figure 4-40. Nested Font Faces

	D0	D1	D1d	D2	D2d	D3	D3d
Family	Helv		Helv		Helv		Helv
DsClass	Sans		Sans		Sans		Sans
DsnSize	12		12		12		12
Weight	5		5	7	7		5
Pos	1	4	4	1	1		1
Struc	1		1	1	1		1
Scores						Rsco	Rsco
Esclass	2		2		2		2
Scrpt	5		5		5		5
Prop	5		5		5		5

Table 4-11. Font References for Nested Fonts

The ODA representation, shown in figure 4-40 and table 4-11, illustrates how nested font manipulations are carried out. Like before, the font references for the document styles are created relative to the root formatting state vector (*D1*, *D2* and *D3*). However, three fonts are calculated for use by the specific logical objects. First, applying the *Emphasis* style associated with the *paragraph* document part results in the font described by *D1d*. *D1d* is added to the "character fonts" attribute of the first specific basic logical object under the *paragraph*. Next, the *text* document part for "aaa" is handled, which applies the document style *Head* to the formatting state vector for the *paragraph*. This means applying the document style to the font described by *D1d*, not *D0* as before. The result is the font described by *D2d*. *D2d* is the font for the first *text* document part, and so is added to the "character fonts" attribute. However, since *D2d* is not the first font in the "character fonts" attribute, it must be invoked explicitly. As shown, the font is invoked by placing an SGR control function into the content information.

In a fashion similar to the calculation of *D2d*, the calculation of *D3d* involves applying the *Emphasis* style to the font described by *D1d*. *D3d* is added to the "character fonts" attribute of the first basic logical object and then invoked in the text object for "bbb" by an SGR control function.

Like margins and paragraph spreads, some font information can be manipulated by relative styles. The document specified in example 4-33 illustrates the use of relative font faces. The ODA representation of this document is shown in figure 4-41 and table 4-12.

```
@LogDoc-root{
  @Paragraph{
     @Phrase-Define{
        @Text-Emphasis{aaa}
        @Text{bbb}
     }
     @Text-Emphasis{ccc}
  }
}
```

Example 4-33: Document with Relative Font Faces

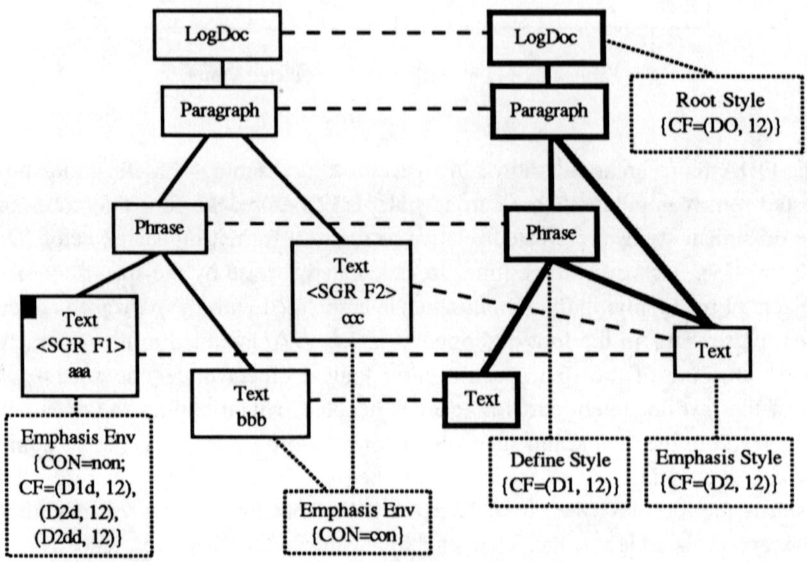

Figure 4-41. Representing Relative Font Faces

	D0	D1	D1d	D2	D2d	D2dd
Family	Helv		Helv		Helv	Helv
DsClass	Sans		Sans		Sans	Sans
DsnSize	12		12		12	12
Weight	5		5		5	5
Pos	1		1	4	4	4
Struc	1		1	4	4	4
Scores		Rsco	Rsco		Rsco	
Esclass	2		2		2	2
Scrpt						
Prop	5		5		5	5

Table 4-12. Font References for Relative Font Faces

Figure 4-41 illustrates how relative document styles that manipulate font faces can affect the calculation of formatting state vectors and their corresponding font references. The *Emphasis* style adds italic to a font face, and is used in two different environments. The first environment is the first *text* object ("aaa"), which is nested in a *phrase*. Here, italic is added to the font reference specified for the *phrase*, which has an underline. This results in the font reference *D2d.* The second place that *Emphasis* is used, for "ccc", adds italic to the root formatting state vector (inherited directly by the *paragraph* document piece since it has no document style associated with it), which results in the font reference denoted *D2dd.* Thus two styles defining character fonts cause five font references to be created.

In addition to font faces, font sizes can be manipulated by relative styles. This is illustrated by the document specified in example 4-34. The ODA representation of this document is shown in figure 4-42 and table 4-13.

```
@LogDoc-root{
  @Paragraph-Incr{
     @Phrase-Incr{
        @Text-Incr{aaa}
        @Text{bbb}
     }
     @Text-Incr{ccc}
     @Text{ddd}
  }
}
```

Example 4-34: Document with Nested Font Sizes

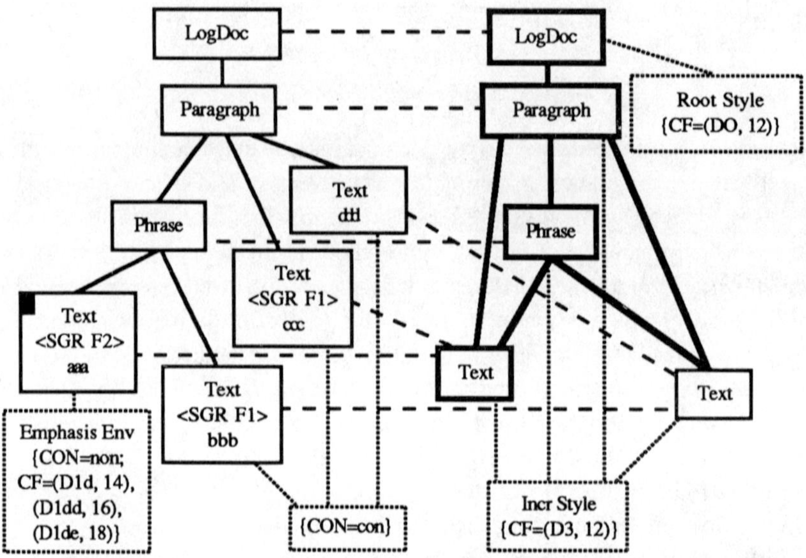

Figure 4-42. Representing Nested Font Sizes

	D0	D1	D1d	D1dd	D1de
Family	Helv		Helv	Helv	Helv
DsClass	Sans		Sans	Sans	Sans
DsnSize	12	14	14	16	18
Weight	5		5	5	5
Pos	1		1	1	1
Struc	1		1	1	1
Scores					
Esclass	2		2	2	2
Scrpt	5		5	5	5
Prop	5		5		5

Table 4-13. Font References for Nested Font Sizes

Figure 4-42 and table 4-13 show how a relative document style, which is repeatedly applied, results in several font references being generated. Specifically, the *Incr* document style increases the font size by 2 points each time it is applied. In the document, there are parts that have *Incr* applied one, two or three times, resulting in different formatting state vectors. Although there is only one font reference for the *Incr* document style (*D1*), there are three font references needed to capture the use of the *Incr* document style at various places in the document (*D1d, D1dd, D1de*).

4.9.8 Content Alignment (Justification)

The EXPRES document model has a single component for aligning content, the *justification* component. This component can have one of four values: left justified, right justified, both sides justified or centered. These values correspond to ODA alignments of 'start-aligned', 'end-aligned', 'justified' and 'centred' respectively. When a document style specifies a justification, the presentation style associated with the corresponding generic logical object will have an "alignment" attribute with the value. Although it should not be necessary, the presentation style associated with the specific logical object may also specify the "alignment" presentation attribute[25]. If the root environment has a

[25]The layout process defined for ODA will ignore the "alignment" attribute associated with any concatenated basic logical object. Since this is the same semantics as the document model used for EXPRES, no special consideration of this interpretation is necessary.

different alignment than the ODA default (which is 'left-aligned'), then an "alignment" attribute must be present on the root (in a default value list).

Our document model permits media besides text to be justified. The interpretation of an alignment on other media is that the block containing the content should be aligned. For example, a raster could be horizontally centered in a page. This is another example where a specific logical structure must override, or more accurately, supplement, the information in the generic structure. For representing style information, the use of the "alignment" attribute is sufficient. However, the "alignment" attribute will have no effect on raster graphics content. Whenever a raster content has a justification other than left justified, the "block alignment" attribute must be specified in the specific logical structure for the basic logical object holding the raster content. The "block alignment" attribute may take the values 'right-hand aligned', 'left-hand aligned', 'centred', which correspond to a left justified, a right justified and a centered raster. A value of both justified for the justification component is represented as a 'centred' value for the "block alignment" attribute.

4.9.9 Line Spacing

The EXPRES document model allows line spacing to be adjusted by a document style. Many systems, such as Andrew, specify this number as either a constant, such as 12 points, or a multiple of the previous spacing (for example, a double spacing style would specify the new spacing as twice the current spacing).

The representation of line spacing is provided by the ODA presentation attribute "line spacing". It is set to the desired value of spacing between baselines. For the "line spacing" attribute to affect formatting, the "proportional line spacing" attribute must have the value 'no', which is its default.

4.9.10 Tab Stops

The EXPRES document model allows the tab stop component of a formatting state vector to contain an ordered list of tab stops, where each tab stop specifies the location and the kind of the tab stop. Possible tab stops are left adjusted, right adjusted, centered and character aligned.

The settings of tab stops are represented using the "line layout table" presentation attribute. When a document style specifies changes to the list of tab stops, that document style is applied to the root environment and the resulting tab stops are represented by a "line layout table" presentation attribute placed in a presentation style on a generic object. ODA requires each tab stop to be identified by an arbitrary reference number. We require that the tab stops be assigned reference numbers from 1 to N, where the reference number is the ordinal position of the tab stop from the left edge of the paper. ODA also uses a different method for measuring the tab stop location than the EXPRES document model. We define the position of the tab stop as an offset from the left edge of the paper. In ODA, the "tab position" parameter is specified as the offset (in scaled measurement units) from the start of the positioning area, which in turn is offset from the basic layout object by the "kerning offset". Thus, the relationship between the value specified by the document model (tab location) and the value needed by the "tab position" parameter is:

```
"tab position" parameter = tab location
                             - kerning offset
                             - left margin
```

Normally, the value of the "kerning offset" attribute is 0, and this equation need only correct for the left margin (recall the distinction between left margin and the offset calculation performed for representing the left margin in section 4.9.4.1). The "alignment" parameter for each tab is set to ‘start-aligned’, ‘end-aligned’, ‘centred’ or ‘aligned-around’, as appropriate. In the case of the ‘aligned-around’ value, the character used in the document model for aligning text should be assigned to the "alignment string" parameter of the "line layout table" attribute.

Like all presentation styles that represent document style information, the presentation style for tab stops is specified on a generic logical object. Typically, the representation calculated for the document style will serve as the value for the formatting state vector as well, in which case nothing more need be done when the document style is used in an environment. If there are nested document styles that make multiple changes to the tab stops, then the tab stops as calculated for an environment’s formatting state vector will likely be different than the representation calculated for the document style. In this case, a presentation style with the correct "line layout table" attribute must be created and attached to the specific logical object.

When a tab is used in a content portion, it must be specified using the STAB control function, not the TAB control function. To preserve imaging fidelity, one needs to simulate the horizontal layout of content to determine which tab stop has been reached, since some content may have extended through several tab stops. We assume that processable documents use tabs only for delimiting fields, and, therefore, the formatting of content will not reach a tab stop without first finding a tabulation control function. Thus, when converting character content that uses TABs to content that uses STABs, we assume that the ordinal count of tabs since the last explicit line break or beginning of paragraph will correspond to the same ordinal position of the tab stop. Thus, the Nth tab found since that last explicit line break of paragraph start will be represented as STAB n.

4.9.11 Page Breaks

When a document style (and its use in an environment) specifies a page break, the attribute "new layout object" is used with the value ‘object type page’. This attribute is placed on the layout style of the corresponding logical object for the document piece specifying the page break. Unlike other attributes, this attribute may apply to a composite object. Therefore, when a document style associated with a composite logical object specifies a page break, a layout style for the generic object (and possibly for the specific object) needs to specify the "new layout object" attribute in a style specified by the logical object, not on the object’s default value list. The specification of the attribute will force a single new page to be generated. If the attribute were placed on the default value list for basic logical objects, like other layout directives, a new page would be generated for each basic object as it was formatted.

4.9.12 Representation of Page Layout Components

When a document style specifies a change that affects page layout, such as a new top margin, a new kind of page must be created that has new values for appropriate attributes. Unlike most other formatting state vector components, page layout components are translated to attributes that are placed in the generic *layout* structure. Sometimes, changes to page layout components require changes to be made both in the logical structure and in the generic layout structure.

The general strategy is to create a new template page for the new page layout, insert it into the generic layout structure, and direct the content from the logical structure into the new page. Each of these processes is discussed in turn, followed by the details of how to make changes in the generic layout structure for each of the page layout components.

4.9.12.1 Design and Insertion of a New Page

The design of a generic page was described in section 4.7, which discussed the generic layout objects that need to be created and how to organize them to define a new page. As explained, the newly constructed page is appended to the sequence of pages subordinate to the pageset. However, defining a new page and placing it into the generic layout structure is not enough. One must also direct the layout process to use the correct page at the right time.

To force the use of a new page format, the logical object to which a new page layout format will be associated should have the attribute "new layout object" in a layout style that is placed directly on the object (not in the default value list). The value of the attribute refers to the object identifier of the new generic page. This process will be illustrated by an example.

A document that requires a new generic layout change is shown in example 4-35. All we need to know about the *TopMargin* document style is that it changes the top page margin component of the formatting state vector.

The ODA representation of the document in example 4-35 is shown in figure 4-43. The figure contains two generic trees, one for the generic logical structure (rooted at *LogDoc*), the other for the generic layout structure (rooted at *LayDoc*).

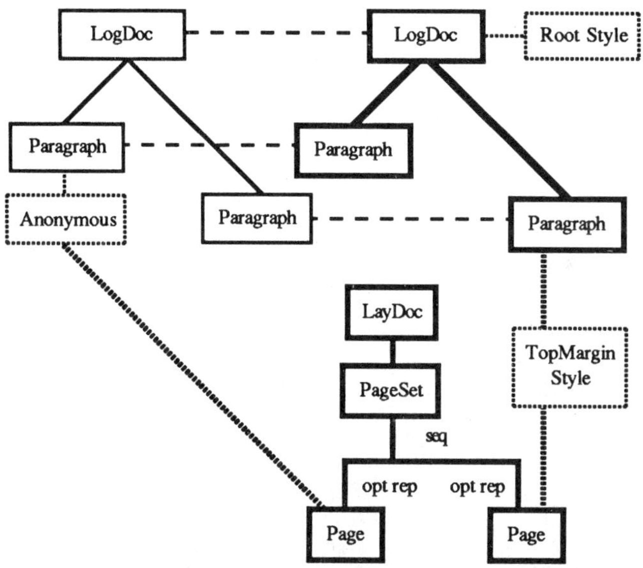

Figure 4-43. Using New Layout Object to Direct Layout Process

```
@LogDoc{
   @Paragraph{...}
   @Paragraph-TopMargin{..}
}
```

Example 4-35: Document that Requires a New Page Template

In example 4-35, we assumed that the first *paragraph* is formatted in the first (default) page format while the second *paragraph* uses a new format, one that specifies a new the top margin. The "new layout object" reference from the first *paragraph* (shown as a heavy striped line from an *anonymous* layout style) ensures that layout begins with the first possible page format. The generic logical object for the second *paragraph* refers to a layout style that has the attribute "new layout object", that refers to the new type of page in the generic layout structure. Thus, the second *paragraph* will be formatted as specified by the second page template. Recall that the layout styles here are specified directly by the logical objects, and are not in the default value list of the objects—the notation in the diagram does not distinguish these cases.

However, the representation presented so far still does not capture all of the intended meaning. As shown, the representation ensures that the first page is formatted as specified by the first generic page, and the beginning of the second *paragraph* is formatted as specified by the second generic page. There is no guarantee that any intermediate pages formatted between the first page of the first *paragraph* and the first page of the second *paragraph* will be in the format of the first generic page. ODA only requires that the second type of page follow the first, but does not dictate when to start using the second type of page. To guarantee that all pages until the second *paragraph* are formatted as specified by the first generic page, a layout category must be defined that directs its stream into the appropriate page type. In particular, the two frames used for text and footnotes must be set to allow only appropriate basic logical objects to be formatted in them. By convention, each *FrameC* (for main-body content) and *FrameF* (for footnote content) in the body frame (*BodyFrame1*) has a "permitted categories" attribute that contains the position of its enclosing page in the sequence of generic pages. All basic logical objects specify explicitly, via the "layout category" attribute, the kind of frame into which they can be formatted, which in turn forces an appropriate page format to be used. This is illustrated by the document in example 4-36, which is essentially the same as the document in example 4-35, except that some basic logical objects are shown as well. The ODA representation is shown in figure 4-44.

```
@LogDoc{
   @Paragraph{
      @Text{...}
   }
   @Paragraph-TopMargin{
      @Text{...}
   }
}
```

Example 4-36: Document Directing Content to New Page Format

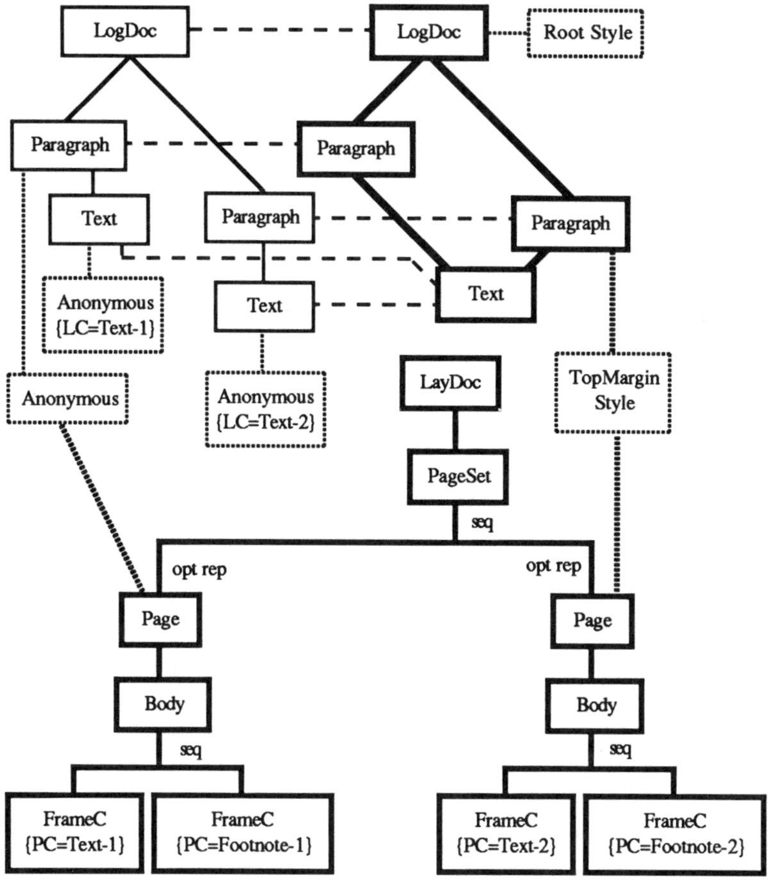

Figure 4-44. Explicit Representation of Formatting Categories

Figure 4-44 shows more of the generic layout structure than in figure 4-43 (although some parts that are not relevant to the discussion are still missing). The generic *FrameC*'s and generic *FrameF*'s are labeled with "permitted categories" attributes. The "permitted categories" attribute specifies which basic logical objects may use them for formatting their content. The figure also shows how each basic logical object specifies, via a layout style, the "layout category" attribute. The layout styles are labeled *anonymous* to show that they are not the representation of any document style. The "layout category" attribute is used to specify the kind of frame in which the basic logical object can be formatted. With both "permitted categories" and "layout category" attributes specified, the semantics of ODA will require that the first page format will be used for all pages until the second *paragraph* is reached, and all content after the second *paragraph* will be formatted with the second page format.

One problem with the conventions, as stated, is that gratuitous generic pages are specified. Consider the document specified in example 4-37. All we need to know about the *BottomMargin* document style is that it changes the bottom page margin component of the formatting state vector.

In the document specified in example 4-37, two changes to a page format are carried out between the formatting of basic blocks. The second *paragraph* changes the top margin while the first *phrase* in the second *paragraph* changes the bottom margin. The result of applying the conventions are illustrated in figure 4-45.

Strictly applying the conventions, the logical object for the second *paragraph* must specify a new layout object. The new layout object for the second

```
@LogDoc{
   @Paragraph{...}
   @Paragraph-TopMargin{
      @Phrase-BottomMargin{...}
   }
}
```

Example 4-37: Document with Two Adjacent Page Changes

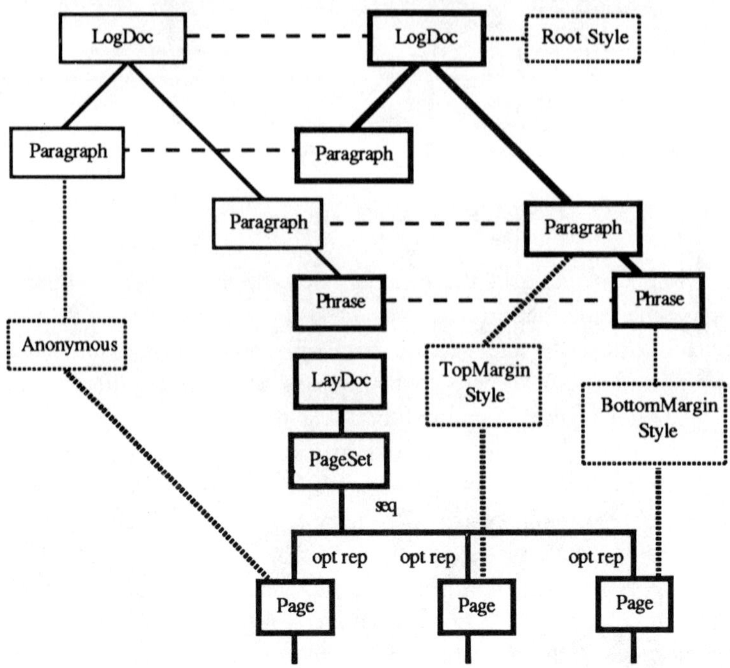

Figure 4-45. Needless Blank Pages for Adjacent Page Changes

paragraph (the middle page in the generic layout structure) would be different from the layout object specified by the *phrase* (the third page specified in the generic layout structure). To avoid this situation, changes made to a page format between the formatting of basic blocks are carried out on the same generic page. The result is shown in figure 4-46. Note that even though both logical objects require a new page, the same page is used.

However, the representation has lost some information. By combining the two changes to the page format into a single generic page, there is no way to recover which document style caused which change. This situation is analogous to the problem of representing multiple changes to a font, where different document styles can combine to create a font reference different than either document style specified. In this situation, separate font references are created to represent the document style information and the formatting state vector used in the environment. The place to hold page formatting information is a generic page. Thus, the way to represent both document styles and the formatting state vectors of environments that require changes to page formats is by having multiple generic pages, some pages being used to hold document style information (and referred to by the generic logical structure) and other generic pages used to hold environment information (and referred to by the specific logical structure). In figure 4-47, the generic logical structure refers to generic pages that contained

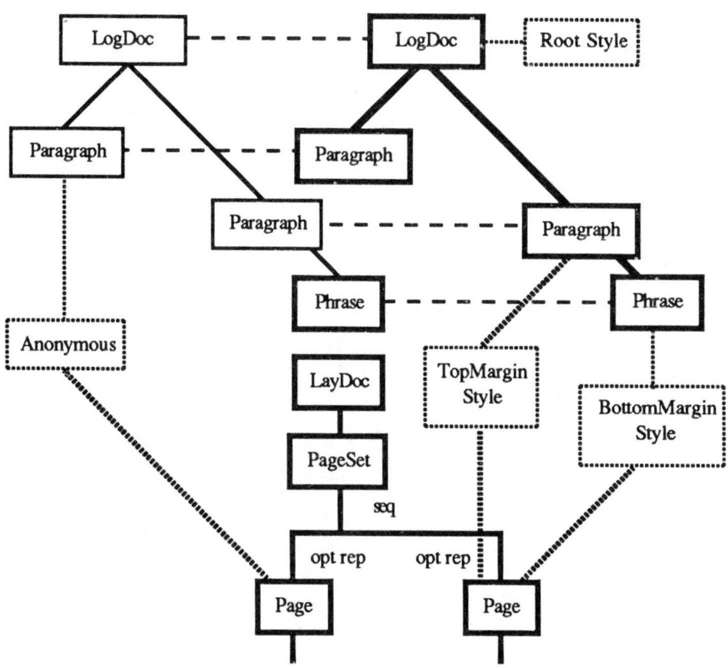

Figure 4-46. Concise Representation of Adjacent Page Changes

the information required for the *TopMargin* and *BottomMargin* styles, while another generic page is used by the specific logical structure that combines all of the necessary page formatting information. Because the *opt-rep* expression is used in the "generator for subordinates" attribute for the sequences of pages in the pageset, the generic pages used by the generic logical structure need never be instantiated, and thus, will be skipped over during the layout process.

We apply the general idea of new generic pages for each of the formatting state vector components that affect page layout. We now consider the representation of each formatting state component in detail.

4.9.12.2 Top and Bottom Margins

In this section, we consider the representation of top and bottom margins. These margins are set by positioning the generic frames appropriately in a generic page. A list of the affected attributes was given in table 4-5 on page 112. Here, we consider the values that are assigned to those attributes.

The representation of page layout components assumes a particular representation of generic page. Specifically, the top margin is the boundary line between the *Header* frame and the *BodyFrame1* frame, and the bottom margin is

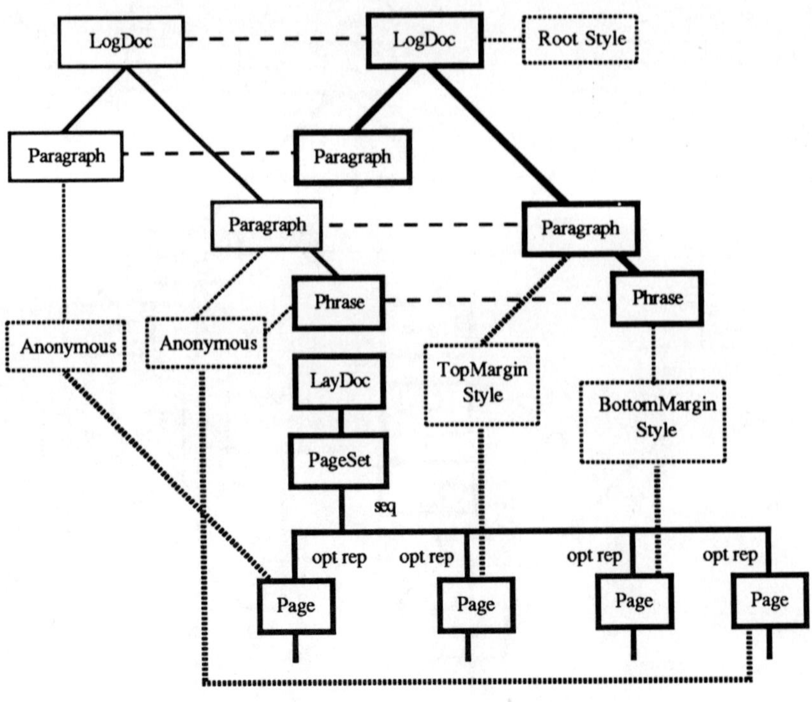

Figure 4-47. Modified Generic Layout for New Page Margins

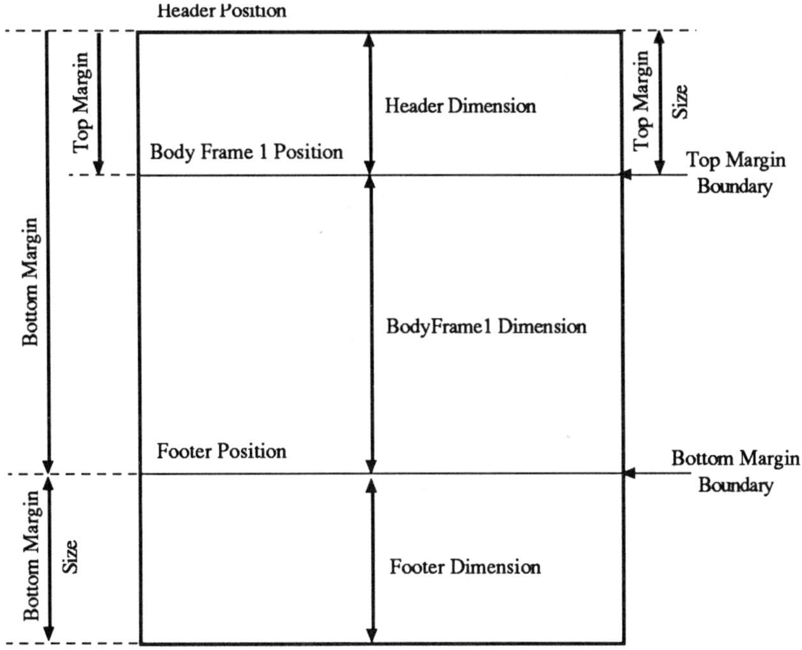

Figure 4-48. Frame Layout on a Page

the boundary line between the *Footer* frame and the bottom of the *BodyFrame1* frame. Thus, changes to the top margin affect the vertical dimension of the *Header* frame, and the position of the *BodyFrame1* frame. Further, if the position of the *BodyFrame1* changes, the vertical dimension of the *BodyFrame1* may also have to change to ensure that the boundary between the *BodyFrame1* and the *Footer* frame does not move. Similarly, when changes to the bottom margin affect the position of the *Footer* frame, the vertical dimension of the *Footer* frame must be adjusted to fit on the page, and the vertical dimension of the *BodyFrame1* frame must be adjusted to match the new *Footer* position. Figure 4-48 shows how the various frames are used with respect to the formatting state vector.

A summary of the values for various parameters and sub-parameters in the generic frames is given in table 4-14. Note that when changing some of these attributes for one margin, others must change as well to maintain the old value of other components. Following the table is a brief description of what is involved in changing each margin.

Frame	Position	Dimension
Header	Fixed Horizontal=0 Vertical=0	Horizontal=Maximum Vertical=Top Margin Boundary
Footer	Fixed Horizontal=0 Vertical=Bottom Margin Boundary	Horizontal=Maximum Vertical=Bottom Margin Size
BodyFrame1	Fixed Horizontal=0 Vertical=Top Margin Boundary	Horizontal=Maximum Vertical=Bottom Margin Boundary - Top Margin Boundary
FrameC	Variable Fill Order=Normal	Horizontal=Maximum Vertical=Rule B
FrameF	Variable Fill Order=Reverse	Horizontal=Maximum Vertical=Rule B

Table 4-14. Values for Generic Frame Attributes

Top Margin

The top margin is represented using the "position" attribute on the *BodyFrame1* frame. The "fixed position" parameter specifies the "vertical position" sub-parameter, which measures where the top of the *BodyFrame1* frame is on the page. Hence, the value for the "vertical position" sub-parameter is exactly the same as the the top margin in the abstract document model. If the top margin alone changes, then the "fixed dimension" sub-parameter of the "vertical dimension" parameter of the "dimensions" attribute must also be adjusted, so that the calculation of the bottom margin is maintained correctly. In addition, the "fixed dimension" sub-parameter of the "vertical dimension" parameter of the "dimensions" attribute of the *Header* frame (immediately above the *BodyFrame1* frame on a page) should be set to the top margin size so that the header and body area do not overlap.

Bottom Margin

The bottom margin is represented using the "fixed dimension" sub-parameter of the "vertical dimension" parameter of the "dimensions" attribute on the *BodyFrame1* frame. This value measures the distance from the top of the *BodyFrame1* frame to the bottom of the *BodyFrame1* frame. Since the bottom of the *BodyFrame1* frame is defined to be the same as the bottom margin, setting the specified sub-parameter would result in setting the bottom margin. If one measures the bottom margin from the bottom of the page, then given a page size P, a top margin T and a bottom margin B, the value for the sub-parameter would be P-(T+B). If one measures the bottom margin from the top of the page, then the value for the sub-parameter would be B-T. In addition, the "vertical position" sub-parameter of the "fixed position" parameter of the "position" attribute of the *Footer* frame should be set so that the footer frame does not overlap with the body frame, and the "fixed dimension" sub-parameter of the "vertical dimension" parameter of the "dimensions" attribute of the *Footer* frame (immediately below the *BodyFrame1* frame on a page) should be set to the bottom margin size so that the footer lays entirely on the page.

4.9.12.3 Headers and Footers

The NIST document application profile allows a variety of content and structure in a header or footer. For EXPRES, a header or footer is described by a generic basic logical object that is either a *Text* object or a *Number* object. To use a particular header or footer in a generic page, the basic logical object is referenced by the "logical source" attribute of the *FrameH* inside of the *Header* frame or *Footer* frame for the page. Although the generic object may be referred to by more than one "logical source" attribute, it is recommended that no other generic logical object and no specific logical object refer to this generic object.

4.9.12.4 Page Numbers

Page numbers may only be used as headers or footers. Therefore, they can only be specified as a *Number* generic basic logical object that is the value of a "logical source" attribute for a generic *FrameH*. Changes to page numbers, either in representation (numerals or roman) or value (changing a page number) are handled the same way as changing a header or footer. The details of how the *Number* generic basic logical object is used are beyond the scope of the EXPRES conventions.

4.10 Document Profile

The document profile in ODA provides information about the document as a whole. For example, the choice of document application profile and presence of certain logical and layout structures are defined in the document profile.

Both ODA and the NIST document application profile require several document profile attributes to be defined, such as "ODA version" and "generic logical structure". For the purposes of EXPRES, most of these attributes were unneeded. We were going to perform limited interchange with other EXPRES participants who were using the same ODA manuals, following these conventions, and with systems configured for use in the United States. Therefore, we were sure which version of ODA we were using, we knew which document structures were going to be present, and so on. Therefore, we omitted nearly all document profile attributes.

Two exceptions were "fonts list" and "unit scaling". One needs the "fonts list" attribute to get a list of the font references used in the document. There is no other place to hold information about a font's family, face codes, weight and other characteristics. The use of the "fonts list" attribute was discussed in section 4.9.7. The "unit scaling" attribute is used to allow translators to use their native system measurements when calculating values for various attributes. Had all participants used the default unit scaling values in the ODA representation, then translators in both directions would have had to perform extraneous calculations for scaling their units to 1/1200 of an inch[26]. By using the "unit scaling" attribute, only the translators from ODA to the native system had to perform scaling to retrieve the correct information.

4.11 Review

In this chapter, we presented the ODA representation of the document model we used for translating between document systems for the EXPRES project. Our goal was to provide high editing fidelity with a lesser concern for imaging fidelity. Therefore, we had to make simplifications in the NIST document application profile model while layering our own set of conventions for the representation of document styles. The result is a strategy that is straightforward to implement and provides a sufficient level of functionality to evaluate ODA as an interchange medium. In the next chapter, we discuss the translators we built using these conventions, describe how we used those translators for interchange and evaluate how well ODA met our needs.

---

[26]Technically, 1/1200 of 25.4 mm. See section 3.10/1 of ISO IS 8613.

5
Experience using ODA

This chapter discusses our experiences using ODA as an interchange medium between document processing systems. Following the conventions in chapter 4, the EXPRES participants and their collaborators built a collection of translators between a variety of multi-media formats and ODA. The EXPRES participants, their collaborators and a group from the Technical University of Berlin connected their systems together and interchanged documents at several demonstrations. In this chapter, we discuss each of the participating systems and translators. We then discuss the scenarios presented at the demonstrations, including screen pictures showing the document as it was interchanged among systems. We close the chapter with an evaluation of the translation efforts: how well ODA served our purpose, how well the translators worked and what problems are intrinsic to the translation process.

5.1 EXPRES Document Systems

The systems shown in interchange demonstrations were the Andrew Toolkit, Diamond, Interleaf and troff. In this section, we provide some background on these systems and their translators. For each system, we give an overview of that system and discuss its translators.

5.1.1 The Andrew Toolkit (ATK)

The Andrew Toolkit is a multi-media user interface tool kit. The system was designed and implemented at the Information Technology Center (ITC) at Carnegie Mellon University as part of an IBM-funded project. The Andrew Toolkit runs on a variety of Unix systems under the X Window System [49, 50]. As bundled with the X Window System distribution (see section K), the Andrew Toolkit comes with a collection of subroutine libraries, dynamic objects for various media and a collection of applications built with the tool kit. The applications include a generic multi-media editor, a multi-media mail and bulletin board system, a graphical shell and a multi-media help system. Many different kinds of document systems have been built with the Andrew Toolkit. For the EXPRES project, we chose to experiment with the most widely used Andrew Toolkit document system: the generic multi-media editor and the *text* object, which we refer to collectively as *ATK*.

5.1.1.1 Document Model

The general document model provided by ATK is both simple and powerful. A document is a sequence of characters, each of which can be a conventional glyph or an *embedded* object. Embedded objects provide media other than text, such as rasters, structured graphics, tables and equations. The power of ATK is provided through the use of a general style system. Styles can manipulate any component of the formatting state vector using both relative and absolute calculations. Furthermore, styles can extend the formatting state vector to provide information beyond the predefined formatting specifications.

Styles can be applied to any consecutive sequence of characters, including embedded objects. Typically, a style is applied to the logical part of a document. For example, a style may be applied to a word to make it italic or to a paragraph to change its indentation. However, a style may start in the middle of one paragraph and finish in the middle of another and not correspond to any document structure. In addition, more than one style may be applied to a sequence of characters. This permits, for example, the application of a style to a paragraph to change its indentation, and the application of another style within the paragraph to italicize a word.

Although the model provided by ATK is general, the implementation of that model is oriented towards display in a window and not on a printed page. Unlike a printed page, a window can be resized, which causes ATK to reformat the visible part of the document. This may change line breaks and scale embedded objects to fit into the available area. In the vertical dimension, ATK provides a view of an infinite scroll and does not divide a document into pages, or provide any of the page layout associated with printed copy. Instead, ATK assumes that alternate information is provided. For example, a window title instead of a page header provides information about a document, a scroll bar instead of page numbers shows location in a document and dynamic *links* and *bookmarks* connect parts of a document instead of static cross references. Similarly, ATK has no predefined notions of page footers, page breaks, top margins or bottom margins[27].

---

[27]Special accommodation for these features can be made when a document is printed, but the addition of these facilities came after the EXPRES developments.

Another special property of the Andrew Toolkit is arbitrary embedding of media. Just as ATK permits other media to be embedded in text, other media permit further embedding. For example, one can put a table in some text, a structured graphic in one cell of the table and an equation as one object in the graphic. Unlike ODA, content is not restricted to the leaves of a document, but can appear anywhere in a document structure.

5.1.1.2 Translator Implementation

Two ATK translators were written for the EXPRES experiment at the ITC: one from ATK format to ODIF[28], the other from ODIF to ATK format. The ATK translators had two unique tasks: synthesis of layout information and synthesis of document organization. As described in the previous section, ATK is display-oriented and lacks any specifications for page layout. However, *some* values for vertical margin information must be present according to both the conventions in chapter 4 and the NIST document application profile[29]. The ATK translators used predefined page-layout values when producing ODIF and ignored incoming values when reading ODIF. With some more work, the ATK translators could have reused information instead of synthesizing it. For example, the design of the translators included facilities for keeping page-layout information that was provided in ODIF instead of throwing it away. When the document was translated back into ODIF, the same page-layout information could be sent instead of providing predefined page-layout information. However, we were not able to complete the implementation of this feature.

The ATK translators also synthesized a document organization. Recall that the ATK document model does not have the concept of a paragraph, let alone the concepts of passages, phrases or figures. But when translating a document into ODIF, an ATK document has to be partitioned into these entities. Two different algorithms were chosen: one for paragraphs, the other for passages and phrases. The translators break a document into paragraphs whenever two consecutive new-line characters are found. Recognizing passage and phrase structure is not as obvious. The only readily available document structure within ATK is the application of styles, so we used the structure imposed by style application to guide the use of passages and phrases: styles that applied to more than one paragraph were used to delimit passages, while styles used within a paragraph were used to delimit phrases. When receiving an ODIF document, the reverse transformations were performed; two new lines were added between

---

[28]Recall that ODIF, Office Document Interchange Format, is the ASN.1 representation of an ODA document.

[29]Omission of the information is not possible, since the defaulting algorithm of ODA will provide a value for all attributes.

paragraphs and styles were applied to the content associated with subtrees specified by passages and phrases.

The translators are two separate programs that took approximately six person-months to write using the EXPRES conventions, the CMU ODA Tool Kit and the Andrew Toolkit, by people who were developers of all three. The translators are about 15,000 lines of C code, which does not include supporting software from either the CMU ODA Tool Kit or the Andrew Toolkit, both of which provide substantial services to the translators.

One interesting result of writing the ODA translators was that we were able to separate the construction of the translators for different media among different people. They did not need to know how the rest of the translation worked in order to write their own section. Therefore, the separation of structure and content in ODA allowed us to partition the work among several people who worked in parallel.

5.1.2 Diamond

The Diamond system is a multi-media document preparation and mail system. The system was designed and implemented at Bolt, Beranek and Newman, Incorporated (BBN) as a DARPA-funded project. It runs on a variety of window systems, including the X Window System and Suntools [59]. The document system provides users with multi-font and multi-national text, raster images, graphics, voice and spreadsheets. The document system is integrated with a mail system that provides X.400 message delivery. For the EXPRES project, the Center for Information Technology Integration (CITI) at the University of Michigan added equations to Diamond, and released the system as UM EXPRES[30]. For interchange purposes, the group at the University of Michigan used the document facilities of Diamond with their added equation facility, but did not use the X.400 transfer abilities for interchange demonstration.

5.1.2.1 Document Model

Diamond provides a straightforward document model, in which a document is a sequence of *paragraphs*, and a paragraph can be either text or another media, such as a raster image or a table[31]. Regardless of medium, each paragraph has

[30]BBN has continued development of Diamond as well and sells a commercial version called BBN/Slate.

[31]Voice can be attached anywhere in the document rather than being a separate paragraph. Since neither ODA nor the EXPRES conventions considered audio media, this difference is unimportant.

its layout described by a style, which can change fonts, horizontal margins, alignment and other formatting properties. Paragraphs are formatted onto pages, and the user may control page-layout information, such as top and bottom margins, headers, footers and page numbers.

Diamond contains a rich style system with three interesting properties. First, styles can specify absolute and relative calculations. However, any particular component of the formatting state vector will be manipulated in only one way, absolutely or relatively. The second interesting property is that there is structure among styles. Diamond allows one style to be defined in terms of another by providing a style inheritance facility. Unlike Andrew's and ODA's styles, Diamond's inherited styles are problematic for interchange. To achieve a high-fidelity translation, one needs some way to retain the structure among styles. However, in the absence of the proposed "variant of" attribute discussed in section 4.5, there is no easy way to represent the style inheritance relationship.

The third interesting property of Diamond styles is their partitioning by location of use in a document. There are three classes of styles, called *global styles*, *paragraph styles* and *font deltas*. Each class of style limits both the formatting state vector components that can be manipulated, and the places in the document where the class can be applied. For example, a single global style provides defaults for all components of the formatting state vector and can be applied only to the whole document. Another class is the paragraph style, which can change all but page-layout components, and, as its name implies, is applied to a single paragraph. The third class, font deltas, contains styles that are applied within a paragraph to change only font components of the formatting state vector. The separation of styles into three classes also controls their application nesting. For each piece of text content, there is an application of the global style, followed by an optional paragraph style, followed by an optional font delta.

5.1.2.2 Translator Implementation

Two translators were written at CITI for the EXPRES experiment. They translate between the Diamond internal representation and ODIF. The Diamond translators were built as an integral part of the Diamond system.

The Diamond translators performed some simple transformations to styles during conversion. To eliminate any inherited styles, the translator that produced ODA collapsed all inherited style definitions into a single equivalent definition, throwing away all style structure. Similarly, no inherited styles were created when reading an ODIF file. Thus, the Diamond translators lost the structure of styles.

Most of the time, little rearranging of style application was necessary. For example, global styles, paragraph styles and font deltas match well with the use of default value lists on the *LogDoc*, *Paragraph* and *Phrase* logical objects in ODA. However, when a *Passage* containing a style was received by Diamond, the translator converted the style into a paragraph style, broke the *Passage* into several paragraphs and applied the style to each paragraph. Thus, the Diamond translators lost all passage structure in an interchanged document.

There was one other transformation performed by the Diamond translator for interchange: the simulation of centered figures. This was necessary because the style system in Diamond ignores alignment requests for media other than text. Hence, a specification in ODA for a figure to be centered could not be provided directly. Instead, the Diamond translator calculated an appropriate left margin that would cause the figure to appear to be centered on the page. This means that an interchanged document could lose centering information.

The translators were about 5,000 lines of additional C code for the Diamond 3.0 system. They took about three person-months to write by someone familiar with the internals of the Diamond system.

5.1.3 Interleaf

Interleaf [46] is a document production system developed and sold by the Interleaf Corporation. Interleaf provides facilities for producing high-quality technical documentation with a variety of media, including graphics, rasters, equations and tables. Interleaf also provides a large collection of layout facilities including *micro-documents*, which allow an author to place a document within a document. The system runs on a variety of hardware, operating systems and window systems, including Unix, VMS and X Windows.

5.1.3.1 Document Model

The basic document model in Interleaf is a sequence of *components*. Each component can range from a structure as simple as a paragraph of text, a drawing or a table, to one as complicated as a micro-document with its own structure and formatting. In addition, Interleaf contains a great variety of embellishments, such as multiple page formats, automatically-generated tables of contents and tight control over detailed formatting decisions, such as kerning and intraline spacing.

Formatting in Interleaf is specified through the use of styles. Like Diamond, Interleaf provides an inherited style facility, where styles specify both relative and absolute calculations. But like Andrew, the styles can be applied to components as a whole, or within a component.

There are two special properties of Interleaf that affect its interoperability using ODA. The first is that Interleaf, like the Andrew Toolkit, allows content to be nested. For example, a table can contain a micro-document, which can contain a graphic. Since ODA has no nested content, translation to ODA can be problematic. The second property of Interleaf is its sheer detail. There are many special effects that Interleaf provides, such as paragraph profiles and multiple initial-lines indentation, that cannot be specified directly using the ODA character content model. As a result, documents that exploit many of Interleaf's features may lose substantial amounts of formatting information during translation.

5.1.3.2 Translator Implementation

Two translators were written by a group at the Aerospace Information Systems of McDonnell Douglas Corporation (MDC) for the EXPRES experiment: one from Interleaf ASCII Reader format to ODIF, the other from ODIF to Interleaf ASCII Reader format. These are separate programs that used the CMU ODA Tool Kit for support but did not use Interleaf's ASCII Reader tool kit[32].

Unlike the Andrew and Diamond translators, the Interleaf translators did not attempt to retain style information. When generating ODA, only the information in the formatting state vector associated with an environment was translated. When receiving ODA, the document style information and document structure within generic logical structure was ignored. Hence, style information was lost when interchanging with either of the Interleaf translators. The translators, including code to support parsing and generation of the Interleaf ASCII format, were about 4000 lines of C code: 2000 lines for each translator. The two translators took about 4 person-months to write by people who were familiar with ODA but not familiar with the ASCII representation of the Interleaf structure.

5.1.4 Troff

Troff is the base of a family of programs for typesetting technical documents. It is derived from the original RUNOFF program by J. E. Saltzer at MIT. Troff was developed by J. F. Ossanna around 1973 in assembly language for the PDP-11. It was rewritten in C around 1975, and until late 1977, it evolved slowly to support multiple output formats and provide better performance. The

---

[32]Interleaf's ASCII Reader tool kit was unavailable during development of MDC's translators.

version used by EXPRES is integrated with utilities to produce PostScript documents.

The main difference between troff and other low-level formatters is the degree to which normal formatting features are predefined by the system. Instead, troff is programmable: it has variables (called registers), arithmetic, and conditional tests on variables. Thus, troff has no special commands to deal with footnotes, paragraphs, tables of contents or titles.

5.1.4.1 Document Model

As a formatting engine, troff does not have a document model, but instead, it has a page model. One can simulate a document model through the use of *commands*, *requests*, *macros* and *traps*.

Troff defines about 80 commands and about 40 requests. It is possible to define macros to bundle various commands and requests, or to hold text. It is also possible to set traps, which can invoke macros depending on the position of the output. These mechanisms, for example, can be used for dividing the output into pages, adding running heads with page numbers, printing multiple columns and imaging footnotes.

A macro package does not only hide the formatting details, it can also function to give a logical structure to the document. For example, the first widely available macro package, the `-ms` [39] macros, provides definitions of paragraphs, section headers, footnotes and abstracts.

In addition to macro packages, one can build preprocessors to provide additional document structure and media, and postprocessors to provide some formatting and imaging semantics. For example, eqn [37] defines a simple language for describing mathematical expressions. Similarly, tbl [40] provides a language for formatting tabular material while pic [36] and ideal [62] are preprocessors for defining line drawings.

5.1.4.2 Translator Implementation

An ODIF to troff translator was written at the ITC for the EXPRES project. We decided that writing converse the translator, from troff to ODIF, would be too difficult. Although the use of various macro packages and preprocessors make troff appear to have a multi-media document structure, no one in practice uses *only* the macro commands and preprocessor commands. Therefore, a troff to ODIF translator could work only by emulating the entire low-level semantics of troff. Emulating all of the troff semantics was far more work than we wanted, and would result in losing the high-level document structure and multi-media content we were seeking. Hence, we restricted ourselves to translating from ODIF into troff.

Because troff is used as an imaging system, the ODIF to troff translator was concerned with presenting an acceptable image and not with providing an editable document. Therefore, the translator made no attempt to recover style information (as macro definitions) or provide a logical representation for content. For example, raster media was not converted into a high-level representation such as pic, but instead converted into a printable image [4].

The translator was about 3,000 lines of C code and required about 1.5 person-months by people who were expert in troff and the CMU ODA Tool Kit.

5.2 ISOTEXT Document System

One interchange demonstration took place between the EXPRES systems and *ISOTEXT*, an ODA system developed outside of the EXPRES project. In this section, we provide some background on ISOTEXT and its capabilities.

ISOTEXT [12] is a native ODA document processing system developed by TELES GmbH in conjunction with the Technical University of Berlin under subcontract from Nixdorf. ISOTEXT implements the ODA document model through the use of a WYSIWYG editor. By its nature, ISOTEXT supports a document organization facility similar to ODA's logical structure and a formatting facility based on ODA's layout structure and layout directives.

Although ODA defines no content for tables or equations, there is work underway in standards committees for adding these media and the ISOTEXT system contains prototype implementations of early versions of the standards.

ISOTEXT adds two important editing features to its use of the ODA document model. First, ISOTEXT adds a *macro* facility that simplifies editing. Second, ISOTEXT contains a Lisp-like extension language for manipulating documents. These features would be lost during interchange, since they can not be represented using ODA.

ISOTEXT was one of several ODA systems developed as part of the *Piloting ODA* (PODA) project [19] sponsored by the *European Strategic Programme for Research and Development in Information Technology* (Esprit), which in turn is directed by the Commission of the European Communities. Since it is part of the PODA project, ISOTEXT does not use the EXPRES interchange conventions. The version of ISOTEXT used for the interchange demonstration at the ODA Workshop held in San Jose assumed that documents would be following the *Q112* document application profile discussed in section 3.4. This profile is a lower profile than *Q113*, which is aligned with the NIST document application profile upon which the EXPRES conventions are based. As a practical matter, the documents generated by the EXPRES translators were subsets of the NIST document application profile that closely match *Q112* requirements except for the use of the "default value lists" attribute and font definitions, so ISOTEXT was able to successfully read these documents. ISOTEXT had no problem with default value lists, since it could process them. The two problems that were apparent were related to font information and raster contents. ISOTEXT did not support the ISO draft font standard for font definitions, so we were not able to exchange font information. The raster problems were caused by the EXPRES participants' shortcuts, as explained in section 5.3.1: we assumed all rasters were 72 pixels per inch while ISOTEXT tried to derive the raster information from the appropriate ODA attributes.

5.3 Implemented Document Features

The previous sections provided details about the systems that participated in the EXPRES document interchange demonstrations. In this section, we discuss, in detail, those document features that were implemented and those that were not. Our description of a feature as implemented implies that all of the translators were able to translate the feature between ODA and each document model.

The implemented features are classified into the three parts of a document model, as defined in section 2.1: organization, formatting instructions and content. In addition, we discuss the editing information retained by the translators. We start by describing the document features that were not implemented completely by all systems or not implemented at all.

5.3.1 Common Limitations of Translators

As is clear from previous sections, we did not implement many of the features defined in the EXPRES conventions. Features were omitted for one or more of the following reasons:

- some document systems had insufficient functionality to support the feature,
- we lacked the resources (time) to implement the feature,
- ODA did not support the feature.

Table 5-1 lists those features that we were not able to interchange universally.

In addition, all of the EXPRES translators made common assumptions when converting between ODA and their native formats:

Logical Object	Use
Number	An integer
FNote	Collection of two numbers and footnote body
FNBody	Content of footnote
Geometric Drawing	Picture composed of geometric descriptions
Equation	Description of a mathematical equation
Table	A collection of columns of information
Line Spacing	The amount of space from one baseline of text to the next
Baseline Offset	The amount of space from baseline to Y coordinate of content
Paragraph Spread Above	The minimum amount of space from current paragraph to the preceding
Paragraph Spread Below	The minimum amount of space from current paragraph to the following
Tab Stops	A list of tabs stops
Header	Current header value
Footer	Current footer value
Footnote	Content targeted for the footnote page area

Table 5-1. Unimplemented Document Model Features

- The only media in a native document would be text, rasters, tables and equations.
- No media would be nested (only applies to systems that support nested media).
- Rasters, tables and equations would not be placed within a line, but would appear in a paragraph by themselves.
- Only ASCII characters would be present in text, no national or graphical character sets would be used.
- All rasters would be represented at 72 dots per inch within the ODIF document.

- Tables and equations would be represented in ODA using their ATK representations.

Each of these assumptions is discussed in turn.

All of the EXPRES systems could support more media than specified in the EXPRES conventions. Rather than defining how a translator should accommodate the presence of other media, we defined the scope of the translators to include native documents with only text, rasters, tables and equations. Similarly, we insisted that native documents not exploit any content nesting features. Thus, Interleaf documents containing micro-documents and Andrew documents containing nested tables would not be interchanged. Although some systems permit non-text media to be embedded within a line of text, the translators were written assuming that no raster, table or equation would appear within a paragraph of text.

ODA permits many different character sets to be used. Although experiments have been made with various national character sets, ATK only handles ASCII characters well. In addition, the versions of Interleaf and Diamond we used for the interchange demonstrations were configured with American character sets. Since all systems were limited to ASCII coded text, we narrowed the scope of the translators to ASCII character codings.

Another restriction we adopted was limiting raster resolution to 72 pixels per inch. All systems in the interchange had the ability to scale rasters, and the default resolution of rasters within ODA[33] did not match any system's choice. The initial interchange attempts exchanged raster content at 200 pixels per inch, which caused all systems to perform scaling when producing and accepting raster content. We found that the image degraded substantially during the scaling process, regardless of which pair of translators were being used. We tried substituting a variety of scaling algorithms, but none worked well. As a result, we decided to limit interchange to rasters that were 72 pixels per inch and not perform scaling during the translation process. A side effect of this agreement is that no translator produced the required ODA information in the raster content portions describing the raster density. As explained before, this prevented ISOTEXT from reading raster information produced by EXPRES translators.

Because no ODA content standards existed for tables and equations in ODA, some representation was needed for these contents. ATK uses formats similar to Excel [43] for tables and eqn [37] for equations. Because these formats are well known and documented, they were used by the translators that implemented tables and equations (ATK and Diamond). The translators im-

[33]The default value is 200 pels per inch. Other values for "pel transmission density" are 240, 300, 400, 600 and 1200 pels per inch.

plemented a related convention: tables and equations are exchanged as character content, not as private content. To conform with the ODA standard, one should define a new content architecture for a new medium, but we adopted the expedient of reusing the character content since our representations for tables and equations were textual. We distinguished the special contents through the value of the "application comments" attribute, as specified in the EXPRES conventions.

Despite these restrictions, we were able to exchange substantial amounts of logical and layout structure. The next sections discuss the document structures that we exchanged.

5.3.2 Organization

Recall that the organization of a document consists of its logical structure, that is, the way that an author organized the document (independent of its content). Section 4.4.1 lists those features of the NIST document application profile that our conventions cover. Due to restrictions in several of the document systems, we were forced to omit some of these features. The common features that were implemented are listed in table 5-2.

5.3.3 Formatting Instructions

Section 4.8 describes those features of a formatting state vector specified by the EXPRES conventions. Again, the lack of supported features in several of the

Logical Object	Use
LogDoc	The entire document
Passage	Collection of paragraphs, figures and other passages
Paragraph	Collections of text, figures and phrases
Phrase	Collection of text and other phrases
Figure	A raster
Text	Collection of characters
Raster	Picture composed of pixels

Table 5-2. Implemented Logical Objects From NIST Document Application Profile

FSV Component	Meaning
Left Margin	The amount of space from left edge of paper to left edge of content in the body area
Right Margin	The amount of space from left edge of paper to right most edge of content in the body area
Top Margin	The amount of space from top edge of paper to top most edge of text area
Bottom Margin	The amount of space from top edge of paper to bottom most edge of footnote area
Paragraph Indentation	The amount of space the first line of the paragraph should be indented
Justification	How content should be aligned: left, right, both or centered

Table 5-3. Implemented Formatting State Vector Components

document systems forced us to eliminate several components of the formatting state vector. The features we implemented are listed in table 5-3.

5.3.4 Content

Also listed in section 4.8 are features related to content that are defined in the EXPRES conventions. The restricted subset of those features that we implemented are listed in table 5-4.

5.3.5 Editing

As described in section 1.1.3, the retention of effective processability during translation was a primary goal of EXPRES. Section 2.3 explained that this capability is embodied in style sheet mechanisms and chapter 4 discussed the EXPRES conventions for style sheets. Diamond, ATK and Interleaf have style sheet mechanisms that are rich enough to allow us to implement the scheme as defined in chapter 4, though the Interleaf translators did not exploit Interleaf's style sheet mechanism.

FSV Component	Meaning
Font Family	The family of typeface to use for characters, restricted to Times, Courier and Helvetica
Font Face	Modifications of the typeface, restricted to subsets of {bold, italic}
Font Presentation	Modifications of the characters appearance, restricted to subsets of {underline}
Font Size	Size of typeface in points
Line Break	Whether the content should start a new line
Paragraph Break	Whether the content should start a new paragraph

Table 5-4. Implemented Formatting State Vector Content Components

5.4 EXPRES Demonstrations

The translators constructed by the EXPRES participants were demonstrated several times to hundreds of people at several venues. The first public demonstration was at the ACM Conference on Document Processing Systems held in Santa Fe, New Mexico during December 5-9, 1988 [1]. The same demonstration, with minor variations to be described, was repeated at the *Workshop on Compound Document Interchange Using ODA* sponsored by IBM and the National Science Foundation, held June 1-2, 1989 at the IBM Almaden Research Center in San Jose, California. This demonstration was seen by an estimated 25 people. Finally, the translators were put through their paces at several of the regular monthly demonstrations given at the Information Technology Center. The audiences at the monthly demonstrations were primarily upper-level managers from around the world.

The demonstration given in Santa Fe was run on three workstations connected by a local Ethernet. The workstations and associated operating and document systems are listed in table 5-5.

The demonstration consisted of taking a document created in ATK and translating the document so that it was editable on the other document systems. In each case, after the document was translated into the new document format, the document was edited to demonstrate its processability. The first page of the original document is shown in figure 5-1 on page 177.

Workstation	OS	Document System
IBM RT-PC	AOS 4.3 (BSD Unix-like)	X11, WM, Andrew
Sun 3/50	Sun 2.0 (BSD Unix-like)	Suntools, Diamond
DEC Vaxstation 2000	VMS 4.7	Interleaf

Table 5-5. Demonstration Systems

The demonstration consisted of the following steps:

1 The ATK document (on the RT) was translated into ODIF.

2 The ODIF representation was translated into troff format (on the RT), run through troff and displayed on the screen.

3 The ODIF representation was transferred to the Sun and Vax machines using FTP.

4 On the Sun, the ODIF representation was translated into Diamond. The Diamond document was edited to replace the phrase "Center for Information Technology Integration" with "CITI". In addition, the raster on the first page was replaced with a raster indicating the path of this document. The first page of the Diamond document can be seen in figure 5-2 on page 178.

5 The Diamond document was translated to ODIF and the ODIF representation was transferred to the RT. This representation was then translated to ATK and edited. Unfortunately, we do not have a figure for this document.

6 On the Vax, the ODIF representation from step 3 was translated into the Interleaf ASCII format. The Interleaf document was edited in a similar fashion to the Diamond document. Figure 5-3 on page 179 shows the first page of the Interleaf document.

A demonstration similar to this one was given at the workshop held in San Jose, with one major addition. Some people from the Technical University of Berlin (TUB) were present and were demonstrating ISOTEXT. The people from TUB volunteered to take the ODA translation of the original ATK document and attempt to read it into their editor, which accepted ODIF datastreams. Several minor changes had to be made to the ODIF stream produced by our translator to eliminate the use of font definitions. Once this was done, the

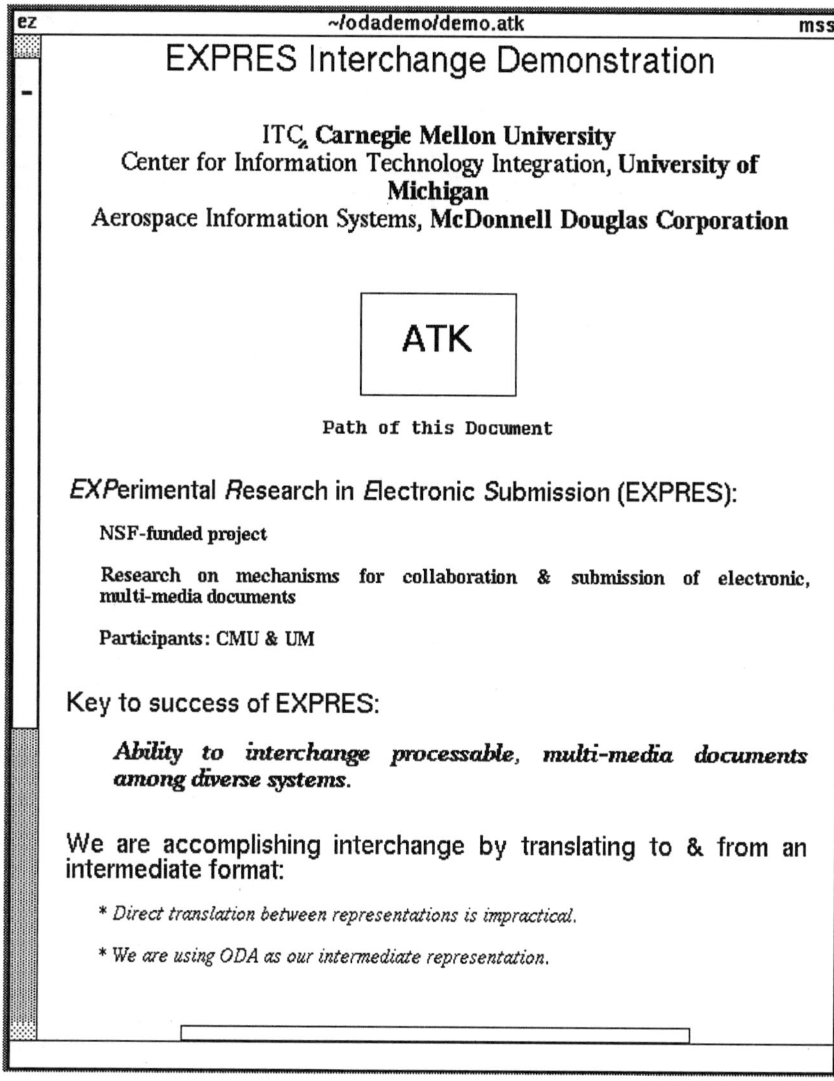

Figure 5-1. Original ATK Document

ISOTEXT system was able to import the document, display it and edit it. ISOTEXT was unable to process the raster information since the ATK translator assumed 72 pixel per inch resolution while the ISOTEXT system tried to to deduce the raster size from ODA information. We were unable to obtain a screen dump of the ISOTEXT document.

The demonstrations at the Information Technology Center were similar to the one in Sante Fe, except that troff was run on a NeXT machine, generating PostScript that was displayed on the screen. A picture of the first page of the troff version of the document is shown in figure 5-4 on page 180.

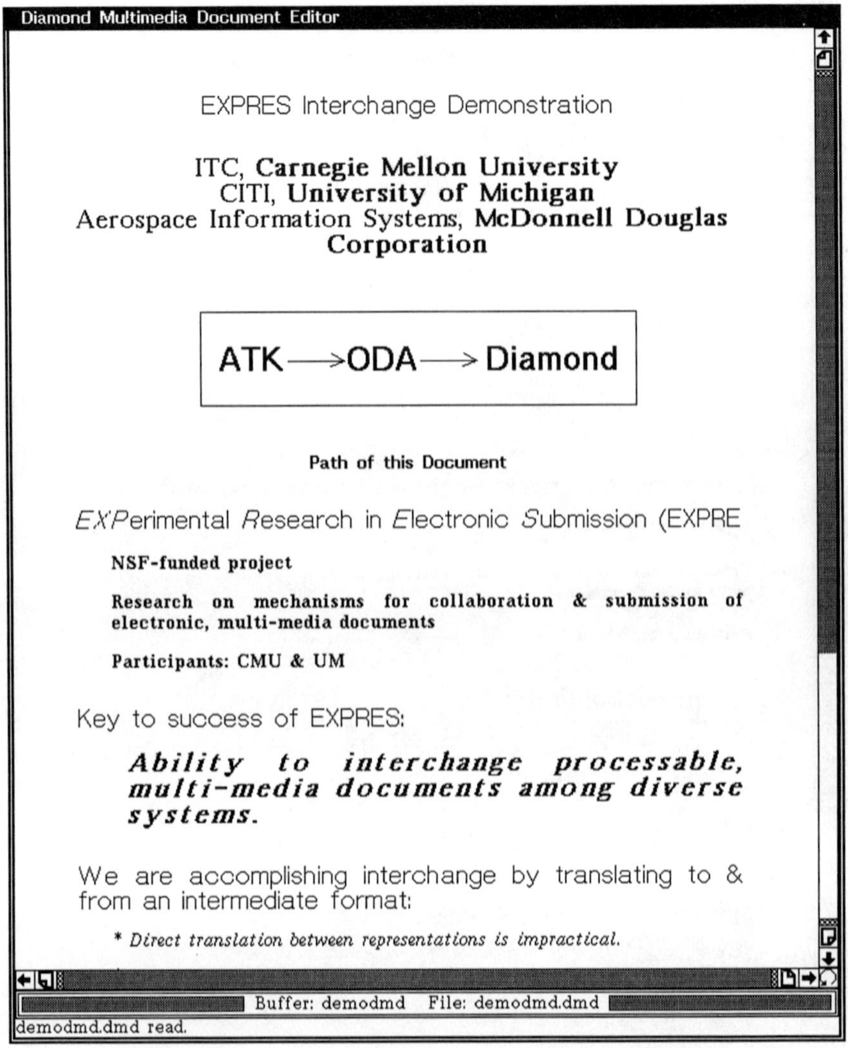

Figure 5-2. Diamond Document: ATK to ODA to Diamond

Peoples' reactions to the demonstrations varied widely. Some thought that we had accomplished our goals of interchanging processable, multi-media documents among diverse systems. More discriminating observers saw minor differences in the documents and thought that we had shown the basic concept through prototype-quality systems, but were not up to production-quality format translation. Others, who insisted on perfect image fidelity, believed the demonstration to be a failure, since the documents did not look the same on all systems: there were visible differences caused by changes in (soft) line breaks, (soft) page breaks and font substitutions. We take the intermediate position, that

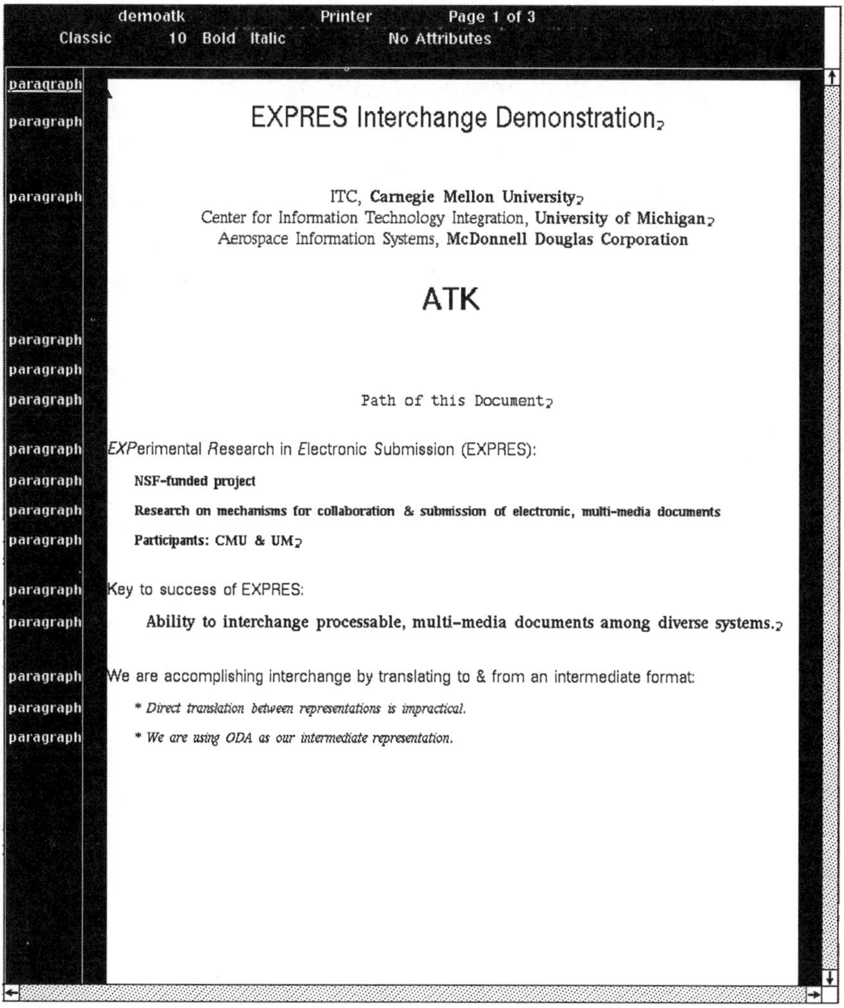

Figure 5-3. Interleaf Document: ATK to ODA to Interleaf

the demonstrations show the approach to be viable, though the translators we built were only prototypes. This position is elaborated in the next section, where we analyze, in detail, our experiences with document interchange.

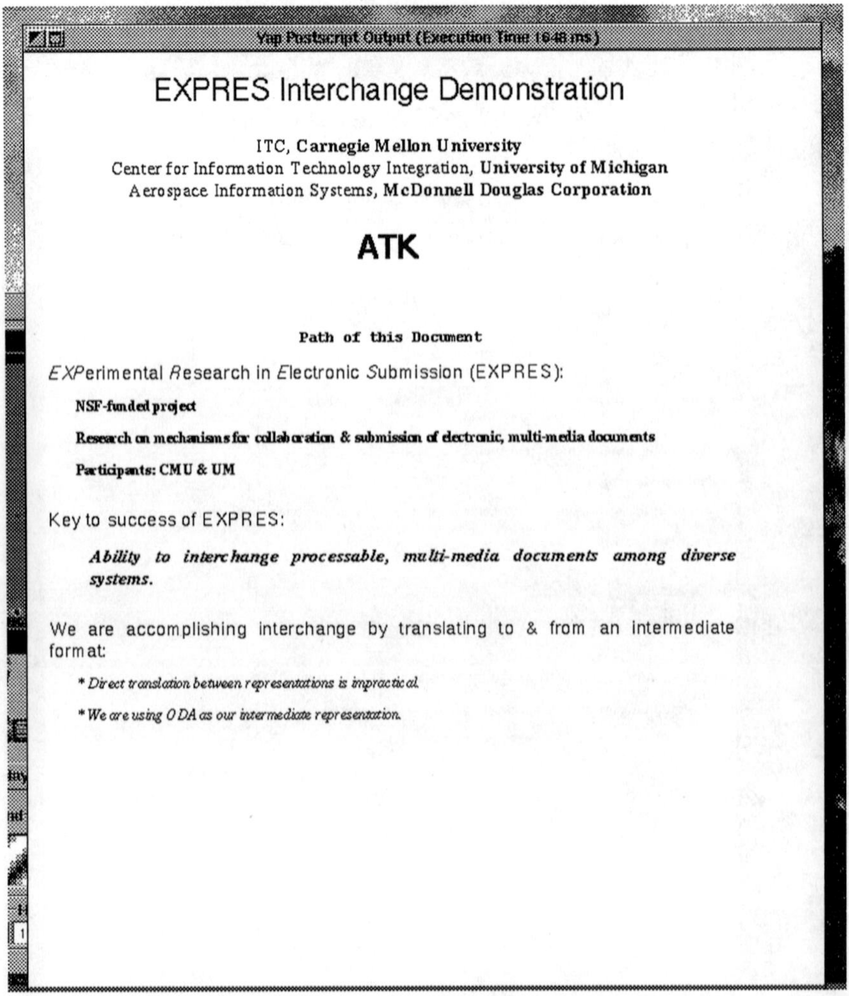

Yap Postscript Output (Execution Time 1648 ms)

EXPRES Interchange Demonstration

ITC, **Carnegie Mellon University**
Center for Information Technology Integration, **University of Michigan**
Aerospace Information Systems, **McDonnell Douglas Corporation**

ATK

Path of this Document

*EXP*erimental *R*esearch in *E*lectronic *S*ubmission (EXPRES):

NSF-funded project

Research on mechanisms for collaboration & submission of electronic, multi-media documents

Participants: CMU & UM

Key to success of EXPRES:

Ability to interchange processable, multi-media documents among diverse systems.

We are accomplishing interchange by translating to & from an intermediate format:

** Direct translation between representations is impractical.*

** We are using ODA as our intermediate representation.*

Figure 5-4. Troff Document: ATK to ODA to troff

5.5 Experiences with Interchange

By performing interchanges, we found several classes of problems that lowered the fidelity our translation of a document from one system to another. We broadly classify the problems as lost functionality, detail mismatch, display artifacts and ODA difficulties. Each is discussed below.

5.5.1 Lost Functionality

One of the more difficult problems we faced was the lack of functionality, either in one of the document systems, in ODA or in the document application profile. We have already discussed the limitations of individual systems (in section 5.1) and of document application profiles (in section4.4). We postpone discussing the limitations of ODA until section 5.5.4.

The loss of a system-specific formatting feature during translation is not overwhelming. Most of the desired information to be communicated is provided in all systems. What causes distress to users is the loss of information when a document returns to its originating system after being translated. The problem with both ODA and most host systems is their inability to hold information that is foreign to them. For example, ODA may not understand links that impose arbitrary structures on contents, but there should be a way to express the links so that if the document were returned to the originating system, the links could be recovered. Similarly, although ATK did not support footnotes, if it received one, it should be able to send it back out when the document is translated back into ODA. Because most systems are not written with interchange in mind, they usually contain no facilities for recording extra-system information.

5.5.2 Detail Mismatch

A less obvious problem to the designer of interchange programs is detail mismatch. A detail mismatch occurs when a source system, a target system and an intermediate representation all define a superficially identical feature. Because the resemblance is only superficial, subtle differences in the feature's implementation or interpretation affect the appearance and structure of a document. Several mismatches that we found are discussed below.

5.5.2.1 Style Systems

Different style systems were the most pervasive mismatch we found. Nearly every system had some notion of bundling formatting information into a style, but the differences between them made interoperation among style systems very difficult. Some of the differences we found included structure between styles, types of styles and applicability of styles. We elaborate each of these differences in turn.

Structure Among Styles

The relationships among styles offered by systems vary a great deal. ODA provides a paired, flat style system, where styles are grouped into pairs, a layout style and a presentation style, but where there is no relationship between any two styles. Andrew also has no structure on styles. However, Diamond and Interleaf allow one style to be defined in terms of another, adding structure to the styles. We found no good way to interchange this information in our demonstrations.

Partitioning of Style Function

We were not able to preserve the partitioning of function among styles. In representations like ODA, the fact that style functions are partitioned into different classes (layout and presentation styles) does not matter, since any particular function can fit in only one place. However, systems like Diamond allow the same kind of editing change to be in many different style classes. For example, a font change can be made in a global style, a property sheet or a font delta. Therefore, when translating between Diamond and ODA, it may not be possible to know which of the various kinds of styles should be used to represent some change to the formatting state vector.

Partitioning of Style Application

A third problem we had when interchanging style information was style applicability. Part of a style's specification is where it may be used in a document. Some systems, like Andrew, permit any style to be applied anywhere, assuming they properly (lexically) nest. Some systems permit styles to be applied only to a paragraph or only within paragraphs. Like the partitioning of style function, we found no good way to exchange information about partitioning of style application.

5.5.2.2 Rasters

Rasters represented another detail mismatch in our translation. Two manifestations of the problem were inappropriate scaling and image degradation.

All our systems support raster graphics, and all systems perform scaling when necessary. What was not clear was when a translator should perform scaling. Some systems provide the ability to store an image at a different resolution than when displayed, in which case one does not want the translator to provide any scaling—any necessary scaling would be performed when the raster is displayed. If a translator producing ODA believed that the raster image would be scaled on receipt, the producer may attempt to perform the scaling before encoding into ODA to take advantage of locally known information, such as whether the raster was a line drawing, a photograph or a dithered gray-scale image. The result of the mismatch is indecision of when to perform raster scaling.

Even if we knew when to do scaling, the process of changing the raster would cause problems. As explained in section 5.3.1, all translators ignore ODA raster semantics to ensure acceptable interchange. As a result of our experience, we consider the raster scaling problem to be one of the more severe problems waiting for future ODA translator writers.

5.5.2.3 Line Spacing

Our document systems and ODA interpret line spacing differently. This resulted in another minor mismatch that affected interchange.

ODA provides a collection of predefined line-spacing values along with an undefined "proportional line spacing" attribute. Andrew, Diamond and Interleaf also provide proportional line spacing. However, all four interpretations differ in how proportional spacing should be calculated. Sometimes, it is by the largest font in a line, sometimes it is by a base font size that is then scaled for double or triple spacing, and sometimes it is controlled by other content in the line, such as a raster image. As a result, the systems all claim to provide proportional line spacing, but the results can appear quite different.

5.5.2.4 Fonts

Because fonts are pervasive in modern document production systems, we tried to exchange meaningful font information. Although the font standard [34] provides a lot of functionality, we still had problems with high-fidelity exchange.

The first problem we had was a simple lack of fonts on all systems. As documents moved from system to system, fonts were substituted, causing both the format and the general appearance to change. As mentioned in section 5.5.1, some of the problem could be alleviated if the target system performed substitution only on display, keeping the true value of the font so that correct information could be encoded in further translations. Only Andrew provided this ability, so translated documents tended to drift towards the common set of fonts supported by all systems.

We encountered another minor detail mismatch with font representations. All systems had superscripts[34], subscripts, underling and font families, but they implemented them differently: the spacing relative to the baseline was different, the location and thickness of the score mark was different and the look of fonts with the same name was different. A similar problem occurred with font faces, such as bold and italic. Although all systems supported changes to the font face, some systems had more changes than others. For example, the Interleaf translator permitted only bold or italic changes to a font, but not both, while Andrew, Diamond and troff all permitted fonts to be both bold and italic. We ignored that particular problem by limiting changes to fonts.

5.5.2.5 Document structure

Finally, we found mismatched document structures. Although all systems claimed to provide some kind of document structure, the details varied. For example, Andrew's document structure is a sequence of characters while Diamond and Interleaf provide a sequence of a paragraphs. The NIST document application profile provides another abstract structure based on passages and phrases. As a result, the attempt to translate a document's structure usually requires a transformation of the structure. For example, paragraphs are created when going from Andrew to ODA, while phrases are eliminated when going from ODA to Interleaf. Because today's systems provide meager manipulation of document structure, this may not be a problem. The receiving system dictates a single document structure, so interchange parties always see their system's native structure. A returning document from an interchange appears to have lost no structure since the receiver's unique structure is imposed on the incoming ODA. However, when document structuring features such as Microsoft Word's outlining and manipulation facilities become commonplace, the difference in document structures among systems may result in the inability to retain the high-level, document-structure manipulation features as a document is translated.

---

[34]Technically, superscripts and subscripts are baseline offsets, but we consider them here because they were represented in ODA using font definitions.

5.5.3 Display Artifacts

A display artifact is the appearance of a document system feature that is not part of a document. Because document systems are intended to edit documents as well as display them, they contain visual cues for the editor. At first glance, these are indistinguishable from the document itself. Thus, as a document moves from system to system, one sees these additional display artifacts come and go. Thus, it appears that a lower fidelity of translation has occurred. Some of the display artifacts we encountered are discussed below.

5.5.3.1 Displayed Margins

One display artifact is the amount of white space displayed for a margin. Although all systems may claim that a document has a left margin of one inch, the information is applied only when printing a document. When displayed on the screen, the system uses some amount of white space between the content and the left side of the window to make editing easy and to conserve screen space. For example, Andrew always places a quarter-inch white boarder on the left side of the document, while Interleaf displays exactly the specified white space. Such a document appears quite differently on each system, and thus, one might believe that the margin information had been lost. To minimize this effect, we exchanged documents that specified a left margin value of 0 inches.

5.5.3.2 Figure Demarcation

Most systems view documents as collections of text with occasional inclusions of other material, such as a figure. To distinguish the figure from the text, document systems frequently provide visual guidelines. For example, placing a box around a figure or including grid lines are ways that a document system aids the editor in document manipulation. However, the additional boxes or lines are not part of a final document, though they may always appear on the screen. When a document with these additional boxes or lines is moved to another system that does not provide figure demarcation, it may appear that part of a document is lost. When translated back to the source system, the boxes reappear. Although this happened in our demonstrations, we ignored it.

5.5.3.3 Line Wrapping

Nearly all systems provide some kind of line wrapping, that is, fitting a contiguous collection of text into a box by breaking the text into lines. However, neither ODA nor most document systems provide precise semantics for calculating where lines should be wrapped. Some factors that affect line wrapping include:

- Available fonts and their corresponding font size tables;
- Wrapping based on logical page margins or wrapping based on window size;
- Algorithm for applying character and word spacing.

For example, all of our systems (as well as ODA) provide automatic line wrapping, yet lines wrapped differently during the demonstration. We decided that our concentration on processable forms of documents implied that documents would be edited at the receiving side, and therefore lines would be reformatted under normal use. If the exact position where line breaks occurred had to be preserved, then a page description interchange should have been used. Hence, we made no attempt to preserve the exact line wrapping induced by the text formatting algorithm of each system.

The lack of functionality, detail mismatch and display artifacts we have been outlining are generic. The problems occur regardless of intermediate representation. We also had some problems with ODA itself, which are discussed in the next section.

5.5.4 Difficulties with ODA

As we applied ODA to our document translation problem, we encountered several problems with ODA. We consider some broad classes of these problems in this section.

5.5.4.1 Size

The first problem we had, and probably the first problem everyone has with ODA, is its size. The main standard is over 750 pages. As a result, a comprehensive knowledge of ODA requires the mastery of a great deal of information. Further, the standard contains references to many subsidiary standards that must be understood to use ODA. Besides increasing the volume of material to read, there are interactions between the ODA standard and subsidiary standards. For example, certain attributes in ODA, such as "graphic rendition" are defined as being synonyms for parts of other standards—here, the SGR control function in ISO 6429 [28]. In addition, there are interactions between the standards. For example, not all of the parameters defined for the SGR control function in ISO 6429 are permitted for the "graphic rendition" attribute in ODA. Mastering the immense number of details and interactions is a substantial effort.

After learning ODA for EXPRES interchange purposes, we found several problems with applying the standard. We discuss these problems, starting with redundancy in ODA.

5.5.4.2 Redundancy

Redundancy in ODA means that more than one way to describe a document feature is either possible or required to effect an interchange. In the former case, the ease of selecting an alternative makes translation into ODA easier, since a producer can select one of several representations. However, this ease comes at the expense of the recipient, who must determine which of several ODA facilities were used to express a document feature. In the latter case, the producer of ODA must generate and the recipient must verify that redundant information is present in an ODA document to ensure that a particular document feature is being exchanged. We consider each case in turn.

Selecting One of Several Representations

There are several places in ODA where we had to choose one of many representations for a document feature. Because we used a relatively limited document model, we certainly have not explored all of the possibilities. For purposes of discussion, we separate two kinds of choices we had to make: low-level and high-level.

Low-Level Representation Choice. Low-level choices are made for small or isolated parts of documents. Usually, the representation of a low-level choice involves a single attribute rather than a combination of features. Four specific low-level redundancies gave us problems.

First, there are several representations for horizontal margins. For example, the ones we considered for the left margin were:

"offset" — One can express a left margin by using the "right-hand offset" parameter of the "offset" attribute in the logical structure. This applies to content to be formatted in a frame extending to the edge of the imaging area. The needed white space is provided by offsetting blocks by the specified amount within the frame. This is the method we chose for representing left margins.

"indentation" — When using character content, one can specify the amount of white space from the interior edge of a block to the text through the use of the "indentation" attribute in the logical structure. If one uses a frame extending to the edge of the imaging area and places contained blocks to the edge of the frame, then the value for "indentation" will be the left margin.

"initial offset" (character content)

If one has formatted character content and wants to position the left margin of the text, one can use the "initial offset" attribute in the layout structure. Again, this assumes that the content block is placed in a frame that extends to the edge of the imaging area and that the block is not otherwise offset.

"initial offset" (raster content)

Like the "initial offset" for character content, the "initial offset" attribute for raster content can be used to shift a raster within a block to provide a left margin.

relative frame "position"

Instead of manipulating how a block is moved within a frame, one can place a nested frame that provides spacing for a left margin. Blocks would be formatted with no offset to the interior frame. The left margin could then be specified in an "offset" sub-parameter within the "variable position" parameter of the "position" attribute of the nested frame.

absolute frame "position"

Using the same strategy as for a relative frame position, one can instead explicitly set the "position" sub-parameters of the "fixed position" of the "position" attribute of the nested frame.

relative block "position"

One need not use a nested frame on the layout side to specify a relative position. Instead, one can use the "position" attribute on a block to specify a relative position to the enclosing frame that generates a left margin. One can do this even without having a specific layout structure by placing an appropriate default value list for blocks on the containing frame in the generic layout structure.

absolute block "position"

Just as one can use "fixed position" parameter for frames, one can use it on the block.

"border" on containing frame

White space between a frame and its contained block can be specified by the "border" attribute on the containing frame. Defining the "border freespace width" of the "right-hand-edge" to the desired left margin size and all of "border line type" sub-parameters to be 'invisible' yields the specified margin.

"border" on contained block

One can place the "border" on the block, either explicitly in a specific layout structure, or implicitly, through the use of a default value list on a frame in the generic logical structure.

One can argue that each of these is acceptable for representing the appearance of a left margin. Some are content specific, some not. Some require formatted content, some not. Most of them can be combined, and thus, the left margin may be an amalgam of several attributes. All of the choices for left margin have implications on the representation of other document features. For example, the use of "offset" for the right margin would naturally fit with use of "offset" for the left margin, while the use of "dimension" for the right margin follows naturally from the use of the "position" for the left margin.

The choice of representation is also affected by the desired document model. For example, the use of "offset" worked well for EXPRES, but we suspect it will not work well in a document that requires multiple columns—a requirement excluded from the EXPRES specifications. Unfortunately, a receiver of an ODA document has no way to know which document model is intended and must search many pieces of the received document to deduce relevant features. Although the details of our example focused on the representation of "left margin", we found the same problem with many abstract document features.

A second source of low-level redundancy was the use of embedded control functions in content, especially character content. Many of the attributes had nearly perfect analogues as embedded control functions, such as "graphic rendition" (SGR) and "graphic character subrepertoire" (IGS). Other attributes, such as "alignment" (JFY), had certain values that could be simulated by embedded control functions. To interpret character content correctly, one had to search for the same information in different locations, as an attribute value and as an embedded function.

A third source of low-level redundancy was with the embedded control functions themselves. Typically, a referenced standard would provide some function, but that function would be perceived to be inadequate. So an additional embedded code would be defined and included. For example partial line movement is provided by reference to ISO 6429 via the control function PLU (partial line up). However, that control function does not specify exactly how much space to move. Hence, an additional control function is defined in ODA, VPB (line position backward[35]), that performs essentially the same function but permits the amount of space to be specified as a parameter.

A fourth source of low-level redundancy was the specification for font information. There are three places that font and character set information is provided: graphic renditions, the "character fonts" attribute and the font definitions in the document profile's "fonts list" attribute. Some aspects of a font definition, such as the use of a shadow font, can be specified only by a font

---

[35]From section 11.1.15/6 of ISO IS 8613: "VPB - Line Position Backward. A control function ... which causes the active position to be moved in the opposite direction to the line progression." The careful reader can deduce that this means superscript.

definition. But others, such as italic or bold, can be specified either with a graphic rendition or through the use of appropriate font definitions. Fortunately, the semantics of these features specify that only one technique can be used—if a font definition is used, then the other graphic rendition codes that affect font characteristics are undefined. However, the size of a font can be recorded in two locations: in the "character fonts" attribute and in the font definition itself. In fact, the size must be specified in the "character fonts" attribute. To our knowledge, ODA does not specify the document semantics if the size in the "character fonts" attribute does not match a size specified in a font definition. We finessed the issue by requiring that the font size information be identical in both locations.

High-Level Representation Choice. High-level choices of representation are made for large or abstract document structures. Because of our explicit document model, we did not face as many problems with these as with the low-level structures. However, from our design studies, we found several sources of problems.

First, there is a generic problem of appearance versus structure. One example of this dilemma is the representation of synchronized columns, that is, columns where every paragraph in each column is lined up horizontally. Presumably, one could specify this format through the use of a frame for each column, and the use of the "synchronization" attribute, resulting in tall, thin columns. However, to achieve the desired visual effect, one could also lay out short, wide frames to collect together matching paragraphs, and then line up the blocks within the rows to form columns. In both cases, one is achieving a "cell" structure where each cell is a paragraph. The use of tall, thin frames for columns matches the logical structure of the document, but the use of short, wide frames for rows might be a more appropriate structure for user manipulation. A third structure might be to simply layout blocks in a tabular structure without any additional row or column information. A system receiving a document would be hard pressed to deduce the existence of synchronized columns without some notion of how that feature might be expressed. Although the receiving system might correctly image the incoming document, regardless of the approach used to represent its feature, the logical editing function associated with that feature would be lost unless the high-level structure had been deduced accurately.

A second example of high-level choices concerns the use of the binding values. Binding values are used to represent a value in a document, such as a page number or section number. One has a choice whether to use a binding value or to place the result of an expression in the appropriate place of a document. For example, the section numbers in a document could be represented by placing the corresponding text in the appropriate place or by placing a reference to an appropriate expression for generated content. Even if one used only expressions, there may be many ways to represent a particular value through references to objects and other binding values. We did not do a detailed study of ODA expressions, but we believe that trying to identify a pattern for arbitrary kinds of counting may prove difficult.

The third example is simply a list of high-level structures that we do not believe have a straightforward representation in ODA. The high-level structures include tables of contents, cross references, lists of figures and indices. We believe that these high-level structures are important for EXPRES purposes, but they were not considered in detail for lack of time. However, we did consider them enough to believe that they require a great deal of careful design and that the design will not be easily recognizable without prior knowledge of the structures it employs. Therefore, we believe that several different approaches will be used and interchange of these high-level structures will not be effective without extra-ODA knowledge.

Needing Several Representations

The previous section discussed how ODA provided redundant mechanisms for representing document features. ODA has another redundancy problem: the need to repeat information in a document. Some examples of this redundancy are given in this section.

We needed to use redundant information in ODA to represent relative style information that was otherwise unavailable. For example, values of attributes in the specific logical structure that represent a nested environment may not match the values in the style placed in the generic structure for document styles. This is because the attributes in the generic structure represent document styles and are calculated relative to a global formatting state vector while the values in the specific logical structure reflect the environment and use their own formatting state vector. The inability of ODA to represent relative style information forces us to include the additional style in the specific logical structure.

A second place where redundancy appears in a document is with fonts. Font definitions group information. Unlike attributes, such as "offset", the different parts of a font definition are not independently defaultable. Therefore, all aspects of a font must be defined when it is used—unspecified aspects cannot be inherited from previous font definitions. For example, the font to be used in a specific place of the document is determined by the formatting state vector and may be, say, a 10 point, italic, Helvetica font. If the current document style specifies that a 12 point font should be used, the formatting state vector in the environment would automatically retain the italic and Helvetica information. Since the definition of the font in the document style would specify only the 12 point requirement, the corresponding font reference used in the ODA generic structure would also only specify the font size. However, the font reference is unusable in the specific structure. Because a font definition is considered complete, the family and face of the font in the generic structure would be undefined and could not be used by the specific logical structure to describe the proper font for imaging the content. Therefore, a second font definition that represents the information in the formatting state vector of the environment must be provided to specify a 12 point and italic and Helvetica font. Hence, two font definitions are needed to represent a change, even when an absolute style is specified.

A third place where redundant information appears in a document is for global document information. For example, one may want to specify that the margins for the entire document are 1 inch, or that the font for the entire document is Helvetica. The defaulting mechanism in ODA works fine with a single logical structure, but a document could have several logical structures: the generic logical root and the roots of each tree used in a "logical source" attribute in the generic layout structure. To provide a default value for the entire document means providing the same default value at the root of each of these trees. Naturally, a recipient system has no way of knowing whether the values in all the roots should be interpreted as a global default or not.

There is also one minor example of redundant information that was already mentioned: font sizes. Even if a font size is specified in the font definition, the size must also be specified in the "character fonts" attribute.

For EXPRES purposes, the redundancy problems were annoying but not difficult to handle, largely because we agreed on the conventions in this book. In addition to the redundancy problems with ODA, we found several problems with the content architecture. These are discussed in the next section.

5.5.4.3 Content Architecture

Most of our problems with ODA's content architectures have been discussed elsewhere. We summarize them here.

Limited Collection of Content

ODA has a limited collection of content types. We found the requirements of defining a private content architecture too onerous and instead decided simply to use a specially marked version of the character content for representing tables and equations. This served us quite well. We assume that new content types will be added over time and will alleviate this problem.

Standards Inconsistency

ODA contents make use of other standards to define their representation. Unfortunately, the content architectures usually start with the other standards and then make amendments as necessary to fit them into ODA. As mentioned, not all graphic rendition codes specified in ISO 6429 are permitted in ODA. Similarly, the meaning of the TAB character is replaced with a new control function: STAB. Each difference by itself is small, but together constitute a substantial learning cost.

No Content Structure

ODA views documents as having a structure with content at the leaves of the structure. However, several systems permit content to have structure that interacts with the structure of the document. For example, both Andrew and Interleaf permit arbitrary content within a table cell, where a table is a kind of content. Similarly, there are relationships between content that ODA cannot represent. For example, a chart in a document may represent the same underlying data that is in a table. In ODA, there is no way to tie together the two content portions. Arbitrary cross references between contents, including such notions as hypertext, are also unsupported.

These problems are being studied in ISO working groups. Their true impact on interoperability is unclear. We prohibited structured content during the EXPRES experiment.

5.5.4.4 Attribute Semantics

We found several problems with the attribute semantics defined by ODA. We do not have any strong criticism about the semantics of any particular attribute, but we found ourselves continually returning to some part of the standard to investigate some subtlety that escaped us on the first several readings. We enumerate some of our pitfalls in this section.

Inconsistent Defaulting Rules

ODA provides a defaulting mechanism for specifying the values of attributes. We found the mechanism useful for both factoring information and representing style structure. Although the rules appear homogeneous in section 5.1.2.4/2 of ISO IS 8613, their actual application is riddled with special cases. Eleven special cases are identified in an informative appendix, section E/2 of ISO IS 8613.

In addition to the rules listed in section 5.1.2.4/2 of ISO IS 8613, there are other interactions when determining the default value of an attribute. First, some, but not all, parameterized attributes can have "independently defaultable" parameters. In essence, this means that the defaulting rules must be applied to each parameter rather than to the attribute as a whole. Second, some coding attributes and presentation attributes do not follow the same rules for finding default values as used in section 2 of ISO IS 8613. One sees examples of the difference in the "initial offset" attribute for character content or the "clipping" attribute in raster content. In both cases, the default value depends on information present elsewhere in the basic object for the content or in the content information.

Semantically Meaningless Attributes

Even if one were able to get easily the value associated with an attribute for a particular constituent, the attribute may not affect document semantics in any way. This is because the attribute may be ignored outright. For example, the presence of a "type of coding" attribute is ignored when a "content architecture" attribute is present.

Other cases of semantically meaningless attributes are more subtle. For example, concatenated paragraphs affect the applicability of attributes. As specified in section 7.5/6 of ISO IS 8613, the values of attributes such as "character fonts" and "line layout table" are ignored on concatenated blocks. Thus, if one wishes to employ different fonts in different NIST document application profile *phrases*, one must collect them together in the "character fonts" attribute for the first content portion for the paragraph rather than with each piece of content. Another subtle example of where an attribute is meaningless occurs when embedded control functions are present in character content. If the content associated with a basic object starts with an embedded control function that covers the same function as an attribute (for examples, see section 5.5.4.2), then the attribute has no meaning in the document.

A related example is the interaction between SGR functions for controlling the weight (bold) and posture (italic) of a font and the SGR functions for invoking a font. If a font is invoked explicitly, then the definition of that font's weight and posture are taken from the font definition and not by any graphic rendition that specified a weight or posture. The extra SGR control functions would be meaningless.

The last set of semantically meaningless attributes comes from their lack of use when processing the document. We provide two examples. First, when determining the content associated with a content portion, the presence of "content information" attribute overrides (and makes meaningless) the presence of "content generator" attribute. Second, during layout, various layout directives take precedence. The exact order is specified in section 5.7.12/2 of ISO IS 8613, but the effect is that some attributes are ignored in favor of other attributes, which makes the interpretation of a document structure difficult in isolation.

Attribute placement

Another problem we found with attributes was their placement. In particular, we would have liked to have a truly global default for some values. In its absence, we were forced to specify redundantly some information (see section 5.5.4.2).

Attribute structure

We found several problems with attribute structure. The first was simply that the structure was not fine grained enough. The introduction of independently defaultable parameters was a start to fixing the problem, but its introduction solved some problems by introducing more mechanism, and hence more implementation. Unfortunately, independently defaultable parameters do not go far enough. As mentioned in section 5.5.4.2, the inability to specify individual parts of a font definition caused extraneous font definitions to be generated.

Another previously mentioned problem is the inability to specify relative attributes. This is more than just the ability to define a margin movement rather than a margin setting. One should be able to specify relationships among attributes. For example, one might want to define the relationship between the left margin, right margin and the page width. Needless to say, ODA does not provide such a facility.

A minor problem we had with attribute structure was the distinction between no attribute and the 'null' value for an attribute. For example, the default value of the "layout style" attribute is null[36]. If one could apply the default value rules for the "layout style" attribute, encountering an explicit null value for the attribute would end the search, while not encountering any attribute would continue the search. For the most part, this was simply a technical problem with no real impact on our work.

5.5.4.5 Layout (Formatting) Semantics

One area of ODA that we found demanding was the formatting (or layout) semantics. Our concern was twofold. First, we believe that accurate interchange required layout to be performed on an otherwise processable-only document. Second, we found the description of layout insufficient for a high-fidelity interchange. Each problem is discussed in turn.

---

[36]It is also unclear why an attribute that is not defaultable has a default value defined for it. We believe the reasoning is that every constituent has a value for every associated attribute. If that attribute is not defaultable and does not appear on a constituent, then the default value specified in the standard is used.

Need for Layout

Two factors influenced our belief in the need to perform layout in order to interchange processable documents. First, the combination of different ways of representing common document information makes isolated inspection of the logical structure insufficient. For example, one may need to perform layout to determine a left margin. In section 5.5.4.2, we noted several ways to represent a left margin—we probably missed several alternatives. As a result, the only effective way to determine the left margin of a document is to format it and see how far the content is from the edge of the paper, ignoring which of the dozen or so mechanisms were used to format the content. A similar argument can be made for tracking how tabs affect the layout of character content.

Second, several aspects of document structure are apparent only when formatting. For example, page breaks and page numbers may only have meaning to some systems after they have been resolved. But content may not be formatted the same on different generic pages—the selection of a generic page cannot be done without formatting the document and following the specifications of layout categories, permitted categories and various generators for subordinate expressions in the generic layout structure.

Complexity of Layout

Given that layout is necessary, we briefly considered its complexity. We quickly decided to avoid using aspects that would require layout because:

- Backtracking is necessary;
- No clear protocol is defined between content and document;
- Open (extra-ODA) issues affect layout.

Because of the nature of construction expressions in ODA, one could be faced with formatting a group of pages only to find that no appropriate page is available for formatting the next piece of content. One could be forced to backtrack arbitrarily through previous formatting in order to make another choice for formatting earlier content. There are no guidelines as to how many iterations to try, nor how to order the available choices for formatting content. In section 6.3.1/2 of ISO IS 8613, the standard states that backtracking should not be performed except in cases where certain attributes (which typically do not cross page boundaries) are concerned, but the standard sheds no light on how one should select new pages from a construction expression to ensure that a document can be formatted without backtracking.

The lack of a clear protocol between the content architecture and the document structure also makes layout difficult. Text can be broken into two (or more) pieces and rasters can be scaled, so there is latitude in formatting content.

The standard does not provide a clear algorithm for determining which part of the formatting process ultimately decides how much space each piece of content receives. Since the space allocated for each content controls the selection of frames and formatting of other content, the lack of precision makes the construction of a formatter difficult.

Open Issues Affecting Layout

There are critical decisions that ODA lets an implementation make when performing layout, particularly when formatting character content. These are enumerated in section 12.2.1.3.2/6 of ISO IS 8613. For example, different font metrics and different hyphenation algorithms can cause text to be formatted differently by formatters. Each layout process is legal, but could use a different amount of space on the page, thus affecting the layout of the entire document.

5.5.4.6 Document Application Profiles

Originally, we wanted to define our own document application profile for use in EXPRES. In the process of trying to do this, we found that the provisions for writing a document application profile did not meet our needs. In ODA, the only real mechanism for defining a document application profile is subsetting. One can restrict certain attributes to be used on certain kinds of constituents and one can limit the values of attributes, but one cannot specify higher-level information. For example, one may not specify attribute interactions, such as we required for specifying paragraph indentations and left margin. One cannot provide document model semantics beyond ODA's document model, as we tried to do for left margins. As a result, one cannot use a document application profile to enforce a particular document model; one uses a document application profile only to specify a subset of the ODA document model.

5.5.4.7 ODA as an Interchange Medium

Our last set of criticisms with ODA concerns generic problems with using it as an interchange medium, rather than as a document model. We have two: missing high-level functionality and difficult interchange formats.

Missing Functionality

ODA is a low-level document representation. To achieve a high fidelity of document interchange, it is important to describe and retain high-level information about a document. However, one must translate concepts such as a table of contents, an index or a style sheet system into lower level terms when using

ODA. The recovery of such information is ad hoc at best or requires special extra-ODA agreements at worst. Although document application profiles hold some hope of bridging this need, it is not clear if one can put enough information in the constituent restrictions to capture the desired abstractions.

Difficult Interchange Formats

The interchange formats required by ODA are difficult to use. The most prevalent one, ODIF, is based on ASN.1, which is a context-sensitive binary format. As discussed in 7.3.3.2, we found the creation of ODIF/ASN.1 support to be the single most resource-consuming part of the project. We found debugging these streams very hard—when some format is in error, it is difficult to hand parse the representation to locate the mistake. Further, these streams are difficult to exchange by many systems. For example, within the United States, most electronic mail systems are character based. Most systems will not transmit binary data and there does not seem to be an accepted binary-to-ASCII encoding scheme. Thus interchanging ODIF streams by electronic mail is problematic.

5.6 Summary

In this chapter, we described the translators we built and the interchanges that we performed. From those experiences, we drew a variety of conclusions about document translation in general and about using ODA in particular. To perform our experiments with ODA, we built tools to help us manipulate ODA structures. In the next chapter, we discuss our tools and the effort to build them.

6
Experience Implementing ODA

We spent significant resources, in personnel and in time, on the CMU ODA Tool Kit, a portable subroutine library for application programs manipulating ODA documents. The tool kit effort involved specification, design, implementation, testing and distribution.

This chapter discusses all aspects of this work, beginning with our reasons for investing our resources and a description of our goals for the tool kit. This discussion is followed by sections giving an overview of the functionality provided by the tool kit and functionality *not* provided by the tool kit. We then review some of the major design decisions that we made and a brief description of our implementation process. The last two sections of this chapter look back on our implementation effort and discuss the lessons we learned. We finish the chapter with some statistics about the size of the tool kit and the resources required to build it.

6.1 The CMU ODA Tool Kit: Motivation and Goals

During our early investigations into ODA, we realized that a common tool kit for manipulating ODA documents would be helpful, if not essential. We were motivated by the complexity of ODA and the fact there would be at least two distinct groups implementing translators.

Unfortunately, we were able to locate only one ODA tool kit [53], and its availability was unclear for several years[37]. We decided, therefore, that we should build our own tool kit. Besides serving as a common tool kit for the EXPRES participants, we hoped that, if done carefully, the tool kit would be of value to the community at large after the project was completed.

---

[37]We believe that the tool kit, known as the Stored ODA (SODA) Interface, is now being sold.

To these ends, we set the following goals for the CMU ODA Tool Kit:

- The tool kit should be comprehensive, covering as much of ODA as resources permitted. In addition, the tool kit should be as general as possible, and not limit its support to any particular document application profile or set of application programs.
- The tool kit should hide the most difficult parts of ODA: attribute manipulation, the defaulting rules and reading and writing ODIF.
- To attain maximum benefit after the project had terminated, the tool kit should be as portable as possible. The tool kit should be distributed in a form that enabled it to be used immediately on several common systems, and it should be easy to port to a new environment.

While there were many advantages to having such a tool kit, there was one disadvantage of which we were aware. The potential problem was that having all EXPRES participants use the same code did not provide an opportunity to check our implementation for correctness to the ODA standard. A common means for checking such correctness is by interchanging with another implementation. It is possible that the two implementations will agree and still be incorrect. Nonetheless, this is a useful method for informal validation, and we did not make explicit provisions for it[38].

6.2 Functionality Provided by the Tool Kit

The CMU ODA Tool Kit is a subroutine library (along with include files and a manual) for use in application programs that manipulate ODA documents. The tool kit is faithful to the low-level semantics of ODA and, though it hides much of the complexity of ODA, use of the tool kit requires a thorough knowledge of the standard. The tool kit is intended for application developers and not end users.

---

[38]We were able, however, to interchange a document produced by our tool kit with a system using another implementation (section 5.2).

6.2.1 Data Types

The tool kit defines and exports a large number of data types for use by the application program. These types include:

- ***Primitive portable types***: these types are provided as replacements for the predefined C types `int` and `char`, to enhance portability[39].
- ***Constituent***: this type is used to represent all of the constituents defined by ODA.
- ***Document***: this type represents an ODA document in memory, and allows multiple documents to be manipulated.
- ***Attribute types***: these are the types for ODA attributes, and include integers, sequences, font definitions and many structured types.
- ***Sequences***: these types represent arbitrary-length sequences of other types. Sequences are common to many attribute values.
- ***Iterator***: this type is used to allow application programs to iterate over parts of an ODA document.

6.2.2 Constants and Pseudo-constants

The tool kit exports a number of identifiers that are to be treated by the application program as constant values. Some of these identifiers are implemented as C macros, which means they are effectively constants. Other identifiers are, in fact, variables and the programmer must be careful not to modify the underlying value. These identifiers denote:

- ***Attribute identifiers***: each attribute in the tool kit is represented by a unique identifier, which is an integer constant.
- ***Tag values***: are used to discriminate among the types in a C `union`. A tag value may be one of several identifiers, each of which maps to a constant integer.
- ***Distinguished values***: for the value of some attributes, ODA defines a set of distinguished values, which can be thought of as defining an enumerated type. A distinguished value is represented by an (integer valued) identifier in the tool kit.

---

[39]section 6.4.2 discusses our portability concerns

- ***Locally-defined system parameters***: there are a number of constants that may be modified by the local site installer to affect the behavior of the tool kit. For example, the length of the buffer used to hold human-readable error messages is such a constant.

6.2.3 Variables

In the interest of good software engineering, there are only two variables exported by the tool kit. The two user-visible variables are part of the error handling capabilities and contain the latest error code and a human-readable explanation of the error. These must be variables so that they can be modified, if necessary, by locally-defined routines.

6.2.4 Routines

The bulk of the tool kit is formed by user-visible routines that serve a variety of functions:

- initialization and finalization of the tool kit,
- machine-dependent actions, such as memory allocation, input and output, and swapping,
- manipulation (such as creation & destruction) of documents and constituents,
- useful operations on certain type values (sequence and font manipulations, for example),
- handling attributes (setting and getting a value, and obtaining a printable name for an attribute),
- evaluating ODA expressions (such as string expressions and object identifier expressions),
- reading and writing ODIF,
- support for programmer-defined iteration over parts of a document,
- modifying the tool kit's run-time behavior (these routines were intended to allow the user to improve the search time required for finding attribute values, but the routines are all unimplemented),

- debugging support (providing a human-readable representation of parts of a document).

6.2.5 Summary

The CMU ODA Tool Kit provides support for ODA in a general fashion by directly implementing ODA semantics. We avoided the temptation to tailor the tool kit to the specific needs of the EXPRES participants. This has resulted in a tool kit that hides many of the difficult parts of ODA, but does not provide a high level of functionality in some areas, as discussed in the next section.

6.3 Functionality Not Provided by the Tool Kit

Although we found the CMU ODA Tool Kit to be invaluable, there are a number of capabilities that it lacks that would have proved useful[40]. For example, because the tool kit mimics the ODA model faithfully, there is no support for high level document concepts, such as margins and paragraphs. This meant that each translator had to include its own code for dealing with these features.

The tool kit treats the information in a content portion (the value of the "content information" attribute) as a sequence of bytes. This leaves the interpretation of those bytes, which can be quite messy, to the application. The character content, for example, can contain embedded control codes, which are tedious to manipulate. Once again, the lack of a capability in the tool kit caused the translator writers to duplicate what could have been common functionality.

Although we made a decision not to provide support in the tool kit for any specific document application profile, it would have been useful to have had a subroutine layer on top of the tool kit for this purpose. This layer would have understood the NIST Document Application Profile [48] and the EXPRES conventions, and provided support for manipulating passages, paragraphs and phrases.

Lastly, the complexity of the application code required for utilizing some parts of the tool kit was daunting. For example, the creation of even a simple construction expression required many lines of complex structure-field selections,

[40]These features were unimplemented either because we lacked the resources, or because we did not perceive the need when we began our work.

assignments and routine invocations. If we had perceived this problem early enough, we would have implemented routines that accepted a concise notation specifying the desired operations.

In addition, there is functionality missing that did not cause us to suffer, but which appears useful to us for potential applications of the tool kit. In particular, the tool kit provides no support for the ODA layout or imaging processes. These are complex processes and the tool kit provides a natural base on which to implement them. Another place where the tool kit falls short, is that it provides no support for the ODL or SDIF datastreams. Although this was not a capability that we required, it would be a useful, and obvious, addition to the tool kit.

Having specified the functionality of the tool kit, we needed to make decisions that affected the tool kit implementation. These are described in the next section.

6.4 Implementation Decisions

During the design stage, we made a number of important decisions regarding the implementation of the CMU ODA Tool Kit. These decisions covered a range of topics: managing the size of the tool kit description, portability concerns, information hiding, and storage management and error handling. These topics are discussed in the following sections.

6.4.1 Managing the Size of the Tool Kit Description

We believed that the description of the CMU ODA Tool Kit (the Application Programmer's Interface Manual [57], now part 3 of this book) should be as small as possible. It was clear that we were going to have to devote a significant amount of text to the descriptions of each ODA attribute (there are more than 150). We were, therefore, interested in limiting the number of routines, since their descriptions would form the bulk of the interface, after the attribute descriptions.

The number of user-visible routines became a concern when we began to specify the routines for manipulating attributes. The design required two kinds of routines: one for associating an attribute and its value with a constituent and one for obtaining the value of an attribute. The obvious form of declaration for the assignment routines is something like[41]

```
SetAttr(constituent, attr, value)
   CONSTITUENT constituent;
   int         attr;
   some-type   value;
```

and similarly for the routines that obtain a value. (Within the tool kit, attributes are designated by integers.)

The problem is that there are 40 distinct types required for the attributes. Because C has no form of generic routines, this would have meant the definition and documentation of 80 routines (40 each for assignment and value access). As there are only around 100 routines exported by the tool kit, this would have increased the number of user-visible routines by more than 50%.

We could have avoided actually describing all 80 routines by describing the common format and listing the distinct routines and associated types. Nonetheless, this would still have required the user to determine the appropriate routine, from the 40 available, for each attribute.

Instead of this, we decided to implement the equivalent of generic routines, by defining the assignment routine as

```
SetAttr(constituent, attr, value)
   CONSTITUENT    constituent;
   int            attr;
   POINTER_type   value;
```

The type `POINTER_type` is a type that can be cast to or from any other pointer type, and this allows the routine to accept an arbitrary pointer as its second argument. This pointer will reference an object of the type appropriate for the specified attribute (denoted by `attr`). This object will contain the value that is to be assigned to the attribute. The routine can determine the type of the referenced object by examining the specified attribute.

For example, to assign the value 5 to the attribute "indentation", the following code might be used:

```
int ind;

ind = 5;
SetAttr(constituent, cc_INDENTATION, &ind);
```

[41]For illustrative purposes, we provide simplified definitions of these routines. See section 16.1.2 for the actual definition.

This technique allowed us define two routines, instead of 80[42]. It turned out to be rather tricky to describe the appropriate method for calling the routine to obtain an attribute's value. Despite this, we are happy with our decision to reduce the complexity of the interface.

We used a similar technique to limit the number of routines that are necessary for manipulating sequences. This was necessary because the sequence type in the tool kit can have elements of one of more than 20 distinct types.

6.4.2 Portability Concerns

From the beginning, we were concerned that the EXPRES project yield some long-term benefit to the problem of multi-media document interchange. We believed that our best hope was to make the CMU ODA Tool Kit useful to a wide audience. It was for this reason that we designed the tool kit to provide general support for ODA.

In addition, we took great care to make the tool kit portable, so that it would be usable in many environments. Making the tool kit portable involved programming conventions and a careful design of a methodology for handling operating system dependent operations.

6.4.2.1 Programming Conventions

We chose to implement the tool kit in the C programming language because C is widely supported on a large number of platforms and operating systems. A tool kit written in C would afford the widest possible coverage.

It is well known, however, that number of dialects of C support different subsets of the language. To enhance portability, we investigated many popular C compilers to determine their restrictions. We then specified a subset of C that would be acceptable to most of these compilers and, yet, not be so restrictive as to hamper programming. A listing of the restrictions we imposed is given in appendix E.

A common problem that inhibits portability is assumptions about the representation of types that are not guaranteed by the language. It is common on Unix workstations, for example, to program as if the size of an integer is 32 bits, since this is true on many machines. But, this may cause problems if the program is to be run on an IBM PC. On many C compilers for the PC, the default represen-

---

[42]Because the assignment of integers and certain other types to attributes was so common, we ended up defining several additional routines specific to these types.

tation for an integer (the int type) is 16 bits. Similar problems can occur with the character type and pointer types.

Our solution to this was to forbid the use of the C types int and char. Instead, the tool kit uses the types INT_type and CHAR_type, which are defined by the tool kit installer at each site. The tool kit manual (part 3 of this book) provides guidelines for each of these types, describing constraints on, and the uses of, each type. For example, on an IBM PC, the installer could define the type INT_type to be the C type long int, which provides a 32 bit integer.

The C language does not define a pointer type that may point to any type of object. But, such a pointer type is needed in many programs, and it is common to use the type char * for this purpose and to cast between this type and other pointer types. Unfortunately, it is not universal among C compilers that the type char * can serve in this fashion. Therefore, installation of the tool kit also requires the definition of the type POINTER_type[43].

Some C compilers restrict the number of characters of an identifier that will be treated as significant. Our investigations turned up 32 as the smallest maximum, and so we restrict the length of any identifier in the tool kit to 32 characters. In addition to a compile-time restriction, some operating systems place restrictions on link-time identifiers. Such restrictions are usually on length and the characters that may be used. For this reason, we have restricted all link-time identifiers to six, case-independent, alphanumeric characters.

6.4.2.2 Operating System Dependent Operations

When designing portable code, it is important to consider how to handle operating system operations, such as input/output and memory allocation. If you are limiting target environments to variants of Unix, the problem becomes slightly more manageable, since the Unix standard i/o package is a de facto standard.

However, we were targeting the tool kit to a variety of non-Unix environments (MS-DOS and VAX/VMS), so this technique would not work. Instead, we identified a set of operations that were dependent on the particular installation environment. These operations were

- input and output (for bytes and characters),
- memory management (allocation, reallocation and deallocation),
- swapping (between primary and secondary memories).

All of these operations are implemented by routines defined by the site installer. The tool kit manual describes the semantics required of these routines.

---

[43]This type is normally defined to be either char * or void *.

6.5 Design Decisions

At the start of our design of the tool kit, we made several major design decisions that affected the implementation effort. These decisions covered storage management, error handling and information hiding. These are discussed in the remainder of this section.

6.5.1 Memory Management

The C language provides no support for dynamic memory management[44]. But, because the tool kit makes heavy use of dynamic memory allocation (all constituents and attributes are allocated dynamically), it was important for us to devise a strategy for dealing with memory management.

We decided to keep the dynamic memory space used by the tool kit separate from any such space used by the application. Thus, the tool kit can not make use of memory allocated by the application program, but must allocate its own memory and must copy values as needed.

For example, the routine for setting the value of an attribute (`SetAttr`, section 16.1.2), accepts a pointer to an object that contains the value to be assigned to the attribute. To maintain the distinction between memory spaces, the tool kit routine will always copy the specified value into a new object, allocating memory as necessary[45]. Note that if the application program had already dynamically allocated the object, it will have to deallocate it after the call.

There are tool kit routines that return (a pointer to) dynamically allocated memory to the application program. These routines create objects for manipulation by the program, such as sequences and iterators. In these cases, it is the responsibility of the program to indicate when the object is no longer required, by invoking an appropriate tool kit defined deallocation routine.

The primary disadvantage of the careful partitioning made by the tool kit is that it can be inefficient in time and memory. In writing the translators, we found that frequently the program would dynamically allocate an object, fill in a value and pass it to the `SetAttr` routine. At this point, the `SetAttr` routine would dynamically allocate an identical object and copy the value from the

---

[44]It is standard practice for C libraries to provide routines (`malloc`, `realloc` and `free`) that allow user control of dynamic memory. But, these routines are not part of the C language definition and no garbage collection is available.

[45]There is an exception to this rule: constituents are not copied because their representations are hidden and are treated as pointer values.

original object. Upon return, the program would immediately deallocate its object. This felt wasteful, since the tool kit could have just used the program's object. We decided that the simplification in semantics afforded by our scheme outweighed the potential inefficiencies.

6.5.2 Error Handling

We decided to impose a uniform technique for handling errors within the tool kit that would ensure that all errors were reported, in as helpful a fashion as possible, to the application program. The salient features of this methodology are:

- All routines are value-returning functions whose possible return values must contain a distinguished error value.
- The execution is never aborted in the case of an error, but the error is returned to the application program, which is free to deal with it in an appropriate manner.
- For each routine, there is a single error return value. Further information about the actual error can be found in two global variables. The first contains one of a fixed number of generic error codes, indicating the general class of error that occurred. The other variable contains a string that provides a human-readable description of the error. For example, a call of `SetAttr` to set the value of the attribute "same layout object" may return an error indication. The global variable will be set to `BADATTR_err`, which indicates that the error involves the invalid use of an attribute. In this case, the global error message might be

  ```
  Attribute "same layout object" must go
      on a layout style
  ```

indicating that the program has attempted to assign the attribute to a constituent on which it does not apply.

6.5.3 Information Hiding

In the interests of maintainability and good programming practice, we designed the tool kit to use information hiding wherever feasible. For this reason, the internals of several of the most important types in the tool kit are hidden from the application programmer.

For example, the types `DOCUMENT_type` and `CONSTITUENT`, which are used to represent ODA documents and constituents, are exported as pointers. The internal representations, which are structures, are manipulated through a number of routines. The same is true of the type `ITERATOR_type`, which is used to iterate over portions of a document.

But, there are many types whose representations are not hidden, including all of the attribute value types (except for the type `CONSTITUENT`). We would have preferred to also hide the representations of these types. But this was impractical, as it would have required hundreds of routines to allow the application program to access and set the values. Instead, we chose to open up the representations and allow the programmer access to the values directly.

We decided to provide some support and enforcement for our information hiding by specifying a complex set of conventions for include files. These conventions, which are described in appendix J, allow an include file to provide exported definitions at several levels of visibility.[46]

6.6 Construction Process

We carefully planned the construction process for the CMU ODA Tool Kit so that the implementation would match the manual and so that the tool kit could be built incrementally. We insisted that the implementation match the specification because there were other groups we expected to use the tool kit, and it was crucial that they have accurate specifications. Also, our work was reviewed by a committee selected by the NSF; complete documentation facilitated their review, and enabled us to demonstrate progress.

We began by specifying completely the application programmer's interface to the tool kit [57] before any implementation started. Once the first version of the manual was written, it occurred to us that we might be able to generate some code and definitions directly from the manual. One of our first steps was to write several programs to scan through the text of the manual and to automatically generate source code. This source code included:

- type definitions for all of the user-visible types,
- constant definitions for all ODA attributes and distinguished values,
- stub routine definitions for all user-visible routines (each of these stubs consisted of code that returned an error indicating that the routine was unimplemented).

[46]These conventions are to be followed by the tool kit implementer and not by the applications programmer.

Besides the fact that this saved us work and effort, it also encouraged the developers to change the manual before modifying the source code.

Because our time was short, it was important that the tool kit provide appropriate functionality for the translators as they were constructed. This requirement, and the complexity of ODA, dictated that we implement the tool kit in an incremental fashion.

To facilitate this, we first created the *empty tool kit*. The empty tool kit contained every routine, type and identifier definition specified by the manual. The type and identifier definitions were complete, but the routine bodies were all empty. This allowed us to begin forming a run-time library of the tool kit and to begin linking it with test application programs. The first versions of the translators could, thus, be constructed with an appropriate skeleton for utilizing the tool kit.

At this point, we could begin making incremental additions to the tool kit implementation in two ways: implementing routines or adding support for unimplemented attributes. This flexibility allowed us to track the needs of the translators so that forward progress proceeded on both the tool kit and translators.

We began our specification and implementation of the tool kit with the Draft International Standard version of ODA, because that was all that was available. During the implementation of the tool kit, the International Standard became available. We decided that that the tool kit would only have future value if it adhered to the final version of ODA, so we halted development while we converted the documentation and implementation.

6.6.1 The CMU ODA Tool Kit Tester

During the earliest phases of the tool kit construction, we considered appropriate methodologies for testing the implementation. There were two obvious strategies for driving the tool kit: write small driver programs or create test ODIF datastreams. Either of these choices promised to be painful. Testing the more than 150 attributes would require dozens of driver programs. The thought of creating binary ODIF datastreams by hand was even more discouraging.

We decided that the appropriate strategy was to create a *tester* program. This would be an interactive program that would accept text commands dictating the tool kit actions to perform. The tester would execute the commands, check the results and report to the user on the validity of the results along with a printable form of them. This program could be built incrementally along with the tool kit.

The tester also provided an easy manner of creating ODIF datastreams for further testing. The user could construct interactively an ODA document and then direct the tester to call `WriteODIF` to produce the ODIF representation.

6.6.2 The CMU ODA Tool Kit Validation Suite

Because we intended from the beginning to distribute the tool kit widely, planning for such a distribution was a constant part of our plans. It was suggested to us that it would be helpful to include an installation validation program with the tool kit source code. This program would ensure that the tool kit had been built properly at the local site. (Note that the program was not intended to check the conformance of our implementation against the ODA standard.)

We agreed that this was a useful idea and began construction of a validation suite, which consisted of a driver program and a number of ODIF datastreams. The datastreams came in pairs: an input datastream and an output datastream. The driver program was a portable program that would perform the following actions for each input datastream:

1 Read the datastream.

2 Perform optional processing on the ODA document (this processing was specified in a table that drove the entire process).

3 Write out the ODA document in ODIF format to a new file.

4 Compare the new datastream with its corresponding output datastream, byte for byte. If the datastreams differed in any manner, it was an error, and a message would be printed.

The output datastreams provided in the suite were generated by us as part of the distribution creation process. The proper operation of the validation suite depended on the ODIF datastreams created at the local site being identical to the datastreams generated at the ITC. It is possible that a tool kit could generate different datastreams from the same ODA document. This could happen, for example, if the current date and time were expressed as the value of the attribute "document date and time". We did not use this attribute, and our tool kit was deterministic, so that we were sure that differing datastreams implied an error.

6.6.3 The Configuration Builder

Because we intended the tool kit to be used on a number of dissimilar platforms, we were concerned with distributing the tool kit in a form that would be amenable to easy building. The Unix *make* facility [20] provides a common, and easily accessible, mechanism for controlling the construction and installation of a software package. All of the systems we were targeting provided a similar construction facility, and we wanted to provide support for each of these facilities.

All of the facilities were quite similar, differing only in minor ways: delimiter syntax, order of statements and means for invoking compilers and link editors. This similarity begged for the creation of a program that could create automatically the source files necessary to drive the facilities. In addition, one of the most painful parts of creating appropriate source files is the necessity of specifying file dependencies, which usually indicate the dependence of a source file on several include files. We also used this program, known as *gmake*, to create this dependency list.

6.7 Lessons

We learned a number of lessons from the specification, design, implementation and distribution of the CMU ODA Tool Kit. Looking back, we believe that we did many things appropriately, but that we also made some incorrect decisions.

In the list of things we got right, the foremost is that we took the time to specify the application programmer's interface to the tool kit completely before we began any implementation. In addition to providing for a cleaner design, the written specification proved to be useful as a publicity tool for the project. We were also able to use the specification to generate automatically a significant amount of source code.

Our time worrying about portability was well spent, as the tool kit is highly portable and was used on several dissimilar environments. The coding conventions we imposed on ourselves were a little difficult to get used to, but became second nature eventually.

Lastly, the validation suite turned out to be a valuable tool, but not for the purpose we had originally envisioned. (The suite was implemented too late in the project for any external sites to make use of it.) However, the suite was invaluable as a regression tester, as it allowed us to check the evolving implementation of the tool kit against a base set of capabilities. In addition, the suite earned its way by serving as a test for the portability of the tool kit on non-Unix machines.

The primary mistake we made was in the specification of the site-dependent input and output facilities. These specifications described the semantics of the routines that have to be implemented at each site to provide input and output routines for the tool kit. The idea of providing local routines was good, but we made the mistake of attempting to define a minimal set of routines as needed by the tool kit internally. In particular, we defined a single read routine and a single write routine for characters (and a pair for bytes). There was no provision for other activities frequently associated with input and output, such as file opening and closing or buffer flushing. All such activities had to be performed as part of the read and write routines and this led to complex and insufficient semantics. We would have been better off defining more routines.

Finally, although we did a reasonable job of providing portable distribution support, we did not go far enough. The distribution building procedure is not a one-step process, but requires the installer to follow a series of steps. This is rather tedious and prone to error. With a small amount of extra effort we could have provided a simpler procedure. The distribution building procedure is also missing support for some common package construction actions, such as actual installation of the appropriate parts of the package and aid for cleaning up the results of the construction.

6.8 Statistics

The time we spent working on the tool kit can be broken down into four categories. The first is the time we spent learning ODA. A total of five people were involved in this activity and spent a total of 8.5 person months. The second category includes the time we spent designing the tool kit. Three people worked on this and spent six person months. Implementing the tool kit is the third class of time. Six people worked on the tool kit for a total of 36 person months. The fourth category includes the time spent testing the tool kit, and porting it to systems other than BSD UNIX. A total of four people worked on this, spending a total of 2.5 person months. Thus, the entire CMU ODA Tool Kit effort required 53 person months of effort.

It is also interesting to consider how much code was written. Table 6-1 contains a break down of this by module. There is a total of about 83,000 of lines of code in the tool kit. The largest single module was the attribute module with about 38,000 lines of code. Other large modules are those for `ReadODIF` and `WriteODIF` (12,000 and 9,000 lines respectively—and these are partial implementations).

Table 6-2 provides details about the attribute module. The categories considered are the drivers, service routines, defaulting, support for the canonical operations on the unique attribute types, and table initialization support. Note that the lion's share of the code in the tool kit supports the canonical operations (about 30,000 lines). We found dealing with the attributes that have parameters especially tedious. There are seven such attributes in ODA, but we spent a total of one person month concerned with these attributes. The code required to support them was about 450 lines long.

Module name	Total Lines	Lines Source	Lines Include
Attribute	38,150	33,300	4,850
Attr. support	1,800	1,350	450
Constituent	3,150	2,400	650
Debug	3,400	3,100	300
Document	750	550	200
Error hndlg	750	500	250
Expression	400	250	200
Font	800	550	250
Format	350	250	100
Init/final	600	350	300
I/O	1,000	800	200
Iteration	2,150	1,700	450
Label	1,050	900	150
Mem	500	300	200
Memory mgmt	350	200	150
Miscellaneous	850	0	850
Priority	400	250	200
Read ASN.1	5,650	5,200	450
Read ODIF	6,600	6,100	500
Sequence	1,950	1,650	300
String	400	250	150
Stubs	2,950	2,950	0
Write ASN.1	5,250	4,700	550
Write ODIF	3,750	3,250	500
Totals	83,050	70,850	12,200

Table 6-1. CMU ODA Tool Kit Module Size

Function	Total	Code	Include
Canonical ops.	29,700	27,300	2,400
Default value list	800	700	100
Defaulting total	3,100	2,800	300
Document application default	900	800	100
Drivers	3,600	2300	1,300
Initialization	1,700	100	1,600
Service routines	900	800	100

Table 6-2. Size of the Attribute Module, in Lines of Code

7 Conclusions

Chapters 1 through 6 have discussed, in detail, the investigation of multi-media document interchange undertaken by the EXPRES participants at Carnegie Mellon University. The remainder of this book (parts 2 through 5) is reference material for the tool kits constructed by our group. In this chapter, the last in part 1, we provide some general conclusions about our overall experience in five areas: the EXPRES interchange requirements, multi-media document interchange, the Office Document Architecture, the two tool kits we constructed and our implementation strategies for the tool kits and the translators.

7.1 The EXPRES Interchange Requirements

Recall that part of the NSF's mission for the EXPRES project was to investigate the feasibility of electronic document interchange among the scientific research community (see section 1.1.3). The EXPRES participants chose to focus on what we perceived as the key issue: the interchange of processable multi-media documents among diverse systems. To support this, we developed a set of requirements for our investigations, which we can summarize as follows:

- we must support a wide variety of document processing systems,
- we must be able to support a variety of non-text media (including raster graphics, structured graphics, equations and tables) and to support the introduction of new media,
- it must be possible for new document systems to participate without requiring a major upheaval to the existing framework,
- documents must retain processability and formatting information as they move among systems,
- when feasible, accepted standards should be used.

These requirements had a profound effect on the choices we made and on the directions our investigations took. In retrospect, we are satisfied that our requirements do, in fact, represent the needs of authors in heterogeneous computing environments. We have reached this conclusion as a result of our participation in many conferences and workshops where electronic document systems were discussed.

7.2 Multi-media Document Interchange

Early in the project, we determined that the most effective means for achieving document interchange among diverse systems was to perform translation between each document format and an intermediate representation. The advantage of this approach over direct translation between each format is that direct translation requires about N^2 translators to support N document formats. In addition, for a new document system to participate freely in the interchange would require the implementation of $2N$ new translators. This is obviously impractical. Note that if translation to an intermediate representation is used, the addition of a new system only requires that two new translators be built (between the new and intermediate formats).

There is, however, another approach that we could have taken: *daisy-chained* translators. This technique is described in the next section.

7.2.1 Daisy-chained Translators

In this technique, the systems are arranged in a circle so that each translator is responsible for translating from its format to just one other format. For example, if

$$\{ d_1, d_2, \ldots, d_N \}$$

were the set of of participating document formats, then a set of translators

$$\{ t_1, t_2, \ldots, t_N \}$$

would be constructed so that

$$
\begin{aligned}
&t_1\colon d_1 \rightarrow d_2\\
&t_2\colon d_2 \rightarrow d_3\\
&\ldots\\
&t_{N-1}\colon d_{N-1} \rightarrow d_N\\
&t_N\colon d_N \rightarrow d_1
\end{aligned}
$$

Note that only N translators need be built with this approach, as opposed to the $2N$ translators required for the intermediate format method. The addition of a new format, d_{N+1}, would require the implementation of only two new translators:

$$
\begin{aligned}
&t'_N\colon d_N \rightarrow d_{N+1}\\
&t_{N+1}\colon d_{N+1} \rightarrow d_1
\end{aligned}
$$

While daisy-chaining minimizes the number of translators, it suffers from several serious drawbacks.

To translate a document from one format, d_i, to another format, d_j, requires participation by all of the translators t_i to t_{j-1} (this assumes that $i<j$). In the worst case, this is all but one of the translators. With this technique, all translators must be available to an author—either each translator must execute in all environments or the user must have access to the appropriate environments. In either case, this is a severe restriction on the practicality of this method.

In addition, it is inevitable that a document will lose information as it is translated between formats. The daisy-chaining approach may subject a document to the maximum number of translations. Consider that a round trip from one format to another and back will force a document to pass through every translator. The weakest translator in the chain will impose a limit on the translation fidelity of a document.

7.2.2 Translation to an Intermediate Format

We have, thus, concluded that translation to an intermediate representation is the only practical approach. However, it also has some problems, which are inherent in document format translation. The biggest danger is that a document may become unusable as it is translated. We have identified two problems that may befall a document during translation: *explosion* and *loss of structure*.

7.2.2.1 Explosion

The translation of a document from one representation to another may entail the construction of complicated structures to convey style information. For example, the EXPRES translation conventions specify a complex mechanism for conveying style structure between a document format and ODA. When a document is translated into ODA, this may result in a large number of layout styles, presentation styles and uses of the "default value lists" attribute.

A translator for a document format with a rich style system can translate this information into reasonable structures. Unfortunately, a translator for a system with a poor style system (or one that does not match the style system assumed by the EXPRES conventions) may produce a messy, deeply-nested style structure. This may produce a valid document, but one that has grown significantly in size during the translations; the document has suffered an *explosion.* Whether this growth turns out to be a problem depends on several factors. If the document has grown sufficiently, it may have become too large for one of the document systems to manipulate. In addition, the additional structure may consume large amounts of processing time, making the document unwieldy.

There is another potential consequence lurking in this scenario. The additional generated style structure may over-specify the document, making it difficult to edit. The difficulty arises because of the plethora of styles and their applications.

We postulated the over-specification problem before any of the translators were built. We then actually observed this phenomenon with an early version of the ATK and Diamond translators. At one point, the Diamond translator was translating style information in a simplistic manner that created a new style for every paragraph and for every font change (see section 5.1.2 for a description of Diamond's style system). We translated an ATK document to ODA, then to Diamond, back to ODA and finally to ATK again. The result was a document that looked and behaved just fine on the screen. However, the document's style structure was unmanageable as it contained thousands of style definitions, each with a single application.

7.2.2.2 Loss of Structure

We surmise that careful translator implementation can cure most cases of explosion, but the problem of *loss of structure* does not lend itself to any apparent solution. Document structure can be lost during translation due to a number of circumstances. The net effect of these losses is that some of the logical structure intended by the author changes or disappears.

The seriousness of the lost structure depends on the details of the loss and can range from annoying to disastrous. For example, imagine that during translation, a style that was applied to two contiguous paragraphs was translated to the same style being applied once to each paragraph. This might be annoying to the authors, but is unlikely to be a serious hindrance. On the other hand, if the translation turned the single style application into two applications of two equivalent, but distinct, styles, this might prove more than annoying. The document now has a different meaning attached to these paragraphs.

Although we have identified and witnessed both explosion and loss of structure, we do not know whether these problems will make document format translation infeasible. We discuss the question of feasibility in section 7.2.6. The next section describes our experiences with implementing document format translators.

7.2.3 Implementing Translators

During the course of the EXPRES project, the ITC participants built two groups of translators: one group between ATK and ODA (section 5.1.1.2) and another from ODA to troff (section 5.1.4.2). We also had close contact with the groups building the Diamond (section 5.1.2.2) and Interleaf (section 5.1.3.2) translators. Our experience and observation was that a translator for the basic structure and content of a document could be built in one to three person-weeks. To clarify this claim, we need to explain what we mean by *basic structure and content*, and describe the level of expertise required of the programmer.

The basic structure of a document includes those pieces that are represented by the ODA attributes used for specifying logical structure. In particular, the following attributes make up the basic structure: "subordinates", "object class", "generator for subordinates" and "content portions". This implies the identification of *passages*, *paragraphs* and *phrases* as specified by the EXPRES conventions (chapter 4).

In addition, the first translators were able to deal with character content in a simple manner: the text could be extracted, ignoring control functions. Whether raster graphics content could be handled easily depended on the complexity of the local document system's raster representation.

It is important to note that our observations of the rapidity with which a translator could be built may not be applicable generally to any translator utilizing ODA. We believe the design of the EXPRES conventions was a significant factor in simplifying the implementations. The EXPRES conventions allowed the translator writers to partition the task into non-interacting parts. If one were implementing a translator for all of ODA, for example, this property would not hold and the addition of some new features would have severe consequences on existing parts of the implementation.

Another important observation we made regards the expertise required of a translator writer. To implement a translator as quickly as possible, it is desirable that the implementor be an expert in the particular document system. Facility with the local document format turned out to be more important than an in-depth understanding of ODA. In fact, a rather shallow knowledge appeared to be sufficient. Once again, the EXPRES conventions were a major factor, hiding the complexity of ODA by providing a predefined set of rules. To make full use this it was, of course, necessary for the programmer to understand the conventions in some detail and to have a good understanding of the CMU ODA Tool Kit.

7.2.4 The EXPRES Conventions

The EXPRES conventions are a detailed description of the means for using ODA to translate into or out of another document format. The conventions served several necessary purposes for the EXPRES project:

- they hid much of the complexity of ODA,
- they reduced the time before successful interchange could commence among the participants,
- they provided a rich document model,
- they acted as a precise specification of the functionality of our translation model.

We discuss each of these factors in turn.

As mentioned in chapter 4 (and, as should be evident from chapter 3), ODA is a large standard with a complex set of semantics. Besides its size, ODA is redundant: for many document features (such as changing the left margin) there are multiple representations possible in ODA. Therefore, a translator not wishing to limit the use of ODA in interchange would be forced to recognize and cope with all of the possible representations. A problem worse than that of the amount of code required to handle all of the possibilities is the fact that many of the uses of attributes can interact in troubling ways.

For example, under one possible representation, a change to the left margin of a paragraph may also involve drawing a bar at the right edge of the paragraph. The problem is not in detecting the various changes an attribute causes—that is straightforward. It is the interactions among attributes that make it difficult to recognize document features. To recognize a change to a left margin, the translator may have to coalesce information from up to three attributes. The translator would then have to determine that there was, among the changes, a change to the left margin.

This kind of analysis is tedious and error prone, and the interactions make it difficult to construct a translator incrementally. Consider that there is no convenient manner in which an implementor could build a translator in steps, each of which consisted of adding support for a new ODA attribute. The problem is that the consideration of a new attribute may affect much of the existing code for handling attributes. For example, when support for the "border" attribute is added to the translator, all of the code for recognizing the changes to left right, top and bottom margins would have to be modified.

The EXPRES conventions ameliorate this problem by choosing a single ODA representation for each supported document feature. In addition, the conventions avoid interactions among document features. This makes it rather easy to partition the implementation of a translator.

Having carefully specified conventions for translation also reduced greatly the time required before the EXPRES participants could begin interchanging documents. Each translator could be written with full knowledge of how each ODA attribute would be used and how to extract the pertinent information.

The conventions also served the purpose of providing a rich document architecture that would satisfy the EXPRES requirements regarding processability and formatting. This was crucial because ODA's document model was not sufficient to represent the functionality we required.

The EXPRES conventions also functioned as a precise specification of the level of interchange we were attempting. Because the EXPRES project was done under contract from the NSF, the participants were subject to periodic reviews by the NSF and an advisory committee. The conventions provided a concrete means by which progress could be reviewed and technical direction set.

Although the conventions were useful (and crucial), it would have been a tremendous help if we could have produced code to support them. In particular, with appropriate resources, we could have implemented a layer of software above the CMU ODA Tool Kit. This layer would have provided for manipulations of document features, such as margins, font changes and indentation, for example. Because all of these notions are alien to ODA, each implementor ended up producing local implementations for these features.

The primary purpose of the EXPRES conventions, of course, is to preserve the fidelity of documents as they are interchanged among systems. The next section discusses our conclusions about translation fidelity.

7.2.5 Translation Fidelity

The success of a document interchange mechanism will be determined ultimately by the level of translation fidelity that is attained (see section 2.3 for a discussion of translation fidelity). The EXPRES participants were aiming for a high level of fidelity, which retained not only the formatting information associated with a document, but also its organization and editability. We believe that the retention of this information is essential for useful interchange. In this section, we discuss our conclusions as to the level of fidelity that can be obtained during interchange of multi-media documents among diverse systems.

As discussed in section 7.2.3, it is straightforward to retain the organization of a document during translation. If the document system has the appropriate functionality, it is also easy to translate faithfully much of the character content information, such as changes to fonts and subscripts. Assuming proper document system support, the faithful translation of raster images is also possible.

Recall that the interchange of a document from one system to another involves two translations: one into ODA and one out of ODA. Each of these steps provides an opportunity for mishap. Depending on the document systems involved, there may be many details lost or modified during a translation. For example, during the translation from ATK to Interleaf, the specification of inter-line spacing is lost. If the document is eventually translated back to ATK format, default inter-line spacing will be assumed.

Many of the details that are lost are impractical to recover. To retain the inter-line spacing in the example above would require simulating the Interleaf formatting process to determine the largest font on each line. This is necessary because Interleaf defines inter-line spacing relative to the largest glyph that appears on a line.

There is another class of lost information that we do not see a way to recover. Imagine that an ATK document with nested superscripts (like "a^{2^3}") is being translated to Interleaf, which does not allow nested subscripts. It is not at all clear what the ODA to Interleaf translator should do with this document piece. Presumably, the translator could translate the functionality it could handle (yielding "a^2", for example) and tag somehow the location as being only a partial translation. Unfortunately, with this simple-minded scenario, when the document is translated to ODA and back to ATK, the nested superscript would be lost irretrievably.

You might imagine a document format rich enough to allow for the arbitrary annotation of document pieces. The translation could then tag the superscript translation, "a^2", with information indicating that there was another, untranslatable, superscript. Upon translation to ODA, the information would be recovered and the additional superscript restored.

This is a reasonable solution, although it is not clear what should happen if the document piece in question is modified. For example, if the user deletes the "^{2}" character, what should the resultant underlying representation indicate? Either deleting the extra superscript (leaving "a") or moving it down (yielding "a^3") seem reasonable answers. Yet another possibility is that the local system could forbid the editing of such a structure.

However, there is another problem that we have been overlooking: the local document system will have no knowledge of such extra information. Depending on the position of the annotation, the information may be lost without warning. If, for example, the extra superscript is attached to the "^{2}", then deleting the "^{2}" will cause the "^{3}" to disappear without warning. In any case, current document systems do not deal intelligently with features they do not support. The handling of features outside of the built-in functionality of a document system is a research issue.

Although we have had substantial experience with the translation of character and raster graphics contents, we have no experience with the translation of other media. We, therefore, can not say anything meaningful about the feasibility of translating such media.

7.2.6 The Feasibility of Document Interchange

It should be clear from the previous section that complete fidelity during document translation is infeasible. There are simply too many places where information can be lost or modified. As a concrete example, consider the case of translating documents between ATK and Interleaf (using ODA as an intermediate representation). Using the current EXPRES translators, the following document features would be translated correctly:

- organization
- font changes: family, posture, size, superscripts/subscripts (limited to numbers), underlining
- paragraph spread
- rasters (at 72 pixels/inch only)
- alignment
- restricted style sheet definitions (no inheritance)
- restricted style sheet applications (no nesting)

The following are document features that would be lost during translation:

- nested content
- inter-line spacing
- complicated Interleaf formatting semantics

- Interleaf style sheets with inherited definitions
- ATK nested style sheet applications

Looking at these lists, one could decide that such a translation was perfectly suitable for interchange. Conversely, one could argue that the missing functionality is crippling and renders the interchange useless. There is no correct answer. The fact is that we simply do not know whether the level of functionality we have attained is suitable for effective interchange. The answer to this question depends on specific details: the document systems involved, the document features exercised, the kinds of documents (size, security, media) and the revisibility requirements. More importantly, the determination of the feasibility of this kind of interchange can be answered only by deploying complete systems, including translators, and studying their use in a realistic situation. This is something that the EXPRES project did not have the resources to pursue.

7.3 Office Document Architecture

Our choice of the Office Document Architecture (ODA) standard as the intermediate representation for our translations had a profound effect on our work. The use of ODA influenced our conventions, the fidelity we were able to attain and the tools we chose to build. This section looks critically at ODA from three points of view: as a standard for document interchange, as an intermediate representation used by the EXPRES translators and as a specification to be implemented (in the CMU ODA Tool Kit).

7.3.1 The Standard

ODA suffers from many of the maladies common to international standards, which are hampered in their design by the need to satisfy many constituencies. The members of the standards committees and working groups are typically employees of corporations involved in the computer or telecommunications fields. These people represent the interests of their employers, as well as attempting to do technical justice to the standards. Unfortunately, technical judgments often give way to politics. This not infrequently leads to a standard that appears to be a hodge-podge of features without strong guiding principles[47]. It is our opinion that ODA suffers from these blemishes as much as any other international standard.

International standards are also plagued by the desire to be in *alignment* with other related or identical standards. A standard is *aligned* with another standard when the standards are identical except for stylistic differences. The need for alignment is sometimes the result of another organization wishing to have its own version of a standard. This is the case with the Open Document Architecture [14], which is the CCITT version of ODA. The resolution of alignment conflicts is a difficult and time-consuming process and can result in apparently arbitrary changes in an evolving standard.

ODA is also quite large and complex, with many intertwining pieces and some inconsistencies. This makes learning the standard a difficult undertaking; it took the ITC EXPRES participants almost two months of full-time study to get a solid grasp on the basics of the standard (section 6.8). In fact, there are parts of the standard we were never able to grasp fully. For example, we never managed to construct a convincing model for the entire layout process; this led us to surmise that ODA was incomplete in this regard, a fact that others have also concluded.

Another symptom of design by committee is the appearance of inconsistencies in a standard. ODA, for example, has a set of rules for determining the value of an attribute at any place in the document. The basic set of rules are straightforward and intuitive. Unfortunately, there are annoying exceptions to these rules, which are rather messy to understand and implement.

It is our belief that the complexity and inconsistency of ODA are the result of the standardization process and not inherent in a standard for multi-media document interchange. We posit that a carefully designed document architecture model and a well chosen set of orthogonal principles could produce a smaller, but just as capable, standard.

Furthermore, ODA is missing many of the media that we believe are essential for effective scientific document interchange: tables (spreadsheets) and equations, in particular. In all fairness, these are being defined as we write this, although due to the standards process, they will not appear for years.

Of course, we realize that the best technical path is not always possible in the standards world. And although we have been critical of ODA, it was, in fact, the only viable candidate at the time we chose our intermediate representation. The situation has not changed much since then, except for the appearance of

[47]One might argue that this is exactly the purpose of an international standard—to allow interoperation among as many parties as possible. This is certainly true. We believe, however, that this end would be served better by evaluating issues on technical merit and economic issues, as far as possible.

DDIF [18] as a DEC product and the effort to define DSSSL [26] as a document architecture using SGML. Had we the choice to make at this point, we would have to give serious consideration to DDIF and DSSSL as contenders for ODA. ODA still has the advantage, however, in being an international standard and in having widespread acceptance in Europe.

In summary, despite its shortcomings, we do not regret our choice of ODA. In the next section, we discuss our evaluation of ODA specifically in regards to the EXPRES translation conventions.

7.3.2 Use as an Intermediate Representation

One of the nicest aspects of ODA from the EXPRES point of view was the ODA model for logical and layout structures. We found the separation of the two structures to be a useful feature, since this allows a document to say as little or as much as it wants about its organization (logical structure) or its formatting (layout structure).

The ODA logical structure is simple, but should be sufficient for the organization of most documents. Our one criticism is that there is no method for decorating the logical structure of a document with application-specific information. For example, an application might wish to use ODA documents in a hypertext system. An obvious way to do this is to annotate some nodes of the document with pointers to other nodes or documents in the system. Unfortunately, the lack of decoration ability in ODA means there is no clean way to accomplish this.

Unlike the logical structure, ODA's layout structure is rather complex. Despite this, we found the model to be rich in functionality and well thought out. In fact, one of the selling points of ODA for us was that it had a complete layout model as part of the standard. All in all, the layout structure in ODA served the EXPRES project well.

The ODA style system, on the other hand, was not quite as desirable for our use. The style system in ODA appears to have been designed to aid in the factoring of the document representation, presumably to save space and processing time. As discussed in section 5.5.2.1, the paucity of ODA's style system caused us problems when attempting to preserve the style information in our other document formats. These other formats were used by document processing systems and, thus, had style systems oriented towards human editing. In particular, the rules ODA specified for definition and application of styles did not match well with the kind of flexibility one desires when editing.

Another problem we had with ODA's style system was actually caused by the rules for determining the default values of attributes. Recall that some attributes in ODA are composed of parameters. For example, the attribute "border" has a value comprising the four parameters "leading edge", "trailing edge", "left-hand edge" and "right-hand edge". When determining a default value for a particular use of the "border" attribute, a default value will be found for each of the parameters. This means that there is no way to construct a style using "border" that has the following semantics wherever it is applied: change the "left-hand edge" parameter to a specified value, but use the current values for the other parameters.

The ability to construct style definitions with fine control over document features is an important part of the style system of some document processing systems. The EXPRES conventions go to great lengths to preserve as much of the fine-grained information as possible, but much of it is lost anyway.

There are advantages to the ODA restriction that content appear only at the leaves of a document. These advantages include the ease of adding a new content architecture, the separation of the document and content architecture definitions and the simplification of the layout process. Unfortunately, the ability to nest content is a powerful part of some document formats (ATK and Interleaf, for example), and any nesting structure is lost during translation.

Looking back at the entire translation experience, our biggest complaints with ODA for use as an intermediate representation are its size and its style system. Our attempts to preserve style information, which embodies the editing information that we believe essential to effective interchange, were stymied time and again by the ODA style system.

7.3.3 Implementation

We expended a great deal of effort in our construction of the CMU ODA Tool Kit. For the most part, the implementation was tedious, but not difficult. The tedium was caused, almost solely, by the necessity of implementing the semantics of 159 distinct attributes. We did, however, run into two problems: the implementation of the attribute value defaulting rules and the implementation of code to read and write ODIF.

7.3.3.1 Implementing the Defaulting Rules

As mentioned in section 7.3.1, there is a basic rule set specified in ODA for determining the value of an attribute at a particular component. These rules specify the order in which the document is searched, beginning at the component, and proceeding to higher structural levels. The search continues until a value is found for the attribute, or if none is found, then the ODA standard itself defines a value. The basic rules are straightforward and easy to implement.

Unfortunately, these basic rules do not apply directly to the determination of the default value for several attributes. In particular, presentation attributes, attributes for content portions, the "content information" attribute and the "content generator" attribute, each have distinct sets of defaulting rules. These defaulting rules are taken from a subset of the basic rules. Unfortunately, these rules are difficult to understand and messy to implement.

7.3.3.2 Reading and Writing ODIF

Undoubtedly, the part of ODA that gave us the most implementation headaches was the ODIF datastream. The ODIF representation suffers from several problems, including the fact that it is defined using ASN.1 [31, 32]. ASN.1 is an international standard defining an abstract syntax for representing data. ASN.1 also defines a binary encoding for this syntax so that data may be exchanged in a machine-independent manner. (Section 3.5.2 contains a discussion of ODIF and the use of ASN.1, as well as an example.)

Unfortunately, the 8-bit encoding specified for ASN.1 is unreadable by humans and this causes two difficulties: it is extremely tedious to hand-simulate the parse of an ASN.1 datastream, and it is impractical to create test ASN.1 datastreams of any size[48].

We would have been happier had ASN.1 defined an alternate representation that was human readable and interchangeable. It would be a simple matter to specify such a representation and to specify transformations between this representation and the binary encoding. One might argue that the human-readable representation would of necessity consume more space than the binary representation. It is not clear that this is true, given that there is reason to suspect that ASN.1 is not itself space (or time) efficient (the book on OSI protocols by Marshall Rose [55] contains a discussion of this issue).

Part of the reason for the difficulty in human processing of an ODIF datastream is the use of the ASN.1 binary encoding. But, this is not our only complaint about the ODIF representation. Reading a binary encoding is easier if it is possible to find reference points in the datastream. This is virtually impossible for an ODIF datastream due to the overloading of ASN.1 tag values.

To understand the reason for this, it is necessary to know something of the manner in which ASN.1 represents a data object. An ASN.1 data object is composed of three fields: a tag field (identifying the type of the value), a length field (indicating the length of the value) and a field containing the value itself (which may, recursively, contain other ASN.1 data objects). The value of a tag

[48]There is an alternate representation for an ODA datastream, known as ODL, which is represented in human-readable form using SGML (section 3.5.3). Unfortunately, the Document Application Profile we were using specified the use of ODIF.

field may be one of the predefined ASN.1 tag values, which are used to specify common data types, such as integers and strings. In addition, tag values may be defined to denote types that are specific to a standard, as is done by ODA.

The ODIF representation consists of over a hundred of these standard-specific types. Unfortunately, the standard did not choose to assign unique tag values to these types. Instead, virtually all of the tag values are small integers, in the range 0 to 15. These integers are reused to represent all of the types, so that the tag value 4, for example, represents the "dimensions" attribute as well as the "offset" attribute. A parser is able to disambiguate among the overloaded tag values by the context of the parse. This works fine for automated parsers. But humans are not able to determine the meaning of a value in the middle of a datastream without parsing the entire datastream from the beginning. This makes debugging a nightmare.

We presume that the tags were overloaded in this manner in the interest of saving space. If a tag value is too large to be contained in a five bits, multiple bytes will be used. The overloading of tag values ensures that no tag will be large enough to require more than five bits. We would have preferred that ODIF specify unique tag values throughout, as this would greatly ease the implementation task. One might argue that this is wasteful, since once an ODA system is working, there will be no need for humans to look at the ODIF. Unfortunately, we know from experience that while this may be true much of the time, the instability of systems guarantees that such representations will continue to require debugging.

This problem can be alleviated somewhat by tools for manipulating ASN.1 that provide debugging support in the form of a human-readable ASN.1 stream. (Such a tool is provided by the ISODE package, discussed in the next section.) Unfortunately, if one encounters an invalid datastream, the tools are only marginally helpful and one may still be forced to look at the datastream.

7.4 The Tool Kits

The design, specification and implementation of the CMU ODA Tool Kit consumed a large portion of our resources. To a lesser extent, the same is true of the Raster Graphics Tool Kit, although it required significantly less time and only a single programmer. In this section, we discuss our general experiences with building and using these tool kits. Our conclusions on the implementation strategies and software engineering decisions we made are deferred to section 7.5.

7.4.1 The CMU ODA Tool Kit

The decision to design and implement a common tool kit for dealing with ODA was wise. The CMU ODA Tool Kit enabled the EXPRES participants to construct translators that could begin to interchange documents with a minimum of effort. In addition, the CMU ODA Tool Kit has been distributed to well over 100 sites worldwide and may, in fact, prove to have value beyond the length of the EXPRES project.

In looking back on our work in building the tool kit, we believe that we made many good decisions that enhanced the usability of the tool kit. We began by specifying completely the application programmer's interface to the tool kit (the final version of the interface specification is part 3 of this book). While this may seem an obvious thing to do—specify before coding—things do not always proceed in this order. This is especially true in a research environment, where it is common practice to code first and specify later. For the EXPRES project, we found it essential to have a detailed specification available for the other EXPRES participants and the technical review committee, as well as for the programmers working on the tool kit.

As described in section 6.4.1, we were religious in keeping the documentation up to date. We were encouraged in this by the fact that some of the tool kit code and definitions were generated automatically from the documentation. If a programmer succumbed to the temptation to skip the documentation update and dive right into changing the code, it was possible that the next time the tool kit was generated, the changes would disappear. Besides saving tedious coding time, this technique kept us honest.

We were also pleased with the human-readable output provided by the tool kit debugging routines (chapter 22). These routines proved invaluable in debugging both the tool kit and the translators. Our only regret is that we did not allow the debugging form to be used as input to the tool kit. If we had done so, we would have easily been able to create test documents by hand. Instead we were forced to create such documents by writing programs for that sole purpose.

Our design included the principle that, whenever possible, the tool kit should check the ODA semantics of actions specified by an application. For the most part, this meant checking the legality of all attempts to assign attribute values to constituents. Providing code to support this checking was an enormous amount of work. But, we believe such checking to be a valuable aid to users of the tool kit. In addition, this checking turned out to be useful to the tool kit implementors, as it provided a check against calls internal to the tool kit that manipulate attributes.

There are two facets of the tool kit that we are not happy about: the ODIF support and the verbosity required to use the tool kit. As mentioned earlier in this chapter (section 7.3.3.2), support for reading and writing ODIF is difficult to implement due to the use of ASN.1. There was freely available a software package, known as the ISODE package [56], that provides support for parts of the OSI protocol suite, including a parser generator for ASN.1.

The ISODE ASN.1 tools are nice and provide helpful debugging support. Unfortunately, this package did not meet the strict portability guidelines we imposed on our tool kit software.

We, therefore, began by implementing our own ASN.1 support for ODIF. Although our code is portable, it was tedious to write and is not as robust as we would like. In particular, it makes assumptions about the specific format of the datastream in many places where ASN.1 allows for several formats. In addition, the implementation was so difficult that the effort required to add support for additional attributes was discouraging.

We also reimplemented some of the ODIF support using the ISODE software package. This worked quite well and provided ODIF support that was much easier to create, augment and debug.

Another major complaint of ours about the tool kit is that it is verbose, requiring long sequences of statements. For example, the program fragment in example 7-1 (taken from one of the ATK translators) creates a construction expression (section 18.4) of the form

SEQ(*header frame*, *body frame*, *footer frame*)

to be used as the value for the attribute "generator for subordinates".

The code required for the construction of this simple expression is verbose, difficult to read and prone to error during construction. One way of handling

```
TopNode.construction_tag = SEQUENCE_tag;
TopNode.construction_value.construction =
   MakeSequence(SEQUENCE_CONSTRUCTION_TERM_tag,
                (INT_type) 3);
TopNode.construction_value.construction->
 sequence_value.construction_terms[0].term_tag =
   REQ_tag;
TopNode.construction_value.construction->
 sequence_value.construction_terms[0].factor.factor_tag =
   OBJECT_CLASS_ID_tag;
TopNode.construction_value.construction->
 sequence_value.construction_terms[0].
   factor.factor_value.object_class_id =
      ssInfo->headerFrame;
TopNode.construction_value.construction->
 sequence_value.construction_terms[1].term_tag =
   REQ_tag;
/*... 5 similar statements not shown ...*/
SetAttr(ssInfo->curPage, at_GEN_FOR_SUBORDINATES,
        (POINTER_type) &TopNode, ALL_PARMS);
```

Example 7-1: CMU ODA Tool Kit Example

this problem would be to provide tool kit routines that accept strings defining the desired expression and return the equivalent tool kit structures. For example, the tool kit could provide a routine

```
CONSTRUCTION_EXPR_type *MakeExpr(s, args)
   CHAR_type *s;
   CONSTITUENT args[];
```

which would be called something like

```
e = MakeExpr("SEQ(0, 1, 2)", args);
```

Where the `args` argument contains constituents that are to be substituted for occurrences of integers in the string. During construction of the translators, we found that the handling of font definitions also required verbose code. We added several routines to the tool kit that provided functionality similar to the hypothetical `MakeExpr` routine.

7.4.2 Raster Graphics Tool Kit

We did not begin implementation of the Raster Graphics Tool Kit until the summer of 1989, too late for use in any of the translators. We also implemented the Raster Graphics Tool Kit so that it could be used as part of the CMU ODA Tool Kit, or as a stand alone package. This was a wise decision, since it allowed the Raster Graphics Tool Kit to be used as part of an undergraduate project to construct a FAX service for the Andrew electronic mail system.

We do not have much experience with the Raster Graphics Tool Kit, but its limited use in the FAX project indicates that it is a useful package.

7.5 Implementation Strategies

The CMU EXPRES group devoted a significant amount of thought to appropriate disciplines and strategies for implementing the CMU ODA Tool Kit. We agreed to a strict documentation discipline, which is discussed in the next section. The section after that discusses the programming conventions we adopted—covering naming, formatting, modularization and portability. This is followed by three sections that discuss auxiliary software we built: a program for testing the CMU ODA Tool Kit, a validation suite and a program for generating site-dependent installation packages.

7.5.1 Documentation and Specification

As mentioned in section 7.4.1, we specified completely the application programmer's interface to the CMU ODA Tool Kit (and to the Raster Graphics Tool Kit) before any code was written. This is, of course, just good practice. In addition, we imposed conventions on our use of the document processing system, Scribe [61], that allowed us to generate some code and definitions directly from the documentation. The standard procedure we used to compile the tool kit would ensure that this source was generated automatically. This motivated us to keep the documentation up to date.

Had we realized the full power of this ability to generate source code automatically, we would have been even more careful in our use of Scribe. Additional conventions would have enabled us to extract further code from the documentation.

7.5.2 Programming Conventions

In preparing to implement the CMU ODA Tool Kit, we agreed to adhere to a set of programming conventions. These conventions covered identifier names, source code formatting, modularization and portability. This section discusses our experiences with these conventions.

Our conventions for creating names used in source code are described in section 6.4. We found that these conventions worked well, providing immediate visual recognition of the class of an identifier. This allowed implementors and application programmers to distinguish quickly among type names and variables, for example. These conventions also alleviated much of the pressure programmers sometimes feel when forced to create unique names for thousands of identifiers.

The naming conventions were also intended to allow the identifiers associated with the myriad ODA attributes to be determined in a well-defined manner. This would have allowed a programmer to determine the type name of an attribute value without consulting the documentation, for example. Unfortunately, we we did not define precise rules for abbreviating identifiers that were too long[49]. This led to inconsistent abbreviations and meant that a programmer had to consult the documentation more often than we had hoped.

---

[49]We restricted the length of identifiers for portability, since some C compilers impose a maximum length on identifiers.

We also decided to adhere to a set of program formatting conventions, which would specify the layout of source code for the tool kit. These conventions determined such things as indentation, positioning of pieces of C syntax and a standard ordering for declarations (placing macro definitions before variable definitions, for example). We believe the conventions to have been worthwhile, as they insured that all of the source code had a consistent appearance. This contributed to the maintainability of the tool kit.

Our modularization conventions (described in section 6.4) were a complex set of rules intended to break the tool kit into a set of modules, each of which was implemented as one or more files. The conventions specified many aspects of the construction of these source code files, including file name, exported routine names, scoping conventions and handling of include files. The primary benefit of these conventions was that they enabled us to utilize C include files in a manner that avoided many of the problems frequently associated with a large software system. In particular, each source file had only to specify a single include file for each module from which it intended to import routines or definitions. (The name of an include file was determined easily by knowing the name of the module.) The conventions ensured that all appropriate subordinate include files were included. In general, when including files that contain nested include files, one needs to be careful about inserting the same file more than once. Our conventions avoided this problem, freeing programmers from concern.

Another important set of conventions were designed to make the tool kit as portable as possible. These conventions included the definition of a subset of the C language, avoidance of all predefined C primitive types and link-time identifier length restrictions (section 6.4.2 contains a discussion of these conventions). The specification of these rules required a lot of thought, investigation and effort and demanded significant coding discipline.

The effort was worthwhile, as the tool kit has proven to be highly portable. We estimate that it requires less than one day's effort to bring the tool kit up on a new machine or operating system. We base this on the observation that it took two of us one day to port the tool kit to a VAX running VMS—at the time, we were quite unfamiliar with that environment.

Although the portability conventions worked well in most instances, there was a concern that we overlooked. The problem was caused by a lack of restrictions pertaining to global data. In particular, some systems place restrictions on global data. For example, some systems limit the total amount of space that may be allocated to such data. To aid applications program debugging, the CMU ODA Tool Kit uses hundreds of strings for such things as attribute names and error messages. These strings are all allocated in the global data space, which became too large for the Macintosh linker. We eventually overcame this limitation by upgrading the compiler.

Similarly, on the VAX/VMS system, we were unable to create a sharable segment for our static global data. This was because we had no conventions for partitioning global data into read-only and read-write segments.

All of our conventions had one negative effect: they greatly increased the size of the source code. This not only made the task of maintaining the code more difficult, but increased the compile time required. In particular, our technique for handling include files required the compiler to process over a hundred include files for many compilations.

7.5.3 The CMU ODA Tool Kit Tester

The complexity of ODA demanded that we have a strategy for testing incrementally the CMU ODA Tool Kit. This problem was especially acute since the obvious strategies for testing required either the construction of ODIF documents or the continual writing of test programs. Neither of these possibilities was attractive.

Instead, we decided to create a *tester* that would be extended incrementally along with the tool kit. The *tester* (section 6.6.1) is an interactive program that is linked with the tool kit and accepts text commands dictating tool kit routines to invoke. The *tester* prompts for appropriate arguments, invokes the routines, checks the result for validity and outputs a printable form of the return value to the user.

The *tester* required significant implementation effort but was worth the trouble. It allowed rigorous testing of the tool kit without requiring the construction of driver programs or ODIF streams. In addition, the *tester* provided an arena for first attempts at implementing tool kit features. Many of ODA's semantics had to be implemented in the *tester* as well as the tool kit so that the *tester* could validate the tool kit's actions. The programmer could experiment with implementation techniques in the tester. Because the *tester* was never meant to be distributed, the code in the *tester* could act as the *throw-one-away* implementation. Thus, the implementation of a feature in the tool kit was the second, presumably better, implementation.

Finally, the semantics embedded in the *tester* provided a check against the semantics implemented in the tool kit. While this does not provide the same level of validity as checking against an external implementation, it did serve to catch a number of inconsistencies and errors in the tool kit.

7.5.4 The CMU ODA Tool Kit Validation Suite

Early in our design of the CMU ODA Tool Kit, it was suggested to us by a colleague that we should include an *installation validation suite* in our external distribution. The validation suite would attempt to ensure that a local installation of the tool kit was correct[50]. We were particularly concerned about this because installation of the tool kit requires the specification of a number of site-dependent definitions and routines (appendix 10).

We did not think that the implementation of the validation suite would consume much time (see section 6.6.2 for a further discussion) and we felt it would be worth the effort. Unfortunately, the suite was not available for use by the EXPRES participants or any of the early recipients of the tool kit. We therefore have no data on the use of the validation suite for its intended purpose.

The suite, however, provided a mechanism for us to test the tool kit as we modified it. The suite provided a base set of functionality that the tool kit could be checked against automatically. We found this capability especially useful when testing the portability of the tool kit on non-Unix machines.

7.5.5 The Configuration Builder

One of our primary goals for the CMU ODA Tool Kit was that it be highly portable. For this reason, the CMU ODA Tool Kit distribution contained support for a number of common environments (BSD Unix, System V Unix, VAX/VMS, Macintosh MPW and MS-DOS). All of these environments provide a mechanism similar to the Unix Make facility [20], which allows the specification of rules for constructing systems. The rules are specified in a file that is consulted by the facility, which takes into account the dependencies among pieces of the system and will force compilations, links and other actions when necessary. These dependencies are usually caused by the fact that a source file is dependent on include files. Whenever one of these include files changes, it is important to recompile the source file.

In addition to the site-dependent code for the environments, we wished to provide the appropriate configuration information. Although the various specification files for the Make-like facilities are quite similar, they differ in a number of annoying ways. For example, the environments use a common syntax for the specification files, with minor variations, such as separator and

---

[50]Note carefully that the validation suite is not intended to validate the correctness of the implementation with respect to the ODA definition.

end-of-line characters. Furthermore, each system has different methods for invoking the C compiler and utilizing the link editor.

It would have been a tremendous headache to construct the specification file for each environment by hand. In addition, as the tool kit was evolving, dependency information changed, and tracking such dependencies is tedious to the point of being impractical.

Our answer to this was to build a program, *gmake*, for the sole purpose of generating the specification files for each environment. (There is a discussion of *gmake* in section 6.6.3.) *Gmake*, which scans each source file and determines automatically dependencies, has knowledge of the peculiarities of each environment and builds complete specification files.

Gmake was a wonderfully useful tool and accomplished a task that was so tedious that we probably would have not attempted it without support.

7.6 Summary

In the two years that we have been investigating the use of ODA as an interchange medium, we have learned some of the advantages and disadvantages of performing multi-media document interchange using an intermediate representation. In a research environment, we found that ODA provided a rich enough description for a document, but only when coupled with a high-level document model. We also believe that achieving high-quality imaging fidelity for a processable document is difficult. The fine typographic control and sophisticated graphics are difficult to translate among systems that have different, even if similar, capabilities.

Our investigations were inconclusive in a number of areas. The translators we implemented are prototypes and have only been used for demonstrations. We do not, therefore, know whether multi-media document interchange will be effective in a work environment. The utility of such interchange can be determined only by people using document systems and translators in realistic situations, over long periods of time.

Our interchange work was limited to text and raster graphics. Although we did not implement translations for structured graphics, at first glance, it appears that translating between ODA's structured graphics format and other such formats may be difficult. This is because structured graphics typically demand a high level of imaging fidelity and the precise semantics of common graphics operations have subtle, but important, differences.

Our investigations into interchanging style sheet information were also inconclusive. Although we were able to exchange some of the style sheet structure and information, we found many features of style systems that could not be captured.

Another area that we did not examine carefully was the relationship between ODA and SGML. Although an SGML-based interchange format for ODA is defined (ODL), we did not investigate whether that could be exploited in an effective way by current SGML-based systems.

The EXPRES effort has started blazing a trail by which the document systems of tomorrow can work together to allow people in diverse environments to collaborate in the developing global electronic communication village that is our world. We look forward to further investigation and development by others in these directions.

2 Introduction to the Tool Kits

8
Introduction to the Toolkits

Parts 3 and 4 of this book contain the application programmer's interface manuals for the CMU ODA Tool Kit and the Raster Graphics Tool Kit, which were specified, designed and implemented by the ITC EXPRES group. The tool kits are built so that each may be used independently, although it is also possible to use them together in a single application program.

This part of the book (part 2) provides information common to both tool kits. Chapter 9 is a discussion of the notational conventions used throughout parts 3 and 4 and is required reading for anyone desiring to use either tool kit. Chapter 10 is a detailed description of the site-dependent types, routines and definitions used by the tool kits. These entities must be defined by the site installer. You should skim this chapter if you intend to use either tool kit. You need only read this chapter carefully if you are going to install either tool kit at your site.

Detailed building and installation instructions for both tool kits are in appendix B. Several annotated example programs illustrating the use of the tool kits can be found in appendix A.

9
Notational Conventions

This chapter describes the notational conventions used in the remainder of this book to designate various entities (attribute names, reserved words and program identifiers, for example) and naming conventions used for identifiers exported from a tool kit. The next two sections describe the formats used to describe user-visible routines and ODA attributes. Section 9.3 then discusses naming conventions used by the tool kits.

9.1 Routine Descriptions

Each routine description has its own section number and uses a stylized format. An example of the format, with explanatory text, appears below:

9.1.0.1 Brief Routine Description: *routine alias name*

```
C declaration of routine and parameters.
```

Real Name

The real name of this routine (see section 9.3.1).

Synopsis

A one or two sentence description of the purpose of this routine.

Parameters

Descriptions of each parameter.

Return Value

Descriptions of each possible return value.

Side Effects

Any unexpected side effects of this routine. In general, obvious side effects (the allocation of storage or the passage of time, for example) are not mentioned.

Semantics

This describes additional semantics not covered by the brief descriptions above.

Errors

Each possible value for `TKError` (section 12.3).

Implementation Status

Any differences between the description of this routine and its implementation.

If this section consists of the word *Implemented*, then the routine is implemented as described.

9.2 Attribute Descriptions

Each attribute description in chapter 17 has its own section number, and its title is the identifier used to designate the attribute within a tool kit. Each entry uses a stylized format, an example of which, with explanatory text, appears below.

9.2.1 "The Attribute Identifier"

Applicability

This part details the constituents to which the attribute applies and the conditions under which the attribute is mandatory, non-mandatory or defaultable. This part is omitted when all of the attributes within a section are applicable in the same circumstances. In such cases, the beginning of the section will explain the applicability.

Attribute Value Type

This part contains a C type definition of the type for values of the attribute and also provides the name of the type. For example, in the definition

```
typedef SEQUENCE_CONSTITUENT_type
    at_BALANCE_type;
```

the type of values for the attribute "balance" is defined, and a name for this type, `at_BALANCE_type`, is provided.

This section also lists any restrictions on the values of the attribute or its parameters. In the case that the value of an attribute, or one of its parameters, is a distinguished value, then the allowed identifiers and the corresponding ODA values are listed in a table, as follows:

Identifier	Meaning
`at_MEDIUM_TYPE_recto`	'recto'
`at_MEDIUM_TYPE_verso`	'verso'
`at_MEDIUM_TYPE_unspecified`	'unspecified'

Attribute Parameters

This section describes the parameters, if any, of the attribute. It provides a mapping between the parameters and fields in the structure defined in the **Attribute Value Type** entry (attribute values that have parameters are always represented by structured types). In addition, identifiers are provided that are used to specify the presence or absence of optional parameters (section 9.3.2.4).

For example, the attribute "offset" has the following parameter description:

Parameter	Field Name	Parameter Name
"leading offset"	`leading_offset`	`LEADING_OFFSET_parm`
"trailing offset"	`trailing_offset`	`TRAILING_OFFSET_parm`
"left-hand offset"	`left_hand_offset`	`LEFT_HAND_OFFSET_parm`
"right-hand offset"	`right_hand_offset`	`RIGHT_HAND_OFFSET_parm`

If the phrase *none* is present in the last column, then there is no name for the corresponding parameter, and it is mandatory.

If this section is not present, then there are no parameters for this attribute.

Default Value

If there is a default value for this attribute, then its value is designated in this part.

Semantics

This part describes any semantics not covered by the previous parts.

9.3 Naming Conventions for the Tool Kits

There are a number of identifiers that are defined by the tool kits. The identifiers can be categorized by their usage as follows:

- routine names,
- attribute names,
- structure field names,
- attribute parameter names,
- types used for values of attributes,
- other types,
- tag fields and values for union types,
- distinguished values of some attributes.

The remainder of this section describes the notational conventions used for these identifiers. In order to ensure portability across as many C compilers as possible, we have imposed a 31-character maximum length for all identifiers. We have endeavored to provide every identifier with a readable name within this limit.

Unfortunately, in many systems the link editor will only recognize the first six characters and ignore the case of a symbol used as an external definition (for example, a routine name or a global variable). For this reason, every entity defined by a tool kit that must be available at link-time has a *real name* that is restricted to six characters and is case-insensitive. The longer, more readable name is known as the *alias*, as described in the next section.

9.3.1 Routine Names

Each exported routine within a tool kit has two names. The first name is the *real name* of the routine and is the name seen by the compiler and link editor. The alias of a routine defined by a tool kit is a longer name that starts with an upper-case letter and has the first character of each embedded word (or partial word) capitalized.[51]. For example, the routine used for creating a component (section 15.4) has the real name

[51]The alias is actually a macro definition that expands to the real name of the routine.

```
mkcomp
```

but its alias is

```
MakeComponent
```

The alias of a site-dependent routine (whose implementation is defined by the site installer, see section 10.2) follows these conventions, but begins with the string "LOCAL_".

9.3.2 Identifiers Associated with Attributes

Because the tool kits use generic routines for the manipulation of attributes, several conventions are necessary for dealing with attributes and their values. Each attribute is referenced by a unique identifier, as discussed in section 9.3.2.1. In addition, all of the parameters of a structured attribute must be defined. The tool kit uses a structure definition, as described in section 9.3.2.3. Some attributes make use of *distinguished values.* Distinguished values, which are similar to enumerated values in programming languages, are used in ODA to define the values of certain attributes or attribute parameters. For example, the attribute "alignment" must have one of the values 'start-aligned', 'end-aligned', 'centred' or 'justified'. The tool kit identifiers for these values are described in section 9.3.5. Finally, the presence or absence of each optional parameter in an attribute's value must be specified. The technique for doing this is presented in section 9.3.2.4.

9.3.2.1 Attribute Names

Within a tool kit, each attribute is represented by a unique integer value. The include files define a macro for each attribute that expands to its integer. It is essential that the application programmer use only the predefined identifiers and not the corresponding integer values to represent attributes; the integer value for an attribute may change between releases of the tool kit.

Except for document profile attributes, presentation attributes and coding attributes (discussed below), an attribute identifier is obtained by taking the attribute name used in ODA, capitalizing all letters, replacing spaces and hyphens with underscores ("_") and prepending "at_". For example, the identifier used to represent the attribute "layout path" is

```
at_LAYOUT_PATH
```

There are several attributes that apply to the document profile whose names are in conflict with other attributes. In order to distinguish these attributes, and for consistency, all document profile attributes are prefixed with the string "dp_" (instead of "at_"). For example, the attribute "layout path" is both a document profile attribute and a layout attribute. Its identifier for use as a document profile attribute is

```
dp_LAYOUT_PATH
```

In order to avoid conflicts among presentation attributes and coding attributes for the content architectures, the names for these attributes are distinguished by a prefix that is unique for each content:

Content Architecture	Attribute Prefix
Character	`cc_`
Raster Graphics	`rg_`
Geometrics Graphics	`gg_`

Due to the 31-character identifier limit, characters may need to be removed from an attribute name to construct the tool kit identifier. In these cases, the name of the attribute is not formed using the above rules, but is abbreviated to a readable form.

9.3.2.2 Attribute Value Types

For each attribute, there is an identifier that is the name of the type used for values of that attribute. These types are designated by taking the identifier for the attribute and appending the string "_type". For example, the type for values of the attribute "medium type" has the name

```
at_MEDIUM_TYPE_type
```

9.3.2.3 Structures

Many of the types defined by the tool kits are structures. The names of fields in these structures are composed of lower-case letters, digits and underscores ("_").

There is a further convention for field names that correspond to parameters of attributes. In these cases, the field names are obtained from the parameter names used in ODA by replacing spaces and hyphens with underscores (abbreviating to obey the 31-character restriction). For example, the field names in the structure definition for the type for values of the attribute "offset" are

```
typedef struct {
    INT_type    leading_offset;
    INT_type    trailing_offset;
    INT_type    right_hand_offset;
    INT_type    left_hand_offset;
} at_OFFSET_type;
```

and these field names correspond to the parameters "leading offset", "trailing offset", "right-hand offset" and "left-hand offset".

It will frequently be necessary to provide a structure type definition with a name of the form[52]

`struct` *structure-name*

In these cases, the name will be formed by replacing the suffix "_type" of the type name with "_struct". For example, the structure name of the type `SEQUENCE_type` is `SEQUENCE_struct`.

9.3.2.4 Attribute Parameter Names

Many attributes in ODA have values that are structured into *parameters* (and *sub-parameters*). Some parameters are optional and need not be specified. To allow the application programmer to determine the presence or absence of specific parameters within an attribute, each parameter is represented by an identifier.[53]. This identifier is obtained by taking the corresponding field name used in the structure type , capitalizing all letters, replacing spaces and hyphens by the underscore character ("_") and appending "`_parm`". For example, the parameters for the attribute "Offset" are named

```
LEADING_OFFSET_parm
TRAILING_OFFSET_parm
RIGHT_HAND_OFFSET_parm
LEFT_HAND_OFFSET_parm
```

[52]This is necessary to allow recursive references to a structure.

[53]Section 16.1.1 contains a complete discussion of the use of parameter names.

9.3.3 Other Types

There are, of course, types defined by the tool kits that are not covered by any of the above categories. These types have names that are upper-case letters concatenated with "_type". For example, the type used for arbitrary length character strings (described in section 13.14) is named

```
STRING_type
```

9.3.4 Tag Fields and Values

There are a large number of union type definitions exported by the tool kits. For example, the attribute "generator for subordinates" has as its value one of the following:

- a construction expression, or
- a simple construction.

The type for this value is represented conveniently using a structure with an embedded union type that can be either a construction expression or a simple construction. Because C does not provide tagged union types, it is necessary to include a *tag field* within the structure. This tag field will contain an integer value that indicates which union component is present. The names for such tag fields are composed of lower-case letters and underscores, and end in "_tag". The values that the tag field may contain are integers, represented by predefined identifiers. The type of a tag field is the site-defined type

```
TAG_type
```

which is defined to be some integer type, according to the rules in section 10.1. These identifiers are composed of upper-case letters and underscores and are terminated by the string "_tag".

For example, the type for values of the attribute "generator for subordinates" is

```
typedef struct {

    TAG_type generator_tag;
       /* CONSTRUCTION_EXPR_tag,
          SIMPLE_CONSTR_tag */

    union {
       CONSTRUCTION_EXPR_type     expr;
       SEQUENCE_CONSTITUENT_type simple_const;
    } generator_body;

} at_GEN_FOR_SUBORDINATES_type;
```

9.3.5 Distinguished Values

There are some attributes (and parameters of some attributes) that may take a value that is a member of a set of distinguished values defined for the attribute. For example, the attribute "protection" may have one of the values 'protected' or 'unprotected'. Each of these identifiers is represented within the tool kits by a unique integer. The include files define a macro for each distinguished value that expands to its integer value. The type used to represent distinguished values within the tool kits is

```
VALUE_type
```

This type is defined by each site to be some integer type, according to the rules in section 10.1. It is essential that the application programmer use only the predefined identifiers and not the corresponding integer values to represent attributes, as the integer value may change between releases of the tool kits.

The identifier for most distinguished values is obtained by taking the name in the ODA manual, replacing spaces and dashes by underscores ("_") and prepending the attribute name. For example, the data elements for the attribute "protection" are represented by the identifiers

```
at_PROTECTION_protected
at_PROTECTION_unprotected
```

9.3.5.1 The Value 'null'

Within ODA, the value 'null' is used in the values of many attributes. In many of these cases, there is no explicit identifier for the value 'null' used within the C type. Instead, the value 'null' may be represented using a number of different methods. The application programmer should be aware of this special treatment and should read carefully the **Attribute Value Type** and **Semantics** parts of each attribute description (chapter 17) carefully.

10
Site Dependencies

To make the tool kits portable for as many environments as possible, there are a number of definitions that must be provided by the local installer. These definitions include

- Types to be used for integers, characters, bytes and arbitrary pointers. The corresponding predefined C types sometimes used for these purposes—`int`, `char`, `unsigned char` and `char *`, for example—are not portable. C compilers are notorious for providing different representations and semantics for predefined types.
- Several compile-time definitions relating to these types: for example, whether a one's-complement or two's-complement representation is used for signed integers[54].
- Routines whose implementation is dependent on the host machine architecture, operating system or C compiler (for example, input-output and storage allocation).

Type and macro definitions for site-dependent types are described in the next section, 10.1. Routines for allocating and manipulating primary memory are covered in section 10.2.5. The final section of this chapter, 10.3, describes the site-dependent routines necessary to allow data structures to be swapped between primary and secondary memory.

10.1 Site-dependent Types

There are a number of types, used both internally in the tool kits and exported to the application programmer interface, that are site-dependent. This section

[54]The representation for signed integers is important because the ASN.1 external representation of an ODA document requires that integers be written in two's-complement notation.

describes the types that must be defined by the local installer before the tool kits can be compiled and installed. The last part of this section, 10.1.12, describes the semantics that these types must obey.

10.1.1 INT_type

The type `INT_type` is used by the tool kits in many places for the declaration of signed integer variables and parameters. In particular, `INT_type` is used in the following cases:

- The integers representing attributes (section 9.3.2.1) are stored in variables of type `INT_type`.
- Errors returned to the applications programmer (section 12.3) are of type `INT_type`.
- The maximum and minimum integers that will be handled without error by the routines that read external representations (chapter 20) are integers that can be represented in a variable of type `INT_type`.
- Attribute and parameter values indicated as integers by ODA are stored as type `INT_type`.

The usual definition for `INT_type` is

```
typedef int INT_type;
```

The type used to represent `INT_type` must be able to represent signed integers.

10.1.1.1 The ONES_COMPLEMENT Macro

The macro `ONES_COMPLEMENT` should be defined to indicate whether the target system for the tool kits uses a one's-complement representation for objects of type `INT_type`. The usual definition for `ONES_COMPLEMENT` is

```
#define ONES_COMPLEMENT (BOOL_false)
```

10.1.1.2 The MAXIMUM_INT and MINIMUM_INT Macros

The macros `MAXIMUM_INT` and `MINIMUM_INT` must be defined to yield the maximum and minimum integers that may be stored in an integer represented as an `INT_type` object.

10.1.2 CHAR_type

The type CHAR_type is used to represent characters by the tool kits. This type must be able to represent at least 8 bits. The underlying type used for CHAR_type must be such that the C type "array of CHAR_type" is the same as the type of string literals. So, for example, the following declaration

```
static CHAR_type[] = "This is a string";
```

must be legal with your C compiler.

The default declaration for CHAR_type is

```
typedef char CHAR_type;
```

10.1.2.1 The MakeChar Macro

The MakeChar macro is used to convert an integer value (of any size) into a value of type CHAR_type. This might be as simple as a type cast, or it might involve manipulations to avoid sign extension.

10.1.3 BYTE_type

The type BYTE_type is used by the tool kit to represent 8-bit octets. The type used to represent BYTE_type must be a scalar type capable of representing 8-bit values. In particular, the installer has only four choices for the representation of the BYTE_type type. The usual (and recommended) definition is

```
typedef unsigned char BYTE_type;
```

Some C compilers have trouble producing code for dealing correctly with unsigned characters. In such cases, the definition

```
typedef char BYTE_type·
```

is recommended.

Finally, some C compilers have a short integer type that is represented as a single byte. In these cases, one of the following definitions may be used:

```
typedef unsigned short int BYTE_type;
typedef short int BYTE_type;
```

10.1.3.1 The MakeByte Macro

The `MakeByte` macro is used to convert an integer value (of any size) into a value of type `BYTE_type`. This might be as simple as a type cast, or it might involve manipulations to avoid sign extension.

10.1.4 BOOL_type

The type `BOOL_type` is used by the tool kits to represent boolean values. The type used to represent `BOOL_type` must be an integer type capable of representing the site-dependent values `BOOL_false` and `BOOL_true`.

The usual definition for `BOOL_type` is

```
typedef INT_type BOOL_type;
```

Possible alternative definitions are

```
typedef BYTE_type BOOL_type;
typedef char BOOL_type;
typedef unsigned char BOOL_type;
typedef short int BOOL_type;
```

10.1.4.1 The BOOL_false and BOOL_true Macros

The site-dependent macros `BOOL_true` and `BOOL_false` are used by the tool kits as boolean values. In particular, the value for `BOOL_true` must be such that

```
BOOL_true
```

is a legal logical expression in C and evaluates to true. Likewise, the value for `BOOL_false` must be such that

```
BOOL_false
```

is a legal logical expression that evaluates to false. In other words, in the statement

```
if (BOOL_true)
    s1;
else
    s2;
```

the statement s_1 must be executed.

The usual definitions for BOOL_true and BOOL_false are

```
#define BOOL_false                    (0)
#define BOOL_true                     (1)
```

10.1.5 VALUE_type

The type VALUE_type is used by the tool kits to represent distinguished values. Distinguished values are used within ODA to represent values that are unique to a particular attribute (see section 9.3.5 for a discussion of the denotation and use of distinguished values). The type used to represent VALUE_type objects must be an integer type that provides at least 8 bits.

The usual definition for this type is

```
typedef INT_type VALUE_type;
```

The primary purpose for this type is to improve the readability of the application program.

10.1.6 TAG_type

The type TAG_type is used by the tool kits within structure definitions to represent a tag value, used to distinguish between variants of an embedded union object (see section 9.3.4 for a discussion of the denotation and use of tag fields). The type used to represent TAG_type objects must be able to represent integers in the range [0, 100].

The usual definition for this type is

```
typedef INT_type TAG_type;
```

The primary purpose for this type is to improve readability of the application program.

10.1.7 PARM_MASK_type

The type PARM_MASK_type is used as a bit mask to indicate the absence or presence of optional parameters in certain attributes (section 16.1.1). The type used to represent PARM_MASK_type must be an integer type that is at least 32 bits long.

Likely definitions for this type are

```
typedef int PARM_MASK_type;
typedef long int PARM_MASK_type;
```

10.1.7.1 Predefined PARM_MASK_type Identifiers

There are two special predefined `PARM_MASK_type` identifiers, `ALL_PARMS` and `NO_PARMS`. A description of `PARM_MASK_type` objects and the use of these two identifiers can be found in section 16.1.1. The remainder of the `PARM_MASK_type` identifiers can be found throughout chapter 17, with the descriptions of the attributes to which they apply.

10.1.8 POINTER_type

There are several routines within the tool kits that deal with pointers that must be able to reference an object of any type. The type `POINTER_type` is used by the tool kits to represent this type. The type used to define the `POINTER_type` type must be a type that can be cast to be any other pointer type and to which any other pointer type can be cast.

On many systems the definition

```
typedef char *POINTER_type;
```

is known to work. There are systems, however, where the following definition is necessary:

```
typedef void *POINTER_type;
```

10.1.8.1 Predefined POINTER_type Identifiers

There is one predefined `POINTER_type` identifier that is used to denote errors. The value `LCL_ERROR_POINTER` is returned from certain routines when an error occurs.

10.1.9 SWAP_INFO_type

The type `SWAP_INFO_type` is used to record site-dependent swapping information about an object for the site-dependent swapping routines (section 10.3). For example, the type may be represented as a structure with fields for indicating a particular disk file and a byte offset into that file.

The type SWAP_INFO_type need be defined only if swapping is enabled at installation time (section B.7.3).[55]

10.1.10 The REGISTER Macro

Many C compilers accept a register declaration for some types of variables and will attempt to allocate the designated variable to a register. In general, careful use of such declarations can reduce the execution time of a program. The code in the tool kits makes use of such declarations by using the REGISTER macro, which is used in place of the reserved word register. Should you wish to disable the designation of any variables as register, you should define the macro REGISTER to be null:

```
#define REGISTER
```

10.1.11 Controlling the Formatting of Error Messages

There are several switches that may be used to control the formatting of human readable error messages by the tool kits.

10.1.11.1 The DONT_FORMAT_ERRORS Macro

This macro controls the formatting of human-readable error messages by the tool kits' routines. See section B.7.2 for full details.

10.1.11.2 The ERROR_MESG_SIZE Macro

When human-redable error messages are constructed by the tool kits, they are formatted into a static buffer whose length is specified by the macro ERROR_MESG_SIZE. The usual value for this macro is 200, but it may be changed as desired, and the tool kits' routines will automatically compensate for the length.

---

[55]The Raster Graphics Tool Kit include files contain a dummy definition for SWAP_INFO_type, which is nessary due to the construction of that tool kit. If you wish to define this type locally, do not forget to remove this definition.

10.1.12 Assignment Semantics

This section describes the semantics that must be obeyed by the site-dependent types. The descriptions in the remainder of this section are phrased in terms of assignment. Because parameter passing in C can be defined as assignment of the actual parameter to the formal parameter, the restrictions stated also apply to parameter passing.

The assignment constraints are as follows:

1 Assignment of an integer value to a variable of type `INT_type`, `CHAR_type`, `BYTE_type`, `BOOL_type`, `VALUE_type`, `TAG_type` or `PARM_MASK_type` must result in the low-order bytes of the value being represented in the least significant bits of the variable. It is not legal to truncate bits that fall within the range of required values for the target type.

- The results of an assignment of a value that is outside of the required range of the type are unspecified.
- The result of an assignment of a long integer constant (`2345667231L`, for example) is unspecified.
- The semantics of sign extension during assignment is unspecified.

2 Assignment of a character constant to a variable of type `CHAR_type` must result in all 8 bits of the character being represented in the least significant 8 bits of the variable. If the host character set is a 7-bit set, then the value of the high-order bit is unspecified.

- The semantics of sign extension during assignment is unspecified.

10.2 Site-dependent Routines

There are a number of routines that must be defined by each site. These routines perform actions that depend on characteristics of the host site, such as the operating system or the hardware configuration.

The installer at each site is responsible for providing object code for each of the routines listed in this section. The installer should be careful that the routines follow the semantics described.

10.2.1 Initialization and Finalization

It may be necessary to perform some actions prior to the invocation of any site-dependent routines. For example, it may be necessary to initialize buffers for for input and output. For this reason, the routine `LOCAL_Init` should be defined by the site installer. If no initialization actions are required, the routine should simply return successfully.

The site installer may also find it desirable to perform certain actions once the program has indicated that it will no longer be using the site-dependent routines. The routine `LOCAL_Final` is provided for this purpose.

It is the responsibility of the application program to invoke the site-dependent initialization routine before the initialization routines of either of the tool kits. In addition, the site-dependent finalization routines must be called after any tool kit finalization routines.

10.2.2 Initialize Site Dependencies: *LOCAL_Init*

```
INT_type LOCAL_Init()
```

Real Name

lclini

Synopsis

This is a site-dependent routine used to perform any local initialization necessary for the suite of site-dependent routines.

Return Value

`LCL_ERROR_INT`	An error has occurred. `LCLError` will contain an indication of the exact error (see the **Errors** part below).
otherwise	Initialization was successful.

Side Effects

The side effects of this routine will be site dependent.

Semantics

This routine must be called by the application program before any of the initialization routines are called. For example, if the application is using the CMU ODA Tool Kit, `LOCAL_Init` must be called before `InitToolKit` is called.

Errors

`LCL_LOCAL_err`	A locally-defined error, not covered by any other available error, has occurred.

10.2.3 Finalize Site Dependencies: *LOCAL_Final*

```
INT_type LOCAL_Final()
```

Real Name

lclfin

Synopsis

This is a site-dependent routine used to perform any local finalization necessary for the suite of site-dependent routines.

Return Value

`LCL_ERROR_INT`	An error has occurred. `LCLError` will contain an indication of the exact error (see the **Errors** part below).
otherwise	Finalization was successful.

Side Effects

The side effects of this routine will be site dependent.

Semantics

This routine must be called by the application program after any of the tool kits' finalization routines are called. For example, if the application is using the CMU ODA Tool Kit, `LOCAL_Final` must be called after `FinalToolKit` is called.

Errors

`LCL_LOCAL_err`	A locally-defined error, not covered by any other available error, has occurred.

10.2.4 Input and Output Routines

All input and output within the tool kits are done by calls to site-dependent routines. Input and output is performed on *input sources* and *output sources.* An input source may be a keyboard, a file, an area of memory or any other source that the site installer is willing to support. Likewise, an output source may be a screen, paper, a disk file or some other locally-supported source.

Each input and output source is represented by an *input descriptor* or an *output descriptor*, respectively. Descriptors are objects of type `INT_type`. The manner in which the integer descriptor maps to a source is not defined by the tool kits. An obvious representation would be to use the integer as an index into an array that provides site-dependent information about the source.

The descriptors are passed as parameters during calls to tool kit routines. For example, the CMU ODA Tool Kit routine that is used to read an ODIF representation of an ODA document, `ReadODIF` (section 20.1.1), is passed an input descriptor. The `ReadODIF` routine will subsequently pass this descriptor to the site-dependent routine for reading octets.

There is no method for routines within the tool kits to control actions other than reading and writing that may be necessary for the proper use of input and output sources. Examples of such actions are those associated with the opening and closing of a file on many systems. It is the responsibility of the site-dependent routines to perform these actions at appropriate times. It is recommended that actions associated with the opening of an input source be performed when the first call of `LOCAL_ReadBytes` or `LOCAL_WriteBytes` is made for a source descriptor. Likewise, actions associated with the closing of a file should be performed during the last call to `LOCAL_ReadBytes` or `LOCAL_WriteBytes` for a source. The last call to `LOCAL_ReadBytes` can be detected because it is the call that returns a 0 (section 10.2.4.1). The last call to `LOCAL_WriteBytes` contains a value of -1 for the length (section 10.2.4.2).

10.2.4.1 Read Bytes: *LOCAL_ReadBytes*

```
INT_type LOCAL_ReadBytes(id, buffer, length)
   INT_type id;
   BYTE_type buffer[];
   INT_type length;
```

Real Name

lrdbyt

Synopsis

This is a site-dependent routine used by the tool kits to read 8-bit bytes.

Parameters

`id`	The input descriptor.
`buffer`	The address of a `BYTE_type` array that will contain the bytes read.
`length`	The maximum number of bytes desired. Must be $\geq$-1.

Return Value

`LCL_ERROR_INT`	An error has occurred. `LCLError` will contain an indication of the exact error (see the **Errors** part below).
otherwise	The number of bytes placed into `buffer`. Must be non-negative.

Side Effects

The side effects of this routine are site dependent.

Semantics

This routine is called by a tool kit to read bytes from an input source.

It is legal for the number of bytes read to be less than the maximum number requested, in which case the next *N* bytes must be read, where *N* is the return value.

A return of 0 indicates that there is no more input available from that source.

If the parameter `length` is -1, this indicates that a tool kit is done with the input descriptor and the routine should perform any local cleanup actions necessary (such as closing a file). At most one call will be made with this value of `length`. In addition, if `length` is -1, then the value of the parameter `buffer` need not be a valid address.

If an error occurs, this routine should ensure that an appropriate value is in `TKError` (see **Errors** part, below).

Errors

`LCL_BADPARM_err`	One of the parameters is improper.
`LCL_LOCAL_err`	A locally-defined error, not covered by any other available error, has occurred.

10.2.4.2 Write Bytes: *LOCAL_WriteBytes*

```
INT_type LOCAL_WriteBytes(od, buffer, length)
   INT_type od;
   BYTE_type buffer[];
   INT_type length;
```

Real Name

lwrbyt

Synopsis

This is a site-dependent routine used by a tool kit to write 8-bit bytes.

Parameters

`od`	The output descriptor.
`buffer`	The address of a `BYTE_type` array that contains the bytes.
`length`	The maximum number of bytes to be written. Must be ≥-1. The value -1 indicates that no more bytes will be written.

Return Value

`LCL_ERROR_INT`	An error has occurred. `LCLError` will contain an indication of the exact error (see the **Errors** part below).
otherwise	The number of bytes written from `buffer`. Must be non-negative.

Side Effects

This routine has site-dependent side effects.

Semantics

This routine is called by a tool kit to write 8-bit bytes to an output source.

It is legal for the number of bytes written to be less than the maximum number requested, in which case the leading bytes have been written.

A return of 0 (indicating 0 bytes were written) is illegal.

If the `length` parameter is -1, then no more bytes will be written to this output descriptor. At most one call with this value of `length` will be made. In addition, if `length` is -1, then the value of parameter `buffer` need not be a valid pointer.

If an error occurs, this routine should ensure that an appropriate value is in `TKError` (see **Errors** part, below).

Errors

`LCL_BADPARM_err`
One of the parameters is improper.

`LCL_LOCAL_err` A locally-defined error, not covered by any other available error, has occurred.

10.2.4.3 Read Characters: *LOCAL_ReadChars*

```
INT_type LOCAL_ReadChars(id, buffer, length)
    INT_type id;
    CHAR_type buffer[];
    INT_type length;
```

Real Name

lrdchr

Synopsis

This is a site-dependent routine used by a tool kit to read characters.

Parameters

`id`	The input descriptor.
`buffer`	The address of a `CHAR_type` array that will contain the characters read.
`length`	The maximum number of characters desired. Must be ≥-1.

Return Value

`LCL_ERROR_INT`	An error has occurred. `LCLError` will contain an indication of the exact error (see the **Errors** part below).
otherwise	The number of characters placed into `buffer`. Must be non-negative.

Side Effects

The side effects of this routine are site-dependent.

Semantics

This routine is called by a tool kit to read characters from an input source.

It is legal for the number of characters read to be less than the maximum number requested, in which case the next N characters must be read, where N is the return value.

A return of 0 indicates that there is no more input available from that source.

If the parameter `length` is -1, this indicates that the tool kits are done with the input descriptor and the routine should perform any local cleanup actions necessary (such as closing a file). At most one call should be made with this value of `length`. In addition, if `length` is -1, then the value of the parameter `buffer` need not be a valid address.

If an error occurs, this routine should ensure that an appropriate value is in `TKError` (see **Errors** part, below).

Errors

`LCL_BADPARM_err`
: One of the parameters is improper.

`LCL_LOCAL_err`
: A locally-defined error, not covered by any other available error, has occurred.

10.2.4.4 Write Characters: *LOCAL_WriteChars*

```
INT_type LOCAL_WriteChars(od, buffer, length)
   INT_type od;
   CHAR_type buffer[];
   INT_type length;
```

Real Name

lwrchr

Synopsis

This is a site-dependent routine used by a tool kit to write characters.

Parameters

`od`	The output descriptor.
`buffer`	The address of a `CHAR_type` array that contains the characters.
`length`	The maximum number of characters to be written. Must be ≥-1. The value -1 indicates that no more characters will be written.

Return Value

`LCL_ERROR_INT`	An error has occurred. `LCLError` will contain an indication of the exact error (see the **Errors** part below).
otherwise	The number of characters written from `buffer`. Must be non-negative.

Side Effects

This routine has site-dependent side effects.

Semantics

This routine is called by a tool kit to write characters to an output source.

It is legal for the number of characters written to be less than the maximum number requested, in which case the leading characters have been written.

A return of 0 (indicating 0 characters were written) is illegal.

If the `length` parameter is -1, then no more characters will be written to this output descriptor. At most one call with this value of `length` will be made. In addition, if `length` is -1, then the value of parameter `buffer` need not be a valid pointer.

If an error occurs, this routine should ensure that an appropriate value is in `TKError` (see **Errors** part, below).

Errors

`LCL_BADPARM_err`
: One of the parameters is improper.

`LCL_LOCAL_err`
: A locally-defined error, not covered by any other available error, has occurred.

10.2.5 Primary Memory: Allocation and Manipulation

The tool kits make extensive use of dynamic storage allocation. All dynamic allocation performed by a tool kit is through a set of site-dependent routines: `LOCAL_Malloc`, `LOCAL_Realloc` and `LOCAL_Free`.

The usual assumption for dynamic allocation is that the storage manipulated by these routines is allocated on demand. That is, the program will acquire more storage from the operating system only as required by the tool kits. In addition, storage that has been allocated dynamically may be freed, that is, released for future use.

The installer is under no obligation to actually implement these routines under these assumptions. For example, it is feasible for the storage to be allocated from a block of storage allocated at link time. The tool kits do make heavy use of dynamic storage and it is highly recommended that these routines have access to a large block of storage.

In all cases, routines that allocate storage must obey the following restrictions:

- The pointer returned must be to the lowest address of storage that is available in the memory block.

- The pointer returned must be to a block of storage that is aligned to the same constraints as a C structure on the target system.

10.2.5.1 Allocate Primary Memory: *LOCAL_Malloc*

```
POINTER_type LOCAL_Malloc(nbytes)
   INT_type nbytes;
```

Real Name

lclmal

Synopsis

This is a site-dependent routine used by the tool kits to allocate a block of primary memory.

Parameters

`nbytes`	The number of bytes of primary memory to allocate. Will be non-negative.

Return Value

`LCL_ERROR_POINTER`
: An error has occurred. `LCLError` will contain an indication of the exact error (see the **Errors** part below).

otherwise
: A pointer to the allocated block of storage (see restrictions in section 10.2.5).

Side Effects

This routine has site-dependent side effects.

Semantics

This routine must allocate a contiguous block of memory of size `nbytes` or greater.

If `nbytes` is zero, this routine must still return a valid pointer to a block of zero or more bytes. Note that this allows calls of `LOCAL_Realloc` (section 10.2.5.2) to be performed.

Errors

`LCL_BADPARM_err`
: One of the parameters is invalid.

`LCL_INTERNAL_err`
: An error internal to the structure or operation of the tool kits has occurred.

`LCL_LOCAL_err`	A locally-defined error, not covered by any other available error, has occurred.
`LCL_NOMEM_err`	No memory available.

10.2.5.2 Reallocate a Block of Primary Memory: *LOCAL_Realloc*

```
POINTER_type LOCAL_Realloc(old, newsize)
   POINTER_type old;
   INT_type newsize;
```

Real Name

lclrlc

Synopsis

This site-dependent routine is used by the tool kits to reallocate a block of previously-allocated memory.

Parameters

`old`	A pointer to the block of memory to be reallocated.
`newsize`	The number of bytes the block should be after the call. Must be non-negative.

Return Value

`LCL_ERROR_POINTER`
: An error has occurred. `LCLError` will contain an indication of the exact error (see the **Errors** part below).

otherwise
: A pointer to the new block of storage (see restrictions in section 10.2.5).

Side Effects

The old block of storage should be released if a new block of storage was allocated.

This routine has site-dependent side effects.

Semantics

This routine must provide a contiguous block of storage of size `newsize` or greater.

The contents of the first N bytes must be the same as the first N bytes of the storage pointed to by `old`, where

N = min(*current size of the block*, `newsize`)

The first N bytes are those that would be accessed in C by the indices 0 to N-1.

The routine need not release any storage in the case that `newsize` is less than the current size, although it is recommended that it be done.

Errors

`LCL_BADPARM_err`
: One of the parameters is invalid.

`LCL_INTERNAL_err`
: An error internal to the structure or operation of the tool kits has occurred.

`LCL_LOCAL_err`
: A locally-defined error, not covered by any other available error, has occurred.

`NOMEM_err`
: No more memory was available for dynamic allocation.

10.2.5.3 Free a Block of Primary Memory: *LOCAL_Free*

```
INT_type LOCAL_Free(block)
   POINTER_type block;
```

Real Name

lclfre

Synopsis

This site-dependent routine is called by the tool kits to free a previously-allocated block of memory.

Parameters

`block`
: The block of memory to be freed. This block will have been created by a call of `LOCAL_Malloc` or `LOCAL_Realloc`.

Return Value

`LCL_ERROR_INT`
: An error has occurred. `LCLError` will contain an indication of the exact error (see the **Errors** part below).

otherwise
: The block of memory has been freed.

Side Effects

This block of storage may be considered to be available for reallocation or other use. All references to this block are to be considered undefined.

Semantics

This routine should release any storage allocated to the block pointed to by `block`. Although it is not required that the storage actually be released for future use, it is highly recommended that this be done.

Errors

`LCL_BADPARM_err`
: Bad parameter value.

`LCL_INTERNAL_err`
: An error internal to the structure or operation of the tool kits has occurred.

`LCL_LOCAL_err`
: A locally-defined error, not covered by any other available error, has occurred.

10.2.5.4 Copy a Block of Bytes: *LOCAL_Bcopy*

```
INT_type LOCAL_Bcopy(from, to, length)
   POINTER_type from;
   POINTER_type to;
   INT_type length;
```

Real Name

lclbcp

Synopsis

This routine is called by a tool kit's routines to perform an efficient copy of a contiguous block of bytes.

Parameters

`from`	The address of the first byte to copy.
`to`	The first address to which to copy the bytes.
`length`	The number of bytes to copy. Will be non-negative.

Return Value

`LCL_ERROR_INT`	An error has occurred. `LCLError` will contain an indication of the exact error (see the **Errors** part below).
otherwise	The specified number of bytes have been copied.

Side Effects

The contents of the first `length` bytes at address `from` have been copied to the bytes starting at address `to`.

Semantics

Providing an implementation for this routine is optional (section B.7 contains a discussion of the compile-time switch that is set to control the need for this routine).

The caller must guarantee that the two `length` blocks at `from` and `to` will not overlap.

This routine must not modify any of the first `length` bytes at location `from`.

Errors

`LCL_BADPARM_err`
: One of the parameters is invalid.

`LCL_INTERNAL_err`
: An error internal to the structure or operation of the tool kits has occurred.

`LCL_LOCAL_err`
: A locally-defined error, not covered by any other available error, has occurred.

10.3 Managing Secondary Memory

The set of routines defined in this section is meant to allow for the swapping from primary memory to secondary memory of objects. At the current time, these routines should be considered as informational only, as none of these routines has been written. In addition, the necessary tool kits' support has never been tested.

10.3.1 Swap Out a Block of Memory: *LOCAL_SwapOut*

```
SWAP_INFO_type *LOCAL_SwapOut(address,
                              length)
   POINTER_type address;
   INT_type length;
```

Real Name

lclswo

Synopsis

This routine is called by the tool kits to have a contiguous chunk of memory copied from primary memory to secondary memory.

Parameters

`address`	Base address of a contiguous block of primary memory that was allocated by a call to `LOCAL_Malloc` or `LOCAL_Realloc` (sections 10.2.5.1 and 10.2.5.2).
`length`	The number of bytes that were requested for the block when it was allocated.

Return Value

`ERROR_SWAP_INFO`
: An error has occurred. `LCLError` will contain an indication of the exact error (see the **Errors** part below).

otherwise
: The address of a `SWAP_INFO_type` object that contains site-dependent information to allow the block to be swapped back in.

Side Effects

The side effects of this routine are site dependent.

Semantics

This routine is responsible for copying the designated block of memory in a manner that allows it to be swapped back in to another address at a later time. This routine is not responsible for deallocating the block of memory—that will be handled by the caller.

The block of memory pointed to by `address` will have been allocated by a call to `LOCAL_Malloc` or `LOCAL_Realloc` (sections 10.2.5.1 and 10.2.5.2), and so will obey the alignment restrictions stated in section 10.2.5.

This routine returns a pointer to an object of type `SWAP_INFO_type`. A copy of this pointer, and not a copy of the pointed-to information, is associated with the in-memory information of the object that was swapped out. Thus, it is the responsibility of the local implementation to ensure that the data referenced by this pointer remains valid in primary memory until a corresponding call of `LOCAL_SwapIn` (section 10.3.2).

Errors

LCL_BADPARM_err
: One of the parameters is invalid.

LCL_INTERNAL_err
: An error internal to the structure or operation of the tool kits has occurred.

LCL_LOCAL_err
: A locally-defined error, not covered by any other available error, has occurred.

10.3.2 Swap In a Block of Memory: *LOCAL_SwapIn*

```
INT_type LOCAL_SwapIn(swapinfo, address,
                      available)
   SWAP_INFO_type *swapinfo;
   POINTER_type address;
   INT_type available;
```

Real Name

lclswi

Synopsis

This routine is called by the tool kits to have data copied from secondary memory back to a contiguous chunk of primary memory.

Parameters

swapinfo
: A pointer to the site-dependent swap information for the desired memory. This pointer was the result of the LOCAL_SwapOut call (section 10.3.1) that caused this block to be swapped out.

address
: Base address of a contiguous block of primary memory, allocated by a call to LOCAL_Malloc or LOCAL_Realloc, into which the swapped-out data will be copied.

available
: The number of bytes available starting at address into which to swap.

Return Value

LCL_ERROR_INT
: An error has occurred. LCLError will contain an indication of the exact error (see the **Errors** part below).

otherwise
: The data has been swapped in.

Side Effects

The contents of `available` bytes of primary memory starting at `address` may have been modified.

The `SWAP_INFO_type` object referenced by `swapinfo` may be deallocated by this routine.

Other side effects of this routine are site dependent.

Semantics

This routine is responsible for restoring the data that was swapped out by the call of `LOCAL_SwapOut` that returned the pointer `swapinfo` (section 10.3.1). This routine is not responsible for allocating the block of memory—that will have been done by the caller.

The block of memory pointed to by `address` will have been allocated by a call to `LOCAL_Malloc` or `LOCAL_Realloc` and, so, will obey the alignment restrictions stated in section 10.2.5.

The value of the parameter `available` is the number of bytes of primary memory starting at `address` that are available for swapping in. The value may be used by this routine to check that the tool kit has allocated enough space.

The pointer `swapinfo`, which references a `SWAP_INFO_type` object, will not be retained by the tool kit after this routine is called. Thus, this object may be deallocated by this routine.

Errors

`LCL_BADPARM_err`
: One of the parameters is invalid.

`LCL_NOMEM_err`
: Not enough memory was allocated by the tool kit.

`LCL_INTERNAL_err`
: An error internal to the structure or operation of the tool kits has occurred.

`LCL_LOCAL_err`
: A locally-defined error, not covered by any other available error, has occurred.

3 The CMU ODA Tool Kit

11
Introduction to the CMU ODA Tool Kit

This chapter presents an overview of the architecture of the CMU ODA Tool Kit. We have chosen to structure this discussion around an example application, that of document translation. We have used document translation because we designed the tool kit with this application in mind. This does not imply, however, that this is the only application for which the tool kit is suitable; layout and imaging are other obvious applications. Please note that the model of translation is illustrative only: we are not presenting a general discussion of document translation.

This chapter is intended for readers with an understanding of ODA. Throughout, we relate ODA constructs to their counterparts in the tool kit. The details of the tool kit interface are not to be found here; there are other chapters of this book that present such detail. Extensive references to these other chapters are provided.

In section 11.1, we present our model for document translation. Sections 11.1.1 and 11.1.2 elaborate on different aspects of document translation and the relationship of these activities to the functionality provided by the tool kit. Finally, the last section of this chapter discusses the manner in which we have structured the source code for the tool kit.

11.1 Overview of Document Translation and Interchange

One practical technique for achieving interchange between systems using different document formats is by translating between each format and a single intermediate format. Figure 11-1 is a picture of an interchange between two document formats, *A* and *B*, using ODIF, an interchange format for ODA, as the intermediate representation. We assume that formats *A* and *B* are different, and that neither is ODIF.

In our model, a document in format *A* is to be translated to yield a document in format *B*. Thus, the model is divided into two major parts. First, it is necessary

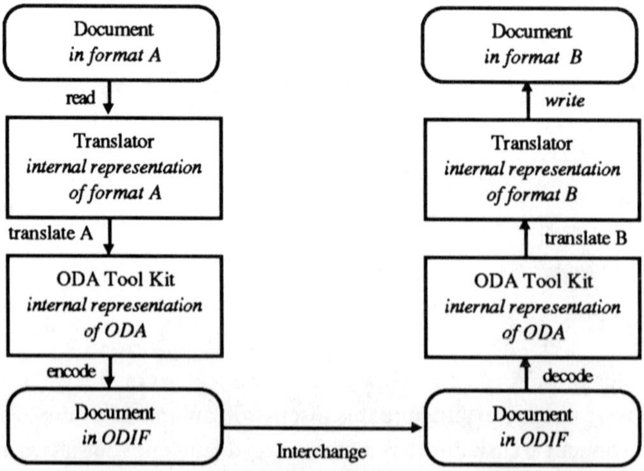

Figure 11-1. Document Format Translation: *A* to *B*

to convert the format *A* document into ODIF. After interchange of the ODIF document, a translator at the receiving site converts this document into format *B*.

We make no assumptions about the physical locations of the format *A* and *B* documents nor about how the interchange is accomplished. When only one machine is involved, the document may be interchanged through a file or main memory. When two machines are involved, interchange requires use of some communications medium.

In the next two sections, we present the tool kit architecture by discussing, in broad terms, the two translators. Presumably, these translators would attempt to maintain a high level of equivalence between the translated documents. We do not address how the translators achieve equivalence. Rather, we give outlines of the translators solely to illustrate use of the tool kit.

11.1.1 Translator from *A* to ODIF

An *A*-to-ODIF translator will take as input a document in format *A* and will produce an ODIF datastream. The translator is a tool kit application that makes calls to tool kit subroutines and uses definitions in the tool kit include files.

The first thing the translator must do is read the format *A* document and produce an internal representation of the document. We are not concerned with the manner in which this is done, as this is strictly a local operation.

Second, the tool kit must be initialized (section 14.1), before any tool kit subroutine can be called or any definition used. The tool kit initialization routine assigns values to counters and data structures, and performs other actions to allow proper functioning of the tool kit.

Next, the tool kit must be instructed to create a new document (section 15.1). The routine that creates a new document returns a value that must be used to refer to this document in subsequent tool kit routine calls. Because the application program may be manipulating more than one document at a time, it is important that the application retain this value in order to designate this document in further routine calls. Code to accomplish this might be:

```
DOCUMENT_type document;

document = MakeDocument();
```

The translator may now begin to traverse the internal representation of the format *A* document and to create the corresponding ODA document by calling tool kit subroutines to create constituents and to attach appropriate attribute-value pairs.

When a constiutent is successfully created, using a call to an appropriate routine (chapter 15), the tool kit returns a value of type CONSTITUENT (section 13.2 and chapter 15). To create the specific document logical root, the code is:

```
CONSTITUENT SpecificDocLogicalRoot;

SpecificDocLogicalRoot = MakeComponent(
        document,
        SPECIFIC_COMPONENT,
        at_OBJECT_TYPE_doc_logical_root );
```

The translator application program must retain this returned value as it is the only manner of designating the constituent in future routine calls. The single type, CONSTITUENT, is used for constituents of all kinds and generic routines are provided for manipulating CONSTITUENT objects. When there is a restriction on the actions that can be performed on a particular class of constituent, the generic routines will perform the appropriate check at run time. The inner details of the type CONSTITUENT are hidden from the application and all manipulations of a constituent (including creation and destruction) are performed using routine calls.

Once a constituent has been created, tool kit routines may be called to associate attribute-value pairs with the constituent (chapter 16). Although there are a modest number of routines for dealing with attributes, many type definitions are required to represent the different kinds of attribute values.

For an attribute whose value is not structured (contains no parameters), the value of the attribute is either an integer or a distinguished value (section 9.3.5). An example of an attribute with an integer value is "layout category". The following type definition defines both a name for the type of values for this attribute (at_LAYOUT_CATEGORY_type) and an underlying type definition (INT_type):

```
typedef INT_type at_LAYOUT_CATEGORY_type;
```

Note that to achieve portability the tool kit uses its own type INT_type and not the predefined C type int (see section 10.1 for details on site dependencies and portability).

An example of an attribute whose value is a distinguished value is "fill order". The ODA standard defines its value as one of the distinguished values 'normal order' or 'reverse order'. In the tool kit, the type definition for the "fill order" attribute is:

```
typedef VALUE_type at_FILL_ORDER_type;
```

The tool kit uses VALUE_type (sections 9.3.5 and 10.1.5) to represent attributes having distinguished values. In addition, the tool kit defines identifiers to use for the distinguished values for the "fill order" attribute:

- the identifier at_FILL_ORDER_normal_order is used for 'normal order',
- the identifier at_FILL_ORDER_reverse_order is used for 'reverse order'.

There are a number of attribute values that are not scalar. These include strings, sequences of other values and structures. For example, the attribute "user visible name" has a value that is a string. In the tool kit, strings are represented by the type STRING_type (section 13.14), and the type definition for this attribute is

```
typedef STRING_type
    at_USER_VISIBLE_NAME_type;
```

The special type for character strings, STRING_type, is necessary so that characters from arbitrary character sets may be included in strings (section 13.14)[56]. Objects with type STRING_type are created using calls to predefined routines. When a new string is created, the tool kit returns a STRING_type value that should be retained by the application to allow further manipulation of the string. Unlike the type CONSTITUENT, however, the structure of the type STRING_type is not hidden from the application programmer. Instead, the structure is visible and the programmer is expected to obey a set of restrictions when manipulating objects of this type. A new STRING_type object can be created using the tool kit call:

---

[56]The C convention of representing character strings as pointers to null-terminated arrays of char objects is insufficient because the null character can not be embedded.

```
STRING_type NewString;

NewString = MakeString( (INT_type) 10);
```

This call will create a string with length 10; note that the parameter to `MakeString` is the length which must be an `INT_type`.

Some attributes have values that may be represented conveniently as a sequence of values. The tool kit supports the type `SEQUENCE_type` for such values (section 13.13.1). The sequencing facility used by the tool kit is not a general purpose sequencing mechanism, but rather supports homogeneous sequences in which each element is a value from a predetermined set of element types. The sequence facility consists of the type `SEQUENCE_type` and a set of routines for performing operations on sequences.

The value of the attribute "subordinates" is defined as a sequence of one or more non-negative integers. Each of these integers is used to form an identifier that then designates uniquely another constituent. Rather than using this representation for the value of the attribute "subordinates", the tool kit defines the value as a sequence of `CONSTITUENT` values. The type definition for the "subordinates" attribute is:

```
typedef SEQUENCE_CONSTITUENT_type
        at_SUBORDINATES_type;
```

Each `CONSTITUENT` value is a reference to another constituent. This representation provides faster access to the constituents than a string of integers. This method for representing constituents that appear in the attribute values of other constituents is used uniformly throughout the Tool Kit.[57]

Objects of type `SEQUENCE_type` are created, destroyed and manipulated by predefined routines (section 13.13.1). As with constituents and strings, when a new sequence is created the tool kit returns a value of the appropriate sequence type that must be retained by the application. The structure of the type `SEQUENCE_type` is not hidden from the application programmer. Instead, the structure is visible and the programmer is expected to obey a set of restrictions when manipulating objects of this type. A new sequence can be made by calling the tool kit as:

```
SEQEUNCE_INT_type NewSequence;

NewSequence = MakeSequence( SEQUENCE_INT_tag,
                (INT_type) 1);
```

[57]Although the internal representation of a reference to a constituent is a value of type `CONSTITUENT`, the external formats defined by ODA require that these references be denoted by strings of integers. The tool kit routines that label a document will perform this transformation automatically (section 20.1.2).

which creates a new sequence of integers. Note that this sequence will have length one, and that the length parameter must be passed as an `INT_type`.

Some attributes have structured values; their values are composed of a number of *parameters* (and, possibly, *sub-parameters*) each of which has its own set of permissible values. For these attributes, the value is represented as a structure.

For some attributes, some or all of the parameters may be *optional*. The value of an optional parameter need not be specified. For some parameters, a default value is defined. For others, there is no value for an unspecified optional parameter. To indicate which parameters are present, a bit mask is associated with each attribute-value pair.[58] A bit mask is of type `PARM_MASK_type` (section 10.1.7) and its value is specified by using a predefined identifier (section 13.1.7.1) or by combining values that designate specific optional parameters. For this purpose, each optional parameter of an attribute has an associated *parameter name*, represented by a predefined identifier (section 9.3.2.4).

An example of an attribute with optional parameters is "offset" and its type definition is

```
typedef struct {
    INT_type        leading_offset;
    INT_type        trailing_offset;
    INT_type        left_hand_offset;
    INT_type        right_hand_offset;
} at_OFFSET_type;
```

The fields `leading_offset` and `trailing_offset` correspond to the parameters "leading offset" and "trailing offset". The identifiers for these parameters are `LEADING_OFFSET_parm` and `TRAILING_OFFSET_parm`. An expression of type `PARM_MASK_type` that specifies only the parameters "leading offset" and "trailing offset" can be made by forming the bit-wise inclusive or of the two parameter names:

```
LEADING_OFFSET_parm | TRAILING_OFFSET_parm
```

When the translator has finished traversing the format *A* document and has built the ODA document, the translator will write the document in ODIF format. A tool kit routine will, when invoked, automatically generate identifiers for constituents and then write the ODIF representation of the document (section 20.1.2). For example, the code to accomplish this is:

---

[58]There are attributes with optional parameters that use techniques other than parameter bit masks to specify the presence or absence of parameters. These attributes are designated clearly in chapter 17.

```
INT_type OutFile;
INT_type ReturnCode;

ReturnCode = WriteODIF( OutFile,
                document,
                dp_INTERCHANGE_FMT_CLASS_a );
```

This call writes the ODIF form of the document to the file designated by OutFile (which must be opened in the appropriate system-specific way) in the interchange format known as format A[59].

Lastly, the application calls the tool kit finalization routine (section 14.2) to allow the tool kit to clean up. For example, dynamically-allocated storage will be released and input/output routines will be given a chance to flush buffers, close files and perform other necessary activity. This can be accomplished by:

```
INT_type ReturnCode;

ReturnCode = FinalToolKit();
```

11.1.2 Translator from ODIF to *B*

The translator from ODIF to *B* is an application program that takes as input a document encoded in ODIF and produces a format *B* document. An outline of possible actions performed by this translator is given in the remainder of this section.

As with the *A*-to-ODIF translator, the ODIF-to-*B* translator starts by initializing the tool kit (section 14.1).

Next, the tool kit is instructed to read the ODIF document using a single call to a tool kit routine (section 20.1.1). This routine builds an internal copy of the ODA document. Upon successful return, this routine will return a value designating a new ODA document. This value must be used by the application program to reference this document in later calls. For example, an ODIF document can be read as:

```
DOCUMENT_type document;
INT_type InFile;

Document = ReadODIF( InFile );
```

where the InFile designates the input file and has been opened in the appropriate system-specific way.

[59]The ODA standard also provides a B interchange format for compatibility with older formats.

The application can now begin to traverse the ODA document by using the tool kit's document traversal mechanism (chapter 19). This mechanism allows the application program to iterate through the constituents of the document in a flexible manner according to various criteria. An `ITERATOR_type` object can be created as:

```
ITERATOR_type iterator;

CONSTITUENT SpecificDocLogicalRoot;

SpecificDocLogicalRoot = FindDocumentRoot(
        document,
        SPECIFIC_DOC_LOGICAL_ROOT);

iterator = MakeIterator(
            SpecificDocLogicalRoot,
            PARENTS_FIRST, DEPTH_FIRST );
```

Here the document's specific logical root is first located by calling the tool kit's `FindDocumentRoot` routine. Then the iterator is created to perform the depth first traversal with parents before their children.

While traversing the document, the application program will visit each node and perform some processing. This processing will obtain the values of attributes to discern structure and formatting information. Attributes in ODA are divided into three classes. A *mandatory* attribute must appear on applicable constituents; its absence is an error. A *non-mandatory* attribute need not appear and its absence on a particular constituent simply indicates that it is unspecified. A *defaultable* attribute, like a non-mandatory attribute, need not be specified; when unspecified, however, ODA specifies a sequence of rules that will determine a default value for the attribute.

To relieve the application program from the burden of implementing this mechanism, the tool kit provides a routine that will yield the value of a specific attribute by obeying the defaulting rules specified in the standard. For example, to obtain the value of the "fill order" attribute when visiting a given constituent, use the tool kit routine `GetAttr`:

```
at_FILL_ORDER_type FillOrder;
CONSTITUENT SomeConstituent;
INT_type ReturnCode;

ReturnCode = GetAttr( SomeConstituent,
            at_FILL_ORDER,
            (POINTER_type) &FillOrder,
            /* defaulting */ BOOL_true,
            /* no parameters */
                (PARM_MASK_type *) 0);
```

This will cause the tool kit to obtain the value for the "fill order" attribute for SomeConstituent and return it in the FillOrder variable which is passed by reference. Note that the tool kit will automatically perform the ODA defaulting to obtain this value and may return the default value if needed. Also note that the last argument instructs the tool kit to not return the parameter mask information; this is irrelevant as the "fill order" attribute does not have parameters.

When the traversal is completed and the format *B* document has been built, the application must finalize the tool kit interface (section 14.2), just like the translator from format *A* to ODIF.

12
Using the Tool Kits

This chapter provides detailed information enabling a programmer to write application programs that make use of the CMU ODA Tool Kit and Raster Graphics Tool Kit. Section 12.1 explains how to make use of the include files provided by the tool kits. This is, of course, strictly a syntactic issue of putting the proper #include preprocessor statements in the appropriate places. The remaining sections of this chapter describe semantic issues. Section 12.2 describes the manner in which the routines exported by the tool kits should be used: how to call them and how to check for execution errors. Section 12.3 describes the global variables exported by the tool kits, the conditions under which these variables will contain meaningful information and what that information will be. Section 12.4 provides instructions for linking the appropriate tool kit libraries with application programs.

12.1 Include Files

In order to use the tool kits' routines or definitions from an application program, it is necessary for the source file to obtain access to the necessary include files. An application program wishing to use the CMU ODA Tool Kit, without the routines provided by the ***isode*** parser[60], should place the preprocessor directive

```
#include <odatk.h>
```

somewhere near the beginning of the program. If one of the ***isode*** routines is to be used, then two preprocessor directives are needed, as shown below.

```
#include <odatk.h>
#include <odaiso.h>
```

If an application program wishes to use the Raster Graphics Tool Kit, then it

---

[60]The routines provided by the ***isode*** parser are `isode_PrintODIF`, `isode_ReadODIF` and `isode_WriteODIF`.

should include the following preprocessor directive near the beginning of the program:

```
#include <rgtk.h>
```

If both the CMU ODA Tool Kit and Raster Graphics Tool Kit are used, then the preprocessor statements for both are necessary, though they may appear in either order. The include files ensure that all definitions are available.

This technique will cause the inclusion of a large number of definitions, not all of which may be needed by a particular application. Although it is possible to individually include only those definitions necessary for a particular source file, the structure of the include files is rather complicated and we do not recommend this practice.

12.2 Exported Routines

The primary functionality of the tool kit is obtained by an application program through the invocation of a set of exported C routines. These routines are described using a common format, which is described in section 9.1. The definitions of these routines are scattered throughout this manual, although an alphabetical list of the routines cross-referenced to their defining sections is available in appendix F.

12.3 Global Variables: TKError, RGError, LCLError and TKErrorMesg

All of the routines defined by a tool kit indicate an error by returning a distinguished value (the description for each routine describes these return values). Once an error has been detected, the application program may obtain further information about the nature of the error by inspecting the appropriate global variables for each tool kit or the site specific support module. Thus, for routines called in the CMU ODA Tool Kit, the application program may examine `TKError` and `TKErrorMesg`. For routines called in the Raster Graphics Tool Kit, the application program may examine `RGError`. Routines that are implemented in the site specific modules report errors through the `LCLError` global variable.

After a routine has returned an error indication, and before the next tool kit routine has been invoked, a global `INT_type` variable contains an indication of the kind of error that occurred. For the CMU ODA Tool Kit, the variable TKError is set; for the Raster Graphics Tool Kit, the variable RGError is set; and for site specific routines (that have the prefix LOCAL), the variable LCLError is set. The value indicates a fairly broad error classification, such as "one of the parameters is invalid" or "an attribute was used incorrectly". Appendix C contains an alphabetized listing of the error identifiers and a generic description of the error. In addition, each routine description in this manual contains an **Errors** part that lists the error codes that may be found in the appropriate global variable if an error occurs during the execution of the routine.

If the error occurred in a call on the CMU ODA Tool Kit, then a detailed description of the error that occurred may be found as the value of the global variable TKErrorMesg. The value of this variable addresses a null-terminated array of CHAR_type objects, which contains a formatted string of ASCII characters that describes the error condition. For example, the following call of the routine MakeComponent (section 15.4)

```
MakeComponent(doc, at_OBJECT_TYPE_frame,
              DAPD_COMPONENT);
```

may return the error value ERROR_CONSTITUENT. The value of TKError might be

```
BADPARM_err
```

indicating that one of the parameters to this routine was invalid. In this example, the contents of the string TKErrorMesg might be "[MakeComponent]: type and kind are inconsistent, i.e. kind of dapd disallows some types" indicating that the values of the parameters type and type could not be used together.

There are two ways in which the formatting of the error message in TKErrorMesg may be inhibited. The site installer may build the CMU ODA Tool Kit such that the code for constructing the error messages is not compiled (section B.7). In this case, the value of TKErrorMesg will always be

```
"Error message formatting is not compiled"
```

In addition, if error message formatting is compiled into the CMU ODA Tool Kit, then the application program may turn the formatting on and off at run time by modifying the appropriate field of the initialization structure described in section 14.1. In the case that the formatting is turned off at run time, the value of TKErrorMesg will be undefined.

12.4 Linking with the Tool Kits

This section describes the commands necessary for building a program that uses the CMU ODA Tool Kit, Raster Graphics Tool Kit or both tool kits. Recall that the CMU ODA Tool Kit and the Raster Graphics Tool Kit are independent of each other, but that both are dependent on the local library built in the ***site*** directory. Thus, you must always include the local library when building a program. It is, however, possible to include either the CMU ODA Tool Kit or the Raster Graphics Tool Kit.

The alternate ODIF support built in the ***isode*** directory need only be included if you are using the CMU ODA Tool Kit and wish to call the alternate routines.

The following is a list of commands for each configuration for linking programs using various combinations of the libraries. In each case, we assume that the libraries have been installed in directory <lib>, that <objs> is a list of object files for the specific application and that <prog> is the desired name for the resulting executable. Be aware that on many systems, the order of items during the link phase is important.

BSD Unix

Just using the CMU ODA Tool Kit:

```
cc <objs> <lib>/liboda.a <lib>/liblocal.a \
   -o <prog>
```

Just using the Raster Graphics Tool Kit:

```
cc <objs> <lib>/libraster.a \
   <lib>/liblocal.a \
   -o <prog>
```

Using the Raster Graphics Tool Kit and the CMU ODA Tool Kit:

```
cc <objs> <lib>/liboda.a <lib>/libraster.a \
   <lib>/liblocal.a -o <prog>
```

Using the CMU ODA Tool Kit and the ***isode*** parser:

```
cc <objs> <lib>/libisode.a <lib>/liboda.a \
   <lib>/liblocal.a -o <prog>
```

System V Unix

Identical with BSD Unix.

VAX VMS

Just using the CMU ODA Tool Kit:

```
link/exec=<prog> <objs>, -
   <lib>liboda.olb/library, -
   <lib>liblocal.olb/library, -
   SYS$LIBRARY:VAXCRTL.OLB/library
```

Just using the Raster Graphics Tool Kit:

```
link/exec=<prog> <objs>, -
   <lib>libraster.olb/library, -
   <lib>liblocal.olb/library, -
   SYS$LIBRARY:VAXCRTL.OLB/library
```

Using the Raster Graphics Tool Kit and the CMU ODA Tool Kit:

```
link/exec=<prog> <objs>, -
   <lib>liboda.olb/library, -
   <lib>libraster.olb/library, -
   <lib>liblocal.olb/library, -
   SYS$LIBRARY:VAXCRTL.OLB/library
```

Using the CMU ODA Tool Kit and the ***isode*** parser:

```
link/exec=<prog> <objs>, -
   <lib>libisode.olb/library, -
   <lib>liboda.olb/library, -
   <lib>liblocal.olb/library, -
   SYS$LIBRARY:VAXCRTL.OLB/library
```

MS-DOS

Just using the CMU ODA Tool Kit:

```
cl -o <prog> <objs> <lib>liboda.lib \
   <lib>liblocal.lib -link /SE:1000 /ST:50000
```

Just using the Raster Graphics Tool Kit:

```
cl -o <prog> <objs> <lib>libraster.lib \
   <lib>liblocal.lib -link /SE:1000 /ST:50000
```

Using the Raster Graphics Tool Kit and the CMU ODA Tool Kit:

```
cl -o <prog> <objs> <lib>liboda.lib \
   <lib>libraster.lib \
   <lib>liblocal.lib -link /SE:1000 /ST:50000
```

Using the CMU ODA Tool Kit and the ***isode*** parser:

```
cl -o <prog> <objs> <lib>liboda.lib \
   <lib>libisode.lib \
   <lib>liblocal.lib -link /SE:1000 /ST:50000
```

Macintosh MPW

Just using the CMU ODA Tool Kit:

```
Link -p -t MPST -c "MPS " ∂
   <objs> <lib>:liboda.a <lib>:liblocal.a ∂
   "{Libraries}"Interface.o ∂
   "{CLibraries}"CRunTime.o ∂
   "{CLibraries}"StdCLib.o ∂
   "{CLibraries}"CInterface.o ∂
   -o <prog>
```

Just using the Raster Graphics Tool Kit:

```
Link -p -t MPST -c "MPS " ∂
   <objs> <lib>:libraster.a  ∂
   <lib>:liblocal.a ∂
   "{Libraries}"Interface.o ∂
   "{CLibraries}"CRunTime.o ∂
   "{CLibraries}"StdCLib.o ∂
   "{CLibraries}"CInterface.o ∂
   -o <prog>
```

Using the Raster Graphics Tool Kit and the CMU ODA Tool Kit:

```
Link -p -t MPST -c "MPS " ∂
   <objs> <lib>:liboda.a <lib>:libraster.a ∂
   <lib>:liblocal.a ∂
   "{Libraries}"Interface.o ∂
   "{CLibraries}"CRunTime.o ∂
   "{CLibraries}"StdCLib.o ∂
   "{CLibraries}"CInterface.o ∂
   -o <prog>
```

Using the CMU ODA Tool Kit and the ***isode*** parser:

```
Link -p -t MPST -c "MPS " ∂
   <objs> <lib>:libisode.a <lib>:liboda.a ∂
   <lib>:liblocal.a ∂
   "{Libraries}"Interface.o ∂
   "{CLibraries}"CRunTime.o ∂
   "{CLibraries}"StdCLib.o ∂
   "{CLibraries}"CInterface.o ∂
   -o <prog>
```

13 Types

The types discussed in this chapter are used in many application programs. Types whose representations are defined by each site, such as `INT_type` and `BOOL_type`, are defined in section 10.1. Types that are particular to the values of specific attributes are defined in chapter 17. Section 13.1 contains brief descriptions of the site-dependent types, which are described in detail in section 10.1. Sections 13.2 through 13.11 describe some predefined types that are necessary for the manipulation of documents and attributes. Section 13.12 contains a number of miscellaneous types and routines expected to be useful in many circumstances. The next section, 13.13, defines a set of type definitions and routines that are used for creating sequences of objects, where each object is of the same type. Although a character string is simply a sequence of `CHAR_type` objects, we expect that character strings will be of widespread use and so we provide a type definition and several routines that provide a specialized string interface. These routines are described in section 13.14.

13.1 Site-dependent Types

There are a set of types that are site-dependent. These types are not predefined by the tool kit, but are defined locally by the site installer. This is done in order to ensure portability of the tool kit to many sites. All of these types are defined in detail in section 10.1. For completeness, however, a brief description of each of the site-dependent types, as well as any predefined identifiers for each type, is provided in this section. The reader is advised to read carefully the sections in chapter 10 for the detailed semantics of these types.

13.1.1 INT_type

The type INT_type, defined in section 10.1.1, is used to represent integer values. This type is not necessarily the same as the C type int.

13.1.1.1 Predefined INT_type Identifiers

There are three predefined identifiers that are used to indicate an error return from a routine that returns an INT_type value. The names of these identifier are

```
ERROR_INT
RG_ERROR_INT
LCL_ERROR_INT
```

which are used by the CMU ODA Tool Kit, the Raster Graphics Tool Kit and the site dependent routines respectively.

There are two site-defined identifiers that may be used to obtain the minimum and maximum integers that may be represented by a variable of type INT_type: they are MINIMUM_INT and MAXIMUM_INT.

13.1.2 CHAR_type

The type CHAR_type, defined in section 10.1.2, is used to represent 7-bit or 8-bit character values. This type is not necessarily the same as the C type char. C character constants may be assigned to variables of type CHAR_type.

13.1.3 BYTE_type

The type BYTE_type, defined in section 10.1.3 is used to represent values that are 8-bit bytes, or octets.

13.1.4 BOOL_type

The type BOOL_type, defined in section 10.1.4, is used to represent C logical values. The result of a C logical expression, or one of the predefined

BOOL_type expressions (defined below), may be assigned to a BOOL_type variable. In addition, a BOOL_type value may be used directly as a C logical expression.

13.1.4.1 Predefined BOOL_type Identifiers

There are two site-defined BOOL_type identifiers, BOOL_true and BOOL_false, that are used to designate the boolean values true and false.

13.1.5 VALUE_type

The site-dependent type VALUE_type, defined in section 10.1.5, is used to represent distinguished values (distinguished values are described in section 9.3.5).

13.1.5.1 Predefined VALUE_type Identifiers

There are a large number of predefined VALUE_type identifiers, most of which correspond to designated values specified in the ODA standard. These identifiers can be found in chapters 16 and 17.

13.1.6 TAG_type

The site-dependent TAG_type, defined in section 10.1.6, is used to represent tag values for certain types, to distinguish between variants of a union field of a structure. Section 9.3.4 describes the use of TAG_type values. Only predefined identifiers should be assigned to variables of type TAG_type.

13.1.7 PARM_MASK_type

The site-dependent PARM_MASK_type, defined in section 10.1.7, is used to represent *parameter masks*, which specify the presence of optional parameters in certain attributes. Identifiers for use in constructing PARM_MASK_type values are defined in section 9.3.2.4, and the use of parameter masks is described in section 16.1.1.

13.1.7.1 Predefined PARM_MASK_type Identifiers

In addition to all of the identifiers defined for specific attribute parameters, there are two special predefined `PARM_MASK_type` identifiers for use in constructing parameter masks. These two identifiers, `ALL_PARMS` and `NO_PARMS`, are described in section 16.1.1.

13.1.8 POINTER_type

The site-dependent type `POINTER_type`, defined in section 10.1.8, is used to represent values that are pointers to an object of arbitrary type. This type is guaranteed to be such that it may be cast into any other pointer type, and such that any pointer type may be cast into the type `POINTER_type`.

13.1.8.1 Predefined POINTER_type Identifiers

The predefined identifiers `ERROR_POINTER`, `LCL_ERROR_POINTER` and `NULL_POINTER` are values of type `POINTER_type` that are guaranteed to be unique: not equal to any other `POINTER_type` objects that may be constructed. Equality between a `POINTER_type` object and these values may be tested using the C equality operator ("=="). The value `ERROR_POINTER` is used as a routine return value to indicate an error.

The value `NULL_POINTER` is used to denote the absence of a POINTER_type object where appropriate.

13.2 The Type CONSTITUENT_type

An ODA document is composed of constituents, each of which is a set of attribute-value pairs. Constituents may be divided into classes according to various sets of criteria. The operations that may be applied to a constituent may depend on the constituent's classification. For example, the attribute "content generator" only applies to constituents that are basic object class descriptions.

To limit the number of routines, types and macros that are necessary for dealing with constituents, there is a single type for all constituents. In addition, there are generic routines for dealing with constituents (these routines perform run-time checks to verify the legality of operations). The single type for declaring all constituents is

```
CONSTITUENT_type
```

For readability, the type CONSTITUENT_type is also available with the name

```
CONSTITUENT
```

despite the fact that this violates the convention that all types exported by the tool kit end with the string "_type".

The structure of the CONSTITUENT type is hidden from the programmer and all manipulations must be made through predefined routines. The introduction to chapter 15 provides a further discussion of the use of CONSTITUENT objects.

13.2.1 Predefined CONSTITUENT_type Identifiers

The predefined identifiers NULL_CONSTITUENT and ERROR_CONSTITUENT are values of type CONSTITUENT that are guaranteed to be unique, that is, not equal to each other or to any other CONSTITUENT objects that may be constructed. Equality between a CONSTITUENT object and either of these values may be tested using the C equality operator ("==").

The value NULL_CONSTITUENT is used to denote the absence of a constituent where appropriate. The value ERROR_CONSTITUENT is used as a routine return value to indicate an error.

13.3 The Type DOCUMENT_type

At any time, the application program may be manipulating more than one ODA document. In order to distinguish between the documents, each document is identified by a value of type DOCUMENT_type. The structure of this type is hidden from the programmer and values are created and destroyed by tool kit routines. Several routines take a DOCUMENT_type value parameter to allow specification of the document to which the operation applies.

13.3.1 Predefined DOCUMENT_type Identifiers

The predefined identifier ERROR_DOCUMENT is a value of type DOCUMENT_type that is guaranteed to be unique, that is, not equal to any other DOCUMENT_type objects that may be constructed. Equality between a DOCUMENT_type object and this value may be tested using the C equality operator ("==").

The value ERROR_DOCUMENT is used as a routine return value to indicate an error.

13.4 The Type ITERATOR_type

The type ITERATOR_type, which is described in chapter 19, is used to maintain the state necessary to allow an application program to traverse portions of an ODA document.

The structure of this type is hidden from the programmer and values are created, destroyed and manipulated by tool kit routines. All of the routines pertaining to this type are described in chapter 19.

13.4.1 Predefined ITERATOR_type Identifiers

The predefined identifier ERROR_ITERATOR is a value of type ITERATOR_type that is guaranteed to be unique, that is, not equal to any other ITERATOR_type objects that may be constructed. Equality between an ITERATOR_type object and this value may be tested using the C equality operator ("==").

The value ERROR_ITERATOR is used as a routine return value to indicate an error.

13.5 The Type INT_PAIR_type

The type INT_PAIR_type is used where a pair of integers is necessary. For example, the value of the attribute "page position" (section 17.1.2.11) is "an ordered pair of integers that specify the horizontal and vertical distances of the reference point ..." [30]. The type for this attribute is of type INT_PAIR_type, shown below.

```
typedef struct {
    INT_type   first, second;
} INT_PAIR_type;
```

When used within the type value for an attribute, the mapping from parameters to the fields of this structure is defined in the specific description for the attribute.

13.6 The Type PERSONAL_NAME_type

The type PERSONAL_NAME_type is used in attributes that describe an agent who has some relationship with a document.

```
typedef struct PERSONAL_NAME_struct {
    SEQUENCE_BYTE_type    surname;
    SEQUENCE_BYTE_type    givenname;
    SEQUENCE_BYTE_type    initials;
    SEQUENCE_BYTE_type    title;
} PERSONAL_NAME_type;
```

The characters in each of these fields must be from the document profile character set.

13.6.1 Specifying Parameters for the PERSONAL_NAME_type Type

The parameter "surname" is mandatory and the field surname must contain a SEQUENCE_BYTE_type value that is not NULL_SEQUENCE.

The remaining parameters, "givenname", "initials" and "title" are optional, and their absence is indicated by the corresponding fields, givenname, initials and title, having the value NULL_SEQUENCE.

13.7 The Type NAME_ORG_type

The type NAME_ORG_type is used in attributes that describe the organization to which an agent belongs.

```
typedef struct NAME_ORG_struct {
    BOOL_type              name_is_present;
    PERSONAL_NAME_type     name;
    STRING_type            organization;
} NAME_ORG_type;
```

The type PERSONAL_NAME_type is defined in section 13.6.

The characters in the field organization, if present, must be from the document profile character set.

When used within the type value for an attribute, the mapping from parameters to the fields of this structure is defined in the specific description for the attribute.

13.7.1 Specifying Parameters for the NAME_ORG_type Type

If the value of the field name_is_present is BOOL_false, then the parameter corresponding to the field name is missing.

The method for specifying the absence or presence of the parameters for the field name is described in section 13.6.1.

The absence of the parameter corresponding to the field organization is specified by the field having the value NULL_STRING.

13.8 The Type ASN1_OBJECT_ID_type

ASN.1 object identifiers, defined in the ISO ASN.1 standards [31, 32], are represented by a sequence of INT_type objects:

```
typedef SEQUENCE_INT_type
    ASN1_OBJECT_ID_type;
```

Each of these integers corresponds to a component of the object identifier. The tool kit will not verify the syntactic or semantic correctness of any user-constructed object of type ASN1_OBJECT_ID_type.

13.8.1 Predefined ASN1_OBJECT_ID_type Identifiers

There are several predefined identifiers that can be used whenever applicable for values of type ASN1_OBJECT_ID_type. These identifiers correspond to ASN.1 object identifiers for the content architecture classes (section 17.1.1.11) and values of the "type of coding" attribute (section 17.1.7.3). The programmer should refer to the appropriate sections for descriptions of these identifiers.

13.9 The Type LAYOUT_OBJECT_type

The type LAYOUT_OBJECT_type is used in the value types for several layout directive attributes: at_INDIVISIBILITY, at_NEW_LAYOUT_OBJECT and at_SAME_LAYOUT_OBJECT. The type definition is shown below.

```
typedef struct {
    TAG_type                   layout_object_tag;
        /*
            LAYOUT_OBJ_DESC_tag,
            LAYOUT_CATEGORY_tag,
            OBJECT_TYPE_PAGE_tag, NULL_tag
        */
    union {
        STRING_type            layout_category;
        CONSTITUENT            constituent;
    } layout_object_value;
} LAYOUT_OBJECT_type;
```

If layout_object_tag is LAYOUT_OBJ_DESC_tag, then the field layout_object_value.constituent contains a reference to a layout object class description.

If layout_object_tag is LAYOUT_CATEGORY_tag, then the field layout_object_value.layout_category must contain characters from the minimum subrepertoire of ISO 6937-2.

If layout_object_tag is OBJECT_TYPE_PAGE_tag, then the value of the attribute is 'object type page'.

If layout_object_tag is NULL_tag, then the value of the attribute is 'null'.

13.10 The Type STRING_OR_ID_type

The type STRING_OR_ID_type is used in several attribute value types to designate a document. Its definition is shown below.

```
typedef struct STRING_OR_ID_struct {
    TAG_type                   string_or_id_tag;
        /* STRING_tag, ID_tag */
    union {
        STRING_type            string;
        ASN1_OBJECT_ID_type id;
    } string_or_id_value;
} STRING_OR_ID_type;
```

If the value of the field string_or_id_tag is STRING_tag, then the field string is present.

If the value of the field string_or_id_tag is ID_tag, then the field id is present.

13.11 The Type FONT_REFERENCE_type

The type FONT_REFERENCE_type is used to define the ISO attributes of a font reference, as defined in ISO DIS 9541 [34].

The method for indicating which parameters are specified is described in section 13.11.1. There are also a large number of predefined identifiers for use in manipulating font descriptions. These are defined in section 13.11.2.

Note that parameters that are specified as having real values in ISO 9541 are represented in this type as fields of type INT_PAIR_type. This corresponds to the ASN.1 representation as a pair of integers (x, y) (where the first field denotes x and the second field denotes y) representing the rational number x/y.

The first field of many structures is named dfnpre, but is currently ignored by the tool kit.

```
typedef struct {
    SEQUENCE_ISO9541_ID_type        dfnpre;
    STRUCTURED_NAME_type            fontname;
    STRUCTURED_NAME_type            difntnam;
    INT_type                        propdata;
    STRUCTURED_NAME_type            typeface;
    STRUCTURED_NAME_type            family;
    INT_type                        posture;
    INT_type                        weight;
    INT_type                        propwdth;
/*  ISO/CHARCOMP unimplemented */
    INT_type                        numchars;
/*  The field ISO/LIGATURS is unimplemented */
    INT_PAIR_type                   dsnsize;
    INT_PAIR_type                   minsize;
    INT_PAIR_type                   maxsize;
    SEQUENCE_INT_type               dsclass;
    INT_type                        structur;
    INT_PAIR_type                   mnfeatsz;
    WRITING_MODES_type              modes;
    STRUCTURED_NAME_type            mode;
/*  ISO/TECHS is unimplemented */
} FONT_REFERENCE_type;
```

```
typedef struct {
   BOOL_type                          specified;
   SEQUENCE_ISO9541_ID_type           dfnpre;
   MODAL_FONT_ATTR_type               iso_0;
   MODAL_FONT_ATTR_type               iso_180;
   MODAL_FONT_ATTR_type               iso_270;
/* non-ISO-modes not implemented */
} WRITING_MODES_type;

typedef struct {
   BOOL_type                          specified;
   SEQUENCE_ISO9541_ID_type           dfnpre;
   INT_PAIR_type                      escapdir;
   INT_type                           esclass;
   INT_PAIR_type                      avgesc;
   INT_PAIR_type                      avglcesc;
   INT_PAIR_type                      avgcpesc;
   INT_PAIR_type                      avgwtesc;
   struct tabesc_struct {
      BOOL_type                       specified;
      INT_PAIR_type                   tx;
      INT_PAIR_type                   ty;
   } tabesc;
   INT_PAIR_type                      mfx;
   INT_PAIR_type                      mlx;
   INT_PAIR_type                      mbx;
   INT_PAIR_type                      mrx;
/* ISO/CLPESCAD is unimplemented */
   INT_PAIR_type                      minampl;
   INT_PAIR_type                      maxampl;
/* ISO/TRACKS & ISO/TRACK unimplemented */
   SCORES_type                        scores;
   VARIANT_SCRIPTS_type               varscrpt;
   INT_PAIR_type                      minl2lsp;
   INT_PAIR_type                      minanasc;
   INT_PAIR_type                      maxanasc;
/* ISO/MODCHATT is not implemented */
/* Non-ISO-attributes unimplemented */
} MODAL_FONT_ATTR_type;

typedef struct {
   BOOL_type                          specified;
   SEQUENCE_ISO9541_ID_type           dfnpre;
   SCORE_type                         rscore;
   SCORE_type                         lscore;
   SCORE_type                         tscore;
/* non-ISO-scores is not implemented */
} SCORES_type;
```

```
typedef struct {
   BOOL_type                         specified;
   INT_PAIR_type                     scorpos;
   INT_PAIR_type                     scorthik;
} SCORE_type;

typedef struct {
   BOOL_type                         specified;
   SEQUENCE_ISO9541_ID_type          dfnpre;
   VARIANT_SCRIPT_type               lscript;
   VARIANT_SCRIPT_type               rscript;
/* non-ISO-scripts is not unimplemented */
} VARIANT_SCRIPTS_type;

typedef struct {
   BOOL_type                         specified;
   INT_PAIR_type                     vxoffset;
   INT_PAIR_type                     vyoffset;
   INT_PAIR_type                     vxscale;
   INT_PAIR_type                     vyscale;
} VARIANT_SCRIPT_type;

typedef struct {
   TAG_type  structured_name_tag;
      /*
         STRUCTURED_NAME_ID_tag,
         STRUCTURED_NAME_SEQ_tag,
         STRUCTURED_NAME_UNSPEC_tag
      */
   union {
      ISO9541_ID_type                id;
      SEQUENCE_ISO9541_ID_type       idseq;
   } structured_name_value;
} STRUCTURED_NAME_type;

/*
   The type ISO9541_ID_type is the element
   type of SEQUENCE_ISO9541_ID_type
*/
typedef struct ISO9541_ID_struct {
   TAG_type  iso9541_id_tag;
   /* ID_INT_tag, ID_STR_tag */
   union {
      INT_type                 integer;
      SEQUENCE_BYTE_type       octet_string;
   } iso9541_id_value;
} ISO9541_ID_type;
```

13.11.1 Specifying FONT_REFERENCE_type Parameters

Due to the large number of parameters and sub-parameters present for the font reference type, this type utilizes a non-standard method for denoting which parameters are specified. For most parameters, the unspecified case is indicated by a unique value as part of the corresponding field's value. Some of these values make use of predefined identifiers, which are defined in section 13.11.2. Several structured parameters indicate the unspecified state with an additional field within the structure.

Tables 13-1 through 13-5 list the ISO parameter name, the corresponding field name and the manner for denoting that the parameter is unspecified. Note that a consistent rule is that all parameters that have real values are represented by `INT_PAIR_type` fields and indicate being unspecified by the value of the `second` subfield being 0[61].

[61]This works because division by 0 is undefined and, hence, illegal.

ISO Name	Unspecified
ISO/FONTNAME	`fontname.structured_name_tag == STRUCTURED_NAME_UNSPEC_tag`
ISO/DIFNTNAM	`difntnam.structured_name_tag == STRUCTURED_NAME_UNSPEC_tag`
ISO/PROPDATA	`propdata == ISO_PROPDATA_unspec`
ISO/TYPEFACE	`typeface.structured_name_tag == STRUCTURED_NAME_UNSPEC_tag`
ISO/FAMILY	`family.structured_name_tag == STRUCTURED_NAME_UNSPEC_tag`
ISO/POSTURE	`posture == ISO_POSTURE_unspec`
ISO/WEIGHT	`weight == ISO_WEIGHT_unspec`
ISO/PROPWDTH	`propwdth == ISO_PROPWDTH_unspec`
ISO/CHARCOMP	*unimplemented*
ISO/NUMCHARS	`numchars == -1`
ISO/DSNSIZE	`dsnsize.second == 0`
ISO/MINSIZE	`minsize.second == 0`
ISO/MAXSIZE	`maxsize.second == 0`
ISO/DSCLASS	`dsclass == NULL_SEQUENCE`
ISO/STRUCTUR	`ISO_STRUCTUR_unspec`
ISO/MNFEATSZ	`mnfeatsz.second == 0`
ISO/MODES	`modes.specified == BOOL_false`
ISO/MODE	`mode.structured_name_tag == STRUCTURED_NAME_UNSPEC_tag`

Table 13-1. FONT_REFERENCE_type Parameters

ISO Name	Unspecified
ISO/0	`iso_0.specified == BOOL_false`
ISO/180	`iso_180.specified == BOOL_false`
ISO/270	`iso_270.specified == BOOL_false`

Table 13-2. WRITING_MODES_type Parameters

ISO Name	Unspecified
ISO/RSCORE	`rscore.specified == BOOL_false`
ISO/LSCORE	`lscore.specified == BOOL_false`
ISO/TSCORE	`tscore.specified == BOOL_false`
ISO/SCORPOS	`scorpos.second == 0`
ISO/SCORTHIK	`scorthik.second == 0`

Table 13-3. SCORES_type and SCORE_type Parameters

ISO Name	Unspecified
ISO/LSCRPT	`lscrpt.specified == BOOL_false`
ISO/RCSRPT	`rscrpt.specified == BOOL_false`
ISO/VXOFFSET	`vxoffset.second == 0`
ISO/VYOFFSET	`vyoffset.second == 0`
ISO/VXSCALE	`vxscale.second == 0`
ISO/VYSCALE	`vyscale.second == 0`

Table 13-4. VARIANT_SCRIPTS_type and VARIANT_SCRIPT_type Parameters

ISO Name	Unspecified
ISO/ESCAPDIR	`escapdir.second == 0`
ISO/ESCLASS	`esclass == ISO_ESCLASS_unspec`
ISO/AVGESC	`avgesc.second == 0`
ISO/AVGLCESC	`avglcesc.second == 0`
ISO/AVGCPESC	`avgcpesc.second == 0`
ISO/AVGWTESC	`avgwtesc.second == 0`
ISO/TABESC	`tabesc.specified == BOOL_false`
ISO/TX	`tabesc.tx.second == 0`
ISO/TY	`tabesc.ty.second == 0`
ISO/MFX	`mfx.second == 0`
ISO/MLX	`mlx.second == 0`
ISO/MBX	`mbx.second == 0`
ISO/MRX	`mrx.second == 0`
ISO/CLPESCAD	*unimplemented*
ISO/MINAMPL	`minampl.second == 0`
ISO/MAXAMPL	`maxampl.second == 0`
ISO/TRACKS	*unimplemented*
ISO/TRACK	*unimplemented*
ISO/SCORES	`scores.specified == BOOL_false`
ISO/VARSCRIPT	`varscrpt.specified == BOOL_false`
ISO/MINL2LSP	`minl2lsp.second == 0`
ISO/MINANASC	`minanasc.second == 0`
ISO/MAXANASC	`maxanasc.second == 0`
ISO/MODCHATT	*unimplemented*

Table 13-5. MODAL_FONT_ATTR_type Parameters

13.11.2 Predefined Identifiers for Fonts

The following identifiers are predefined by the tool kit for use in dealing with fonts.

13.11.2.1 Sub-parameter Values

ISO/PROPDATA

`ISO_PROPDATA_unspec`	parameter is unspecified
`ISO_PROPDATA_no`	this font does not include proprietary data
`ISO_PROPDATA_yes`	this font includes proprietary data

ISO/POSTURE

`ISO_POSTURE_unspec`	parameter is unspecified
`ISO_POSTURE_upright`	upright
`ISO_POSTURE_clockwise`	obliquely slanted clockwise, with no other design adjustments
`ISO_POSTURE_counter`	obliquely slanted counter-clockwise, with no other design adjustments
`ISO_POSTURE_italic`	slanted clockwise and the design then adjusted for better appearance
`ISO_POSTURE_other`	other

ISO/WEIGHT

`ISO_WEIGHT_unspec`	parameter is unspecified
`ISO_WEIGHT_ultra_light`	ultra light
`ISO_WEIGHT_extra_light`	extra light
`ISO_WEIGHT_light`	light
`ISO_WEIGHT_semi_light`	semi light
`ISO_WEIGHT_medium`	medium
`ISO_WEIGHT_semi_bold`	semi bold
`ISO_WEIGHT_bold`	bold
`ISO_WEIGHT_extra_bold`	extra bold
`ISO_WEIGHT_ultra_bold`	ultra bold

ISO/PROPWDTH

`ISO_PROPWDTH_unspec`	parameter is unspecified
`ISO_PROPWDTH_ultra_condensed`	ultra condensed
`ISO_PROPWDTH_extra_condensed`	extra condensed
`ISO_PROPWDTH_condensed`	condensed
`ISO_PROPWDTH_semi_condensed`	semi condensed
`ISO_PROPWDTH_medium`	medium
`ISO_PROPWDTH_semi_expanded`	semi expanded
`ISO_PROPWDTH_expanded`	expanded
`ISO_PROPWDTH_extra_expanded`	extra expanded
`ISO_PROPWDTH_ultra_expanded`	ultra expanded

ISO/STRUCTUR

`ISO_STRUCTUR_unspec`	parameter is unspecified
`ISO_STRUCTUR_solid`	solid
`ISO_STRUCTUR_outline`	outline
`ISO_STRUCTUR_inline`	inline
`ISO_STRUCTUR_shadow`	shadow
`ISO_STRUCTUR_patterned`	patterned

ISO/ESCLASS

`ISO_ESCLASS_unspec`	parameter is unspecified
`ISO_ESCLASS_fixed`	fixed pitch
`ISO_ESCLASS_variable`	variable pitch

13.11.2.2 ISO/DSCLASS Values

The identifiers specified in this section correspond to names in appendix A of part 5 of ISO DIS 9541.

- `DSCLASS_ERROR`: used in `Ints2Dsclass` (section 13.12.3).
- `DSCLASS_NOMATCH`: used in `Ints2Dsclass` (section 13.12.3).
- 1 `DSCLASS_UNCIALS`: UNCIALS
  - 1.1 `DSCLASS_UNCIALS_1`: Single Alphabet
    - 1.1.1 `DSCLASS_UNCIALS_1_sans`: sans serif *(SIMPLEX)*

- 1.1.2 `DSCLASS_UNCIALS_1_serif`: serif *(LIBRA)*
- 1.2 `DSCLASS_UNCIALS_2`: Duplex Alphabet
  - 1.2.1 `DSCLASS_UNCIALS_2_sans`: sans serif
  - 1.2.2 `DSCLASS_UNCIALS_2_serif`: serif *(AMERICAN UNCIAL)*
- 2 `DSCLASS_INSCR`: INSCRIPTIONALS
  - 2.1 `DSCLASS_INSCR_solids`: Solids
    - 2.1.1 `DSCLASS_INSCR_solids_sans`: sans serif
    - 2.1.2 `DSCLASS_INSCR_solids_serif`: serif *(SISTINA)*
  - 2.2 `DSCLASS_INSCR_inlines`: Inlines
    - 2.2.1 `DSCLASS_INSCR_inlines_sans`: sans serif
    - 2.2.2 `DSCLASS_INSCR_inlines_serif`: serif *(AUGUSTEA INLINE)*
  - 2.3 `DSCLASS_INSCR_outlines`: Outlines
    - 2.3.1 `DSCLASS_INSCR_outlines_sans`: sans serif
    - 2.3.2 `DSCLASS_INSCR_outlines_serif`: serif *(COLUMNA)*
- 3 `DSCLASS_BLACK`: BLACKLETTERS
  - 3.1 `DSCLASS_BLACK_formal`: Formal Style
    - 3.1.1 `DSCLASS_BLACK_formal_sans`: sans serif *(ITC HONDA)*
    - 3.1.2 `DSCLASS_BLACK_formal_serif`: serif *(GOUDY TEXT)*
    - 3.1.3 `DSCLASS_BLACK_formal_sans_eng`: sans serif engraved
    - 3.1.4 `DSCLASS_BLACK_formal_serif_eng`: serif engraved *(DEUTSCHE ZEIRSCHRIFT)*
  - 3.2 `DSCLASS_BLACK_round`: Round Style
    - 3.2.1 `DSCLASS_BLACK_round_sans`: sans serif
    - 3.2.2 `DSCLASS_BLACK_round_serif`: serif *(GOUDY THIRTY)*
    - 3.2.3 `DSCLASS_BLACK_round_sans_eng`: sans serif engraved
    - 3.2.4 `DSCLASS_BLACK_round_serif_eng`: serif engraved

- 3.3 `DSCLASS_BLACK_hybrid`: Hybrid Style
  - 3.3.1 `DSCLASS_BLACK_hybrid_sans`: sans serif
  - 3.3.2 `DSCLASS_BLACK_hybrid_serif`: serif *(WALLAU)*
  - 3.3.3 `DSCLASS_BLACK_hybrid_sans_eng`: sans serif engraved
  - 3.3.4 `DSCLASS_BLACK_hybrid_serif_eng`: serif engraved
- 3.4 `DSCLASS_BLACK_inform`: Informal Style
  - 3.4.1 `DSCLASS_BLACK_inform_sans`: sans serif
  - 3.4.2 `DSCLASS_BLACK_inform_serif`: serif *(RHAPSODIE)*
  - 3.4.3 `DSCLASS_BLACK_inform_sans_eng`: sans serif engraved
  - 3.4.4 `DSCLASS_BLACK_inform_serif_eng`: serif engraved

- 4 `DSCLASS_SERIFS`: SERIFS
  - 4.1 `DSCLASS_SERIFS_oldstyle`: Oldstyle
    - 4.1.1 `DSCLASS_SERIFS_oldstyle_vene`: Venetian *(EUSEBIUS)*
    - 4.1.2 `DSCLASS_SERIFS_oldstyle_garalde`: Garalde *(GARAMOND)*
    - 4.1.3 `DSCLASS_SERIFS_oldstyle_dutch`: Dutch/English *(JANSON CASLON)*
  - 4.2 `DSCLASS_SERIFS_trans`: Transitional
    - 4.2.1 `DSCLASS_SERIFS_trans_direct`: Direct Line *(BASKERVILLE)*
    - 4.2.2 `DSCLASS_SERIFS_trans_modif`: Modified *(PALATINO / GOUDY OLDSTYLE)*
  - 4.3 `DSCLASS_SERIFS_modern`: Modern
    - 4.3.1 `DSCLASS_SERIFS_modern_italian`: Italian *(BODONI)*
    - 4.3.2 `DSCLASS_SERIFS_modern_fat`: Fat Face *(THOROWGOOD ROMAN)*
  - 4.4 `DSCLASS_SERIFS_cont`: Contemporary

- 4.4.1 `DSCLASS_SERIFS_cont_eclectic`: Eclectic *(PERPETUA DEROOS)*
- 4.4.2 `DSCLASS_SERIFS_cont_fine`: Fine serif *(QUOROM ELAN)*
- 4.4.3 `DSCLASS_SERIFS_cont_lettering`: Lettering *(UNIVERSITY)*

• 4.5 `DSCLASS_SERIFS_legi`: Legibility

- 4.5.1 `DSCLASS_SERIFS_legi_rounded`: Rounded *(TIMES ROMAN)*
- 4.5.2 `DSCLASS_SERIFS_legi_square`: Square *(MELIOR)*

• 4.6 `DSCLASS_SERIFS_square`: Square Serif

- 4.6.1 `DSCLASS_SERIFS_square_monotone`: Monotone *(STYMIE)*
- 4.6.2 `DSCLASS_SERIFS_square_clarendon`: Clarendon *(FORTUNE)*
- 4.6.3 `DSCLASS_SERIFS_square_french`: French Clarendon *(P T BARNUM)*
- 4.6.4 `DSCLASS_SERIFS_square_short`: Short *(CHELTENHAM)*
- 4.6.5 `DSCLASS_SERIFS_square_type`: Typewriter *(BULLETIN TYPEWRITER)*
- 4.6.6 `DSCLASS_SERIFS_square_dot`: Dot Matrix

• 4.7 `DSCLASS_SERIFS_latin`: Latin

- 4.7.1 `DSCLASS_SERIFS_latin_solid`: Solid *(WIDE LATIN)*
- 4.7.2 `DSCLASS_SERIFS_latin_incised`: Incised *(CHISEL)*

• 4.8 `DSCLASS_SERIFS_engraving`: Engraving

- 4.8.1 `DSCLASS_SERIFS_engraving_barbed`: Barbed Serif *(NICHOLAS COCHIN)*
- 4.8.2 `DSCLASS_SERIFS_engraving_strait`: Straight Serif *(COPPERPLATE GOTHIC)*

• 4.9 `DSCLASS_SERIFS_art`: Art Nouveau

- 4.9.1 `DSCLASS_SERIFS_art_french`: French *(AURIOL)*
- 4.10 `DSCLASS_SERIFS_computer`: Computer
  - 4.10.1 `DSCLASS_SERIFS_computer_ocr`: OCR
  - 4.10.2 `DSCLASS_SERIFS_computer_dig`: Digital
- 4.11 `DSCLASS_SERIFS_misc`: Miscellaneous *(CALSON ANTIQUE)*

- 5 `DSCLASS_SANS`: SANS SERIFS
  - 5.1 `DSCLASS_SANS_gothic`: Gothic
    - 5.1.1 `DSCLASS_SANS_gothic_grot`: Grotesque *(FRANKLIN GOTHIC)*
    - 5.1.2 `DSCLASS_SANS_gothic_neo`: Neo-grotesque *(HELVETICA)*
    - 5.1.3 `DSCLASS_SANS_gothic_type`: Typewriter
- 6 `DSCLASS_SCRIPTS`: SCRIPTS
- 7 `DSCLASS_ORNAM`: ORNAMENTALS

13.12 Miscellaneous Type Routines

This section contains the specifications of a number of routines for manipulating objects of the types defined so far in this chapter.

13.12.1 Make a Font Have Unspecified Parameters: *UnspecifyFont*

```
INT_type UnspecifyFont(font)
   FONT_REFERENCE_type *font;
```

Real Name

unsfnt

Synopsis

This routine is used to fill in the fields of a `FONT_REFERENCE_type` object so that all parameters are unspecified.

Parameters

`font`	The address of a `FONT_REFERENCE_type` object.

Return Value

`ERROR_INT`	An error has occurred. `TKError` will contain an indication of the exact error (see the **Errors** part below).
otherwise	The font's fields have been filled in.

Side Effects

The object referenced by `font` has been modified.

Semantics

The current values of fields will be ignored.

The object will contain values such that all fields and sub-fields contain values indicating the unspecified state. For example, not only will the field `font->iso_0.specified` be equal to `BOOL_false`, but all of the fields of the `iso_0` field will be marked as unspecified.

Errors

`BADPARM_err`	One of the parameters is invalid.

Implementation Status

Implemented.

13.12.2 Get Integers Corresponding to an ISO/DSCLASS: *Dsclass2Ints*

```
SEQUENCE_INT_type Dsclass2Ints(value)
   VALUE_type value;
```

Real Name

ds2int

Synopsis

This routine obtains the sequence of integers corresponding to a predefined set of values for the ISO/DSCLASS parameter.

Parameters

`value`	One of the predefined values described in section 13.11.2.2.

Return Value

`ERROR_SEQUENCE`
: An error has occurred. `TKError` will contain an indication of the exact error (see the **Errors** part below).

otherwise
: The `SEQUENCE_INT_type` object corresponding to the specified value.

Side Effects

None.

Semantics

The value of the `dsclass` field will be a sequence of integers as defined in Appendix A of ISO DIS 9541.

The sequence of integers has been created with a call to `MakeSequence` and the application should deallocate it (via `DeleteSequence`) when it is no longer needed.

Errors

`BADPARM_err`	One of the parameters is invalid.
`NOMEM_err`	No more memory was available for dynamic allocation.
`INTERNAL_err`	An error internal to the structure or operation of the tool kits has occurred.

Implementation Status

All of the font values listed in section 13.11.2.2 are implemented by this routine. The listing in section 13.11.2.2 does not include all design classes defined in Appendix A of ISO DIS 9541.

13.12.3 Get ISO/DSCLASS Corresponding to Integers: *Ints2Dsclass*

```
VALUE_type Ints2Dsclass(ints)
   SEQUENCE_INT_type ints;
```

Real Name

int2ds

Synopsis

This routine obtains the value for the ISO/DSCLASS parameter corresponding to a sequence of integers.

Parameters

`ints` A sequence of integers designating a value of the ISO/DSCLASS parameter.

Return Value

`DSCLASS_ERROR` An error has occurred. `TKError` will contain an indication of the exact error (see the **Errors** part below).

`DSCLASS_NOMATCH`
The sequence of integers does not match any predefined ISO/DSCLASS value defined by the tool kit.

otherwise The `VALUE_type` (from section 13.11.2.2) corresponding to the specified value.

Side Effects

None.

Errors

`BADPARM_err` One of the parameters is invalid.

`INTERNAL_err` An error internal to the structure or operation of the tool kits has occurred.

Implementation Status

All of the font values listed in section 13.11.2.2 are implemented by this routine. The listing in section 13.11.2.2 does not include all design classes defined in Appendix A of ISO DIS 9541.

13.12.4 Fill Structured Name from Strings: *Array2StrctName*

```
INT_type Array2StrctName(NameArray, NumNames,
                         StructuredName)
   CHAR_type **NameArray;
   INT_type NumNames;
   STRUCTURED_NAME_type *StructuredName;
```

Real Name

ary2sn

Synopsis

This routine assigns the name specified in the name array to the structure defining a structured name.

Parameters

`NameArray`	A pointer to an array of C strings, i.e., null-terminated strings.
`NumNames`	Number of strings in `NameArray`.
`StructuredName`	A pointer to the structure to be filled in based on the `NameArray`.

Return Value

`ERROR_INT`	An error has occurred. `TKError` will contain an indication of the exact error (see the **Errors** part below).
otherwise	The structure specified by `StructuredName` has been filled with the names given in the name array.

Side Effects

None.

Semantics

The current values of fields in the structure referred to by `StructuredName` will be ignored.

The array must contain between 2 and 72 strings. Each string must contain between 1 and 72 characters. Only alphanumerics and the character '-' are permitted.

Errors

`BADPARM_err`	One of the parameters is invalid.
`INTERNAL_err`	An error internal to the structure or operation of the tool kits has occurred.

Implementation Status

Implemented.

13.12.5 Fill in a structured name from a string: *String2StrctName*

```
INT_type String2StrctName(NameString,
                          StructuredName)
   CHAR_type *NameString;
   STRUCTURED_NAME_type *StructuredName;
```

Real Name

str2sn

Synopsis

This routine assigns the name specified in the name string to the structure defining a structured name.

Parameters

`NameString` A pointer to a null-terminated C string that specifies the identifiers to be placed in the structured name.

`StructuredName`
A pointer to the structure to be filled in with a structured name based on the `NameString`.

Return Value

`ERROR_INT` An error has occurred. `TKError` will contain an indication of the exact error (see the **Errors** part below).

otherwise The structure specified by `StructuredName` has been filled with the identifiers given in the name string.

Side Effects

None.

Semantics

The current values of fields in the structured referred to by `StructuredName` will be ignored.

The generated name consists of a sequence of identifiers.

The string consists of identifiers separated by the character "/". The string must contain between 2 and 72 identifiers. Each identifier must contain between 1 and 72 characters. Only alphanumerics and the character '-' are permitted in identifiers.

Errors

`BADPARM_err` One of the parameters is invalid.

`INTERNAL_err` An error internal to the structure or operation of the tool kits has occurred.

Implementation Status

Implemented.

13.12.6 Fill in a Font Name from an Array: *SetFontFamilyArray*

```
INT_type SetFontFamilyArray(FontReference,
                            FamilyArray,
                            NumNames)
  FONT_REFERENCE_type *FontReference;
  CHAR_type **FamilyArray;
  INT_type NumNames;
```

Real Name

stffar

Synopsis

This routine assigns the name specified in the array of strings to the family field for the specified font reference.

Parameters

`FontReference`	A pointer to a font reference structure that should have its family component set.
`NameArray`	A pointer to an array of C strings that specifies the identifiers to be placed in the structured name for the family.
`NumNames`	The number of strings in the `NameArray` array.

Return Value

`ERROR_INT`	An error has occurred. `TKError` will contain an indication of the exact error (see the **Errors** part below).
otherwise	The family component of the structure specified by `FontReference` is assigned the value of a structured name specified by the identifiers given in the name array.

Side Effects

None.

Semantics

The current values of the family field in the structure referred to by `FontReference` will be ignored.

The generated family name consists of a sequence of identifiers.

The array must contain between 2 and 72 strings. Each string in the array specifies one identifier. Each identifier must contain between 1 and 72 characters. Only alphanumerics and the character '-' are permitted in identifiers.

Errors

`BADPARM_err`	One of the parameters is invalid.
`INTERNAL_err`	An error internal to the structure or operation of the tool kits has occurred.

Implementation Status

Implemented.

13.12.7 Fill in a Font Family Name from a String: *SetFontFamily*

```
INT_type SetFontFamily(FontReference,
                       NameString)
   FONT_REFERENCE_type *FontReference;
   CHAR_type *NameString;
```

Real Name

stffst

Synopsis

This routine assigns the name specified in the name string to the family component of the passed font reference.

Parameters

`FontReference`	A pointer to a font reference structure that should have its family component set.
`NameString`	A pointer to a null-terminated C string that specifies the identifiers to be placed in the structured name for the font's family.

Return Value

`ERROR_INT`	An error has occurred. `TKError` will contain an indication of the exact error (see the **Errors** part below).
otherwise	The family component of the structure specified by `FontReference` has been filled with the identifiers given in the name string.

Side Effects

None.

Semantics

The current values for family in the structure referred to by `FontReference` will be ignored.

The generated name consists of a sequence of identifiers.

The string consists of names separated by the character "/". The string must contain between 2 and 72 names. Each name must contain between 1 and 72 characters. Only alphanumerics and the character '-' are permitted in names.

Errors

`BADPARM_err`	One of the parameters is invalid.
`INTERNAL_err`	An error internal to the structure or operation of the tool kits has occurred.

Implementation Status

Implemented.

13.12.8 Look Up an ASN.1 Object Identifier: *LookUpObjectId*

```
VALUE_type LookUpObjectId(id)
   ASN1_OBJECT_ID_type id;
```

Real Name

lkobid

Synopsis

Determine if an `ASN1_OBJECT_ID_type` object is the same object identifier as one of the predefined ASN.1 object identifiers, given below, and return a value indicating which.

Parameters

`id`	The `ASN1_OBJECT_ID_type` object whose value the application is interested in.

Return Value

`ERROR_id`	An error has occurred. `TKError` will contain an indication of the exact error (see the **Errors** part below).
otherwise	A `VALUE_type` indicating the predefined object identifier matched (see **Semantics** below for list of values).

Side Effects

None.

Semantics

The returned value will be one of the following:

Identifier	Meaning
`FORM_CC_id`	'formatted character content architecture'
`PROC_CC_id`	'processable character content architecture'
`FP_CC_id`	'formatted processable character content architecture'
`FORM_RG_id`	'formatted raster graphics content architecture'
`FP_RG_id`	'formatted processable raster graphics content architecture'
`FORM_GG_id`	'formatted geometric graphics content architecture'
`CC_CODING_id`	*character content encoding*
`RG_T6_CODING_id`	*raster graphics* 'T6 encoding'
`RG_T4_1_CODING_id`	*raster graphics* 'T4 one dimensional encoding'
`RG_T4_2_CODING_id`	*raster graphics* 'T4 two dimensional encoding'
`RG_BITMAP_CODING_id`	*raster graphics* 'bitmap encoding'
`GG_CODING_id`	*geometric graphics encoding*
`UNKNOWN_id`	*unknown object identifier*

Errors

`BADPARM_err`	One of the parameters is invalid.
`INTERNAL_err`	An error internal to the structure or operation of the tool kits has occurred.

Implementation Status

Implemented.

13.12.9 Unspecify Default Value Lists: *UnspecifyDefaultValueLists*

```
INT_type UnspecifyDefaultValueLists(dvl)
   at_DEFAULT_VALUE_LISTS_type *dvl;
```

Real Name

unsdvl

Synopsis

This routine is used to initialize an `at_DEFAULT_VALUE_LISTS_type` so that no default values are specified.

Parameters

`dvl`	`dvl` should contain the address of an `at_DEFAULT_VALUE_LISTS_type` object.

Return Value

`ERROR_INT`	An error has occurred. `TKError` will contain an indication of the exact error (see the **Errors** part below).
otherwise	The structure's fields have been filled in with NULL_CONSTITUENT values.

Side Effects

The object referenced by `dvl` is modified.

Semantics

The current values of fields will be ignored.

The object will contain values such that no defaults are specified for any kind of constituent.

Errors

`BADPARM_err`	One of the parameters is invalid.

Implementation Status

Implemented.

13.12.10 Initialize Default Value Lists: *InitializeDefaultValueLists*

```
INT_type
InitializeDefaultValueLists(doc, dvl,
                            constituent_mask)
   DOCUMENT_type doc;
   at_DEFAULT_VALUE_LISTS_type *dvl;
   PARM_MASK_type constituent_mask;
```

Real Name

intdvl

Synopsis

Fills in the fields of an at_DEFAULT_VALUE_LISTS_type object with a NULL_CONSTITUENT or an appropriate default constituent value for the specified field.

Parameters

doc	The document in which new constituents should be created.
dvl	dvl should contain the address of an at_DEFAULT_VALUE_LISTS_type object.
constituent_mask	A parameter mask value indicating which fields should be initialized with new default constituents. The mask values are specified below. NO_PARMS and ALL_PARMS are also legitimate values.

Subroutine Parameters

Parameter	Field Name	Parameter Name
"page attributes"	page_attributes	PAGE_ATTRIBUTES_parm
"frame attributes"	frame_attributes	FRAME_ATTRIBUTES_parm
"block attributes"	block_attributes	BLOCK_ATTRIBUTES_parm
"composite logical attributes"	composite_logical_attributes	COMP_LOGICAL_ATTRIBUTES_parm
"basic logical attributes"	basic_logical_attributes	BAS_LOGICAL_ATTRIBUTES_parm

Return Value

ERROR_INT	An error has occurred. TKError will contain an indication of the exact error (see the **Errors** part below).
otherwise	The structure's fields have been filled in with either NULL_CONSTITUENT values or appropriate new, default constituent values.

Side Effects

The object referenced by dvl is modified.

Semantics

The current values of fields will be ignored.

For each specified parameter, a new constituent will be created and assigned to that field in the structure. All other fields will be set to NULL_CONSTITUENT. If NO_PARMS is passed as the parameter_mask, then all fields will have the value NULL_CONSTITUENT. If ALL_PARMS is passed as the parameter_mask, then all fields will have new default constituents assigned.

Errors

BADPARM_err One of the parameters is invalid.

Implementation Status

Implemented.

13.13 Sequences

There are many instances within the tool kit where it is necessary to represent an unbounded sequence of objects of a particular type. For example, the value of the attribute "permitted categories" is a sequence of strings, each representing a layout category (section 17.1.2.7). To provide a consistent interface for manipulating such sequences, the Tool Kit defines a single type, SEQUENCE_type, and a set of routines. This type is defined so as to be able to represent an unbounded sequence of objects of a specified set of types. The type SEQUENCE_type is defined in the next three sections, 13.13.1 through 13.13.3, and the routines for manipulating sequence objects are defined in sections 13.13.4 through 13.13.9.

13.13.1 The Type SEQUENCE_type

```
typedef struct SEQUENCE_struct {

    POINTER_type                    private_info;
    TAG_type                        sequence_tag;
    INT_type                        length;
```

```
    union {
        struct SEQUENCE_struct          **asn1_object_ids;
        struct ATOMIC_STRING_EXPR_struct
                                     *atomic_string_exprs;
        struct BINDING_PAIR_struct      *binding_pairs;
        BYTE_type                       *bytes;
        CHAR_type                       *chars;
        CONSTITUENT                     *constituents;
        CONSTRUCTION_TERM_type        *construction_terms;
        struct COPYRIGHT_INFO_struct    *copyright_infos;
        struct FONT_PAIR_struct         *font_pairs;
        INT_type                        *ints;
        INT_PAIR_type                   *int_pairs;
        struct ISO9541_ID_struct        *iso9541_ids;
        at_MEDIUM_TYPE_type             *medium_types;
        struct NAME_ORG_struct          *name_orgs;
        struct PERSONAL_NAME_struct     *personal_names;
        struct RESOURCE_MAP_struct      *resource_maps;
        struct REVISION_HISTORY_struct
                                     *revision_histories;
        struct REVISER_struct           *revisers;
        struct SEQUENCE_struct          **strings;
        struct STRING_OR_ID_struct      *string_or_ids;
        struct TAB_STOP_struct          *tab_stops;
        VALUE_type                      *values;
    } sequence_value;

} *SEQUENCE_type;
```

The tag field sequence_tag is a read-only field that is set by the tool kit routines. Its value specifies which of the possible union fields is present in this SEQUENCE_type object. Table 13-6 lists the possible tag values and the corresponding union fields. The programmer is warned to avoid assigning a value to the sequence_tag field. This field is set during the call to MakeSequence and must not change during the lifetime of the resultant SEQUENCE_type object.

The field length, which is set by the tool kit routines, is a read-only field that specifies how many sequence elements there are room for in this SEQUENCE_type object. This field does not specify how many of these elements have actually been given a value by the programmer. It is the programmer's responsibility to keep track of the number of valid values in a sequence. The programmer is warned to avoid assigning a value to the length field; this field may be manipulated indirectly by use of the routines ExpandSequence (section 13.13.7) and ContractSequence (section 13.13.8).

The field private_info is used to store a pointer to information that is necessary for the tool kit in manipulating SEQUENCE_type objects[62]. The application programmer must not modify the value of this field.

---

[62]Currently, the data referenced by the private_info field is used to allow the value of a sequence to be moved between primary and secondary memory (section 21.1).

The appropriate sub-field of the union field sequence_value can be determined by inspecting the value of the sequence_tag field (as described in table 13-6). So, if the value of the sequence_tag field is SEQUENCE_INT_PAIR_tag, then the field sequence_value.int_pairs is present and the sequence's elements are of type INT_PAIR_type.

Each sub-field is to be treated as a pointer to an array of contiguous objects of the corresponding type. For example, the value of the field sequence_value.int_pairs is a pointer to a contiguous array of objects of type INT_PAIR_type. The number of such objects can be determined by the value of the field length. Like all arrays in C, the indices run between 0 and length-1.

The programmer is free to assign and modify the contents of the array pointed to by the appropriate union field within the constraints imposed by the maximum length of the sequence (specified by the field length). The programmer is warned to avoid assigning any value to the pointer field (the field asn1_object_ids, bytes, ... as appropriate), however, as this field is manipulated by the tool kit routines. In particular, the programmer should be aware that the value of the union sub-field may change during calls to ExpandSequence (section 13.13.7) and ContractSequence (section 13.13.8).

13.13.2 Additional Sequence Type Names

To enhance readability there is, for each possible sequence element type, a type defined for a sequence of that element type. The application programmer should be aware that each of these types is actually identical to the type SEQUENCE_type. (These additional sequence types are defined in table 13-6 on page 334.)

```
typedef SEQUENCE_type SEQUENCE_ASN1_OBJECT_ID_type;
typedef SEQUENCE_type SEQUENCE_ATOMIC_STR_EXPR_type;
typedef SEQUENCE_type SEQUENCE_BINDING_PAIR_type;
typedef SEQUENCE_type SEQUENCE_BYTE_type;
typedef SEQUENCE_type SEQUENCE_CHAR_type;
typedef SEQUENCE_type SEQUENCE_CONSTITUENT_type;
typedef SEQUENCE_type SEQUENCE_CONSTRUCTION_TERM_type;
typedef SEQUENCE_type SEQUENCE_COPYRIGHT_INFO_type;
typedef SEQUENCE_type SEQUENCE_FONT_PAIR_type;
typedef SEQUENCE_type SEQUENCE_INT_type;
typedef SEQUENCE_type SEQUENCE_INT_PAIR_type;
typedef SEQUENCE_type SEQUENCE_ISO9541_ID_type;
```

```
typedef SEQUENCE_type SEQUENCE_at_MEDIUM_TYPE_type;
typedef SEQUENCE_type SEQUENCE_NAME_ORG_type;
typedef SEQUENCE_type SEQUENCE_PERSONAL_NAME_type;
typedef SEQUENCE_type SEQUENCE_RESOURCE_MAP_type;
typedef SEQUENCE_type SEQUENCE_REVISION_HISTORY_type;
typedef SEQUENCE_type SEQUENCE_REVISER_type;
typedef SEQUENCE_type SEQUENCE_STRING_type;
typedef SEQUENCE_type SEQUENCE_STRING_OR_ID_type;
typedef SEQUENCE_type SEQUENCE_TAB_STOP_type;
typedef SEQUENCE_type SEQUENCE_VALUE_type;
```

13.13.3 Predefined SEQUENCE_type Identifiers

The identifiers NULL_SEQUENCE and ERROR_SEQUENCE are each values of type SEQUENCE_type that are guaranteed to be unique, that is, not equal to each other or to any other SEQUENCE_type objects that may be constructed. Equality between a SEQUENCE_type object and either of these values may be tested using the C equality operator ("==").

The value ERROR_SEQUENCE is used as a routine return value to indicate an error. The value NULL_SEQUENCE, which is not the same as an empty sequence, is used to denote the absence of a sequence where appropriate.

13.13.4 Create a Sequence Object: *MakeSequence*

```
SEQUENCE_type MakeSequence(type, length)
   TAG_type type;
   INT_type length;
```

Real Name

mkseq

Synopsis

This routine is used to create a sequence of a specified initial length.

Parameters

type	A value indicating the type of objects that will make up the sequence. The value must be one of the tag values listed in table 13-6.
length	The initial length of the sequence. This value must be non-negative.

Type Name	Tag	Element Type
SEQUENCE_ASN1_OBJECT_ID_type	SEQUENCE_ASN1_OBJECT_ID_tag	ASN1_OBJECT_ID_type
SEQUENCE_ATOMIC_STR_EXPR_type	SEQUENCE_ATOMIC_STR_EXPR_tag	ATOMIC_STRING_EXPR_type
SEQUENCE_BINDING_PAIR_type	SEQUENCE_BINDING_PAIR_tag	BINDING_PAIR_type
SEQUENCE_BYTE_type	SEQUENCE_BYTE_tag	BYTE_type
SEQUENCE_CHAR_type	SEQUENCE_CHAR_tag	CHAR_type
SEQUENCE_CONSTITUENT_type	SEQUENCE_CONSTITUENT_tag	CONSTITUENT
SEQUENCE_CONSTRUCTION_TERM_type	SEQUENCE_CONSTRUCTION_TERM_tag	CONSTRUCTION_TERM_type
SEQUENCE_COPYRIGHT_INFO_type	SEQUENCE_COPYRIGHT_INFO_tag	COPYRIGHT_INFO_type
SEQUENCE_FONT_PAIR_type	SEQUENCE_FONT_PAIR_tag	FONT_PAIR_type
SEQUENCE_INT_type	SEQUENCE_INT_tag	INT_type
SEQUENCE_INT_PAIR_type	SEQUENCE_INT_PAIR_tag	INT_PAIR_type
SEQUENCE_ISO9541_ID_type	SEQUENCE_ISO9541_ID_tag	ISO9541_ID_type
SEQUENCE_at_MEDIUM_TYPE_type	SEQUENCE_at_MEDIUM_TYPE_tag	at_MEDIUM_TYPE_type
SEQUENCE_NAME_ORG_type	SEQUENCE_NAME_ORG_tag	NAME_ORG_type
SEQUENCE_PERSONAL_NAME_type	SEQUENCE_PERSONAL_NAME_tag	PERSONAL_NAME_type
SEQUENCE_RESOURCE_MAP_type	SEQUENCE_RESOURCE_MAP_tag	RESOURCE_MAP_type
SEQUENCE_REVISION_HISTORY_type	SEQUENCE_REVISION_HISTORY_tag	REVISION_HISTORY_type
SEQUENCE_REVISER_type	SEQUENCE_REVISER_tag	REVISER_type
SEQUENCE_STRING_type	SEQUENCE_STRING_tag	STRING_type
SEQUENCE_STRING_OR_ID_type	SEQUENCE_STRING_OR_ID_tag	STRING_OR_ID_type
SEQUENCE_TAB_STOP_type	SEQUENCE_TAB_STOP_tag	TAB_STOP_type
SEQUENCE_VALUE_type	SEQUENCE_VALUE_tag	VALUE_type

Table 13-6. Sequence Types

Return Value

ERROR_SEQUENCE	An error has occurred. `TKError` will contain an indication of the exact error (see the **Errors** part below).
otherwise	The `SEQUENCE_type` object.

Semantics

This routine will create a `SEQUENCE_type` object of the specified length. This length can be changed by a call to the `ExpandSequence` or `ContractSequence` routines (sections 13.13.7 and 13.13.8).

Errors

`BADPARM_err`	The parameter `type` or `length` is invalid.
`INTERNAL_err`	An error internal to the structure or operation of the tool kits has occurred.
`NOMEM_err`	No more memory was available for dynamic allocation.

Implementation Status

Implemented.

13.13.5 Delete a Sequence Object: *DeleteSequence*

```
INT_type DeleteSequence(sequence)
   SEQUENCE_type sequence;
```

Real Name

delseq

Synopsis

This routine is used to delete a `SEQUENCE_type` object.

Parameters

`sequence`	The `SEQUENCE_type` object to delete. This may not be `NULL_SEQUENCE` or `ERROR_SEQUENCE`.

Return Value

`ERROR_INT`	An error has occurred. `TKError` will contain an indication of the exact error (see the **Errors** part below).
otherwise	The sequence has been deleted.

Semantics

The storage used for the SEQUENCE_type structure and the array of objects pointed to by the sequence_value field is released. Storage associated with objects that are referenced by the sequence is not released (see DeleteSequenceDeeply in section 13.13.6). References to this object are undefined after this call.

Errors

BADPARM_err	The parameter is invalid.
INTERNAL_err	An error internal to the structure or operation of the tool kits has occurred.

Implementation Status

Implemented.

13.13.6 Delete Sequence and Elements: *DeleteSequenceDeeply*

```
INT_type DeleteSequenceDeeply(sequence)
    SEQUENCE_type sequence;
```

Real Name

dlsqd

Synopsis

This routine is like DeleteSequence but will, in most cases, also release storage associated with objects referenced by the sequence.

Parameters

sequence	The SEQUENCE_type object to delete. This may not be NULL_SEQUENCE or ERROR_SEQUENCE.

Return Value

ERROR_INT	An error has occurred. TKError will contain an indication of the exact error (see the **Errors** part below).
otherwise	The storage for the sequence and most associated objects (see **Semantics**) has been deleted.

Side Effects

Storage is released.

Semantics

This routine will release the storage associated with the sequence object `sequence`. In addition, for most sequence types whose objects are dynamically allocated, this routine will also release the storage associated with the objects that comprise the sequence. For example, if the `sequence` object is a sequence of `ASN1_OBJECT_ID` objects, then the storage associated with any of these objects in this sequence will also be released.

The following table specifies the actions of the routine for each type of sequence. Note that this routine is unimplemented for certain types (and will return `UNIMPL_err` in those cases). Also, for objects whose values are not normally dynamically allocated (such as `INT_type` objects), this routine behaves identically to `DeleteSequence`.

Sequence element type	Action on elements
`ASN1_OBJECT_ID`	free storage
`ATOMIC_STRING_EXPR_type`	free storage
`BINDING_PAIR_type`	free storage
`BYTE_type`	*not applicable*
`CHAR_type`	*not applicable*
`CONSTITUENT`	*not applicable*
`CONSTRUCTION_TERM_type`	free storage
`COPYRIGHT_INFO_type`	free storage
`FONT_PAIR_type`	unimplemented
INT_type	*not applicable*
`INT_PAIR_type`	*not applicable*
`ISO9541_ID_type`	unimplemented
`at_MEDIUM_TYPE_type`	*not applicable*
`NAME_ORG_type`	free storage
`PERSONAL_NAME_type`	free storage
`RESOURCE_MAP_type`	free storage
`REVISION_HISTORY_type`	free storage
`REVISER_type`	free storage
`STRING_type`	free storage
`STRING_OR_ID_type`	free storage
`TAB_STOP_type`	free storage
VALUE_type	*not applicable*

Errors

BADPARM_err	The parameter is ERROR_SEQUENCE or a null pointer, or the sequence is badly formed.
UNIMPL_err	The routine is unimplemented for the element type of the sequence.
INTERNAL_err	An error internal to the structure or operation of the tool kits has occurred.

Implementation Status

This routine may not behave properly if a value (other than a constituent) is referred to multiple times within the sequence or its values.

13.13.7 Increase the Length of a Sequence Object: *ExpandSequence*

```
INT_type ExpandSequence(sequence, amount)
    SEQUENCE_type sequence;
    INT_type amount;
```

Real Name

expseq

Synopsis

This routine is used to increase the length of a SEQUENCE_type object.

Parameters

sequence	The SEQUENCE_type object whose length is to change. The value may not be NULL_SEQUENCE or ERROR_SEQUENCE.
amount	The number of additional elements to allow for. This must be non-negative.

Return Value

ERROR_INT	An error has occurred. TKError will contain an indication of the exact error (see the **Errors** part below).
otherwise	The length of the object was changed and the return value is the former length.

Side Effects

The storage allocated for the array of objects may be moved. In other words, the value of the pointer in the sequence_value field may be changed as a result of calling this routine.

Semantics

Upon returning from this routine, the length of the object will be changed as specified. The values of the elements in the original array will be the first elements of the new array. The remainder of the elements in the new array will be uninitialized.

Errors

`BADPARM_err`	The parameter `amount` is negative or the parameter `sequence` is invalid.
`INTERNAL_err`	An error internal to the structure or operation of the tool kits has occurred.
`LOCAL_err`	A locally-defined error, not covered by any other available error, has occurred.
`NOMEM_err`	No more memory was available for dynamic allocation.

Implementation Status

Implemented.

13.13.8 Decrease the Length of a Sequence Object: *ContractSequence*

```
INT_type ContractSequence(sequence, amount)
   SEQUENCE_type sequence;
   INT_type amount;
```

Real Name

conseq

Synopsis

This routine is used to decrease the length of a `SEQUENCE_type` object.

Parameters

`sequence`	The `SEQUENCE_type` object whose length is to change. The value may not be `NULL_SEQUENCE` or `ERROR_SEQUENCE`.
`amount`	The number of elements to remove from the sequence. This must be non-negative.

Return Value

`ERROR_INT`	An error has occurred. `TKError` will contain an indication of the exact error (see the **Errors** part below).
otherwise	The length of the object was changed and the return value is the former length.

Side Effects

The storage allocated for the array of objects may be moved. In other words, the value of the pointer in the `sequence_value` field may be changed as a result of calling this routine.

Semantics

Upon returning from this routine, the length of the object will have been changed as specified. The values of the elements in the new array will be identical to the corresponding elements of the old array.

Errors

`BADPARM_err`	The parameter `amount` is negative, or greater than the length of the sequence, or the parameter `sequence` is invalid.
`INTERNAL_err`	An error internal to the structure or operation of the tool kits has occurred.

Implementation Status

Implemented.

13.13.9 Append Two Sequences: *AppendSequence*

```
INT_type AppendSequence(s1, s2)
   SEQUENCE_type s1;
   SEQUENCE_type s2;
```

Real Name

appseq

Synopsis

This routine is used to append one sequence to the end of another.

Parameters

`s1`	`s2` will be appended to this sequence. The value may not be `NULL_SEQUENCE` or `ERROR_SEQUENCE`.
`s2`	Will be appended to `s1`. The value may not be `NULL_SEQUENCE` or `ERROR_SEQUENCE`.

Return Value

`ERROR_INT`	An error has occurred. `TKError` will contain an indication of the exact error (see the **Errors** part below).
otherwise	The sequence `s1` has the value of `s2` appended to it.

Side Effects

The length and value of `s1` may be modified.

Semantics

The element type of the two sequences must be the same. `s2` will not be modified. `s1` may be expanded.

Errors

`BADPARM_err`	The parameter `s1` or the parameter `s2` is invalid.
`INTERNAL_err`	An error internal to the structure or operation of the tool kits has occurred.
`LOCAL_err`	A locally-defined error, not covered by any other available error, has occurred.
`NOMEM_err`	No more memory was available for dynamic allocation.

Implementation Status

Implemented.

13.13.10 Copy a Sequence: *CopySequence*

```
SEQUENCE_type CopySequence(sequence)
   SEQUENCE_type sequence;
```

Real Name

cpyseq

Synopsis

This routine is used to make a copy of a sequence.

Parameters

`sequence`	The sequence to copy.

Return Value

`ERROR_SEQUENCE`	An error has occurred. `TKError` will contain an indication of the exact error (see the **Errors** part below).
otherwise	The new sequence.

Side Effects

None.

Errors

`BADPARM_err`	One of the parameters is invalid.
`NOMEM_err`	No more memory was available for dynamic allocation.
`INTERNAL_err`	An error internal to the structure or operation of the tool kits has occurred.

Implementation Status

Implemented.

13.14 The STRING_type Type

The ODA standard makes extensive use of unbounded length character strings. The value of many attributes—"user-readable comments", "user-visible name", "alternative representation" and "subject", for example—is a string of characters from some character set. In addition, the tool kit defines routines that deal with arbitrary-length strings of characters.

Character strings are represented as `SEQUENCE_type` objects with `CHAR_type` objects as elements. But character strings are familiar to programmers and it is convenient to provide a specific set of operations for dealing with strings: concatenation, for example. For this reason, there are a set of routines defined strictly for dealing with `STRING_type` objects. In addition, there is a set of predefined identifiers that are specific to the `STRING_type` type. These predefined identifiers are defined in the next section and the routines in the remaining part of this section.

The definition of the type STRING_type is

```
typedef SEQUENCE_type STRING_type;
```

In addition, for a STRING_type object, str, it is guaranteed that

```
str.sequence_tag == SEQUENCE_CHAR_tag
```

Note that while objects and identifiers of type STRING_type may be used as SEQUENCE_type objects, the converse is not true. That is, an arbitrary SEQUENCE_type object may not be passed in place of a STRING_type parameter to a tool kit routine.

13.14.1 Predefined STRING_type Identifiers

There are two predefined STRING_type identifiers: NULL_STRING and ERROR_STRING. Both of these values are unique STRING_type objects that are guaranteed to be different from each other and from any other STRING_type objects that can be created. Equality between a STRING_type object and either of these identifiers may be tested using the C equality operator ("==").

The value NULL_STRING, which is not the same as an empty string, is used to denote the absence of a STRING_type value where appropriate. The value ERROR_STRING is used as a routine return value to indicate an error.

13.14.2 Create a STRING_type Object: *MakeString*

```
STRING_type MakeString(len)
   INT_type len;
```

Real Name

mkstr

Synopsis

This routine is used to create a STRING_type object with an initial value.

Parameters

len	The number of chars in this string (must be non-negative).

Return Value

`ERROR_STRING`	An error has occurred. `TKError` will contain an indication of the exact error (see the **Errors** part below).
otherwise	The return value is an object of type `STRING_type` with the specified length.

Side Effects

None.

Errors

`BADPARM_err`	One of the parameters is invalid.
`INTERNAL_err`	An error internal to the structure or operation of the tool kits has occurred.
`LOCAL_err`	A locally-defined error, not covered by any other available error, has occurred.
`NOMEM_err`	No more memory was available for dynamic allocation.

Implementation Status

Implemented.

13.14.3 Make a STRING_type From a Null-terminated String: *Chars2String*

```
STRING_type Chars2String(str)
   CHAR_type *str;
```

Real Name

chrstr

Synopsis

This routine is used to create a `STRING_type` from a null-terminated C-type character string.

Parameters

`str`	A null-terminated character string. The value of this parameter must not be `(CHAR_type *) 0`.

Return Value

`ERROR_STRING`	An error has occurred. `TKError` will contain an indication of the exact error (see the **Errors** part below).
otherwise	The return value is an object of type `STRING_type` initialized with the specified characters.

Side Effects

None.

Semantics

This routine will return a `STRING_type` object that is initialized to the characters specified in the parameter. Its length will be set to the number of characters in the string. The value `NULL_STRING` will never be returned.

Errors

`BADPARM_err`	One of the parameters is invalid.
`INTERNAL_err`	An error internal to the structure or operation of the tool kits has occurred.
`LOCAL_err`	A locally-defined error, not covered by any other available error, has occurred.
`NOMEM_err`	No more memory was available for dynamic allocation.

Implementation Status

Implemented.

13.14.4 Delete a STRING_type Object: *DeleteString*

```
INT_type DeleteString(string)
   STRING_type string;
```

Real Name

delstr

Synopsis

This routine is used to delete a `STRING_type` object and free any associated storage.

Parameters

`string`	The `STRING_type` object to be deleted.

Return Value

`ERROR_INT`	An error has occurred. `TKError` will contain an indication of the exact error (see the **Errors** part below).
otherwise	The `STRING_type` object has been deleted.

Side Effects

None.

Errors

`BADPARM_err`	One of the parameters is invalid.
`INTERNAL_err`	An error internal to the structure or operation of the tool kits has occurred.
`LOCAL_err`	A locally-defined error, not covered by any other available error, has occurred.

Implementation Status

Implemented

13.14.5 Concatenate Two Strings: *ConcatString*

```
STRING_type ConcatString(s1, s2)
   STRING_type s1;
   STRING_type s2;
```

Real Name

catstr

Synopsis

This routine is used to concatenate two strings and produce a third.

Parameters

`s1`	The first string to concatenate. It must not be either of the values `NULL_STRING` or `ERROR_STRING`.
`s2`	The second string to concatenate. It must not be either of the values `NULL_STRING` or `ERROR_STRING`.

Return Value

`ERROR_STRING`	An error has occurred. `TKError` will contain an indication of the exact error (see the **Errors** part below).
otherwise	A new `STRING_type` object that is the result of concatenating the values of `s1` and `s2`.

Side Effects

None.

Errors

`BADPARM_err`	One of the parameters is invalid.
`INTERNAL_err`	An error internal to the structure or operation of the tool kits has occurred.
`LOCAL_err`	A locally-defined error, not covered by any other available error, has occurred.
`NOMEM_err`	No more memory was available for dynamic allocation.

Implementation Status

Implemented.

14
Initialization and Finalization

To allow proper access to any of the routines, variables or other identifiers exported by the tool kit, the application program must first initialize the tool kit. This is accomplished by calling the routine `InitToolKit`, described in section 14.1. The programmer is warned that the application program should not attempt to gain access to any of the tool kit facilities until this routine has been called and has returned successfully.

When the application program is done with the tool kit—that is, when no more use will be made of any exported facilities—it is recommended that the finalization routine, `FinalToolKit`, be called. This routine will perform cleanup operations such as releasing dynamic storage and indicating termination on input and output sources.

It is not required that the finalization routine be called before the application program terminates. However, in the case that the program may wish to restart use of the tool kit, it is necessary that `FinalToolKit` be called before `InitToolKit` is invoked again. After `FinalToolKit` is called, the application program must call the initialization routine again before using any of the tool kit definitions. The programmer is advised that no information relating to the internal state of the memory managed by the tool kit will be retained after a call to `FinalToolKit`.

14.1 Initialize the Tool Kit: *InitToolKit*

```
INT_type InitToolKit(parms)
   INIT_PARMS_type *parms;
```

Real Name

tkinit

Synopsis

This routine will initialize the tool kit.

Parameters

`params`	A pointer to a structure that contains fields that are used to configure various aspects of the tool kit. Section 14.1.1 describes the type of this parameter.

Return Value

`ERROR_INT`	An error has occurred. `TKError` will contain an indication of the exact error (see the **Errors** part below).
otherwise	The tool kit has been initialized.

Side Effects

After this routine is called, the full functionality of the tool kit will be available.

Semantics

A pointer to the `INIT_PARMS_type` structure will be retained by the Tool kit, and the values of the fields will be checked at various times by the tool kit routines. This means that changing the values of fields may cause changes in the run-time behavior of the tool kit (section 14.1.1 contains a complete description of how the fields are used). The application program must be sure that the structure whose address is passed will be available until the `FinalToolKit` is called.

Errors

`BADPARM_err`	One of the parameters is invalid.
`INTERNAL_err`	An error internal to the structure or operation of the tool kits has occurred.
`NOFINAL_err`	The tool kit was already initialized and hasn't been finalized (by calling `FinalToolKit`, see section14.2)
`NOMEM_err`	No more memory was available for dynamic allocation.

Implementation Status

Implemented.

14.1.1 The Type INIT_PARMS_type

```
typedef struct {
    BOOL_type                  automatic_label;
    INT_type                   error_od;
    BOOL_type                  format_errors;
    BOOL_type                  print_warnings;
    INT_type                   (*output_filter)();
} INIT_PARMS_type;
```

`automatic_label`
: If `BOOL_true`, then the tool kit will label automatically all appropriate constituents when writing an external ASN.1 representation. If `BOOL_false`, then the application will be responsible for labeling constituents or calling explicitly the tool kit labeling routines (section 20.3). The programmer should read the warnings in section 20.3 before attempting any application-specific labeling.

`error_od`
: An output descriptor (section 10.2.4) to which error messages should be written. If this parameter is -1, then error and warning messages will not be written.

`format_errors`
: No formatting of human-readable error messages will be performed by tool kit routines when the value of this field is `BOOL_false`. In addition, when this field is `BOOL_false`, the value of `print_warnings` will be treated as if it were `BOOL_false`.

`print_warnings`
: The tool kit will not print warning messages to the file indicated by `error_od` when the value of this field is `BOOL_false`.

`output_filter`
: A procedure to be called by the tool kit whenever the value of a `STRING_type` object is to be output. This procedure is described in detail in section 14.1.1.1. If this field has the value of the predefined identifier `NULL_FILTER_ROUTINE`, then a default routine will be used. This default routine will assume that the character set is 7-bit ASCII and will write non-printing characters in the format

```
\ooo
```

where `ooo` is the octal representation of all eight bits of the non-printable character.

14.1.1.1 Application-defined Routine to Write a Character String: *output_filter*

```
INT_type output_filter(od, string)
   INT_type od;
   STRING_type string;
```

Real Name

Not applicable.

Synopsis

This routine is called by the tool kit routines when it is necessary to write a character string that may be composed of characters from application-specific character sets.

Parameters

`od`	The output descriptor to which the characters should be written by this routine.
`string`	The `STRING_type` object whose value is tc be written.

Return Value

`ERROR_INT`	An error has occurred. `TKError` will contain an indication of the exact error (see the **Errors** part below).
otherwise	The characters have been written successfully.

Side Effects

The side effects of this routine are application dependent.

Semantics

This routine will be called whenever a tool kit routine needs to write a character string that may contain characters from application-specific character sets. This routine will only be called during debugging (chapter 22) to write the values of the following attributes: "user-readable comments", "user-visible name", "alternative representation", "title", "subject", "document type", "abstract", "status", "user specific codes", "distribution list", "keywords", "document reference", "languages", "authorization", "security classification", "access", "encryption indicator", "password", and "content information".

The value of the parameter `string` will be the uninterpreted value of the attribute. It is up to the application to discern the appropriate character sets involved as defined in ISO 2022 [27].

The application program must not modify the value of the character string.

Errors

`BADPARM_err`	One of the parameters is invalid.
`LOCAL_err`	A locally-defined error, not covered by any other available error, has occurred.

Implementation Status

Any user definition of this routine will be ignored.

14.2 Finalize the Tool Kit: *FinalToolKit*

```
INT_type FinalToolKit()
```

Real Name

tkfinl

Synopsis

This routine is used to indicate that the application is done with the tool kit.

Parameters

Return Value

`ERROR_INT`	An error has occurred. `TKError` will contain an indication of the exact error (see the **Errors** part below).
otherwise	Finalization for the tool kit has been completed successfully.

Side Effects

Any storage dynamically-allocated by the tool kit has been released (via calls to `LOCAL_free`, section 10.2.5.3).

Semantics

Errors

`INTERNAL_err`	An error internal to the structure or operation of the tool kits has occurred.
`NOFINAL_err`	The tool kit was never initialized (via a call to `InitToolKit`) or has not been initialized since the last call to this routine.

Implementation Status

This routine will return success, but it does not perform any actions.

15
Documents and Constituents

An application program using the tool kit may need to manipulate a number of ODA documents. To distinguish between the various documents, values of type `DOCUMENT_type` are used (section 13.3). A new ODA document is created by a call of the routine `MakeDocument` (described in this chapter, section 15.1) or the routine `ReadODIF` (described in section 20.1.1).

The routines defined in sections 15.4 through 15.9 of this chapter are used to create and manipulate the various kinds of *constituents* that can occur in an ODA document. ISO IS 8613 defines a *constituent* as

> A set of attributes that is one of the following types: a document profile, an object description, an object class description, a presentation style, a layout style or a content portion description.

None of the internal structure of a constituent is visible to the application program: all access to a constituent is made through routines. This includes creation, deletion and access to attributes.

There are several ways that constituents are classified by the ODA standard. For example, a constituent may be classified according to whether it is a component description, layout style, presentation style or document profile. Alternatively, a constituent may be classified on the basis of its being a logical or a layout component. The operations that may be applied to a constituent may depend on the constituent's classification. For example, the attribute "content generator" only applies to constituents that are basic object class descriptions.

To limit the number of routines, types and macros that are necessary for dealing with constituents, there is a single type for all constituents. In addition, there are generic routines for dealing with constituents (these routines perform run-time checks to verify the legality of operations on the particular constituents). The single type for declaring all constituents is

```
CONSTITUENT_type
```

For readability, the type `CONSTITUENT_type` is also available with the name

```
CONSTITUENT
```

despite the fact that this violates the convention that all types exported by the tool kit end with the string "_type".

Because of the various orthogonal bases for classifying constituents, a single routine for creating a constituent of any kind would have been unwieldy. Such a routine would have had many parameters to allow for the creation of the various kinds of constituents. Instead, there are four routines used to create constituents: `MakeComponent`, `MakeStyle`, `MakeContent` and `MakeDocumentProfile`. The division into these routines is based on the (arbitrary) classification of constituents into the four groups' component descriptions, layout and presentation styles, content portions and document profiles. This division is illustrated in figure 15-1.

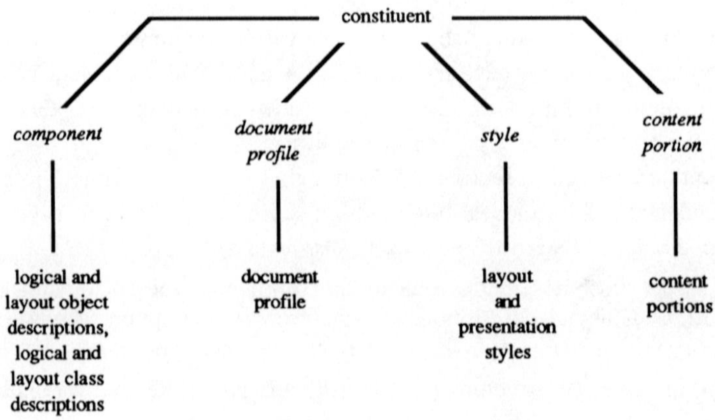

Figure 15-1. Classification of Constituents

15.1 Create a New Document: *MakeDocument*

```
DOCUMENT_type MakeDocument()
```

Real Name

mkdoc

Synopsis

This routine is used to create a new ODA (initially empty) document.

Return Value

`ERROR_DOCUMENT`
: An error has occurred. `TKError` will contain an indication of the exact error (see the **Errors** part below).

otherwise
: A value that can be used to specify an ODA document in subsequent tool kit calls.

Side Effects

None.

Semantics

The document will be empty when it is created.

Errors

`INTERNAL_err` An error internal to the structure or operation of the tool kits has occurred.

`NOMEM_err` No more memory was available for dynamic allocation.

Implementation Status

Implemented.

15.2 Get a Document: *GetDocument*

```
DOCUMENT_type GetDocument(constituent)
   CONSTITUENT constituent;
```

Real Name

getdoc

Synopsis

This routine is used to retrieve the document containing the passed constituent.

Parameters

`constituent` A constituent in a document.

Return Value

`ERROR_DOCUMENT`	An error has occurred. `TKError` will contain an indication of the exact error (see the **Errors** part below).
otherwise	A value that specifies the ODA document containing the passed constituent.

Side Effects

None.

Semantics

The identifying value for the document containing the parameter `constituent` is returned.

Errors

`BADPARM_err`	One of the parameters is invalid.
`INTERNAL_err`	An error internal to the structure or operation of the tool kits has occurred.

Implementation Status

Implemented.

15.3 Delete a Document: *DeleteDocument*

```
INT_type DeleteDocument(document, delete)
   DOCUMENT_type document;
   BOOL_type delete;
```

Real Name

deldoc

Synopsis

This routine is used to delete a document (and all of its constituents, if desired).

Parameters

`document`	The document to delete.
`delete`	If `BOOL_true`, then all of the constituents associated with this document will be deleted, and their storage released. If `BOOL_false`, only the document description information will be freed.

Return Value

`ERROR_INT`	An error has occurred. `TKError` will contain an indication of the exact error (see the **Errors** part below).
otherwise	The document and all of its constituents and attribute values are deleted.

Side Effects

If `delete` == `BOOL_true`, all of the constituents associated with this document, and all of the attribute-value pairs associated with those constituents will be deleted by this routine. Storage for all of these objects will be released (via calls to `LOCAL_Free`, section 10.2.5.3).

Semantics

The `document` is no longer valid and should not be used in any tool kit calls.

Errors

`BADPARM_err`	One of the parameters is invalid.
`INTERNAL_err`	An error internal to the structure or operation of the tool kits has occurred.
`LOCAL_err`	A locally-defined error, not covered by any other available error, has occurred.

Implementation Status

This routine calls `DeleteConstituent` (section 15.9), which should be consulted for status.

15.4 Create a Component: *MakeComponent*

```
CONSTITUENT MakeComponent(document, type,
                          kind)
   DOCUMENT_type document;
   VALUE_type type;
   VALUE_type kind;
```

Real Name

mkcomp

Synopsis

This routine is used to create a logical or layout component description.

Parameters

`document` The document in which to place the new constituent.

`type` An identifier denoting the object type of component desired. `Type` must be one of the legal values for the attribute "Object Type" (section 17.1.1.1).

`kind` The value of this argument determines whether the component will be a specific component or a generic component:

Identifier	Meaning
SPECIFIC_COMPONENT	specific component
GENERIC_COMPONENT	generic component
DVL_COMPONENT	"default value lists" component
DAPD_COMPONENT	"document application profile defaults" component

Return Value

`ERROR_CONSTITUENT`
An error has occurred. `TKError` will contain an indication of the exact error (see the **Errors** part below).

otherwise A constituent of the specified object type and kind.

Side Effects

The attribute `at_OBJECT_TYPE` will be assigned an appropriate value.

Semantics

The newly created constituent will be part of the specified document.

The kinds `DVL_COMPONENT` and `DAPD_COMPONENT` are used to create components to be used in the values of the attributes "default value lists" and "document application profile defaults" (see section 15.10).

If `kind` is `DVL_COMPONENT` or `DAPD_COMPONENT`, then the value of `type` may not be any of the following values:

- `at_OBJECT_TYPE_doc_layout_root`,
- `at_OBJECT_TYPE_page_set`,
- `at_OBJECT_TYPE_doc_logical_root`.

Errors

`BADPARM_err`	One of the parameters is invalid.
`EXISTS_err`	This call was trying to create a document root that already existed.
`INTERNAL_err`	An error internal to the structure or operation of the tool kits has occurred.
`LOCAL_err`	A locally-defined error, not covered by any other available error, has occurred.
`NOMEM_err`	No more memory was available for dynamic allocation.

Implementation Status

Implemented.

15.5 Create a Style: *MakeStyle*

```
CONSTITUENT MakeStyle(document, kind)
   DOCUMENT_type document;
   VALUE_type kind;
```

Real Name

mkstyl

Synopsis

This routine is used to create a layout or presentation style description.

Parameters

`document` — The document in which to place the new constituent.

`kind` — The value of this parameter determines if the style will be a presentation or a layout style:

Identifier	Meaning
`PRESENTATION_STYLE`	presentation style
`LAYOUT_STYLE`	layout style

Return Value

`ERROR_CONSTITUENT`
An error has occurred. `TKError` will contain an indication of the exact error (see the **Errors** part below).

otherwise — A constituent that is constrained to be a layout or presentation style as specified.

Side Effects

None.

Semantics

The newly created constituent will be part of the specified document.

Errors

`BADPARM_err`	One of the parameters is invalid.
`INTERNAL_err`	An error internal to the structure or operation of the tool kits has occurred.
`LOCAL_err`	A locally-defined error, not covered by any other available error, has occurred.
`NOMEM_err`	No more memory was available for dynamic allocation.

Implementation Status

Implemented.

15.6 Create a Content Portion: *MakeContent*

```
CONSTITUENT MakeContent(document, type)
   DOCUMENT_type document;
   VALUE_type type;
```

Real Name

mkcont

Synopsis

This routine is used to create a content portion.

Parameters

`document`	The document in which to place the new constituent.
`type`	An integer denoting the content architecture desired; may be one of the values

Identifier	Meaning
`CHARACTER_CONTENT`	*character content portion*
`GEOMETRIC_GRAPHICS_CONTENT`	*geometric graphics portion*
`RASTER_GRAPHICS_CONTENT`	*raster graphics portion*

Return Value

`ERROR_CONSTITUENT`
: An error has occurred. `TKError` will contain an indication of the exact error (see the **Errors** part below).

otherwise
: A constituent that is constrained to contain a content of the specified content architecture.

Side Effects

None.

Semantics

The newly created constituent will be part of the specified document.

Due to a restriction in the CMU ODA Tool Kit, it is imperative that all content portions have some attribute that allows the particular kind of content (character, raster graphics or geometric graphics) to be determined when the containing document is read by `ReadODIF` (section 20.1.1). This attribute can be "type of coding" (section 17.1.7.3) or a presentation attribute (sections 17.1.9, 17.1.10 and 17.1.12).

Errors

`BADPARM_err`	One of the parameters is invalid.
`INTERNAL_err`	An error internal to the structure or operation of the tool kits has occurred.
`LOCAL_err`	A locally-defined error, not covered by any other available error, has occurred.
`NOMEM_err`	No more memory was available for dynamic allocation.

Implementation Status

Implemented.

15.7 Create the Document Profile: *MakeDocumentProfile*

```
CONSTITUENT MakeDocumentProfile(document)
   DOCUMENT_type document;
```

Real Name

mkdocp

Synopsis

This routine is used to create the document profile constituent.

Parameters

`document`	The document in which to place the new constituent.

Return Value

`ERROR_CONSTITUENT`
: An error has occurred. `TKError` will contain an indication of the exact error (see the **Errors** part below).

otherwise
: The constituent representing the document profile is returned.

Side Effects

None.

Semantics

The new constituent will become part of the specified document.

Only one document profile may be in existence for a document at any time. If a new one is to be created, the previous one, if any, must be deleted, using `DeleteConstituent` (section 15.9.)

Errors

`EXISTS_err`	A document profile for this document currently exists.
`INTERNAL_err`	An error internal to the structure or operation of the tool kits has occurred.
`LOCAL_err`	A locally-defined error, not covered by any other available error, has occurred.
`NOMEM_err`	No more memory was available for dynamic allocation.

Implementation Status

Implemented.

15.8 Copy a Constituent: *CopyConstituent*

```
CONSTITUENT CopyConstituent(document,
                            constituent)
   DOCUMENT_type document;
   CONSTITUENT constituent;
```

Real Name

cpcons

Synopsis

This routine is used to create a copy of a constituent and place it into a designated document.

Parameters

`document`	The document in which to place the new constituent.
`constituent`	The constituent to copy.

Return Value

`ERROR_CONSTITUENT`

	An error has occurred. `TKError` will contain an indication of the exact error (see the **Errors** part below).
otherwise	A copy of the constituent.

Side Effects

None.

Semantics

The new constituent will become part of the specified document.

The effect will be as if a copy of the specified constituent and all of its attributes were created. That is, the attributes are not shared and modifying attributes in the original constituent will not affect the attributes of the copy. The values of all of the attributes will be copied. However, any constituents referenced by the values of attributes will not be copied, but references to them are copied.

Errors

`BADPARM_err`	One of the parameters is invalid.
`INTERNAL_err`	An error internal to the structure or operation of the tool kits has occurred.
`LOCAL_err`	A locally-defined error, not covered by any other available error, has occurred.
`NOMEM_err`	No more memory was available for dynamic allocation.

Implementation Status

Implemented.

15.9 Delete a Constituent: *DeleteConstituent*

```
INT_type DeleteConstituent(constituent)
   CONSTITUENT constituent;
```

Real Name

delcns

Synopsis

This routine is used to delete a constituent.

Parameters

`constituent`	The constituent to delete.

Return Value

`ERROR_INT`	An error has occurred. `TKError` will contain an indication of the exact error (see the **Errors** part below).
otherwise	The constituent has been deleted.

Side Effects

If any storage was dynamically allocated for this constituent, then it will be released (via calls to `LOCAL_Free`, section 10.2.5.3).

Semantics

This routine will only delete the specified constituent. All attributes and their values associated with the constituent are also deleted except for any constituents referenced as attribute values which are not deleted.

It is the responsibility of the application programmer to modify references to this constituent to avoid dangling references.

Errors

`BADPARM_err`	The parameter `CONSTITUENT` had an invalid value.
`INTERNAL_err`	An error internal to the structure or operation of the tool kits has occurred.

Implementation Status

Implemented.

15.10 Default Components

The routine MakeComponent can be used to create two kinds of *default components* that serve a special purpose in the CMU ODA Tool Kit: "default value lists" components and "document application profile defaults" components. These components do not correspond to ODA components but may be used in forming values for the attributes "defaults value lists" (section 17.1.1.17) and "document application profile defaults" (section 17.1.8.11).

The values of these attributes are structures that contain several fields of type CONSTITUENT. In each case, the value of such a field may be NULL_CONSTITUENT indicating that no default values are specified for the corresponding attributes. If a field is not NULL_CONSTITUENT, then its value must be a default component of the appropriate kind, as follows.

For the attribute at_DEFAULT_VALUE_LISTS:

field	kind
page_attributes	at_OBJECT_TYPE_page
frame_attributes	at_OBJECT_TYPE_frame
block_attributes	at_OBJECT_TYPE_block
composite_logical_attributes	at_OBJECT_TYPE_comp_logical_obj
basic_logical_attributes	at_OBJECT_TYPE_bas_logical_obj

For the attribute dp_DOC_APP_PROF_DEFAULTS:

field	kind
cc_presentation	at_OBJECT_TYPE_bas_logical_obj
cc_coding	at_OBJECT_TYPE_bas_logical_obj
rg_presentation	at_OBJECT_TYPE_bas_logical_obj
rg_coding	at_OBJECT_TYPE_bas_logical_obj
gg_presentation	at_OBJECT_TYPE_bas_logical_obj
gg_coding	at_OBJECT_TYPE_bas_logical_obj

When a default component is present, the values assigned to the attributes of this component are taken as the default values for the corresponding kinds of objects. For example, to indicate a set of default values for attributes of basic logical attributes, an object of type at_DEFAULT_VALUE_LISTS_type should be created. The field basic_logical_attributes should be assigned a constituent, created with the following call:

```
MakeComponent(doc,
              at_OBJECT_TYPE_bas_logical_obj,
              DVL_COMPONENT)
```

The values of the attributes applied to this constituent will determine the default values for the same attributes of basic logical objects that will fall within the scope of the "default value lists" attribute.

16
Attribute Routines

The routines defined in this chapter are used to create and manipulate the various kinds of *attributes* that can occur in an ODA document. The ODA standard defines an attribute as

> An element of a constituent of a document that has a name and a value and that expresses a characteristic of this constituent or a relationship with one or more constituents [30].

The routines described in this chapter may be applied to any type of constituent. The routines will, however, return an error if an attempt is made to access an attribute that is illegal for the specific constituent. For example, an attempt to obtain the value of the `at_OBJECT_TYPE` attribute of a layout style constituent will yield an error at run time[63].

16.1 Attributes and Values

Each attribute is specified by the use of a predefined identifier (defined in an include file, see section 12.1). Tables 17-1 through 17-12 in the next chapter list the predefined identifiers and the name of the type for the attribute's value.

The routines used for assigning values to and obtaining values from attributes are described in sections 16.1.2 through 16.1.11.

The routines `SetAttr` and `GetAttr` may be used to assign and access attribute-value pairs for any attribute. For convenience, there are routines that are specific to attributes with `INT_type` and `VALUE_type` values: `SetIntAttr`/`GetIntAttr` and `SetValueAttr`/`GetValueAttr`.

A variety of specialized attribute routines are defined to assist with document manipulation. The routines `SetLayoutAttr` and `SetPresAttr` are used to assign attributes to styles associated with constituents. The `SetStringAttribute` routine is used to convert C strings and attach them

[63]This is because the "object type" attribute is not applicable to layout styles.

as either `SEQUENCE_BYTE_type` or `SEQUENCE_CHAR_type` attributes. Finally, `AppendSubordinate` is used to add a new constituent to the sequence of subordinates found on a constituent.

16.1.1 Parameter Masks

Some attributes have values that are structured and are composed of parameters and sub-parameters. (We will refer to sub-parameters as simply parameters in the remainder of this section.) The value of such an attribute need not have values specified for all of its parameters. In fact, it is not necessary to supply a value for any parameter. The ODA manual gives the default value for any parameter that is unspecified.

Within the tool kit, attributes with structured values have an associated type that is used to represent these values (section 9.3.2.2). In general, there is a field within the structure for each parameter. There is, in general, no method within a structure for indicating the absence of a value for a parameter. Instead, there is associated with each instance of an attribute a *parameter mask.* This mask, which is of type `MASK_type` (section 10.1.7), determines which parameters are considered to have values.

The association between a parameter mask and an attribute value is made at attribute assignment time, within the call of the `SetAttr` (or other) routine (section 16.1.3). In addition, specific parameters may be changed from specified to unspecified after the attribute has been assigned to a constituent using the `UnspecParms` routine (section 16.4).

In addition, the parameter mask is returned when an attribute value is obtained via the `GetAttr` routine (section 16.1.3). The `GetParmMask` routine may be used to obtain the parameter mask for an attribute at any time (section 16.5).

Associated with each parameter that may be unspecified is an identifier that is used to indicate that parameter (section 9.3.2.4).[64] A number of parameters may be specified by combining their identifiers using the C bitwise-or operator. For example, the two parameters "border line width" and "border line type" for the attribute "border" are indicated by the expression

```
BORDER_LINE_WIDTH_parm | BORDER_LINE_TYPE_parm
```

[64]There is no associated parameter identifier for parameters of an attribute that has no default value.

There are two special values that may be used in dealing with parameter masks:

`ALL_PARMS`	Used to indicate all parameters applicable to an attribute.
`NO_PARMS`	Used to indicate no parameters.

It is never an error for a programmer-supplied parameter mask to specify parameters that do not apply to the specific attribute. Parameters that do not apply are simply ignored. This is useful because it allows the programmer to indicate simply that only a single parameter is unspecified. For example, the following C expression is a parameter mask indicating that all of the parameters except "border edges" are specified for the attribute "border":

```
(PARM_MASK_type) ~BORDER_EDGES_parm
```

For an attribute with no parameters, or an attribute that has no default value, any value specified for the parameter mask is legal and will be ignored.

16.1.2 Assign an Attribute: *SetAttr*

```
INT_type SetAttr(constituent, attribute,
                 value, mask)
   CONSTITUENT constituent;
   INT_type attribute;
   POINTER_type value;
   PARM_MASK_type mask;
```

Real name

setatt

Synopsis

This routine is used to assign a value to an attribute. It may be used for any attribute, regardless of the value type. It will always replace any existing value for the specified attribute.

Parameters

`constituent`	The constituent to which to attach the specified attribute-value pair.
`attribute`	The integer specifying the desired attribute.
`value`	A reference to the value to be assigned to this attribute (see **Semantics** below for details).
`mask`	A bit mask indicating which parameters of this attribute are present. The special values `ALL_PARMS` and `NO_PARMS` may be used (see section 16.1.1).

Return Value

ERROR_INT	An error has occurred. TKError will contain an indication of the exact error (see the **Errors** part below).
otherwise	The attribute was assigned to the constituent.

Side Effects

None.

Semantics

The value of the parameter value must be of type POINTER_type and be a reference to a value of the appropriate type. The possible values for attributes may be divided into five classes depending on the type of the value: VALUE_type values, INT_type values, values whose underlying type is SEQUENCE_type, CONSTITUENT values and other structured types (those defined in chapters 13 and 17). The following table shows the appropriate type to pass for an attribute value of each class (this value must always be cast to POINTER_type):

Value class	Type of value parameter
VALUE_type	VALUE_type *
INT_type	INT_type *
SEQUENCE_type	SEQUENCE_type
CONSTITUENT	CONSTITUENT
other structured type	*other structured type* *

Notice that for values of type CONSTITUENT and SEQUENCE_type, the value itself must be passed directly. For values of type INT_type, VALUE_type and other structured types, the address of a variable must be passed. Remember to cast to POINTER_type all values passed as the parameter value.

Do not attempt to assign a value for an attribute that is not applicable to the specified constituent. In such a case, the constituent will not be modified and the value of TKError will be BADATTR_err.

For an attribute with no parameters, either of the values ALL_PARMS or NO_PARMS may be used.

Errors

BADATTR_err	The attribute specified is not applicable to this constituent.
NOREPLACE_err	An attempt has been made to modify the value of a read-only attribute.

BADPARM_err	One of the parameters is invalid.
INTERNAL_err	An error internal to the structure or operation of the tool kits has occurred.
NOMEM_err	No more memory was available for dynamic allocation.
UNIMPL_err	The specified attribute is unimplemented.

Implementation Status

This routine has restricted semantic checking of some attributes, as follows:

- The following attributes are not checked for semantic correctness with respect to ODA (although for values of a sequence type, the sequence is checked to see if it has a legal element type): "abstract", "access rights", "additional information", "alternative representation", "alternative representation character sets", "application comments", "authorization", "balance", "code extension announcers", "coding attributes", "comments character sets", "content architecture class", "content identifier layout", "content identifier logical", "content information", "content portions", "creation date and time", "document application profile defaults", "document date and time", "document size", "document type", "expiry date and time", "external document class", "graphic character sets", "graphic rendition", "imaging order", "layout category", "layout object class", "layout style", "layout style identifier", "local filing date time", "logical source", "number objects per page", "number of pages", "object class", "object class identifier", "object identifier", "page dimensions", "page position", "presentation features", "presentation style", "presentation style identifier", "profile character sets", "purge date and time", "release date and time", "resources", "security classification", "start date and time", "status", "subject", "subordinates", "synchronization", "title", "type of coding", "user readable comments", "user visible name".

- The following attributes are not checked at all: "default value lists", "document application profile defaults", "languages", "local file reference", "ODA version", "organizations", "permitted categories", "revision history", "unit scaling", "user specific codes".

- The attribute "fonts list" has limited support. See the **Implementation Status** entry for the routine `WriteConstituentState` (section 22.2.4) for a specification of the supported fields.

16.1.3 Obtain the Value of an Attribute: *GetAttr*

```
INT_type GetAttr(constituent, attribute,
                 value, inherited, mask)
   CONSTITUENT constituent;
   INT_type attribute;
   POINTER_type value;
   BOOL_type inherited;
   PARM_MASK_type *mask;
```

Real name

getatt

Synopsis

This routine is used to obtain the value of any attribute, regardless of the type of the value. Unless otherwise indicated (by the parameter `inherited`), the default value rules of ODA will be used.

Parameters

`constituent`	The constituent whose attribute value is desired.
`attribute`	The integer specifying the desired attribute.
`value`	The address of a variable in which to place the value of the attribute (see the **Semantics** section below for a detailed description of this parameter).
`inherited`	If `BOOL_false`, then the specified attribute must appear directly for this constituent or an error will be returned. If `inherited` is `BOOL_true`, then the ODA attribute inheritance rules will be applied to obtain a value.
`mask`	The address of a `MASK_type` variable into which will be placed a bit mask indicating which parameters of this attribute are present. The value of this parameter may be `((PARM_MASK_type *) 0)` if the value of the mask is not desired or if the attribute has no parameter mask. If the specified attribute has no parameters, then the value of `*mask` will be undefined upon return.

Return Value

`ERROR_INT`	An error has occurred. `TKError` will contain an indication of the exact error (see the **Errors** part below).
otherwise	The attribute value was obtained.

Side Effects

None.

Semantics

The value of the parameter value must be of type POINTER_type and be the address of a variable of the appropriate type. The possible values for attributes may be divided into five classes depending on the type of the value: VALUE_type values, INT_type values, values whose underlying type is SEQUENCE_type, CONSTITUENT values and other structured types (other structured types are the structured types defined in chapters 13 and chapter 17). The following table shows the appropriate type to pass for an attribute value of each class:

Value class	Type of value parameter
VALUE_type	VALUE_type *
INT_type	INT_type *
SEQUENCE_type	SEQUENCE_type *
CONSTITUENT	CONSTITUENT *
other structured type	*other structured type* **

Notice that for values of *other structured types* a pointer to a variable that is itself a pointer must be passed. Remember to cast to POINTER_type all values passed as the parameter value.

Do not request the value of an attribute that is not applicable to the constituent. In such cases, the objects pointed to by the parameters value and mask will not have been modified and the value of TKError will be BADATTR_err.

Do not request the use of the inheritance rules for an attribute that is not defaultable. In this case, the value of TKError will be BADPARM_err.

Errors

BADATTR_err	The attribute specified is not applicable to this constituent.
BADPARM_err	One of the parameters is invalid.
NOMEM_err	No more memory was available for dynamic allocation.
UNIMPL_err	The specified attribute is unimplemented.
NOATTR_err	The attribute is not associated with this constituent.

Implementation Status

This routine may fail to yield the proper value in the case that the attribute value is to be obtained from the document profile attribute "document application profile defaults".

This routine does not check whether the requested attribute is applicable to the specified constituent.

This routine does not support the defaulting of values from resource documents.

16.1.4 Assign an INT_type Attribute: *SetIntAttr*

```
INT_type SetIntAttr(constituent, attribute,
                    value)
   CONSTITUENT constituent;
   INT_type attribute;
   INT_type value;
```

Real name

sinatt

Synopsis

This routine is used to assign an `INT_type` value to an attribute. It will replace any existing value for the specified attribute.

Parameters

`constituent`	The constituent to which to attach the specified attribute-value pair.
`attribute`	The integer specifying the desired attribute.
`value`	An `INT_type` value to be assigned to this attribute.

Return Value

`ERROR_INT`	An error has occurred. `TKError` will contain an indication of the exact error (see the **Errors** part below).
otherwise	The attribute was assigned to the constituent.

Side Effects

None.

Semantics

The value passed to this routine is simply an `INT_type` (and not a pointer to an `INT_type`, as in `SetAttr`).

It is an error to attempt to assign a value for an attribute that is not applicable to the specified constituent. In such a case, the constituent will not be modified and the value of `TKError` will be `BADATTR_err`.

Errors

BADATTR_err	The attribute specified is not applicable to this constituent.
BADPARM_err	One of the parameters is invalid.
INTERNAL_err	An error internal to the structure or operation of the tool kits has occurred.
NOMEM_err	No more memory was available for dynamic allocation.
UNIMPL_err	The specified attribute is unimplemented.

Implementation Status

See the **Implementation Status** section of SetAttr (section 16.1.2).

16.1.5 Obtain the Value of an INT_type Attribute: *GetIntAttr*

```
INT_type GetIntAttr(constituent, attribute,
                    value, inherited)
   CONSTITUENT constituent;
   INT_type attribute;
   INT_type *value;
   BOOL_type inherited;
```

Real name

ginatt

Synopsis

This routine is used to obtain the value of an INT_type attribute. Unless otherwise indicated by the parameter inherited, the default value rules will be used.

Parameters

constituent	The constituent with the desired attribute.
attribute	The integer specifying the desired attribute.
value	The address of an INT_type variable in which the value of the attribute will be placed.
inherited	If BOOL_false, then the specified attribute must appear directly for this constituent or an error will be returned. If inherited is BOOL_true, then the ODA attribute inheritance rules will be applied to obtain a value.

Return Value

`ERROR_INT`	An error has occurred. `TKError` will contain an indication of the exact error (see the **Errors** part below).
otherwise	The attribute value was obtained.

Side Effects

None.

Semantics

Do not request the value of an attribute that is not applicable to the constituent. In such cases, the object pointed to by the parameter `value` will not have been modified and the value of `TKError` will be `BADATTR_err`.

Errors

`BADATTR_err`	The attribute specified is not applicable to this constituent.
`BADPARM_err`	One of the parameters is invalid.
`NOMEM_err`	No more memory was available for dynamic allocation.
`UNIMPL_err`	The specified attribute is unimplemented.
`NOATTR_err`	The attribute is not associated with this constituent.

Implementation Status

See the **Implementation Status** section of `GetAttr` (section 16.1.3).

16.1.6 Assign a VALUE_type Attribute: *SetValueAttr*

```
INT_type SetValueAttr(constituent, attribute,
                      value)
   CONSTITUENT constituent;
   INT_type attribute;
   VALUE_type value;
```

Real name

svlatt

Synopsis

This routine is used to assign a `VALUE_type` value to an attribute. It will replace any existing value for the specified attribute.

Parameters

`constituent`	The constituent to which to attach the specified attribute-value pair.
`attribute`	The integer specifying the desired attribute.
`value`	A `VALUE_type` value to be assigned to this attribute.

Return Value

`ERROR_INT`	An error has occurred. `TKError` will contain an indication of the exact error (see the **Errors** part below).
otherwise	The attribute was assigned to the constituent.

Side Effects

None.

Semantics

The value passed to this routine is simply a `VALUE_type` (and not a pointer to a `VALUE_type`, as in `SetAttr`).

Do not attempt to assign a value for an attribute that is not applicable to the specified constituent. In such a case, the constituent will not be modified and the value of `TKError` will be `BADATTR_err`.

Errors

`BADATTR_err`	The attribute specified is not applicable to this constituent.
`BADPARM_err`	One of the parameters is invalid.
`INTERNAL_err`	An error internal to the structure or operation of the tool kits has occurred.
`NOMEM_err`	No more memory was available for dynamic allocation.
`UNIMPL_err`	The specified attribute is unimplemented.

Implementation Status

See the **Implementation Status** section of `SetAttr` (section 16.1.2).

16.1.7 Obtain the Value of a VALUE_type Attribute: *GetValueAttr*

```
INT_type GetValueAttr(constituent, attribute,
                      value, inherited)
   CONSTITUENT constituent;
   INT_type attribute;
   VALUE_type *value;
   BOOL_type inherited;
```

Real name

gvlatt

Synopsis

This routine is used to obtain the value of a `VALUE_type` attribute. Unless otherwise indicated (by the parameter `inherited`), the attribute value inheritance rules of ODA will be used.

Parameters

`constituent`	The constituent wtih the desired attribute.
`attribute`	The integer specifying the desired attribute.
`value`	The address of a `VALUE_type` variable in which the value of the attribute will be placed.
`inherited`	If `BOOL_false`, then the specified attribute must appear directly for this constituent or an error will be returned. If `inherited` is `BOOL_true`, then the ODA attribute inheritance rules will be applied to obtain a value.

Return Value

`ERROR_INT`	An error has occurred. `TKError` will contain an indication of the exact error (see the **Errors** part below).
otherwise	The attribute value was obtained.

Side Effects

None.

Semantics

Do not request the value of an attribute that is not applicable to the constituent. In such cases, the object pointed to by the parameter `value` will not have been modified and the value of `TKError` will be `BADATTR_err`.

Errors

BADATTR_err	The attribute specified is not applicable to this constituent.
BADPARM_err	One of the parameters is invalid.
NOMEM_err	No more memory was available for dynamic allocation.
UNIMPL_err	The specified attribute is unimplemented.
NOATTR_err	The attribute is not associated with this constituent.

Implementation Status

See the **Implementation Status** section of `GetAttr` (section 16.1.3).

16.1.8 Assign an Attribute to a Layout Style: *SetLayoutAttr*

```
INT_type SetLayoutAttr(constituent,
                       attribute,
                       value, mask)
   CONSTITUENT constituent;
   INT_type attribute;
   POINTER_type value;
   PARM_MASK_type mask;
```

Real name

setlat

Synopsis

This routine is used to assign a value to an attribute on a layout style associated with the constituent. It may be used for any layout attribute, regardless of the value type. It will replace any existing value for the specified attribute.

Parameters

constituent	The constituent that refers to a layout style to which to attach the specified attribute-value pair.
attribute	The integer specifying the desired attribute.
value	A reference to the value to be assigned to this attribute (see **Semantics** below for details).
mask	A bit mask indicating which parameters of this attribute are present. The special values ALL_PARMS and NO_PARMS may be used (see section 16.1.1).

Return Value

`ERROR_INT`	An error has occurred. `TKError` will contain an indication of the exact error (see the **Errors** part below).
otherwise	The attribute was assigned to the constituent's layout style.

Side Effects

None.

Semantics

The procedure is used to place an attribute-value pair on the layout style associated with the constituent using `SetAttr`. If the constituent does not already have a layout style, one is created (using `MakeStyle`) and added to the constituent using `SetAttr`.

The interpretation of the `attribute`, `value` and `mask` parameters are the same as for `SetAttr` (section 16.1.2).

Do not attempt to assign a value for an attribute that is not applicable to a layout style. It is also an error to specify a constituent that cannot specify a layout style attribute. In such cases, the value of `TKError` will be `BADATTR_err`.

For an attribute with no parameters, either of the values `ALL_PARMS` or `NO_PARMS` may be used.

Errors

`BADATTR_err`	The attribute specified is not applicable to a layout constituent or a layout style is not applicable to this constituent.
`NOREPLACE_err`	An attempt has been made to modify the value of a read-only attribute.
`BADPARM_err`	One of the parameters is invalid.
`INTERNAL_err`	An error internal to the structure or operation of the tool kits has occurred.
`NOMEM_err`	No more memory was available for dynamic allocation.
`UNIMPL_err`	The specified attribute is unimplemented.

Implementation Status

See the **Implementation Status** section of `SetAttr` (section 16.1.2).

16.1.9 Assign an Attribute to a Presentation Style: *SetPresAttr*

```
INT_type SetPresAttr(constituent, attribute,
                     value, mask)
   CONSTITUENT constituent;
   INT_type attribute;
   POINTER_type value;
   PARM_MASK_type mask;
```

Real name

setrat

Synopsis

This routine is used to assign a value to an attribute on a presentation style associated with the constituent. It may be used for any presentation attribute, regardless of the value type. It will replace any existing value for the specified attribute.

Parameters

`constituent`	The constituent that refers to a presentation style to which to attach the specified attribute-value pair.
`attribute`	The integer specifying the desired attribute.
`value`	A reference to the value to be assigned to this attribute (see **Semantics** below for details).
`mask`	A bit mask indicating which parameters of this attribute are present. The special values `ALL_PARMS` and `NO_PARMS` may be used (see section 16.1.1).

Return Value

`ERROR_INT`	An error has occurred. `TKError` will contain an indication of the exact error (see the **Errors** part below).
otherwise	The attribute was assigned to the constituent's presentation style.

Side Effects

None.

Semantics

The procedure is used to place an attribute-value pair on the presentation style associated with the constituent using `SetAttr`. If the constituent does not already have a presentation style, one is created (using `MakeStyle`) and added to the constituent using `SetAttr`.

The interpretation of the `attribute`, `value` and `mask` parameters are the same as for `SetAttr` (section 16.1.2).

Do not attempt to assign a value for an attribute that is not applicable to a presentation style. It is also an error to specify a constituent that cannot specify a presentation style attribute. In such cases, the value of `TKError` will be `BADATTR_err`.

For an attribute with no parameters, either of the values `ALL_PARMS` or `NO_PARMS` may be used.

Errors

`BADATTR_err`	The attribute specified is not applicable to a presentation constituent or a presentation style is not applicable to this constituent.
`NOREPLACE_err`	An attempt has been made to modify the value of a read-only attribute.
`BADPARM_err`	One of the parameters is invalid.
`INTERNAL_err`	An error internal to the structure or operation of the tool kits has occurred.
`NOMEM_err`	No more memory was available for dynamic allocation.
`UNIMPL_err`	The specified attribute is unimplemented.

Implementation Status

See the **Implementation Status** section of `SetAttr` (section 16.1.2).

16.1.10 Set a String Attribute: *SetStringAttr*

```
INT_type SetStringAttr(constituent,
                       attribute,
                       value, chars)
   CONSTITUENT constituent;
   INT_type attribute;
   CHAR_type *value;
   BOOL_type chars;
```

Real name

ststat

Synopsis

This routine converts the characters specified by the `value` parameter into an appropriate sequence type and then sets that sequence type as the specified attribute for the passed constituent.

Parameters

constituent	The constituent for which an attribute is being specified.
attribute	An integer value denoting the attribute to be defined for the constituent.
value	A C string: a null-terminated string specifying the characters to be used in building the sequence for the attribute.
chars	A boolean flag indicating whether the value should be converted into a SEQUENCE_CHAR_type or SEQUENCE_BYTE_type. The mnemonic values DATA_CHAR_TYPE and DATA_BYTE_TYPE are provided to specify BOOL_true and BOOL_false respectively.

Return Value

ERROR_INT	An error has occurred. TKError will contain an indication of the exact error (see the **Errors** part below).
otherwise	The specified attribute has been assigned to the specified constituent.

Side Effects

None.

Semantics

If chars is BOOL_true, the character data are converted into a SEQUENCE_CHAR_type. If chars is BOOL_false, the character data are converted into a SEQUENCE_BYTE_type. The converted sequence is used as the attribute value in a call of SetAttr. See section 16.1.2 for a discussion of SetAttr.

The data specified by data are not altered.

Errors

BADATTR_err	The attribute is not applicable to the constituent.
BADPARM_err	One of the parameters is invalid.
INTERNAL_err	An error internal to the structure or operation of the tool kits has occurred.
NOMEM_err	No more memory was available for dynamic allocation.
UNIMPL_err	The specified attribute is unimplemented.

Implementation Status

See the **Implementation Status** section of `SetAttr` (section 16.1.2).

16.1.11 Append a Subordinate Constituent: *AppendSubordinate*

```
INT_type AppendSubordinate(parent, child)
   CONSTITUENT parent;
   CONSTITUENT child;
```

Real name

appsub

Synopsis

This routine is used to change the value of the "subordinates" attribute of the parent by appending the constituent value for the child.

Parameters

`parent`	The constituent that contains (or may contain) a "subordinates" attribute which should specify the child constituent.
`child`	The constituent that should be made subordinate to the parent.

Return Value

`ERROR_INT`	An error has occurred. `TKError` will contain an indication of the exact error (see the **Errors** part below).
otherwise	The "subordinates" attribute associated with the parent has the child constituent appended to it.

Side Effects

None.

Semantics

The procedure is used to enlarge the subordinates associated with a parent constituent by appending the child constituent to the sequence of subordinates for the parent. If the parent constituent has a "subordinates" attribute defined, it will be replaced with a sequence of subordinates expanded to include the child constituent. If the parent constituent has no "subordinates" attribute defined, a sequence will be created, consisting of only the child constituent, and assigned to the parent as its "subordinates" attribute.

Errors

`BADATTR_err`	The "subordinates" attribute is not applicable to the parent constituent or the child attribute may not be specified as a subordinate of the parent.
`BADPARM_err`	One of the parameters is invalid.
`INTERNAL_err`	An error internal to the structure or operation of the tool kits has occurred.
`NOMEM_err`	No more memory was available for dynamic allocation.
`UNIMPL_err`	The specified attribute is unimplemented.

Implementation Status

Implemented.

16.1.12 Determine the Kind of Content for a Basic Object: *GetContentKind*

```
VALUE_type GetContentKind(constituent)
   CONSTITUENT constituent;
```

Real name

gcntkn

Synopsis

This routine is used to determine the kind of content associated with a basic object as described in sections 5.3.4.1 and 5.3.4.2 in part 2 of ISO IS 8613.

Parameters

`constituent`	The basic object to determine the content kind for.

Return Value

`ERROR_kind`	An error has occurred. `TKError` will contain an indication of the exact error (see the **Errors** part below).
otherwise	The content kind (see **Semantics** below for a list of values).

Side Effects

None.

Semantics

The returned value will be one of the following:

ERROR_kind	*an error has occurred (see* **Errors** *below)*
FORM_CC_kind	'formatted character content architecture'
PROC_CC_kind	'processable character content architecture'
FP_CC_kind	'formatted processable character content architecture'
FORM_RG_kind	'formatted raster graphics content architecture'
FP_RG_kind	'formatted processable raster graphics content architecture'
FORM_GG_kind	'formatted geometric graphics content architecture'

Errors

BADPARM_err	One of the parameters is invalid.
INTERNAL_err	An error internal to the structure or operation of the tool kits has occurred.
NOMEM_err	No more memory was available for dynamic allocation.

Implementation Status

Implemented.

16.1.13 Get the Content Associated with a Basic Object: *GetContent*

```
SEQUENCE_BYTE_type GetContent(constituent)
   CONSTITUENT constituent;
```

Real name

gcontt

Synopsis

This routine is used to get the content associated with a basic object as defined in section 5.1.2.5 in part 2 of ISO IS 8613.

Parameters

constituent	The basic object whose content is to be found.

Return Value

`ERROR_SEQUENCE`
: An error has occurred. `TKError` will contain an indication of the exact error (see the **Errors** part below).

otherwise
: The content associated with the specified object.

Side Effects

None.

Errors

`BADPARM_err` One of the parameters is invalid.

`NOMEM_err` No more memory was available for dynamic allocation.

Implementation Status

This routine does not evaluate any string expressions it may encounter, because the routine `EvalStringExpr` (section 18.1.2) is not implemented.

16.2 Test for the Presence of an Attribute: *HasAttr*

```
INT_type HasAttr(constituent, attribute)
   CONSTITUENT constituent;
   INT_type attribute;
```

Real name

hasatt

Synopsis

This routine is used to test for the presence of an attribute on a constituent.

Parameters

`constituent` The constituent to test.

`attribute` The integer specifying the desired attribute.

Return Value

`ERROR_INT`	An error has occurred. `TKError` will contain an indication of the exact error (see the **Errors** part below).
`NO_ATTR`	The attribute is not present.
`HAS_ATTR`	The attribute is present.

Side Effects

None.

Semantics

It is illegal to test for an attribute that is inapplicable to the specified constituent. In such cases the value of `TKError` will be `BADATTR_err`.

Errors

`BADATTR_err`	The attribute specified is not applicable to this constituent.
`BADPARM_err`	One of the parameters is invalid.
`INTERNAL_err`	An error internal to the structure or operation of the tool kits has occurred.

Implementation Status

This routine does not check to see whether the requested attribute is applicable to the specified constituent.

16.3 Remove an Attribute: *RemAttr*

```
INT_type RemAttr(constituent, attribute)
   CONSTITUENT constituent;
   INT_type attribute;
```

Real name

rematt

Synopsis

This routine is used to remove an attribute from a constituent.

Parameters

`constituent`	The constituent.
`attribute`	The integer specifying the attribute to be removed.

Return Value

`ERROR_INT`	An error has occurred. `TKError` will contain an indication of the exact error (see the **Errors** part below).
otherwise	The attribute has been removed.

Side Effects

Any storage allocated for the attribute will be released (via calls to `LOCAL_Free`, section 10.2.5.3).

Errors

`BADATTR_err`	The attribute specified is not applicable to this constituent.
`BADPARM_err`	One of the parameters is invalid.
`NOATTR_err`	The specified attribute does not exist on this constituent.
`INTERNAL_err`	An error internal to the structure or operation of the tool kits has occurred.

Implementation Status

This routine does not check to see whether the requested attribute is applicable to the specified constituent.

16.4 Make Specified Parameters be Unspecified: *UnspecParms*

```
INT_type UnspecParms(constituent, attribute,
                     unspec)
   CONSTITUENT constituent;
   INT_type attribute;
   PARM_MASK_type unspec;
```

Real name

unsprm

Synopsis

This routine is used to make certain parameters of an attribute value become unspecified.

Parameters

`constituent`	The constituent that has the attribute.
`attribute`	An integer representing the desired attribute.
`unspec`	A bit mask indicating the parameters to be unspecified. It is legal to specify parameters that are already unspecified.

Return Value

`ERROR_INT`	An error has occurred. `TKError` will contain an indication of the exact error (see the **Errors** part below).
otherwise	The parameters were unspecified.

Side Effects

None.

Semantics

It is illegal to attempt to modify the parameters of an attribute that is not applicable to the specified constituent. In such cases no parameter masks will be modified and the value of `TKError` will be `BADATTR_err`.

Errors

`BADATTR_err`	The attribute is not applicable to the specified constituent.
`NOATTR_err`	The specified attribute does not exist on the constituent.
`BADPARM_err`	One of the parameters is invalid.
`INTERNAL_err`	An error internal to the structure or operation of the tool kits has occurred.

Implementation Status

This routine does not check to see whether the requested attribute is applicable to the specified constituent.

16.5 Obtain a Parameter Mask: *GetParmMask*

```
INT_type GetParmMask(constituent, attribute,
                     mask)
   CONSTITUENT constituent;
   INT_type attribute;
   PARM_MASK_type *mask;
```

Real name

getprm

Synopsis

This routine is used to obtain the parameter mask for a specified attribute and constituent.

Parameters

`constituent`	The constituent with the attribute.
`attribute`	The identifier specifying the value of the desired attribute.
`mask`	The address of a variable in which the parameter mask for this attribute will be returned.

Return Value

`ERROR_INT`	An error has occurred. `TKError` will contain an indication of the exact error (see the **Errors** part below).
otherwise	The parameter mask has been returned.

Side Effects

None.

Errors

`BADATTR_err`	The attribute specified is not applicable to this constituent.
`NOATTR_err`	The specified attribute does not exist on this constituent.
`BADPARM_err`	One of the parameters is invalid.
`INTERNAL_err`	An error internal to the structure or operation of the tool kits has occurred.

Implementation Status

This routine does not check to see whether the requested attribute is applicable to the specified constituent.

16.6 Get a Printable Name for an Attribute: *GetAttrName*

```
CHAR_type *GetAttrName(attribute)
   INT_type attribute;
```

Real Name

gattnm

Synopsis

This routine is used to obtain a printable string corresponding to an attribute.

Parameters

`attribute`	The identifier specifying the attribute.

Return Value

`(CHAR_type *) 0`	An error has occurred. `TKError` will contain an indication of the exact error (see the **Errors** part below).
otherwise	A null-terminated character string whose value is the ODA name of the attribute.

Side Effects

None.

Semantics

The character string referenced by the return value of this routine is in static storage and must not be modified by the application program.

The ODA name of an attribute is conventionally surrounded by double quote marks ("). However, the name returned by this routine is not.

Errors

`BADPARM_err`	One of the parameters is invalid.
`INTERNAL_err`	An error internal to the structure or operation of the tool kits has occurred.
`NOMEM_err`	No more memory was available for dynamic allocation.

Implementation Status

Implemented.

16.7 Get a Printable Name for a Distinguished Value: *GetValueName*

```
CHAR_type *GetValueName(value)
   VALUE_type value;
```

Real Name

gvalnm

Synopsis

This routine is used to obtain a printable string corresponding to a distinguished value.

Parameters

`value`	The value whose name is desired.

Return Value

`(CHAR_type *) 0`	An error has occurred. `TKError` will contain an indication of the exact error (see the **Errors** part below).
otherwise	A null-terminated character string whose value is the ODA name of the specified value.

Side Effects

None.

Semantics

The character string referenced by the return value of this routine is in static storage and must not be modified by the application program.

The ODA representation of a distinguished value is conventionally surrounded by matched quote marks (''), however, the name returned by this routine is not.

Errors

`BADPARM_err`	One of the parameters is invalid.
`INTERNAL_err`	An error internal to the structure or operation of the tool kits has occurred.

Implementation Status

Implemented.

17
Specific Attribute Descriptions

Each attribute is specified by the use of a predefined identifier (defined in an include file, see section 12.1). Tables 17-1 through 17-12 list the predefined identifiers and the name of the type for the attribute's value.

Identifier	Value type
at_OBJECT_TYPE	VALUE_type
at_OBJECT_IDENTIFIER	STRING_type
at_OBJECT_CLASS_IDENTIFIER	STRING_type
at_GEN_FOR_SUBORDINATES	CONSTRUCTION_EXPR_type
at_CONTENT_GENERATOR	STRING_EXPR_type
at_OBJECT_CLASS	CONSTITUENT
at_SUBORDINATES	SEQUENCE_CONSTITUENT_type
at_CONTENT_PORTIONS	SEQUENCE_CONSTITUENT_type
at_RESOURCE	STRING_type
at_PRESENTATION_STYLE	CONSTITUENT
at_CONTENT_ARCH_CLASS	SEQUENCE_INT_type
at_CONTENT_TYPE	VALUE_type
at_USER_READABLE_COMMENTS	STRING_type
at_APPLICATION_COMMENTS	SEQUENCE_BYTE_type
at_USER_VISIBLE_NAME	STRING_type
at_BINDINGS	SEQUENCE_BINDING_PAIR_type
at_DEFAULT_VALUE_LISTS	at_DEFAULT_VALUE_LISTS_type

Table 17-1. Shared Attributes

Identifier	Value type
at_POSITION	at_POSITION_type
at_DIMENSIONS	at_DIMENSIONS_type
at_BORDER	at_BORDER_type
at_BALANCE	SEQUENCE_CONSTITUENT_type
at_LAYOUT_PATH	VALUE_type
at_LOGICAL_SOURCE	CONSTITUENT
at_PERMITTED_CATEGORIES	SEQUENCE_STRING_type
at_IMAGING_ORDER	SEQUENCE_CONSTITUENT_type
at_TRANSPARENCY	VALUE_type
at_COLOUR	VALUE_type
at_PAGE_POSITION	INT_PAIR_type
at_MEDIUM_TYPE	at_MEDIUM_TYPE_type

Table 17-2. Layout Attributes

Identifier	Value type
at_PROTECTION	VALUE_type
at_LAYOUT_STYLE	CONSTITUENT

Table 17-3. Logical Attributes

Identifier	Value type
at_LAYOUT_STYLE_IDENTIFIER	STRING_type

Table 17-4. Layout Style Attributes

Identifier	Value type
at_BLK_ALGNM	VALUE_type
at_CONCATENATION	VALUE_type
at_FILL_ORDER	VALUE_type
at_INDIVISIBILITY	LAYOUT_OBJECT_type
at_LAYOUT_CATEGORY	STRING_type
at_LAYOUT_OBJECT_CLASS	CONSTITUENT
at_NEW_LAYOUT_OBJECT	LAYOUT_OBJECT_type
at_OFFSET	at_OFFSET_type
at_SAME_LAYOUT_OBJECT	at_SAME_LAYOUT_OBJECT_type
at_SEPARATION	at_SEPARATION_type
at_SYNCHRONIZATION	CONSTITUENT

Table 17-5. Layout Directive Attributes

Identifier	Value type
at_PRESENTATION_STYLE_ID	STRING_type

Table 17-6. Presentation Style Attributes

Identifier	Value type
at_CONTENT_ID_LOGICAL	STRING_type
at_CONTENT_ID_LAYOUT	STRING_type
at_TYPE_OF_CODING	SEQUENCE_INT_type
at_CONTENT_INFORMATION	SEQUENCE_BYTE_type
at_ALT_REP	STRING_type

Table 17-7. Content Portion Attributes

Identifier	Value type
dp_GENERIC_LAY_STRUCT	VALUE_type
dp_SPECIFIC_LAY_STRUCT	VALUE_type
dp_GENERIC_LOG_STRUCT	VALUE_type
dp_SPECIFIC_LOG_STRUCT	VALUE_type
dp_LAYOUT_STYLES	VALUE_type
dp_PRESENTATION_STYLES	VALUE_type
dp_EXTERNAL_DOCUMENT_CLASS	SEQUENCE_INT_type
dp_RESOURCE_DOCUMENT	STRING_OR_ID_type
dp_RESOURCES	SEQUENCE_RESOURCE_MAP_type
dp_DOC_APP_PROFILE	SEQUENCE_INT_type
dp_DOC_APP_PROF_DEFAULTS	dp_DOC_APP_PROF_DEFAULTS_type
dp_DOC_ARCH_CLASS	VALUE_type
dp_CONTENT_ARCH_CLASSES	SEQUENCE_ASN1_OBJECT_ID_type
dp_INTERCHANGE_FMT_CLASS	VALUE_type
dp_ODA_VERSION	dp_ODA_VERSION_type
dp_PROFILE_CHARACTER_SETS	SEQUENCE_BYTE_type
dp_COMMENTS_CHARACTER_SETS	SEQUENCE_BYTE_type
dp_ALT_REP_CHARACTER_SETS	SEQUENCE_BYTE_type
dp_PAGE_DIMENSIONS	SEQUENCE_INT_PAIR_type
dp_MEDIUM_TYPES	SEQUENCE_at_MEDIUM_TYPE_type
dp_LAYOUT_PATHS	SEQUENCE_VALUE_type
dp_PROTECTIONS	SEQUENCE_VALUE_type
dp_BLOCK_ALIGNMENTS	SEQUENCE_VALUE_type
dp_FILL_ORDERS	SEQUENCE_VALUE_type
dp_TRANSPARENCIES	SEQUENCE_VALUE_type
dp_COLOURS	SEQUENCE_VALUE_type
dp_PAGE_POSITIONS	SEQUENCE_INT_PAIR_type
dp_TYPES_OF_CODING	SEQUENCE_ASN1_OBJECT_ID_type
dp_CODING_ATTRIBUTES	SEQUENCE_CONSTITUENT_type
dp_PRESENTATION_FEATURES	SEQUENCE_CONSTITUENT_type
dp_NUMBER_OBJECTS_PER_PAGE	INT_type
dp_UNIT_SCALING	INT_PAIR_type
dp_FONTS_LIST	SEQUENCE_FONT_PAIR_type
dp_TITLE	STRING_type
dp_SUBJECT	STRING_type
dp_DOCUMENT_REFERENCE	STRING_OR_ID_type
dp_DOCUMENT_TYPE	STRING_type

Table 17-8. Document Profile Attributes, I

Identifier	Value type
dp_ABSTRACT	STRING_type
dp_DOCUMENT_DATE_AND_TIME	STRING_type
dp_CREATION_DATE_AND_TIME	STRING_type
dp_LOCAL_FILING_DATE_TIME	STRING_type
dp_EXPIRY_DATE_AND_TIME	STRING_type
dp_START_DATE_AND_TIME	STRING_type
dp_PURGE_DATE_AND_TIME	STRING_type
dp_RELEASE_DATE_AND_TIME	STRING_type
dp_REVISION_HISTORY	SEQUENCE_REVISION_HISTORY_type
dp_ORGANIZATIONS	SEQUENCE_STRING_type
dp_PREPARERS	SEQUENCE_NAME_ORG_type
dp_OWNERS	SEQUENCE_NAME_ORG_type
dp_AUTHORS	SEQUENCE_NAME_ORG_type
dp_COPYRIGHT	SEQUENCE_COPYRIGHT_INFO_type
dp_STATUS	STRING_type
dp_USER_SPECIFIC_CODES	SEQUENCE_STRING_type
dp_DISTRIBUTION_LIST	SEQUENCE_NAME_ORG_type
dp_ADDITIONAL_INFORMATION	SEQUENCE_BYTE_type
dp_REFS_TO_OTHER_DOCS	SEQUENCE_STRING_OR_ID_type
dp_SUPERCEDED_DOCUMENTS	SEQUENCE_STRING_OR_ID_type
dp_LOCAL_FILE_REFERENCE	dp_LOCAL_FILE_REFERENCE_type
dp_DOCUMENT_SIZE	INT_type
dp_NUMBER_OF_PAGES	INT_type
dp_LANGUAGES	SEQUENCE_STRING_type
dp_AUTHORIZATION	STRING_type
dp_SECURITY_CLASSIFICATION	STRING_type
dp_ACCESS_RIGHTS	STRING_type

Table 17-9. Document Profile Attributes, II

Identifier	Value type
cc_ALIGNMENT	VALUE_type
cc_CHARACTER_FONTS	SEQUENCE_INT_PAIR_type
cc_CHAR_ORIENTATION	VALUE_type
cc_CHARACTER_PATH	VALUE_type
cc_CHARACTER_SPACING	INT_type
cc_CODE_EXT_ANNOUNCERS	SEQUENCE_BYTE_type
cc_1ST_LINE_OFFSET	INT_type
cc_GRAPHIC_CHARACTER_SETS	SEQUENCE_BYTE_type
cc_GRAPHIC_CHAR_SUBREP	INT_type
cc_GRAPHIC_RENDITION	SEQUENCE_INT_type
cc_ITEMISATION	cc_ITEMISATION_type
cc_KERNING_OFFSET	INT_PAIR_type
cc_LINE_LAYOUT_TABLE	SEQUENCE_TAB_STOP_type
cc_LINE_PROGRESSION	VALUE_type
cc_LINE_SPACING	INT_type
cc_PAIRWISE_KERNING	VALUE_type
cc_FORMATTING_INDICATOR	VALUE_type
cc_INITIAL_OFFSET	INT_PAIR_type
cc_INDENTATION	INT_type
cc_ORPHAN_SIZE	INT_type
cc_PROP_LINE_SPACING	VALUE_type
cc_WIDOW_SIZE	INT_type

Table 17-10. Character Content Presentation Attributes

Identifier	Value type
rg_CLIPPING	rg_CLIPPING_type
rg_LINE_PROGRESSION	VALUE_type
rg_PEL_PATH	VALUE_type
rg_INITIAL_OFFSET	INT_PAIR_type
rg_PEL_TRANS_DENSITY	VALUE_type
rg_IMAGE_DIMENSIONS	rg_IMAGE_DIMENSIONS_type
rg_PEL_SPACING	rg_PEL_SPACING_type
rg_SPACING_RATIO	INT_PAIR_type
rg_COMPRESSION	VALUE_type
rg_NUMBER_OF_LINES	INT_type
rg_NUMBER_OF_PELS_PER_LINE	INT_type
rg_NUMBER_DISCARDED_PELS	INT_type

Table 17-11. Raster Graphics Content Presentation and Coding Attributes

Identifier	Value type
gg_ENCODING_ANNOUNCER	gg_ENCODING_ANNOUNCER_type
gg_LINE_RENDITION	gg_LINE_RENDITION_type
gg_MARKER_RENDITION	gg_MARKER_RENDITION_type
gg_TEXT_RENDITION	gg_TEXT_RENDITION_type
gg_FILLED_AREA_REND	gg_FILLED_AREA_REND_type
gg_EDGE_RENDITION	gg_EDGE_RENDITION_type
gg_COLOUR_REPS	gg_COLOUR_REPS_type
gg_TRANSPARENCY_SPEC	gg_TRANSPARENCY_SPEC_type
gg_TRANSFORMATION_SPEC	gg_TRANSFORMATION_SPEC_type
gg_REGION_INTEREST_SPEC	gg_REGION_INTEREST_SPEC_type
gg_PICTURE_ORIENTATION	VALUE_type
gg_PICTURE_DIMENSIONS	gg_PICTURE_DIMENSIONS_type

Table 17-12. Geometric Graphics Content Presentation Attributes

17.1 Attribute Descriptions

This section contains a detailed description of each attribute, as supported by the CMU ODA Tool Kit. The format for each attribute description is as described in section 9.2.

17.1.1 Shared Attributes

The attributes defined in this section can be specified for more than one type of constituent.

17.1.1.1 at_OBJECT_TYPE

Applicability

This attribute can be specified only for component descriptions.

It is mandatory for object class descriptions. It is defaultable for an object description that refers to an object class description; otherwise it is mandatory.

Attribute Value Type

```
typedef VALUE_type at_OBJECT_TYPE_type;
```

The value of this attribute must be one of:

Identifier	Meaning
`at_OBJECT_TYPE_doc_layout_root`	'document layout root'
`at_OBJECT_TYPE_page_set`	'page set'
`at_OBJECT_TYPE_page`	'composite or basic page'
`at_OBJECT_TYPE_frame`	'frame'
`at_OBJECT_TYPE_block`	'block'
`at_OBJECT_TYPE_doc_logical_root`	'document logical root'
`at_OBJECT_TYPE_comp_logical_obj`	'composite logical object'
`at_OBJECT_TYPE_bas_logical_obj`	'basic logical object'

Semantics

This is a read-only attribute and it cannot be removed or assigned a value. Its value is set automatically when the component is created (by `MakeComponent`, section 15.4).

17.1.1.2 at_OBJECT_IDENTIFIER

Applicability

This attribute can be specified only for object descriptions.

It is mandatory except for the exceptional case described in section 5.3.1.2, Part 2 of ISO IS 8613.

Attribute Value Type

```
typedef STRING_type
        at_OBJECT_IDENTIFIER_type;
```

Semantics

The characters forming the value of the `STRING_type` should be an object identifier according to the rules in section 5.3.1.2, Part 2 of ISO IS 8613.

The programmer is warned that values assigned to this attribute will not be checked for legality. Read section 20.3 for a discussion of the rules applying to object identifiers.

17.1.1.3 at_OBJECT_CLASS_IDENTIFIER

Applicability

This attribute can be specified only for object class descriptions.

It is mandatory.

Attribute Value Type

```
typedef STRING_type
        at_OBJECT_CLASS_IDENTIFIER_type;
```

Semantics

The characters forming the value of the `STRING_type` should be an object class identifier according to the rules in section 5.3.1.3, Part 2 of ISO IS 8613.

The programmer is warned that values assigned to this attribute will not be checked for legality. Read section 20.3 for a discussion of the rules applying to object class identifiers.

17.1.1.4 at_GEN_FOR_SUBORDINATES

Applicability

This attribute may be specified only for composite object class descriptions.

It is non-mandatory (but see the description in section 5.3.2.1 in Part 2 of ISO IS 8613).

Attribute Value Type

```
typedef CONSTRUCTION_EXPR_type
        at_GEN_FOR_SUBORDINATES_type;
```

The type `CONSTRUCTION_EXPR_type` is defined in section 18.4.

17.1.1.5 at_CONTENT_GENERATOR

Applicability

This attribute may be specified only for basic object class descriptions and basic logical object descriptions.

It is non-mandatory.

Attribute Value Type

```
typedef STRING_EXPR_type
        at_CONTENT_GENERATOR_type;
```

See section 18.1 for a description of the `STRING_EXPR_type` type.

17.1.1.6 at_OBJECT_CLASS

Applicability

This may be specified only for object descriptions.

It is non-mandatory (but see the description in section 5.3.3.1, Part 2 of ISO IS 8613).

Attribute Value Type

```
typedef CONSTITUENT at_OBJECT_CLASS_type;
```

The value of this attribute must be a constituent that is an object class description.

17.1.1.7 at_SUBORDINATES

Applicability

This attribute may be specified only for composite object descriptions.

It is mandatory except for the exceptional case described in section 5.3.3.2, Part 2 of ISO IS 8613.

Attribute Value Type

```
typedef SEQUENCE_CONSTITUENT_type
               at_SUBORDINATES_type;
```

17.1.1.8 at_CONTENT_PORTIONS

Applicability

This attribute may be specified only for basic component descriptions.

It is non-mandatory.

Attribute Value Type

```
typedef SEQUENCE_CONSTITUENT_type
            at_CONTENT_PORTIONS_type;
```

Each CONSTITUENT in the sequence of constituents must be a content portion.

17.1.1.9 at_RESOURCE

Applicability

This attribute applies only to object class descriptions.
It is non-mandatory.

Attribute Value Type

```
typedef STRING_type at_RESOURCE_type;
```

The characters in the string must be from the minimum subrepertoire of ISO 6937-2.

17.1.1.10 at_PRESENTATION_STYLE

Applicability

This attribute may be specified only for basic component descriptions.
It is non-mandatory for object class descriptions and defaultable for object descriptions.

Attribute Value Type

```
typedef CONSTITUENT
        at_PRESENTATION_STYLE_type;
```

Default Value

NULL_CONSTITUENT
The constituent referenced must be a presentation style.

Semantics

The value NULL_CONSTITUENT denotes the value 'null'.

17.1.1.11 at_CONTENT_ARCH_CLASS

Applicability

This attribute may be specified only for basic component descriptions.
It is non-mandatory for object class descriptions and defaultable for object descriptions.

Attribute Value Type

```
typedef ASN1_OBJECT_ID_type
        at_CONTENT_ARCH_CLASS_type;
```

See section 13.8 for a description of the type `ASN1_OBJECT_ID_type`.

The value of this attribute must designate a content architecture class defined in ISO IS 8613.

There are predefined identifiers that designate the content architectures defined in ISO IS 8613:

Identifier	Meaning
`at_CONTENT_ARCH_CLASS_form_cc`	'formatted character content architecture'
`at_CONTENT_ARCH_CLASS_proc_cc`	'processable character content architecture'
`at_CONTENT_ARCH_CLASS_fp_cc`	'formatted processable character content architecture'
`at_CONTENT_ARCH_CLASS_form_rg`	'formatted raster graphics content architecture'
`at_CONTENT_ARCH_CLASS_fp_rg`	'formatted processable raster graphics content architecture'
`at_CONTENT_ARCH_CLASS_form_gg`	'formatted geometric graphics content architecture'

The value of this attribute must be one of these predefined attributes.

Default Value

`at_CONTENT_ARCH_CLASS_form_cc`

Semantics

This attribute will override the value of the `at_CONTENT_TYPE` attribute if both are present.

17.1.1.12 at_CONTENT_TYPE

Applicability

This attribute may be specified only for basic layout component descriptions.

It is non-mandatory for object class descriptions and defaultable for object descriptions.

Attribute Value Type

```
typedef VALUE_type at_CONTENT_TYPE_type;
```

Default Value

at_CONTENT_TYPE_form_rgca

Semantics

The value must be one of

Identifier	Meaning
at_CONTENT_TYPE_form_rgca	'formatted raster graphics content architecture'

The value of this attribute will be ignored if the attribute at_CONTENT_ARCH_CLASS is also specified.

17.1.1.13 at_USER_READABLE_COMMENTS

Applicability

This attribute may be specified only for component descriptions and styles.

It is non-mandatory for object class descriptions and styles. It is defaultable for object descriptions.

Attribute Value Type

```
typedef STRING_type
        at_USER_READABLE_COMMENTS_type;
```

The characters in the string must be from the character sets mentioned in the document profile attribute "comments character sets" (section 17.1.8.17).

Default Value

A STRING_type object whose length is zero.

17.1.1.14 at_APPLICATION_COMMENTS

Applicability

This attribute may be specified only for component descriptions.

It is non-mandatory for object class descriptions. It is defaultable for object descriptions.

Attribute Value Type

```
typedef SEQUENCE_BYTE_type
        at_APPLICATION_COMMENTS_type;
```

Default Value

A SEQUENCE_BYTE_type object whose length is zero.

17.1.1.15 at_USER_VISIBLE_NAME

Applicability

This attribute may be specified only for component descriptions and styles.

It is non-mandatory for object class descriptions and styles. It is defaultable for object descriptions.

Attribute Value Type

```
typedef STRING_type
        at_USER_VISIBLE_NAME_type;
```

The characters in the string must be from the character sets mentioned in the document profile attribute "comments character sets" (section 17.1.8.17).

Default Value

A `STRING_type` object whose length is zero.

17.1.1.16 at_BINDINGS

Applicability

This attribute may be specified only for component descriptions.

It is non-mandatory for object class descriptions and defaultable for object descriptions.

Attribute Value Type

```
typedef SEQUENCE_BINDING_PAIR_type
                at_BINDINGS_type;
```

Where the element type, `BINDING_PAIR_type`, is defined below.

```
typedef struct {
    TAG_type    binding_pair_tag;
      /*
        STRING_EXPR_tag, NUMERIC_EXPR_tag,
        OBJECT_ID_EXPR_tag, INTEGER_tag,
        STRING_tag, CONSTITUENT_tag
      */
    union {
      STRING_EXPR_type      string_expr;
      NUMERIC_EXPR_type     numeric_expr;
      OBJECT_ID_EXPR_type   object_id_expr;
      INT_type              integer;
      STRING_type           string;
      CONSTITUENT           constituent;
    } binding_pair_value;
} BINDING_VALUE_type;

typedef struct BINDING_PAIR_struct {
    STRING_type        binding_identifier;
    BINDING_VALUE_type value;
} BINDING_PAIR_type;
```

The characters in the value of the field binding_identifier must be from the minimum subrepertoire of ISO 6937-2.

The type STRING_EXPR_type is defined in section 18.1, the type NUMERIC_EXPR_type in section 18.2 and the type OBJECT_ID_EXPR_type in section 18.3.

Default Value

NULL_SEQUENCE

Semantics

The value NULL_SEQUENCE indicates that no bindings are specified.

17.1.1.17 at_DEFAULT_VALUE_LISTS

Applicability

This attribute may be specified only for composite component descriptions.

It is non-mandatory.

Attribute Value Type

```
typedef struct {
    CONSTITUENT  page_attributes;
    CONSTITUENT  frame_attributes;
    CONSTITUENT  block_attributes;
    CONSTITUENT
             composite_logical_attributes;
    CONSTITUENT  basic_logical_attributes;
} at_DEFAULT_VALUE_LISTS_type;
```

The value of each field must be `NULL_CONSTITUENT` or a component created via a call to `MakeComponent` with the `kind` having the value `DVL_COMPONENT` (section 15.4).

Semantics

Each field in the structure corresponds to an object for which defaultable attributes may be specified, as follows:

`page_attributes`
: Default values for attributes of pages.

`frame_attributes`
: Default values for attributes of frames.

`block_attributes`
: Default values for attributes of blocks.

`composite_logical_attributes`
: Default values for attributes of composite logical components.

`basic_logical_attributes`
: Default values for attributes of basic logical components.

If the value of a field is `NULL_CONSTITUENT`, then there are no default attribute values specified for components of the kind corresponding to that field. Otherwise, the value of a field is a constituent and the values of attributes attached to that constituent determine the default values of the corresponding attributes for that kind of component. For example, if the field `frame_attributes` is a constituent that has the attribute "layout path" with value '270°', then this specifies a default value of '270°' for that attribute on frame components.

The attributes that may be attached to constituents for this structure must obey the restrictions specified in section 5.3.5.5, part 2 of ISO IS 8613. Note that when a layout style or presentation style is attached to a constituent, the default values specified are for the attributes attached to the style and not for the attributes "layout style" or "presentation style" themselves.

Also, note that although ISO IS 8613 indicates that the attribute "content type" may be defaulted for basic logical components, we believe this to be in error, since the attribute "content type" is not applicable to basic logical components.

See sections 13.12.9 and 13.12.10 for utility routines that manipulate the `at_DEFAULT_VALUE_LISTS_type` structure.

17.1.2 Layout Attributes

The attributes described in this section may be specified only for layout components.

17.1.2.1 at_POSITION

Applicability

This attribute may be specified only for frame and block component descriptions.

It is non-mandatory for object class descriptions and defaultable for object descriptions.

Value Type

```
typedef struct {
    TAG_type            position_tag;
        /*
            FIXED_POSITION_tag,
            VARIABLE_POSITION_tag
        */
    union {
        FIXED_POSITION_type
                        fixed_position;
        VARIABLE_POSITION_type
                        variable_position;
    } position_body;
} at_POSITION_type;
typedef struct {
    INT_type        horizontal_position;
    INT_type        vertical_position;
} FIXED_POSITION_type;
```

```
typedef struct {
    struct {
        INT_type     leading_offset;
        INT_type     trailing_offset;
        INT_type     left_hand_offset;
        INT_type     right_hand_offset;
    } offset;
    struct {
        INT_type     leading_edge;
        INT_type     trailing_edge;
        INT_type     centre_separation;
    } separation;
    VALUE_type     alignment;
    VALUE_type     fill_order;
} VARIABLE_POSITION_type;
```

If the value of the field `position_tag` is `FIXED_POSITION_tag`, then the field `fixed_position` is present. If the value of the field `position_tag` is `VARIABLE_POSITION_tag`, then the field `variable_position` is present.

The values of the fields `horizontal_position` and `vertical_position` (if specified) must be non-negative integers.

The values of the fields `leading_offset`, `trailing_offset`, `right_hand_offset` and `left_hand_offset` (if specified) must be non-negative integers.

The field `alignment` must have one of the values

Identifier	Meaning
at_POSITION_right_hand_aligned	'right-hand aligned'
at_POSITION_centred	'centred'
at_POSITION_left_hand_aligned	'left-hand aligned'

The field `order` must have one of the values

Identifier	Meaning
at_POSITION_normal_order	'normal order'
at_POSITION_reverse_order	'reverse order'

Attribute Parameters

Parameter	Field Name	Parameter Name
"horizontal position"	`fixed_position.horizontal_position`	`HORIZONTAL_POSITION_parm`
"vertical position"	`fixed_position.vertical_position`	`VERTICAL_POSITION_parm`
"leading offset"	`variable_position.offset.leading_offset`	`POS_LEADING_OFFSET_parm`
"trailing offset"	`variable_position.offset.trailing_offset`	`POS_TRAILING_OFFSET_parm`
"right-hand offset"	`variable_position.offset.right_hand_offset`	`POS_RIGHT_HAND_OFFSET_parm`
"left-hand offset"	`variable_position.offset.left_hand_offset`	`POS_LEFT_HAND_OFFSET_parm`
"leading edge"	`variable_position.separation.leading_edge`	`POS_LEADING_EDGE_parm`
"trailing edge"	`variable_position.separation.trailing_edge`	`POS_TRAILING_EDGE_parm`
"centre separation"	`variable_position.separation.centre_separation`	`POS_CENTRE_SEPARATION_parm`
"alignment"	`variable_offset.alignment`	`POS_ALIGNMENT_parm`
"fill order"	`variable_offset.fill_order`	`POS_FILL_ORDER_parm`

Default Value

If a parameter is specified without including a value for one or more of its sub-parameters or sub-sub-parameters then the following defaults are defined for the undefined sub-parameters or sub-sub-parameters:

```
fixed_position.
            horizontal_position: 0
            vertical_position:   0
variable_position.offset.
            leading_offset:      0
            trailing_offset:     0
            left_hand_offset:    0
            right_hand_offset:   0
```

```
variable_position.separation.
            leading_edge:        0
            trailing_edge:       0
            centre_separation:   0
variable_position.alignment:  at_POSITION_right_hand_aligned
variable_position.fill_order:    at_POSITION_normal_order
```

17.1.2.2 at_DIMENSIONS

Applicability

This attribute may be specified only for page, frame or block component descriptions.

It is non-mandatory for object class descriptions and defaultable for object descriptions.

Attribute Value Type

```
typedef struct {
   INT_type   minimum_dimension;
   INT_type   maximum_dimension;
} RULE_type;

typedef struct {
   TAG_type       dimensions_tag;
      /*
         FIXED_DIMENSIONS_tag,
         VARIABLE_PAGE_HEIGHT_tag,
         RULE_A_tag, RULE_B_tag,
         MAX_SIZE_tag
      */
   union {
      INT_type    fixed_dimension;
      INT_type    variable_page_height;
      RULE_type   rule;
   } dimensions_body;
} DIMENSION_type;

typedef struct {
   BOOL_type
            horizontal_dimension_is_null;
   DIMENSION_type  horizontal_dimension;
   BOOL_type vertical_dimension_is_null;
   DIMENSION_type  vertical_dimension;
} at_DIMENSIONS_type;
```

If either of the fields `horizontal_dimension_is_null` or `vertical_dimension_is_null` is `BOOL_true`, then the corresponding parameter ("horizontal dimension" or "vertical dimension") has the value ‘null’.

If the field dimensions_tag has the value MAX_SIZE_tag, then the corresponding parameter ("horizontal dimension" or "vertical dimension") has the sub-parameter "maximum size" with the value 'applies'.

If the value of the field dimensions_tag is RULE_A_tag or RULE_B_tag, then the corresponding parameter ("horizontal dimension" or "vertical dimension") has the sub-parameter "Rule A" or "Rule B".

Attribute Parameters

Parameter	Field Name	Parameter Name
"horizontal dimension"	horizontal_dimension	HORIZONTAL_DIMENSION_parm
"vertical dimension"	vertical_dimension	VERTICAL_DIMENSION_parm
"fixed dimension"	fixed_dimension	*none*
"variable page height"	variable_page_height	*none*
"Rule A"	rule	*none*
"Rule B"	rule	*none*

Default Value

Each of the parameters ("horizontal dimension" and "vertical dimension") is independently defaultable.

For a frame or block: horizontal_dimension and vertical_dimension both with the sub-parameter "fixed dimension", with a value that is the maximum size that can be achieved for the position within the area of the immediately superior object.

For a page: horizontal_dimension and vertical_dimension both with the sub-parameter "fixed dimension", with a value that is the assured reproduction area for ISO A4.

Semantics

The field vertical_dimension.dimensions_tag can have the value VARIABLE_PAGE_HEIGHT_tag only if the field horizontal_dimension.dimensions_tag has the value FIXED_DIMENSIONS_tag.

The field dimensions_tag may only have the value VARIABLE_PAGE_HEIGHT_tag when the attribute applies to basic pages, which are currently unsupported.

If the value of the field dimensions_tag is RULE_A_tag or RULE_B_tag, then the corresponding parameter ("horizontal dimension" or "vertical dimension") has the sub-parameter "Rule A" or "Rule B".

17.1.2.3 at_BORDER

Applicability

This attribute may be specified only for frame and block component descriptions and presentation styles.

It is non-mandatory for object class descriptions and presentation styles and defaultable for object descriptions.

Attribute Value Type

```
typedef struct {
    BORDER_INFO_type     left_hand_edge;
    BORDER_INFO_type     right_hand_edge;
    BORDER_INFO_type     trailing_edge;
    BORDER_INFO_type     leading_edge;
} at_BORDER_type;

typedef struct {
    BOOL_type  is_null;
    INT_type   border_line_width;
    VALUE_type border_line_type;
    INT_type   border_freespace_width;
} BORDER_INFO_type;
```

If the field `is_null` is `BOOL_true`, then the corresponding parameter has the value 'null'.

If the sub-parameter "border line width" is specified, then the field `border_line_width` must contain a non-negative integer.

If the parameter "border line type" is specified, then the field `border_line_type` must contain one of the values

Identifier	Meaning
at_BORDER_solid	'solid'
at_BORDER_dashed	'dashed'
at_BORDER_dot	'dot'
at_BORDER_dash_dot	'dash-dot'
at_BORDER_dash_dot_dot	'dash-dot-dot'
at_BORDER_invisible	'invisible'

If the parameter "border freespace width" is specified, then the field `border_freespace_width` must contain a non-negative integer.

Attribute Parameters

Parameter	Field Name	Parameter Name
"left-hand-edge"	left_hand_edge	LEFT_HAND_EDGE_parm
"right-hand-edge"	right_hand_edge	RIGHT_HAND_EDGE_parm
"trailing-edge"	trailing_edge	BORDER_TRAILING_EDGE_parm
"leading-edge"	leading_edge	BORDER_LEADING_EDGE_parm

Default Value

```
{
  { BOOL_false, 20, at_BORDER_solid, 0 },
  { BOOL_false, 20, at_BORDER_solid, 0 },
  { BOOL_false, 20, at_BORDER_solid, 0 },
  { BOOL_false, 20, at_BORDER_solid, 0 }
}
```

17.1.2.4 at_BALANCE

Applicability

This attribute may be specified only for composite layout component descriptions without immediately subordinate blocks.

It is non-mandatory for layout object class descriptions and defaultable for layout object descriptions.

Attribute Value Type

```
typedef SEQUENCE_CONSTITUENT_type
          at_BALANCE_type;
```

If the value of this attribute is NULL_SEQUENCE, then the value of the attribute is 'null'.

If the value is not NULL_SEQUENCE, then there must be at least two constituents in the sequence.

All constituents specified must be object class descriptions.

Default Value

NULL_SEQUENCE

Semantics

The value NULL_SEQUENCE indicates the ODA distinguished value 'null'.

17.1.2.5 at_LAYOUT_PATH

Applicability

This attribute may be specified only for frame component descriptions.

It is non-mandatory for object class descriptions and defaultable for object descriptions.

Attribute Value Type

```
typedef VALUE_type at_LAYOUT_PATH_type;
```

The value of this attribute must be one of

Identifier	Meaning
at_LAYOUT_PATH_type_0	'0°'
at_LAYOUT_PATH_type_90	'90°'
at_LAYOUT_PATH_type_180	'180°'
at_LAYOUT_PATH_type_270	'270°'

Default Value

at_LAYOUT_PATH_type_270

17.1.2.6 at_LOGICAL_SOURCE

Applicability

This attribute may be specified only for frame object class descriptions.

It is non-mandatory.

Attribute Value Type

```
typedef CONSTITUENT
           at_LOGICAL_SOURCE_type;
```

Semantics

The value of this attribute must be a constituent that is a logical object class.

The value NULL_CONSTITUENT is not allowed.

17.1.2.7 at_PERMITTED_CATEGORIES

Applicability

This attribute may be specified only for frame component descriptions.

It is non-mandatory for object class descriptions and defaultable for object descriptions.

Attribute Value Type

```
typedef SEQUENCE_STRING_type
        at_PERMITTED_CATEGORIES_type;
```

The characters in each of the strings in the attribute value must be from the minimum subrepertoire of ISO 6937-2.

Default Value

NULL_SEQUENCE

Semantics

The value NULL_SEQUENCE indicates the value 'null'.

17.1.2.8 at_IMAGING_ORDER

Applicability

This attribute may be specified only for composite page or frame object descriptions.

It is non-mandatory.

Attribute Value Type

```
typedef SEQUENCE_CONSTITUENT_type
        at_IMAGING_ORDER_type;
```

Each constituent in the attribute value must be one of the immediate subordinates to the constituent to which the attribute applies.

17.1.2.9 at_TRANSPARENCY

Applicability

This attribute may be specified only for page, frame or block component descriptions and for presentation styles.

It is non-mandatory for object class descriptions and presentation styles. It is defaultable for object descriptions.

Attribute Value Type

```
typedef VALUE_type at_TRANSPARENCY_type;
```

The attribute must have one of the values

Identifier	Meaning
at_TRANSPARENCY_transparent	'transparent'
at_TRANSPARENCY_opaque	'opaque'

Default Value

```
at_TRANSPARENCY_transparent
```

17.1.2.10 at_COLOUR

Applicability

This attribute may be specified only for page, frame or block component descriptions and for presentation styles.

It is non-mandatory for object class descriptions and presentation styles. It is defaultable for object descriptions.

Attribute Value Type

```
typedef VALUE_type at_COLOUR_type;
```

The attribute must have one of the values

Identifier	Meaning
`at_COLOUR_colourless`	'colourless'
`at_COLOUR_white`	'white'

Default Value

```
at_COLOUR_colourless
```

17.1.2.11 at_PAGE_POSITION

Applicability

This attribute may be specified only for page component descriptions.

It is non-mandatory for object class descriptions and defaultable for object descriptions.

Attribute Value Type

```
typedef INT_PAIR_type
         at_PAGE_POSITION_type;
```

Each integer must be non-negative.

Attribute Parameters

Parameter	Field Name	Parameter Name
the horizontal distance	`first`	*none*
the vertical distance	`second`	*none*

Default Value

Such that edge losses are minimized.

17.1.2.12 at_MEDIUM_TYPE

Applicability

This attribute may be specified only for page component descriptions.

It is non-mandatory for object class descriptions and defaultable for object descriptions.

Attribute Value Type

```
typedef struct {
    INT_PAIR_type    nominal_page_size;
    VALUE_type       side_of_sheet;
} at_MEDIUM_TYPE_type;
```

The field `side_of_sheet` must have one of the values

Identifier	Meaning
`at_MEDIUM_TYPE_recto`	'recto'
`at_MEDIUM_TYPE_verso`	'verso'
`at_MEDIUM_TYPE_unspecified`	'unspecified'

Attribute Parameters

Parameter	Field Name	Parameter Name
"nominal page size"	`nominal_page_size`	`NOMINAL_PAGE_SIZE_parm`
horizontal dimension	`nominal_page_size.first`	*none*
vertical dimension	`nominal_page_size.second`	*none*
"side of sheet"	`side_of_sheet`	`SIDE_OF_SHEET_parm`

Default Value

{ { *Dimensions for ISO A4* },
`at_MEDIUM_TYPE_unspecified` }

17.1.3 Logical Attributes

17.1.3.1 at_PROTECTION

Applicability

This attribute may be specified only for logical component descriptions.

It is non-mandatory for object class descriptions and defaultable for object descriptions.

Attribute Value Type

```
typedef VALUE_type at_PROTECTION_type;
```

The attribute must have one of the values

Identifier	Meaning
at_PROTECTION_protected	'protected'
at_PROTECTION_unprotected	'unprotected'

Default Value

at_PROTECTION_unprotected

17.1.3.2 at_LAYOUT_STYLE

Applicability

This attribute may be specified only for logical component descriptions.

It is non-mandatory.

Attribute Value Type

```
typedef CONSTITUENT at_LAYOUT_STYLE_type;
```

The value of this attribute must be a reference to a layout style constituent.

Default Value

NULL_CONSTITUENT

Semantics

The value NULL_CONSTITUENT designates the distinguished value 'null'.

17.1.4 Layout Style Attributes

17.1.4.1 at_LAYOUT_STYLE_IDENTIFIER

Applicability

This attribute may be specified only for layout styles.

It is mandatory.

Attribute Value Type

```
typedef STRING_type
          at_LAYOUT_STYLE_IDENTIFIER_type;
```

Semantics

The characters forming the value of the `STRING_type` should be an object identifier according to the rules in section 5.6.1, Part 2 of ISO IS 8613.

The programmer is warned that assignment of a value to this attribute will not be checked for legality. Read section 20.3 for a discussion of the rules applying to identifiers.

17.1.5 Layout Directive Attributes

A layout directive is an attribute of a logical component or a logical style.

17.1.5.1 at_BLK_ALGNM

Applicability

This attribute may be specified only for layout styles and is applicable only to basic logical component descriptions.

It is non-mandatory when specified for layout styles and when applied to logical object class descriptions. It is defaultable when applied to logical object descriptions.

Attribute Value Type

```
typedef VALUE_type at_BLK_ALGNM_type;
```

The attribute must have one of the values

Identifier	Meaning
`at_BLK_ALGNM_right_hand_aligned`	'right-hand aligned'
`at_BLK_ALGNM_left_hand_aligned`	'left-hand aligned'
`at_BLK_ALGNM_centred`	'centred'
`at_BLK_ALGNM_null`	'null'

Default Value

`at_BLK_ALGNM_right_hand_aligned`

17.1.5.2 at_CONCATENATION

Applicability

This attribute may be specified only for layout styles and is applicable only to basic logical component descriptions.

It is non-mandatory when specified for layout styles and when applied to logical object class descriptions. It is defaultable when applied to logical object descriptions.

Attribute Value Type

```
typedef VALUE_type at_CONCATENATION_type;
```

The value of this attribute must be one of

Identifier	Meaning
`at_CONCATENATION_concatenated`	'concatenated'
`at_CONCATENATION_non_concated`	'non-concatenated'

Default Value

`at_CONCATENATION_non_concated`

17.1.5.3 at_FILL_ORDER

Applicability

This attribute may be specified only for layout styles and is applicable only to basic logical component descriptions.

It is non-mandatory when specified for layout styles and when applied to logical object class descriptions. It is defaultable when applied to logical object descriptions.

Attribute Value Type

```
typedef VALUE_type at_FILL_ORDER_type;
```

This attribute must have one of the values

Identifier	Meaning
`at_FILL_ORDER_normal_order`	'normal order'
`at_FILL_ORDER_reverse_order`	'reverse order'

Default Value

`at_FILL_ORDER_normal_order`

17.1.5.4 at_INDIVISIBILITY

Applicability

This attribute may be specified only for layout styles and is applicable only to all logical component descriptions except the document logical root.

It is non-mandatory when specified for layout styles and when applied to logical object class descriptions. It is defaultable when applied to logical object descriptions.

Attribute Value Type

```
typedef LAYOUT_OBJECT_type
           at_INDIVISIBILITY_type;
```

See section 13.9 for a description of the type `LAYOUT_OBJECT_type`.

Default Value

`{ NULL_tag, NULL_CONSTITUENT }`

17.1.5.5 at_LAYOUT_CATEGORY

Applicability

This attribute may be specified only for layout styles and is applicable only to basic logical component descriptions.

It is non-mandatory when specified for layout styles and when applied to logical object class descriptions. It is defaultable when applied to logical object descriptions.

Attribute Value Type

```
typedef STRING_type
           at_LAYOUT_CATEGORY_type;
```

The characters in the string must be from the minimum subrepertoire of ISO 6937-2.

Default Value

NULL_STRING

Semantics

If the value is NULL_STRING, then the value of the attribute is 'null'.

17.1.5.6 at_LAYOUT_OBJECT_CLASS

Applicability

This attribute may be specified only for layout styles and is applicable to all logical component descriptions.

It is non-mandatory when specified for layout styles and when applied to logical object class descriptions. It is defaultable when applied to logical object descriptions.

Attribute Value Type

```
typedef CONSTITUENT
            at_LAYOUT_OBJECT_CLASS_type;
```

Default Value

NULL_CONSTITUENT

Semantics

If the value of this attribute is NULL_CONSTITUENT, then the value of the attribute is 'null'.

17.1.5.7 at_NEW_LAYOUT_OBJECT

Applicability

This attribute may be specified only for layout styles and is applicable only to all logical component descriptions except the document logical root.

It is non-mandatory when specified for layout styles and when applied to logical object class descriptions. It is defaultable when applied to logical object descriptions.

Attribute Value Type

```
typedef LAYOUT_OBJECT_type
            at_NEW_LAYOUT_OBJECT_type;
```

See section 13.9 for a description of the type LAYOUT_OBJECT_type.

Default Value

```
{ NULL_tag, NULL_CONSTITUENT }
```

17.1.5.8 at_OFFSET

Applicability

This attribute may be specified only for layout styles and is applicable only to basic logical component descriptions.

It is non-mandatory when specified for layout styles and when applied to logical object class descriptions. It is defaultable when applied to logical object descriptions.

Attribute Value Type

```
typedef struct {
    INT_type    leading_offset;
    INT_type    trailing_offset;
    INT_type    left_hand_offset;
    INT_type    right_hand_offset;
} at_OFFSET_type;
```

Each field, if specified, must have a value that is a non-negative integer.

Attribute Parameters

Parameter	Field Name	Parameter Name
"leading offset"	leading_offset	LEADING_OFFSET_parm
"trailing offset"	trailing_offset	TRAILING_OFFSET_parm
"left-hand offset"	left_hand_offset	LEFT_HAND_OFFSET_parm
"right-hand offset"	right_hand_offset	RIGHT_HAND_OFFSET_parm

Default Value

Each of the parameters is independently defaultable. The default value for each parameter is 0.

17.1.5.9 at_SAME_LAYOUT_OBJECT

Applicability

This attribute may be specified only for layout styles and is applicable to all logical component descriptions except the document logical root.

It is non-mandatory when specified for layout styles and when applied to logical object class descriptions. It is defaultable when applied to logical object descriptions.

Attribute Value Type

```
typedef struct {
    CONSTITUENT                          parm1;
    LAYOUT_OBJECT_type                   parm2;
} at_SAME_LAYOUT_OBJECT_type;
```

See section 13.9 for a description of the type `LAYOUT_OBJECT_type`.

It is illegal for the value of `parm2.layout_object_tag` to be `NULL_tag`.

Default Value

A value such that the value of the field `parm1` is `NULL_CONSTITUENT`.

Semantics

If this attribute is applied to a logical object description, then the `CONSTITUENT` specified for the field `parm1` must be for another logical object description, or `NULL_CONSTITUENT`.

If this attribute is applied to a logical object class description, then the `CONSTITUENT` specified for the field `parm1` must be for an object description, or `NULL_CONSTITUENT`.

If the field `parm1` is `NULL_CONSTITUENT`, then the value of `parm2` is ignored.

If the field `parm1` is not `NULL_CONSTITUENT`, then the value of `parm2.layout_object_tag` must not be `NULL_tag`.

17.1.5.10 at_SEPARATION

Applicability

This attribute may be specified only for layout styles and is applicable only to basic logical component descriptions.

It is non-mandatory for object class descriptions and layout styles. It is defaultable for object descriptions.

Attribute Value Type

```
typedef struct {
    INT_type    leading_edge;
    INT_type    trailing_edge;
    INT_type    centre_separation;
} at_SEPARATION_type;
```

Each field, if specified, must have a non-negative integer for a value.

Attribute Parameters

Parameter	Field Name	Parameter Name
"leading edge"	`leading_edge`	`LEADING_EDGE_parm`
"trailing edge"	`trailing_edge`	`TRAILING_EDGE_parm`
"centre separation"	`centre_separation`	`CENTRE_SEPARATION_parm`

Default Value

Each of the three parameters is independently defaultable. The default value for each parameter is `0`.

17.1.5.11 at_SYNCHRONIZATION

Applicability

This attribute may be specified only for layout styles and is applicable to all logical component descriptions except the document logical root.

It is non-mandatory when specified for layout styles and when applied to logical object class descriptions. It is defaultable when applied to logical object descriptions.

Attribute Value Type

```
typedef CONSTITUENT
          at_SYNCHRONIZATION_type;
```

Default Value

`NULL_CONSTITUENT`

Semantics

The value `NULL_CONSTITUENT` indicates the value ‘null’.

If this attribute is specified for a logical object description, the value of this attribute must be a reference to another logical object description or `NULL_CONSTITUENT`.

If this attribute is specified for a logical object class description or a layout style, the value of this attribute must be a reference to an object description or `NULL_CONSTITUENT`.

17.1.6 Presentation Style Attributes

All of the attributes in this section are applicable only to presentation styles.

17.1.6.1 at_PRESENTATION_STYLE_ID

Applicability

This attribute may be specified only for presentation styles.

It is mandatory.

Attribute Value Type

```
typedef STRING_type
          at_PRESENTATION_STYLE_ID_type;
```

Semantics

The characters forming the value the `STRING_type` should be an identifier according to the rules in section 5.8.1, Part 2 of ISO IS 8613.

The programmer is warned that values assigned to this attribute will not be checked for legality. Read section 20.3 for a discussion of the rules applying to object identifiers.

17.1.7 Content Portion Attributes

All of the attributes in this section may be specified only for content portions. Note, however, that certain values of these attributes may be further restricted according to the particular content architecture.

17.1.7.1 at_CONTENT_ID_LOGICAL

Applicability

It is non-mandatory, but one of `at_CONTENT_ID_LOGICAL` or `at_CONTENT_ID_LAYOUT` must be specified for a content portion. (See exceptional case described in section 5.9.1, Part 2 of ISO IS 8613.)

Attribute Value Type

```
typedef STRING_type
          at_CONTENT_ID_LOGICAL_type;
```

Semantics

The characters forming the value the `STRING_type` should be an identifier according to the rules in section 5.9.1, Part 2 of ISO IS 8613.

The programmer is warned that values assigned to this attribute will not be checked for legality. Read section 20.3 for a discussion of the rules applying to object identifiers.

17.1.7.2 at_CONTENT_ID_LAYOUT

Applicability

It is non-mandatory, but one of `at_CONTENT_ID_LOGICAL` or `at_CONTENT_ID_LAYOUT` must be specified for a content portion. (See exceptional case described in section 5.9.1, Part 2 of ISO IS 8613.)

Attribute Value Type

```
typedef STRING_type
          at_CONTENT_ID_LAYOUT_type;
```

Semantics

The characters forming the value of the `STRING_type` should be an identifier according to the rules in section 5.9.1, Part 2 of ISO IS 8613.

The programmer is warned that values assigned to this attribute will not be checked for legality. Read section 20.3 for a discussion of the rules applying to object identifiers.

17.1.7.3 at_TYPE_OF_CODING

Applicability

It is defaultable.

Attribute Value Type

```
typedef ASN1_OBJECT_ID_type
            at_TYPE_OF_CODING_type;
```

The value of each of these attributes must be an ASN.1 object identifier as specified in the appropriate section of each content architecture.

The tool kit does not support the exception case described in section 5.9.2, Part 2 of ISO IS 8613.

The following identifiers are predefined by the tool kit for each content architecture. For the character content architecture:

Identifier	Meaning
`at_TYPE_OF_CODING_cc`	*only 1 value is defined*

For the raster graphics content architecture:

Identifier	Meaning
`at_TYPE_OF_CODING_T6`	'T6 encoding'
`at_TYPE_OF_CODING_T4_1`	'T4 one dimensional encoding'
`at_TYPE_OF_CODING_T4_2`	'T4 two dimensional encoding'
`at_TYPE_OF_CODING_bitmap`	'bitmap encoding'

For the geometric graphics content architecture:

Identifier	Meaning
`at_TYPE_OF_CODING_gg`	*only 1 value is defined*

17.1.7.4 at_CONTENT_INFORMATION

Applicability

It is non-mandatory.

Attribute Value Type

```
typedef SEQUENCE_BYTE_type
            at_CONTENT_INFORMATION_type;
```

Semantics

The structure and semantics of the value of this attribute is determined by the content architecture definitions.

17.1.7.5 at_ALT_REP

Applicability

It is non-mandatory.

Attribute Value Type

```
typedef STRING_type at_ALT_REP_type;
```

The characters in the string must be from the character sets specified in the document profile attribute "alternative representation character sets" (17.1.8.18).

17.1.8 Document Profile Attributes

All of the attributes in this section are applicable only to the document profile and are non-mandatory.

17.1.8.1 dp_GENERIC_LAY_STRUCT

Attribute Value Type

```
typedef VALUE_type
            dp_GENERIC_LAY_STRUCT_type;
```

The value of this attribute must be one of

Identifier	Meaning
dp_GENERIC_LAY_STRUCT_fact_set	'factor set'
dp_GENERIC_LAY_STRUCT_part_gen	'partial generator set'
dp_GENERIC_LAY_STRUCT_comp_gen	'complete generator set'

17.1.8.2 dp_SPECIFIC_LAY_STRUCT

Attribute Value Type

```
typedef VALUE_type
          dp_SPECIFIC_LAY_STRUCT_type;
```

The value of this attribute must be

Identifier	Meaning
dp_SPECIFIC_LAY_STRUCT_present	'present'

17.1.8.3 dp_GENERIC_LOG_STRUCT

Attribute Value Type

```
typedef VALUE_type
          dp_GENERIC_LOG_STRUCT_type;
```

The value of this attribute must be one of

Identifier	Meaning
dp_GENERIC_LOG_STRUCT_fact_set	'factor set'
dp_GENERIC_LOG_STRUCT_part_gen	'partial generator set'
dp_GENERIC_LOG_STRUCT_comp_gen	'complete generator set'

17.1.8.4 dp_SPECIFIC_LOG_STRUCT

Attribute Value Type

```
typedef VALUE_type
          dp_SPECIFIC_LOG_STRUCT_type;
```

The value of this attribute must be

Identifier	Meaning
dp_SPECIFIC_LOG_STRUCT_present	'present'

17.1.8.5 dp_LAYOUT_STYLES

Attribute Value Type

```
typedef VALUE_type dp_LAYOUT_STYLES_type;
```

The value of this attribute must be

Identifier	Meaning
dp_LAYOUT_STYLES_present	'present'

17.1.8.6 dp_PRESENTATION_STYLES

Attribute Value Type

```
typedef VALUE_type
            dp_PRESENTATION_STYLES_type;
```

The value of this attribute, if specified, must be

Identifier	Meaning
dp_PRESENTATION_STYLES_present	'present'

17.1.8.7 dp_EXTERNAL_DOCUMENT_CLASS

Attribute Value Type

```
typedef STRING_OR_ID_type
            dp_EXTERNAL_DOCUMENT_CLASS_type;
```

The type STRING_OR_ID_type is described in section 13.10.

Semantics

The value of this attribute must designate a document class description in accordance with the rules specified in Part 2 of ISO IS 8613.

There are no predefined values for this attribute.

17.1.8.8 dp_RESOURCE_DOCUMENT

Attribute Value Type

```
typedef STRING_OR_ID_type
            dp_RESOURCE_DOCUMENT_type;
```

The type STRING_OR_ID_type is described in section 13.10.

17.1.8.9 dp_RESOURCES

Attribute Value Type

```
typedef SEQUENCE_RESOURCE_MAP_type
            dp_RESOURCES_type;
```

Each element is of type

```
typedef struct RESOURCE_MAP_struct {
    STRING_type         name;
    CONSTITUENT         id;
} RESOURCE_MAP_type;
```

The characters in the field `name` must be from the minimum subrepertoire of ISO 6937-2.

The constituents appearing in the sequence of `RESOURCE_MAP_type` objects must be object class descriptions.

17.1.8.10 dp_DOC_APP_PROFILE

Attribute Value Type

```
typedef ASN1_OBJECT_ID_type
          dp_DOC_APP_PROFILE_type;
```

The value of this attribute must designate an application profile as defined in Part 1 of ISO IS 8613.

There are no predefined values for this attribute.

Note that the tool kit does not support the integer form of this attribute.

17.1.8.11 dp_DOC_APP_PROF_DEFAULTS

Attribute Value Type

```
typedef struct {
   ASN1_OBJECT_ID_type content_arch_class;
   BOOL_type           content_type_specified;
   VALUE_type          content_type;
   BOOL_type           dimensions_specified;
   PARM_MASK_type      dimensions_parm_mask;
   at_DIMENSIONS_type  dimensions;
   BOOL_type           transparency_specified;
   VALUE_type          transparency;
   BOOL_type           colour_specified;
   VALUE_type          colour;
   BOOL_type           border_specified;
   PARM_MASK_type      border_parm_mask;
```

```
        at_BORDER_type      border;
        BOOL_type           layout_path_specified;
        VALUE_type          layout_path;
        BOOL_type           page_position_specified;
        INT_PAIR_type       page_position;
        BOOL_type           medium_type_specified;
        PARM_MASK_type      medium_type_parm_mask;
        at_MEDIUM_TYPE_type medium_type;
        BOOL_type           block_alignment_specified;
        VALUE_type          block_alignment;
        ASN1_OBJECT_ID_type type_of_coding;
        CONSTITUENT         cc_presentation;
        CONSTITUENT         cc_coding;
        CONSTITUENT         rg_presentation;
        CONSTITUENT         rg_coding;
        CONSTITUENT         gg_presentation;
        CONSTITUENT         gg_coding;
    } dp_DOC_APP_PROF_DEFAULTS_type;
```

Semantics

Most of the fields of the structure for this type correspond to attributes whose default values in the document profile attribute "document application profile defaults":

Identifier	Attributes
`content_arch_class`	"content architecture class"
`content_type`	"content type"
`dimensions`	"dimensions"
`transparency`	"transparency"
`colour`	"colour"
`border`	"border"
`layout_path`	"layout path"
`page_position`	"page position"
`medium_type`	"medium type"
`block_alignment`	"block alignment"
`type_of_coding`	"type of coding"
`cc_presentation`	*presentation attributes for character content*
`cc_coding`	*coding attributes for character content*
`rg_presentation`	*presentation attributes for raster graphics content*
`rg_coding`	*coding attributes for raster graphics content*

`gg_presentation`	*presentation attributes for geometric graphics content*
`gg_coding`	*coding attributes for geometric graphics content*

The following table shows the method by which the absence of a default value for each field in the structure is designated:

Field	Unspecified
content_arch_class	content_arch_class == NULL_SEQUENCE
content_type	content_type_specified == BOOL_false
dimensions	dimensions_specified == BOOL_false
transparency	transparency_specified == BOOL_false
colour	colour_specified == BOOL_false
border	border_specified == BOOL_false
layout_path	layout_path_specified == BOOL_false
page_position	page_position_specified == BOOL_false
medium_type	medium_type_specified == BOOL_false
block_alignment	block_alignment_specified == BOOL_false
type_of_coding	type_of_coding == NULL_SEQUENCE
cc_presentation	cc_presentation == NULL_CONSTITUENT
cc_coding	cc_coding == NULL_CONSTITUENT
rg_presentation	rg_presentation == NULL_CONSTITUENT
rg_coding	rg_coding == NULL_CONSTITUENT
gg_presentation	gg_presentation == NULL_CONSTITUENT
gg_coding	gg_coding == NULL_CONSTITUENT

7.1.8.12 dp_DOC_ARCH_CLASS

Attribute Value Type

```
typedef VALUE_type dp_DOC_ARCH_CLASS_type;
```

The attribute must have one of the values

Identifier	Meaning
dp_DOC_ARCH_CLASS_formatted	'formatted'
dp_DOC_ARCH_CLASS_processable	'processable'
dp_DOC_ARCH_CLASS_fp	'formatted processable'

17.1.8.13 dp_CONTENT_ARCH_CLASSES

Attribute Value Type

```
typedef SEQUENCE_ASN1_OBJECT_ID_type
          dp_CONTENT_ARCH_CLASSES_type;
```

Semantics

Each ASN.1 object identifier in the sequence must designate a content architecture defined in ISO IS 8613 or other standards. Section 17.1.1.11 defines the set of identifiers that may be used as predefined values for the ASN.1 object identifiers.

17.1.8.14 dp_INTERCHANGE_FMT_CLASS

Attribute Value Type

```
typedef VALUE_type
          dp_INTERCHANGE_FMT_CLASS_type;
```

This attribute must have one of the values

Identifier	Meaning
dp_INTERCHANGE_FMT_CLASS_a	'A'
dp_INTERCHANGE_FMT_CLASS_b	'B'

17.1.8.15 dp_ODA_VERSION

Attribute Value Type

```
typedef struct {
    STRING_type        standard;
    STRING_type        date;
} dp_ODA_VERSION_type;
```

The value of the field `date` must be a complete calendar date character string in accordance with ISO 8601.

It is illegal for the value of either field to be `NULL_STRING`.

Attribute Parameters

Parameter	Field Name	Parameter Name
document architecture standard	standard	*none*
date of document standard	date	*none*

Semantics

The contents of the fields of this attribute will not be checked for syntactic or semantic validity.

17.1.8.16 dp_PROFILE_CHARACTER_SETS

Attribute Value Type

```
typedef SEQUENCE_BYTE_type
            dp_PROFILE_CHARACTER_SETS_type;
```

The value of this attribute consists of the escape sequence(s) used to designate the character set(s) in accordance with ISO 2022 and the register of ISO 2375.

Semantics

The value of this attribute will not be checked for syntactic or semantic validity.

17.1.8.17 dp_COMMENTS_CHARACTER_SETS

Attribute Value Type

```
typedef SEQUENCE_BYTE_type
            dp_COMMENTS_CHARACTER_SETS_type;
```

The value of this attribute consists of the escape sequence(s) used to designate the character set(s) in accordance with ISO 2022 and the register of ISO 2375.

Semantics

The value of this attribute will not be checked for syntactic or semantic validity.

17.1.8.18 dp_ALT_REP_CHARACTER_SETS

Attribute Value Type

```
typedef SEQUENCE_BYTE_type
            dp_ALT_REP_CHARACTER_SETS_type;
```

The value of this attribute consists of the escape sequence(s) used to designate the character set(s) in accordance with ISO 2022 and the register of ISO 2375.

Semantics

The value of this attribute will not be checked for syntactic or semantic validity.

17.1.8.19 dp_PAGE_DIMENSIONS

Attribute Value Type

```
typedef SEQUENCE_INT_PAIR_type
        dp_PAGE_DIMENSIONS_type;
```

Each field must be a positive integer.

Attribute Parameters

Parameter	Field Name	Parameter Name
horizontal page dimension	`first`	*none*
vertical page dimension	`second`	*none*

17.1.8.20 dp_MEDIUM_TYPES

Attribute Value Type

```
typedef SEQUENCE_at_MEDIUM_TYPE_type
        dp_MEDIUM_TYPES_type;
```

Each element of the sequence must be a valid value for the attribute "medium type", as described in section 17.1.2.12, except that all parameters of each value must be specified.

17.1.8.21 dp_LAYOUT_PATHS

Attribute Value Type

```
typedef SEQUENCE_VALUE_type
        dp_LAYOUT_PATHS_type;
```

Semantics

Each element of the sequence must be a valid value for the attribute "layout path", as described in section 17.1.2.5.

17.1.8.22 dp_PROTECTIONS

Attribute Value Type

```
typedef SEQUENCE_VALUE_type
           dp_protections_type;
```

Semantics

Each element of the sequence must be a valid value for the attribute "protection", as described in section 17.1.3.1.

17.1.8.23 dp_BLOCK_ALIGNMENTS

Attribute Value Type

```
typedef SEQUENCE_VALUE_type
           dp_BLOCK_ALIGNMENTS_type;
```

Semantics

Each element of the sequence must be a valid value for the attribute "block alignment", as described in section 17.1.5.1.

17.1.8.24 dp_FILL_ORDERS

Attribute Value Type

```
typedef SEQUENCE_VALUE_type
              dp_FILL_ORDERS_type;
```

Semantics

Each element of the sequence must be a valid value for the attribute "fill order", as described in section 17.1.5.3.

17.1.8.25 dp_TRANSPARENCIES

Attribute Value Type

```
typedef SEQUENCE_VALUE_type
           dp_TRANSPARENCIES_type;
```

Semantics

Each element of the sequence must be a valid value for the attribute "transparency", as described in section 17.1.2.9.

17.1.8.26 dp_COLOURS

Attribute Value Type

```
typedef SEQUENCE_VALUE_type
                    dp_COLOURS_type;
```

Semantics

Each element of the sequence must be a valid value for the attribute "colour", as described in section 17.1.2.10.

17.1.8.27 dp_BORDERS

Attribute Value Type

```
typedef SEQUENCE_VALUE_type
                    dp_BORDERS_type;
```

Semantics

Each element of the sequence must be a valid value for the attribute "border", as described in section 17.1.2.3.

17.1.8.28 dp_PAGE_POSITIONS

Attribute Value Type

```
typedef SEQUENCE_INT_PAIR_type
                dp_PAGE_POSITIONS_type;
```

Semantics

Each element of the sequence must be a valid value for the layout directive attribute "page position", as described in section 17.1.2.11.

17.1.8.29 dp_TYPES_OF_CODING

Attribute Value Type

```
typedef SEQUENCE_ASN1_OBJECT_ID_type
            dp_TYPES_OF_CODING_type;
```

Semantics

Each element of the sequence must be a valid value for the layout directive attribute "type of coding", as described in section 17.1.7.3.

17.1.8.30 dp_CODING_ATTRIBUTES

Attribute Value Type

```
typedef SEQUENCE_CONSTITUENT_type
            dp_CODING_ATTRIBUTES_type;
```

Semantics

The value of this attribute must be a sequence of constituents, which are content portions. The set of coding attributes and corresponding values associated with all of the content portions are considered to specify the non-basic coding attribute values for this attribute.

At the current time, there is no representation for non-basic coding attributes for any of the content architectures. Thus, the only legal value for this attribute is an empty sequence. Note that the value `NULL_SEQUENCE` is not the same as the empty sequence.

17.1.8.31 dp_PRESENTATION_FEATURES

Attribute Value Type

```
typedef SEQUENCE_CONSTITUENT_type
            dp_PRESENTATION_FEATURES_type;
```

Semantics

This attribute is unimplemented.

17.1.8.32 dp_NUMBER_OBJECTS_PER_PAGE

Attribute Value Type

```
typedef INT_type
            dp_NUMBER_OBJECTS_PER_PAGE_type;
```

The value of this attribute must be a positive integer.

17.1.8.33 dp_UNIT_SCALING

Attribute Value Type

```
typedef INT_PAIR_type
            dp_UNIT_SCALING_type;
```

The field `first` of the attribute value will form the numerator of the ratio for scaling BMU's. The field `second` will form the denominator.

Attribute Parameters

Parameter	Field Name	Parameter Name
m	`first`	*none*
n	`second`	*none*

Semantics

There is no default value for this attribute. However, if this attribute is unspecified, then the unit used in the positioning and dimensioning attributes is equal to 1 BMU.

17.1.8.34 dp_FONTS_LIST

Attribute Value Type

```
typedef SEQUENCE_FONT_PAIR_type
            dp_FONTS_LIST_type;
```

Where the type of each element is `FONT_PAIR_type`, described below:

```
typedef struct FONT_PAIR_struct {
    INT_type                  font_id;
    FONT_REFERENCE_type       font_reference;
} FONT_PAIR_type;
```

The type `FONT_REFERENCE_type` is described in section 13.11.

17.1.8.35 dp_TITLE

Attribute Value Type

```
typedef STRING_type dp_TITLE_type;
```

The characters in the string are from the document profile character set.

17.1.8.36 dp_SUBJECT

Attribute Value Type

```
typedef STRING_type dp_SUBJECT_type;
```

The characters in the string are from the document profile character set.

17.1.8.37 dp_DOCUMENT_REFERENCE

Attribute Value Type

```
typedef STRING_OR_ID_type
            dp_DOCUMENT_REFERENCE_type;
```

The type STRING_OR_ID_type is described in section 13.10.

17.1.8.38 dp_DOCUMENT_TYPE

Attribute Value Type

```
typedef STRING_type dp_DOCUMENT_TYPE_type;
```

The characters in the string are from the document profile character set.

17.1.8.39 dp_ABSTRACT

Attribute Value Type

```
typedef STRING_type dp_ABSTRACT_type;
```

The characters in the string are from the document profile character set.

17.1.8.40 dp_DOCUMENT_DATE_AND_TIME

Attribute Value Type

```
typedef STRING_type
            dp_DOCUMENT_DATE_AND_TIME_type;
```

The value of this attribute must be a date character string in accordance with ISO 8601 and, optionally, a time of day character string in accordance with ISO 8601.

Semantics

The value of this attribute will not be checked for syntactic or semantic validity.

17.1.8.41 dp_CREATION_DATE_AND_TIME

Attribute Value Type

```
typedef STRING_type
            dp_CREATION_DATE_AND_TIME_type;
```

The value of this attribute must be a date character string in accordance with ISO 8601 and, optionally, a time of day character string in accordance with ISO 8601.

Semantics

The value of this attribute will not be checked for syntactic or semantic validity.

17.1.8.42 dp_LOCAL_FILING_DATE_TIME

Attribute Value Type

```
typedef STRING_type
        dp_LOCAL_FILING_DATE_TIME_type;
```

The value of this attribute must be a date character string in accordance with ISO 8601 and, optionally, a time of day character string in accordance with ISO 8601.

Semantics

The value of this attribute will not be checked for syntactic or semantic validity.

17.1.8.43 dp_EXPIRY_DATE_AND_TIME

Attribute Value Type

```
typedef STRING_type
        dp_EXPIRY_DATE_AND_TIME_type;
```

The value of this attribute must be a date character string in accordance with ISO 8601 and, optionally, a time of day character string in accordance with ISO 8601.

Semantics

The value of this attribute will not be checked for syntactic or semantic validity.

17.1.8.44 dp_START_DATE_AND_TIME

Attribute Value Type

```
typedef STRING_type
        dp_START_DATE_AND_TIME_type;
```

The value of this attribute must be a date character string in accordance with ISO 8601 and, optionally, a time of day character string in accordance with ISO 8601.

Semantics

The value of this attribute will not be checked for syntactic or semantic validity.

17.1.8.45 dp_PURGE_DATE_AND_TIME

Attribute Value Type

```
typedef STRING_type
         dp_PURGE_DATE_AND_TIME_type;
```

The value of this attribute must be a date character string in accordance with ISO 8601 and, optionally, a time of day character string in accordance with ISO 8601.

Semantics

The value of this attribute will not be checked for syntactic or semantic validity.

17.1.8.46 dp_RELEASE_DATE_AND_TIME

Attribute Value Type

```
typedef STRING_type
         dp_RELEASE_DATE_AND_TIME_type;
```

The value of this attribute must be a date character string in accordance with ISO 8601 and, optionally, a time of day character string in accordance with ISO 8601.

Semantics

The value of this attribute will not be checked for syntactic or semantic validity.

17.1.8.47 dp_REVISION_HISTORY

Attribute Value Type

```
typedef SEQUENCE_REVISION_HISTORY_type
         dp_REVISION_HISTORY_type;
```

The type of each element, REVISION_HISTORY_type, is defined below.

```
typedef struct REVISION_HISTORY_struct {
    STRING_type          revision_date_and_time;
    STRING_type          version_number;
    SEQUENCE_REVISER_type
                         revisers;
    STRING_OR_ID_type version_reference;
    SEQUENCE_BYTE_typeuser_comments;
} REVISION_HISTORY_type;

/* REVISER_type used in SEQUENCE_REVISER_type */
typedef struct REVISER_struct {
    SEQUENCE_PERSONAL_NAME_type names;
    SEQUENCE_BYTE_type          position;
    SEQUENCE_BYTE_type          organization;
} REVISER_type;
```

The type SEQUENCE_PERSONAL_NAME_type is described in sections 13.6 and 13.13.

The value of the field revision_date_and_time must be a date character string in accordance with ISO 8601 and, optionally, a time of day character string in accordance with ISO 8601.

The value of the field user_comments must consist of characters from the document profile character set.

The value of each of the strings comprising each of the fields names must have the format of a personal name as described in Annex C, Part 4 of ISO IS 8613.

The values of each of the fields position and organization must consist of characters from the document profile character set.

Attribute Parameters

The presence or absence of parameters for values of the "revision history" attribute is handled differently from the standard method used in the Tool Kit (see section 16.1.1). For this attribute, the status of parameters is indicated directly within the value.

For each REVISION_HISTORY_type element in the sequence value, the absence of:

- the parameter "revision date and time" is indicated by the field revision_date_and_time having the value NULL_STRING;
- the parameter "version number" is indicated by the field version_number having the value NULL_STRING;
- the parameter "reviser(s)" is indicated by the field revisers having the value NULL_SEQUENCE;

- the parameter "user comments" is indicated by the corresponding field, `user_comments` having the value `NULL_SEQUENCE`;
- the sub-parameters "name(s)", "position" or "organization" is indicated by the corresponding fields `names`, `position` and `organization` having the value(s) `NULL_SEQUENCE`.

Semantics

The programmer should be aware that the values of the fields comprising the value of this attribute will not be checked for syntactic or semantic validity.

17.1.8.48 dp_ORGANIZATIONS

Attribute Value Type

```
typedef SEQUENCE_STRING_type
            dp_ORGANIZATIONS_type;
```

The characters in the strings must be from the document profile character set.

17.1.8.49 dp_PREPARERS

Attribute Value Type

```
typedef SEQUENCE_NAME_ORG_type
          dp_PREPARERS_type;
```

The type `NAME_ORG_type` is defined in section 13.7.

Attribute Parameters

Parameter	Field Name	Parameter Name
"personal name of preparer"	`name`	*none*
"preparer's organisation"	`organization`	*none*

The technique used to designate the presence or absence of parameters for each element of the value of this attribute is described in section 13.7.1.

17.1.8.50 dp_OWNERS

Attribute Value Type

```
typedef SEQUENCE_NAME_ORG_type
          dp_OWNERS_type;
```

The type NAME_ORG_type is defined in section 13.7.

Attribute Parameters

Parameter	Field Name	Parameter Name
"personal name of owner"	name	*none*
"owner's organisation"	organization	*none*

The technique used to designate the presence or absence of parameters for the value of this attribute is described in section 13.7.1.

17.1.8.51 dp_AUTHORS

Attribute Value Type

```
typedef SEQUENCE_NAME_ORG_type
           dp_AUTHORS_type;
```

The type NAME_ORG_type is defined in section 13.7.

Attribute Parameters

Parameter	Field Name	Parameter Name
"personal name of author"	name	*none*
"author's organisation"	organization	*none*

The technique used to designate the presence or absence of parameters for the value of this attribute is described in section 13.7.1.

17.1.8.52 dp_COPYRIGHT

Attribute Value Type

```
typedef SEQUENCE_COPYRIGHT_INFO_type
           dp_COPYRIGHT_type;

/*
   COPYRIGHT_INFO_type is used in
   SEQUENCE_COPYRIGHT_INFO_type
*/
typedef struct COPYRIGHT_INFO_struct {
   SEQUENCE_STRING_type
             copyright_information;
   SEQUENCE_STRING_type
             copyright_dates;
} COPYRIGHT_INFO_type;
```

The characters in each string must be from the document profile character set.

Attribute Parameters

Parameter	Field Name	Parameter Name
"copyright information"	`copyright_information`	*none*
"copyright dates"	`copyright_dates`	*none*

The absence of each parameter is designated by the corresponding field having the value `NULL_SEQUENCE`.

17.1.8.53 dp_STATUS

Attribute Value Type

```
typedef STRING_type dp_STATUS_type;
```

The characters in the string must be from the document profile character set.

17.1.8.54 dp_USER_SPECIFIC_CODES

Attribute Value Type

```
typedef SEQUENCE_STRING_type
          dp_USER_SPECIFIC_CODES_type;
```

The characters in the strings must be from the document profile character set.

17.1.8.55 dp_DISTRIBUTION_LIST

Attribute Value Type

```
typedef SEQUENCE_NAME_ORG_type
          dp_DISTRIBUTION_LIST_type;
```

The type `NAME_ORG_type` is defined in section 13.7.

Attribute Parameters

Parameter	Field Name	Parameter Name
"personal name of recipient"	`name`	*none*
"recipient's organisation"	`organization`	*none*

The technique used to designate the presence or absence of parameters for the value of this attribute is described in section 13.7.1.

17.1.8.56 dp_ADDITIONAL_INFORMATION

Attribute Value Type

```
typedef SEQUENCE_BYTE_type
          dp_ADDITIONAL_INFORMATION_type;
```

The value of this attribute is a sequence of bytes that designates an ASN.1 value constructed according to ISO 8825.

Semantics

The value of this attribute will be uninterpreted by the tool kit: it will not be checked for syntactic or semantic correctness but will be written directly to the output datastream.

17.1.8.57 dp_REFS_TO_OTHER_DOCS

Attribute Value Type

```
typedef SEQUENCE_STRING_OR_ID_type
          dp_REFS_TO_OTHER_DOCS_type;
```

The element type `STRING_OR_ID_type` is defined in section 13.10.

17.1.8.58 dp_SUPERSEDED_DOCUMENTS

Attribute Value Type

```
typedef SEQUENCE_STRING_OR_ID_type
          dp_SUPERSEDED_DOCUMENTS_type;
```

The element type `STRING_OR_ID_type` is defined in section 13.10.

17.1.8.59 dp_LOCAL_FILE_REFERENCE

Attribute Value Type

```
typedef struct {
    STRING_type        filename;
    STRING_type        location;
    STRING_type        user_comments;
} dp_LOCAL_FILE_REFERENCE_type;
```

Each field must consist of characters from the document profile character set.

Attribute Parameters

Parameter	Field Name	Parameter Name
the file name	`filename`	*none*
the location of the document	`.location`	*none*
user comments	`user_comments`	*none*

The absence of each parameter is indicated by the value of the corresponding field being `NULL_STRING`.

17.1.8.60 dp_DOCUMENT_SIZE

Attribute Value Type

```
typedef INT_type dp_DOCUMENT_SIZE_type;
```

The value of this attribute must be a positive integer, which is not less than the number of 8-bit bytes in the document datastream.

17.1.8.61 dp_NUMBER_OF_PAGES

Attribute Value Type

```
typedef INT_type dp_NUMBER_OF_PAGES_type;
```

17.1.8.62 dp_LANGUAGES

Attribute Value Type

```
typedef SEQUENCE_STRING_type
            dp_LANGUAGES_type;
```

The characters in the strings must be from the document profile character set.

17.1.8.63 dp_AUTHORIZATION

Attribute Value Type

```
typedef struct {
    TAG_type            authorization_tag;
        /* NAME_tag, ORG_tag */
```

```
    union {
        PERSONAL_NAME_type
                          name;
        STRING_type       organization;
    } authorization_value;

} dp_AUTHORIZATION_type;
```

If the value of the field `authorization_tag` is `NAME_tag`, then the field `name` is present. The type `PERSONAL_NAME_type` is described in section 13.6.

If the value of the field `authorization_tag` is `ORG_tag`, then the field `organization` is present and must consist of a string of characters from the document profile character set.

17.1.8.64 dp_SECURITY_CLASSIFICATION

Attribute Value Type

```
typedef STRING_type
            dp_SECURITY_CLASSIFICATION_type;
```

The characters in the string are from the document profile character set.

17.1.8.65 dp_ACCESS_RIGHTS

Attribute Value Type

```
typedef SEQUENCE_STRING_type
            dp_ACCESS_RIGHTS_type;
```

The characters in the strings are from the document profile character set.

17.1.9 Character Content Presentation Attributes

All of the attributes described in this section are applicable to basic logical component descriptions that have content portions that are of the character content architecture.

17.1.9.1 cc_ALIGNMENT

Attribute Value Type

```
typedef VALUE_type cc_ALIGNMENT_type;
```

The attribute must have one of the values

Identifier	Meaning
cc_ALIGNMENT_start_aligned	'start-aligned'
cc_ALIGNMENT_end_aligned	'end-aligned'
cc_ALIGNMENT_centred	'centred'
cc_ALIGNMENT_justified	'justified'

Default Value

cc_ALIGNMENT_start_aligned

17.1.9.2 cc_CHARACTER_FONTS

Attribute Value Type

```
typedef SEQUENCE_INT_PAIR_type
          cc_CHARACTER_FONTS_type;
```

All of the integer values must be positive.
There may be at most 10 elements in the sequence.

Attribute Parameters

Parameter	Field Name	Parameter Name
font size	first	*none*
font identifier	second	*none*

Default Value

NULL_SEQUENCE

17.1.9.3 cc_CHAR_ORIENTATION

Attribute Value Type

```
typedef VALUE_type
          cc_CHAR_ORIENTATION_type;
```

The value of this attribute must be one of

Identifier	Meaning
cc_CHAR_ORIENTATION_0	0 degrees
cc_CHAR_ORIENTATION_90	90 degrees
cc_CHAR_ORIENTATION_180	180 degrees
cc_CHAR_ORIENTATION_270	270 degrees

Default Value

cc_CHAR_ORIENTATION_0

17.1.9.4 cc_CHARACTER_PATH

Attribute Value Type

```
typedef VALUE_type cc_CHARACTER_PATH_type;
```

The value of this attribute must be one of

Identifier	Meaning
cc_CHARACTER_PATH_0	0 degrees
cc_CHARACTER_PATH_90	90 degrees
cc_CHARACTER_PATH_180	180 degrees
cc_CHARACTER_PATH_270	270 degrees

Default Value

cc_CHARACTER_PATH_0

17.1.9.5 cc_CHARACTER_SPACING

Attribute Value Type

```
typedef INT_type
          cc_CHARACTER_SPACING_type;
```

The value of this attribute must be a positive integer.

Default Value

The equivalent of 120 SMU's.

17.1.9.6 cc_CODE_EXT_ANNOUNCERS

Attribute Value Type

```
typedef SEQUENCE_BYTE_type
          cc_CODE_EXT_ANNOUNCERS_type;
```

The octets must be the string escape sequences, in accordance with ISO 2022, used to announce code extension features.

Default Value

The escape sequences announcing the use of the G0 and G2 sets, the G2 set being invoked in columns 10 to 15.

17.1.9.7 cc_1ST_LINE_OFFSET

Attribute Value Type

```
typedef INT_type cc_1ST_LINE_OFFSET_type;
```

17.1.9.8 cc_GRAPHIC_CHARACTER_SETS

Attribute Value Type

```
typedef SEQUENCE_BYTE_type
           cc_GRAPHIC_CHARACTER_SETS_type;
```

The value of this attribute consists of the escape sequence(s) used to designate the character set(s) in accordance with ISO 2022 and the register of ISO 2375.

Default Value

The escape sequences and shift functions designating and invoking the primary character set of ISO 6937/2 as the G0 set and the supplementary character set of ISO 6937/2 as the G1 set in columns 10 to 15.

17.1.9.9 cc_GRAPHIC_CHAR_SUBREP

Attribute Value Type

```
typedef INT_type
           cc_GRAPHIC_CHAR_SUBREP_type;
```

Default Value

```
0
```

17.1.9.10 cc_GRAPHIC_RENDITION

Attribute Value Type

```
typedef SEQUENCE_INT_type
            cc_GRAPHIC_RENDITION_type;
```

Default Value

An object, `SI`, of type `SEQUENCE_INT_type` such that

```
SI.sequence_tag == SEQUENCE_INT_tag &&
SI.length == 1 &&
SI.sequence_value.ints[0] == 0
```

17.1.9.11 cc_ITEMISATION

Attribute Value Type

```
typedef struct {
    VALUE_type identifier_alignment;
    INT_type   identifier_start_offset;
    INT_type   identifier_end_offset;
} cc_ITEMISATION_type;
```

The field `identifier_alignment` must have one of the values

Identifier	Meaning
`cc_ITEMISATION_no_itemisation`	'no itemisation'
`cc_ITEMISATION_start_aligned`	'start-aligned'
`cc_ITEMISATION_end_aligned`	'end-aligned'

Default Value

{ `cc_ITEMISATION_no_itemisation`,
distance from start edge of positioning area to line home,
`0` }

17.1.9.12 cc_KERNING_OFFSET

Attribute Value Type

```
typedef INT_PAIR_type
            cc_KERNING_OFFSET_type;
```

The value of each field must be a non-negative integer.

Default Value

```
{ 0, 0 }
```

17.1.9.13 cc_LINE_LAYOUT_TABLE

Attribute Value Type

```
typedef SEQUENCE_TAB_STOP_type
          cc_LINE_LAYOUT_TABLE_type;
```

Where the type of each element, TAB_STOP_type, in the sequence is defined below:

```
typedef struct TAB_STOP_struct {
   STRING_type   tab_reference;
   INT_type      tab_position;
   VALUE_type    alignment;
   STRING_type   alignment_string;
} TAB_STOP_type;
```

The value of the field tab_reference must be a STRING_type object whose value contains from 1 to 4 characters representing decimal digits.

The value of the field tab_position must be a non-negative integer.

The field alignment must have one of the values

Identifier	Meaning
cc_LINE_LAYOUT_TABLE_start_alnd	'start-aligned'
cc_LINE_LAYOUT_TABLE_end_alnd	'end-aligned'
cc_LINE_LAYOUT_TABLE_centred	'centred'
cc_LINE_LAYOUT_TABLE_alnd_arnd	'aligned-around'

Default Value

```
NULL_SEQUENCE
```

which indicates that no tabulation stops are defined.

17.1.9.14 cc_LINE_PROGRESSION

Attribute Value Type

```
typedef VALUE_type
          cc_LINE_PROGRESSION_type;
```

This attribute must have one of the values

Identifier	Meaning
cc_LINE_PROGRESSION_90	90 degrees
cc_LINE_PROGRESSION_270	270 degrees

Default Value

```
cc_LINE_PROGRESSION_270
```

17.1.9.15 cc_LINE_SPACING

Attribute Value Type

```
typedef INT_type cc_LINE_SPACING_type;
```

This attribute must have a value that is a positive integer.

Default Value

The equivalent of 200 SMU's.

17.1.9.16 cc_PAIRWISE_KERNING

Attribute Value Type

```
typedef VALUE_type
          cc_PAIRWISE_KERNING_type;
```

The value of this attribute must be one of

Identifier	Meaning
cc_PAIRWISE_KERNING_yes	'yes'
cc_PAIRWISE_KERNING_no	'no'

Default Value

```
cc_PAIRWISE_KERNING_no
```

17.1.9.17 cc_FORMATTING_INDICATOR

Attribute Value Type

```
typedef VALUE_type
          cc_FORMATTING_INDICATOR_type;
```

The value of this attribute must be one of

Identifier	Meaning
cc_FORMATTING_INDICATOR_yes	'yes'
cc_FORMATTING_INDICATOR_no	'no'

Default Value

```
cc_FORMATTING_INDICATOR_no
```

17.1.9.18 cc_INITIAL_OFFSET

Attribute Value Type

```
typedef INT_PAIR_type
          cc_INITIAL_OFFSET_type;
```

Default Value

The default values for this attribute are determined according to Table 2 in section 7.2.2, Part 6 of ISO IS 8613.

17.1.9.19 cc_INDENTATION

Attribute Value Type

```
typedef INT_type cc_INDENTATION_type;
```

The value of this attribute must be a non-negative integer.

Default Value

0

17.1.9.20 cc_ORPHAN_SIZE

Attribute Value Type

```
typedef INT_type cc_ORPHAN_SIZE_type;
```

The value of this attribute must be a positive integer.

Default Value

1

17.1.9.21 cc_PROP_LINE_SPACING

Attribute Value Type

```
typedef VALUE_type
          cc_PROP_LINE_SPACING_type;
```

The attribute must have one of the values

Identifier	Meaning
cc_PROP_LINE_SPACING_yes	'yes'
cc_PROP_LINE_SPACING_no	'no'

Default Value

```
cc_PROP_LINE_SPACING_no
```

17.1.9.22 cc_WIDOW_SIZE

Attribute Value Type

```
typedef INT_type cc_WIDOW_SIZE_type;
```

The value of this attribute must be a positive integer.

Default Value

```
1
```

17.1.10 Raster Graphics Content Presentation Attributes

All of the attributes described in this section are applicable to basic logical component descriptions that have content portions that are of the raster graphics content architecture.

17.1.10.1 rg_CLIPPING

Attribute Value Type

```
typedef struct {
    INT_PAIR_type coord1, coord2;
} rg_CLIPPING_type;
```

The integers in each of the fields must be non-negative.

Attribute Parameters

Parameter	Field Name	Parameter Name
none	coord1	COORD1_parm
none	coord2	COORD2_parm

Default Value

{ { 0, 0 }, { N-1, L-1 } }

Where N is the number of pels per line and L is the number of lines.

17.1.10.2 rg_LINE_PROGRESSION

Attribute Value Type

```
typedef VALUE_type
        rg_LINE_PROGRESSION_type;
```

This attribute must have one of the values

Identifier	Meaning
rg_LINE_PROGRESSION_90	90 degrees
rg_LINE_PROGRESSION_270	270 degrees

Default Value

```
rg_LINE_PROGRESSION_270
```

17.1.10.3 rg_PEL_PATH

Attribute Value Type

```
typedef VALUE_type rg_PEL_PATH_type;
```

The value of this attribute must be one of

Identifier	Meaning
rg_PEL_PATH_0	0 degrees
rg_PEL_PATH_90	90 degrees
rg_PEL_PATH_180	180 degrees
rg_PEL_PATH_270	270 degrees

Default Value

```
rg_PEL_PATH_0
```

17.1.10.4 rg_INITIAL_OFFSET

Attribute Value Type

```
typedef INT_PAIR_type
        rg_INITIAL_OFFSET_type;
```

Attribute Parameters

Parameter	Field Name	Parameter Name
horizontal coordinate	first	*none*
vertical coordinate	second	*none*

Default Value

The default value of this attribute depends on the values of the attributes "pel path" and "line progression", as specified in Table 2, part 7 of ISO IS 8613.

17.1.10.5 rg_PEL_TRANS_DENSITY

Attribute Value Type

```
typedef VALUE_type
          rg_PEL_TRANS_DENSITY_type;
```

This attribute must have one of the values

Identifier	Meaning
rg_PEL_TRANS_DENSITY_1	1 BMU
rg_PEL_TRANS_DENSITY_2	2 BMU
rg_PEL_TRANS_DENSITY_3	3 BMU
rg_PEL_TRANS_DENSITY_4	4 BMU
rg_PEL_TRANS_DENSITY_5	5 BMU
rg_PEL_TRANS_DENSITY_6	6 BMU

Default Value

```
rg_PEL_TRANS_DENSITY_6
```

17.1.10.6 rg_IMAGE_DIMENSIONS

Attribute Value Type

```
typedef struct {
   TAG_type  image_dimensions_tag;
      /*
         WIDTH_CONTROLLED_tag,
         HEIGHT_CONTROLLED_tag,
         AREA_CONTROLLED_tag,
         AUTOMATIC_tag
      */
   union {
      INT_PAIR_type    width_controlled;
      INT_PAIR_type    height_controlled;
      IMAGE_SIZE_type  area_controlled;
   } image_dimensions_value;
} rg_IMAGE_DIMENSIONS_type;
```

```
typedef struct {
    INT_type              minimum_width;
    INT_type              preferred_width;
    INT_type              minimum_height;
    INT_type              preferred_height;
    VALUE_type            aspect_ratio_flag;
} IMAGE_SIZE_type;
```

If the value of the field `image_dimensions_tag` is `WIDTH_CONTROLLED_tag`, then the field `width_controlled` is present and the first element of the `INT_PAIR_type` object corresponds to "minimum width" and the second element to "preferred width".

If the value of the field `image_dimensions_tag` is `HEIGHT_CONTROLLED_tag`, then the field `image_height` is present and the first element of the `INT_PAIR_type` object corresponds to "minimum height" and the second element to "preferred height".

If the value of the field `image_dimensions_tag` is `AREA_CONTROLLED_tag`, then the field `image_size` is present.

If the value of the field `image_dimensions_tag` is `AUTOMATIC_tag`, then no field is present and the value of the attribute consists of the "automatic" set.

The value of the field `aspect_ratio_flag` must be one of

Identifier	Meaning
`rg_IMAGE_DIMENSIONS_fixed`	'fixed'
`rg_IMAGE_DIMENSIONS_variable`	'variable'

Attribute Parameters

For the "width controlled" set:

Parameter	Field Name	Parameter Name
"minimum width"	`image_dimensions_value.width_controlled.first`	*none*
"preferred width"	`image_dimensions_value.width_controlled.second`	*none*

For the "height controlled" set:

Parameter	Field Name	Parameter Name
"minimum height"	`image_dimensions_value.height_controlled.first`	*none*
"preferred width"	`image_dimensions_value.height_controlled.second`	*none*

For the "area controlled" set:

Parameter	Field Name	Parameter Name
"minimum width"	image_dimensions_value.area_controlled.minimum_width	*none*
"preferred width"	image_dimensions_value.area_controlled.preferred_width	*none*
"minimum height"	image_dimensions_value.area_controlled.minimum_height	*none*
"preferred height"	image_dimensions_value.area_controlled.preferred_height	*none*
"aspect ratio flag"	image_dimensions_value.area_controlled.aspect_ratio_flag	*none*

Default Value

An object, `ID`, of type `rg_IMAGE_DIMENSIONS_type` such that

```
ID.image_dimensions_tag == AUTOMATIC_tag
```

17.1.10.7 rg_PEL_SPACING

Attribute Value Type

```
typedef struct {
    BOOL_type           null;
    INT_type            length;
    INT_type            pel_spaces;
} rg_PEL_SPACING_type;
```

If the field `null` is `BOOL_true`, then the value of the attribute is 'null'. If the field `null` is `BOOL_false`, then the value of the attribute is determined by the values of the fields `length` and `pel_spaces`, which must be positive integers.

Attribute Parameters

Parameter	Field Name	Parameter Name
"length"	length	*none*
"pel spaces"	pel_spaces	*none*

Default Value

```
{ BOOL_false, 4, 1 }
```

17.1.10.8 rg_SPACING_RATIO

Attribute Value Type

```
typedef INT_PAIR_type
        rg_SPACING_RATIO_type;
```

The value of each integer must be positive.

Attribute Parameters

Parameter	Field Name	Parameter Name
"line spacing value"	`first`	*none*
"pel spacing value"	`second`	*none*

Default Value

```
{ 1, 1 }
```

17.1.11 Raster Graphics Content Coding Attributes

All of the attributes in this section are applicable only to content portions of a raster graphics content architecture class.

17.1.11.1 rg_COMPRESSION

Applicability

This attribute is defaultable.

Attribute Value Type

```
typedef VALUE_type rg_COMPRESSION_type;
```

The value of this attribute must be one of:

Identifier	Meaning
`rg_COMPRESSION_compressed`	'compressed'
`rg_COMPRESSION_uncompressed`	'uncompressed'

Default Value

```
rg_COMPRESSION_compressed
```

17.1.11.2 rg_NUMBER_OF_LINES

Applicability

This attribute is mandatory.

Attribute Value Type

```
typedef INT_type rg_NUMBER_OF_LINES_type;
```

The value of this attribute must be a positive integer.

17.1.11.3 rg_NUMBER_OF_PELS_PER_LINE

Applicability

This attribute is defaultable or mandatory depending on the content architecture level. If the attribute is specified as defaultable, then the content architecture level must specify the default value.

Attribute Value Type

```
typedef INT_type
        rg_NUMBER_OF_PELS_PER_LINE_type;
```

The value of this attribute must be a non-negative integer.

Default Value

The default value of this attribute depends on the value of the attribute "pel transmission density", as specified in Table 3 in section 7.2.3, Part 7 of ISO IS 8613.

17.1.11.4 rg_NUMBER_DISCARDED_PELS

Attribute Value Type

```
typedef INT_type
        rg_NUMBER_DISCARDED_PELS_type;
```

The value of this attribute must be a non-negative integer.

Default Value

If the number of pels per line exceeds the line length, then the default value is half the excess number of discarded pels; otherwise it is 0.

17.1.12 Geometric Graphics Content Presentation Attributes

The geometric graphics content presentation attributes are unimplemented in the tool kit. All of the appropriate attribute identifiers are defined, but attempts to assign or obtain the value of this attribute will yield an error (see sections 16.1.2 and 16.1.3).

17.1.12.1 gg_ENCODING_ANNOUNCER

17.1.12.2 gg_LINE_RENDITION

17.1.12.3 gg_MARKER_RENDITION

17.1.12.4 gg_TEXT_RENDITION

17.1.12.5 gg_FILLED_AREA_REND

17.1.12.6 gg_EDGE_RENDITION

17.1.12.7 gg_COLOUR_REPS

17.1.12.8 gg_TRANSPARENCY_SPEC

17.1.12.9 gg_TRANSFORMATION_SPEC

17.1.12.10 gg_REGION_INTEREST_SPEC

17.1.12.11 gg_PICTURE_ORIENTATION

17.1.12.12 gg_PICTURE_DIMENSIONS

18
Expressions

Within ODA, the value of some attributes can be specified by an expression. These attributes are:

- "generator for subordinates" (section 17.1.1.4),
- "content generator" (section 17.1.1.5),
- "same layout object" (section 17.1.5.9),
- "synchronization" (section 17.1.5.11).

There are four kinds of expressions. *String expressions*, which may be used in the value for the "content generator" attribute, are described in section 18.1. *Numeric expressions*, which are used within string expressions, are defined in section 18.2. Section 18.3 describes *object identifier expressions*, which are used in the "same layout object" and "synchronization" attributes. *Construction expressions* may be used as the value of the attribute "generator for subordinates" and are described in section 18.4. In addition, any of these expressions may refer to *binding values*, which are defined in section 18.5.

The reader is advised that the type definitions in this chapter are complex and depend on a complete understanding of the appropriate sections of Part 2 of ISO IS 8613: 5.1.3 (Expressions), 5.1.3.1 (String expressions), 5.1.3.2 (Numeric expressions), 5.1.3.3 (Object identifier expressions), 5.1.3.4 (References to binding values), 5.1.4 (Current instance function) and 5.3.2.1 (Generator for subordinates).

18.1 String Expressions

A string expression, defined in section 5.1.3.1, Part 2 of ISO IS 8613, consists of a sequence of one or more *atomic string expressions*. An atomic string expression is either:

- a string literal,
- a reference to a binding value (section 18.5),
- a string function application.

The types and routines needed to deal with string expressions are defined in the remainder of this section (except for the type `STRING_type`, which is defined in section 13.14).

18.1.1 Types

```
typedef SEQUENCE_ATOMIC_STR_EXPR_type STRING_EXPR_type;
    /* ATOMIC_STRING_EXPR_type is defined below */

typedef struct ATOMIC_STRING_EXPR_struct {

   TAG_type atomic_string_expr_tag;
      /*
         STRING_LITERAL_tag, REF_BINDING_VALUE_tag,
         STRING_FUNCTION_tag
      */

   union {
      STRING_type              string;
      REF_BINDING_VALUE_type   value;
      STRING_FUNCTION_type     function;
   } atomic_string_value;

} ATOMIC_STRING_EXPR_type;

typedef struct {

   TAG_type                 function_tag;
      /*
         MAKE_STRING_tag, UPPER_ALPHA_tag,
         LOWER_ALPHA_tag,
         UPPER_ROMAN_tag, LOWER_ROMAN_tag
      */

   NUMERIC_EXPR_type        arg;

} STRING_FUNCTION_type;
```

If the value of the tag field `atomic_string_expr_tag` is `STRING_LITERAL_tag`, then only the field `string` is present, and the string expression is a string literal.

If the value of the tag field `atomic_string_expr_tag` is `REF_BINDING_VALUE_tag`, then only the field `value` is present, and the string expression is a reference to a binding value. The type `REF_BINDING_VALUE_type` is defined in section 18.5.

If the value of the tag field atomic_string_expr_tag is STRING_FUNCTION_tag, then only the field function is present, and the string expression is a string function application.

If the value of the tag field function_tag is MAKE_STRING_tag, then this is an application of the function MAKE-STRING and the parameter is the numeric expression designated by the field arg.

If the value of the tag field function_tag is UPPER_ALPHA_tag, then this is an application of the function UPPER-ALPHA and the parameter is the numeric expression designated by the field arg.

If the value of the tag field function_tag is LOWER_ALPHA_tag, then this is an application of the function LOWER-ALPHA and the parameter is the numeric expression designated by the field arg.

If the value of the tag field function_tag is UPPER_ROMAN_tag, then this is an application of the function UPPER-ROMAN and the parameter is the numeric expression designated by the field arg.

If the value of the tag field function_tag is LOWER_ROMAN_tag, then this is an application of the function LOWER-ROMAN and the parameter is the numeric expression designated by the field arg.

The type NUMERIC_EXPR_type is defined in section 18.2.

18.1.2 Evaluate a String Expression: *EvalStringExpr*

```
STRING_type EvalStringExpr(expr)
   STRING_EXPR_type *expr;
```

Real name

evlstr

Synopsis

This routine is used to evaluate a string expression according to the rules of ODA (section 5.1.3.1, Part 2 of ISO IS 8613).

Parameters

expr	The address of a string expression to evaluate.

Return Value

ERROR_STRING	An error has occurred. TKError will contain an indication of the exact error (see the **Errors** part below).
otherwise	A STRING_type object representing the value of the expression.

Side Effects

None.

Semantics

The `STRING_type` object returned is as if allocated by a call to `MakeString` (section 13.14.2). It is the user's responsibility to delete this `STRING_type` by a call to `DeleteString` when it is no longer needed.

If the `STRING_type` value returned is `NULL_STRING`, then the value of the string expression is 'null'.

Errors

`BADPARM_err`	One of the parameters is invalid.
`INTERNAL_err`	An error internal to the structure or operation of the tool kits has occurred.
`LOCAL_err`	A locally-defined error, not covered by any other available error, has occurred.
`NOMEM_err`	No more memory was available for dynamic allocation.

Implementation Status

Unimplemented.

18.2 Numeric Expressions

A *numeric expression* is one of the following:

- a numeric literal,
- a reference to a binding value,
- a numeric function application.

The types and routines needed to deal with numeric expressions are defined in the remainder of this section.

18.2.1 Types

```
typedef struct NUMERIC_EXPR_struct {

   TAG_type numeric_expr_tag;
      /*
         NUMERIC_LITERAL_tag, REF_BINDING_VALUE_tag,
         NUMERIC_FUNCTION_tag
      */

   union {
      INT_type                         numeric_literal;
      REF_BINDING_VALUE_type           binding_value;
      NUMERIC_FUNCTION_type            function;
   } numeric_expr_value;

} NUMERIC_EXPR_type;

typedef struct {

   TAG_type function_tag;
      /* INCREMENT_tag, DECREMENT_tag, ORDINAL_tag */

   union {
      struct NUMERIC_EXPR_struct   *incr_decr_parameter;
      OBJECT_REF_type              ord_parameter;
   } numeric_function_value;

} NUMERIC_FUNCTION_type;
```

If the value of the tag field `numeric_expr_tag` is `NUMERIC_LITERAL_tag`, then only the field `numeric_literal` is present and the numeric expression represented is a numeric literal.

If the value of the tag field `numeric_expr_tag` is `REF_BINDING_VALUE_tag`, then only the field `binding_value` is present and the numeric expression represented is a reference to a binding value. The type `REF_BINDING_VALUE_type` is defined in section 18.5.

If the value of the tag field `numeric_expr_tag` is `NUMERIC_FUNCTION_tag`, then only the field `function` is present and the numeric expression represented is an application of a numeric function.

If the value of the tag field `function_tag` is `INCREMENT_tag`, then the `NUMERIC_FUNCTION_type` object is an application of INCREMENT. The parameter to INCREMENT is the numeric expression referenced by the field `incr_decr_parameter`.

If the value of the tag field `function_tag` is `DECREMENT_tag`, then the `NUMERIC_FUNCTION_type` object is an application of DECREMENT. The parameter to DECREMENT is the numeric expression referenced by the field `incr_decr_parameter`.

If the value of the tag field `function_tag` is `ORDINAL_tag`, then the `NUMERIC_FUNCTION_type` object is an application of ORDINAL. The parameter to ORDINAL is the object reference in the field `ord_parameter`. The type `OBJECT_REF_type` is defined in section 18.6.

18.2.2 Evaluate a Numeric Expression: *EvalNumExpr*

```
INT_type EvalNumExpr(expr, value)
   NUMERIC_EXPR_type *expr;
   INT_type *value;
```

Real name

evlnum

Synopsis

This routine is used to evaluate a numeric expression (according to the rules specified in section 5.1.3.2, Part 2 of ISO IS 8613).

Parameters

`expr`	The address of a numeric expression to evaluate.
`value`	The address of an `INT_type` variable into which the value of the evaluated expression will be placed.

Return Value

`ERROR_INT`	An error has occurred. `TKError` will contain an indication of the exact error (see the **Errors** part below).
otherwise	`Value` contains the value of the expression.

Side Effects

The value of the variable referenced by `*value` may be modified.

Semantics

If the return value of this routine is `ERROR_INT`, then the value of `*value` is undefined.

Errors

`BADPARM_err`	One of the parameters is invalid.
`INTERNAL_err`	An error internal to the structure or operation of the tool kits has occurred.
`NOMEM_err`	No more memory was available for dynamic allocation.

Implementation Status

Unimplemented.

18.3 Object Identifier Expressions

Some attributes that apply to object descriptions have values that consist of references to other object descriptions. The value of such an attribute is represented by an *object identifier expression*: an expression that, when evaluated, produces the value of an object identifier. Object identifier expressions are defined in section 5.1.3.3, Part 2 of ISO IS 8613. The types used to represent these expressions and the routines for manipulating these types within the tool kit are described in the remaining parts of this section.

18.3.1 Types

```
typedef struct OBJECT_ID_EXPR_struct {

   TAG_type object_selection_function_tag;
      /*
         CURRENT_OBJECT_tag, CURRENT_INSTANCE_tag,
         SUPERIOR_OBJECT_tag, PRECEDING_OBJECT_tag
      */

   union {
      CURR_INSTANCE_PARMS_type    curr_inst_parms;
      struct OBJECT_ID_EXPR_struct*sup_or_pre_parms;
   } parameters;

} OBJECT_ID_EXPR_type;

typedef struct {

   TAG_type parm1_tag;
      /* OBJECT_CLASS_ID_tag, OBJECT_TYPE_tag */

   union {
      CONSTITUENT object_class_id;
      VALUE_type  object_type;
   } parm1;

   struct OBJECT_REF_struct *parm2;
} CURR_INSTANCE_PARMS_type;
```

If the value of the tag field `object_selection_function_tag` is `CURRENT_OBJECT_tag`, then this is an application of the object selection function CURRENT-OBJECT, which takes no parameters.

If the value of the tag field object_selection_function_tag is CURRENT_INSTANCE_tag, then this is an application of the object selection function CURRENT-INSTANCE, with parameters represented by the field curr_inst_parms.

If the value of the tag field object_selection_function_tag is SUPERIOR_OBJECT_tag, then this is an application of the object selection function SUPERIOR-OBJECT, with parameters represented by the field sup_or_pre_parms.

If the value of the tag field object_selection_function_tag is PRECEDING_OBJECT_tag, then this is an application of the object selection function PRECEDING-OBJECT, with parameters represented by the field sup_or_pre_parms.

If the tag field parm1_tag has the value OBJECT_CLASS_ID_tag, then the field object_type is present and the first parameter to CURRENT-INSTANCE is the object class referenced by the union field object_class_id.

If the tag field parm1_tag has the value OBJECT_TYPE_tag, then the field object_type is present and the first parameter to CURRENT-INSTANCE is an object type. The value of the object_type field must be one of the values for the "object type" attribute (section 17.1.1.1).

18.3.2 Evaluate an Object Identifier Expression: *EvalObjIdExpr*

```
CONSTITUENT EvalObjIdExpr(expr)
   OBJECT_ID_EXPR_type *expr;
```

Real name

evlobi

Synopsis

This routine is used to evaluate an object identifier expression (according to the rules specified in section 5.1.3.3, Part 2 of ISO IS 8613).

Parameters

expr	The address of an object identifier expression to evaluate.

Return Value

`ERROR_CONSTITUENT`
: An error has occurred. `TKError` will contain an indication of the exact error (see the **Errors** part below).

otherwise
: The value of the expression.

Side Effects

None.

Errors

`BADPARM_err`	One of the parameters is invalid.
`INTERNAL_err`	An error internal to the structure or operation of the tool kits has occurred.
`LOCAL_err`	A locally-defined error, not covered by any other available error, has occurred.
`NOMEM_err`	No more memory was available for dynamic allocation.

Implementation Status

Unimplemented.

18.4 Construction Expressions

A *construction expression* is one of the possible values for the attribute "generator for subordinates" (section 17.1.1.4).

18.4.1 Types

```
typedef struct CONSTRUCTION_EXPR_struct {

   TAG_type construction_tag;
      /* TERM_tag, SEQUENCE_tag,
      AGGREGATE_tag, CHOICE_tag */
```

```
   union {

      CONSTRUCTION_TERM_type          term;
      SEQUENCE_CONSTRUCTION_TERM_type construction;
         * Applies to SEQUENCE_tag, AGGREGATE_tag
           & CHOICE_tag */
   } construction_value;

} CONSTRUCTION_EXPR_type;

typedef struct {

   TAG_type                     term_tag;
      /* REQ_tag, OPT_tag, REP_tag, OPT_REP_tag */

   CONSTRUCTION_FACTOR_type  factor;
} CONSTRUCTION_TERM_type;

typedef struct {

   TAG_type factor_tag;    /* OBJECT_CLASS_ID_tag,
                              EXPR_tag */

   union {

      CONSTITUENT                         object_class_id;
      struct CONSTRUCTION_EXPR_struct *expr;
   } factor_value;

} CONSTRUCTION_FACTOR_type;
```

If the value of tag field `construction_tag` is `TERM_tag`, then only the field `term` is present and the construction expression is a construction term.

If the value of tag field `construction_tag` is `SEQUENCE_tag`, `AGGREGATE_tag` or `CHOICE_tag`, then the construction expression is a sequence, aggregate or choice, respectively. The elements of the expression are represented by the elements of the field `construction`. In the case of a sequence, the order of the elements in the sequence is important. For aggregates and choices, the order is arbitrary.

In the type `CONSTRUCTION_TERM_type`, if the value of the tag field `term_tag` is `REQ_tag`, then this construction term is a required construction factor.

If the value of the tag field `term_tag` is `OPT_tag`, then this construction term is an optional construction factor.

If the value of the tag field `term_tag` is `REP_tag`, then this construction term is a repetitive construction factor.

If the value of the tag field `term_tag` is `OPT_REP_tag`, then this construction term is an optional repetitive construction factor.

In the type `CONSTRUCTION_FACTOR_type`, if the value of the tag field `factor_tag` is `OBJECT_CLASS_ID_tag`, then the field object_class_id is present and the construction factor is an object class identifier.

If the value of the tag field factor_tag is EXPR_tag, then the field expr is present and the construction factor is a construction expression.

18.4.2 Delete a Construction Expression: *DeleteConstExpr*

```
INT_type DeleteConstExpr(expr)
   CONSTRUCTION_EXPR_type *expr;
```

Real name

dlcnex

Synopsis

This routine is used to recover the storage associated with a construction expression.

Parameters

expr	The address of a structure of type CONSTRUCTION_EXPR_type. Note that at_GEN_FOR_SUBORDINATES_type is the same as CONSTRUCTION_EXPR_type.

Return Value

ERROR_INT	An error has occurred. TKError will contain an indication of the exact error (see the **Errors** part below).
otherwise	The space associated with the construction expression, including all referenced sequences and expressions, is reclaimed.

Side Effects

None.

Semantics

The space associated with the expression is freed and should not be referenced. In the case that the return value is ERROR_INT, some, but not all, of the of the storage used by the construction expression may have been reclaimed.

It is assumed that no aliasing occurs, that is, that no two subexpressions or subsequences share dyanmically allocated storage.

Errors

BADPARM_err	One of the parameters is invalid.
INTERNAL_err	An error internal to the structure or operation of the tool kits has occurred.

Implementation Status

This routine will not behave properly if the construction expression does not form a tree.

This routine calls DeleteSequence (section 13.13.5), which should be consulted for further status information.

Note that for this routine to work properly, all storage must have been allocated by the tool kit using LOCAL_Malloc (see section 10.2.5.1).

18.5 References to Binding Values

Within a string expression or a numeric expression, it is possible to refer to the value of a binding within the "bindings" attribute of an object. The semantics of a reference to a binding value are defined in section 5.1.3.4, Part 2 of ISO IS 8613.

18.5.1 Types

```
typedef struct {
    OBJECT_REF_type object_ref;
    INT_type        binding_identifier;
} REF_BINDING_VALUE_type;
```

The type OBJECT_REF_type is described in section 18.6.

18.5.2 Evaluate a Reference to a Binding Value: *EvalRefBind*

```
INT_type EvalRefBind(refbind, string,
                         integer)
    REF_BINDING_VALUE_type *refbind;
    STRING_type *string;
    INT_type *integer;
```

Real name

evlrbv

Synopsis

This routine is used to evaluate a reference to a binding value (according to the rules specified in section 5.1.3.4, Part 2 of ISO IS 8613) used within a numeric or string expression.

Parameters

`refbind`	The address of a structure of type `REF_BINDING_VALUE_type`.
`string`	The address of a `STRING_type` object in which to place a `STRING_type` value if the result of this evaluation is a character string.
`integer`	The address of an `INT_type` object in which to place an `INT_type` value if the result of this evaluation is a integer.

Return Value

`ERROR_INT`	An error has occurred. `TKError` will contain an indication of the exact error (see the **Errors** part below).
`STRING_VALUE`	The result of this evaluation is a character string and the corresponding `STRING_type` object has been stored in the address specified by the parameter `string`.
`INTEGER_VALUE`	The result of this evaluation is an integer and the corresponding `INT_type` value has been stored in the address specified by the parameter `integer`.

Side Effects

If the return value is not `ERROR_INT`, then the value of the object referenced by one of `string` or `integer` will have been modified.

Semantics

In the case that the return value is `ERROR_INT`, then neither of the objects referenced by `string` or `integer` has been modified.

If the return value is `STRING_VALUE`, then the object referenced by `string` has been modified, but that referenced by `integer` is untouched.

If the return value is `INTEGER_VALUE`, then the object referenced by `integer` has been modified, but that referenced by `string` is untouched.

Errors

`BADPARM_err`	One of the parameters is invalid.
`INTERNAL_err`	An error internal to the structure or operation of the tool kits has occurred.
`NOMEM_err`	No more memory was available for dynamic allocation.

Implementation Status

Unimplemented.

18.6 Object References

Several of the expressions described in the preceding sections of this chapter allow a reference to an object, which is either an object identifier or an object identifier expression. This is represented by the type `OBJECT_REF_type` within the tool kit.

```
typedef struct OBJECT_REF_struct {

   TAG_type obj_ref_tag;
      /* OBJECT_ID_tag, OBJECT_ID_EXPR_tag */

   union {
      CONSTITUENT               object_id;
      OBJECT_ID_EXPR_type       object_id_expr;
         /* see section 18.3 */
   } object_ref_value;

} OBJECT_REF_type;
```

If the tag field `obj_ref_tag` has the value `OBJECT_ID_tag`, then only the field `object_id` is present and the object reference is an object identifier.

If the tag field `obj_ref_tag` has the value `OBJECT_ID_EXPR_tag`, then only the field `object_id_expr` is present and the object reference is an object identifier expression.

19
Traversing an ODA Document

The routines described in this chapter are used for obtaining access to parts of the current document in various ways.

The first set of routines (section 19.1) is used by the application programmer to locate key constituents of the document: the document logical root(s), the document layout root(s) and the document profile.

The routines described in section 19.2 may be used to create and use *iterators*. An iterator is an object that may be used to traverse the constituents of part of the document in a useful order.

19.1 Locating Constituents

The routines described in this section are used by the application programmer to locate key constituents of the document. These routines are needed in the case that the document was generated by reading an external representation and not by the application program.

These routines may be used in conjunction with the routines for creating iterators in order to traverse the various structures of a document. For example, the call

```
root = FindDocumentRoot(document,
                        SPECIFIC_DOC_LOGICAL_ROOT);
iter = MakeSubgraphIterator(root,
                            PARENTS_FIRST,
                            DEPTH_FIRST);
```

may be used to create an iterator for traversing the specific logical structure in preorder.

19.1.1 Locate a Document Root: *FindDocumentRoot*

```
CONSTITUENT FindDocumentRoot(document, kind)
   DOCUMENT_type document;
   VALUE_type kind;
```

Real Name

fndrt

Synopsis

This routine is used to locate a specified document root.

Parameters

`document` — The document in which to look for the root.

`kind` — Determines the document root to be found which is one of:

Identifier	Meaning
SPECIFIC_DOC_LOGICAL_ROOT	specific document logical root
GENERIC_DOC_LOGICAL_ROOT	generic document logical root
SPECIFIC_DOC_LAYOUT_ROOT	specific document layout root
GENERIC_DOC_LAYOUT_ROOT	generic document layout root

Return Value

`ERROR_CONSTITUENT` — An error has occurred. `TKError` will contain an indication of the exact error (see the **Errors** part below).

otherwise — The specified document root.

Side Effects

None.

Errors

`BADPARM_err` — One of the parameters is invalid.

`INTERNAL_err` — An error internal to the structure or operation of the tool kits has occurred.

`NOROOT_err` — The specified root can not be found.

Implementation Status

Implemented.

19.1.2 Locate a Document Profile: *FindDocumentProfile*

```
CONSTITUENT FindDocumentProfile(document)
   DOCUMENT_type document;
```

Real Name

fnddp

Synopsis

This routine is used to find the document profile of a specific document.

Parameters

`document`	The document whose document profile is to be returned.

Return Value

`ERROR_CONSTITUENT`	An error has occurred. `TKError` will contain an indication of the exact error (see the **Errors** part below).
otherwise	The document profile.

Side Effects

None.

Errors

`BADPARM_err`	The parameter `document` is invalid.
`INTERNAL_err`	An error internal to the structure or operation of the tool kits has occurred.
`NOPROF_err`	The document has no document profile.

Implementation Status

Implemented.

19.2 Iterators

The tool kit provides a mechanism that allows the application program to iterate through the constituents of the ODA document in a variety of orders. When traversing a logical or layout structure, the programmer may choose between generating a constituent before or after its *descendants* and to traverse the graph in breadth-first or depth-first order. (Because there is no order inherent in the layout or presentation styles, these constituents are yielded in an unspecified order.)

In order to search a logical or layout structure, the application program specifies a starting constituent for the search. According to the specified order all of the descendant constituents of the initial constituent are yielded. A constituent is a *descendant* of a designated constituent if:

- it is the designated constituent, or
- it is a child of the designated constituent, or
- it is a child of a descendant of the designated constituent.

The *children* of a constituent depend on the kind of the constituent:

- If the constituent is a composite object class description, then its children are the constituents designated in the value of the attribute "generator for subordinates".
- If the constituent is a composite object description, then its children are the constituents designated in the value of the attribute "subordinates".

19.2.1 Create an Iterator for a Subgraph: *MakeSubgraphIterator*

```
ITERATOR_type
    MakeSubgraphIterator(root,
                         parent_order,
                         walk_order)
   CONSTITUENT root;
   VALUE_type parent_order;
   VALUE_type walk_order;
```

Real Name

mksbgn

Synopsis

This routine is used to create an iterator for traversing a subgraph.

Parameters

`root`	The root of the subgraph to traverse. This constituent must be part of a logical or layout structure.
`parent_order`	Determines the order in which the children of a constituent are generated in relation the constituent itself:

Identifier	Meaning
PARENTS_FIRST	traversal will yield a constituent before any of its subordinates
CHILDREN_FIRST	constituent will be yielded after its subordinates.

walk_order Determines the order in which the subgraph will be traversed:

Identifier	Meaning
BREADTH_FIRST	constituents will be yielded breadth first
DEPTH_FIRST	constituents will be yielded depth first

Return Value

ERROR_ITERATOR
An error has occurred. TKError will contain an indication of the exact error (see the **Errors** part below).

otherwise An iterator for generating the constituents of the graph rooted at root.

Side Effects

None.

Semantics

A normal preorder tree traversal may be obtained (but only for specific structures; the generic structures are graphs) by passing (PARENTS_FIRST, DEPTH_FIRST) as the values for the parameters parent_order and walk_order. A postorder tree traversal is obtained by specifying these parameters as (CHILDREN_FIRST, DEPTH_FIRST).

The combination (CHILDREN_FIRST, BREADTH_FIRST) is not supported.

During the walk of specific structure, the attribute "subordinates" may not be changed on any constituent that has yet to be yielded. It may be changed on constituents that have already been yielded and on constituents that will not be yielded. Changing the attribute "subordinates" of a yielded node will not affect the collection of constituents that are yielded, or the order in which they are yielded. In other words, the yielding of constituents will be as if no change to the attribute "subordinates" had been made.

During the walk of a generic structure, the attribute "generator For subordinates" may not be changed on any constituent that has yet to be yielded. It may be changed on constituents that have already been yielded and on constituents that will not be yielded. Changing the attribute "generator For subordinates" of a yielded node will not affect the constituents that are yielded, or the order that they are yielded. In other words, the yielding of constituents will be as if no change to the attribute "generator For subordinates" had been made.

Errors

`BADPARM_err`	One of the parameters is invalid.
`INTERNAL_err`	An error internal to the structure or operation of the tool kits has occurred.
`NOMEM_err`	No more memory was available for dynamic allocation.
`UNIMPL_err`	Attempt to create an iterator for a breadth first walk with parents after children.

Implementation Status

Implemented.

19.2.2 Create an Iterator for Styles: *MakeStylesIterator*

```
ITERATOR_type MakeStylesIterator(document,
                                  kind)
    DOCUMENT_type document;
    VALUE_type kind;
```

Real Name

mkstgn

Synopsis

This routine is used to create an iterator for traversing the constituents of either the layout styles or the presentation styles for a specified document.

Parameters

`document`	The document in which to look for styles.
`kind`	Determines whether the iterator will be for presentation styles or layout styles. See section 15.5 for a description of the legal values for this parameter.

Return Value

`ERROR_ITERATOR`	An error has occurred. `TKError` will contain an indication of the exact error (see the **Errors** part below).
otherwise	An iterator for generating the constituents of the specified styles.

Side Effects

None.

Semantics

The order in which the constituents of the styles are yielded is unspecified.
An error value is indicated if the document contains no styles.

Errors

`BADPARM_err`	One of the parameters is invalid.
`INTERNAL_err`	An error internal to the structure or operation of the tool kits has occurred.
`NOMEM_err`	No more memory was available for dynamic allocation.
`NO_STYLES_err`	There are no styles in the specified document.

Implementation Status

Implemented.

19.2.3 Yield the Next Constituent: *NextConstituent*

```
CONSTITUENT NextConstituent(iterator)
   ITERATOR_type iterator;
```

Real Name

nxtcns

Synopsis

This routine is used to yield the next constituent in a traversal defined by the specified iterator.

Parameters

`iterator`	The `ITERATOR_type` object to use.

Return Value

`NULL_CONSTITUENT`
: The final constituent was yielded during the last call to this routine. Repeated calls of this routine will produce this return value.

`ERROR_CONSTITUENT`
: An error has occurred. `TKError` will contain an indication of the exact error (see the **Errors** part below).

otherwise
: The next constituent.

Side Effects

The parameter `iterator` has been modified so that the next call of this routine will yield the successor constituent.

Errors

`BADPARM_err`	One of the parameters is invalid.
`INTERNAL_err`	An error internal to the structure or operation of the tool kits has occurred.
`NOMEM_err`	No more memory was available for dynamic allocation.
`WALK_err`	Tree is not structured properly for walking (for example, a loop exists in the specific structure).

Implementation Status

Implemented.

19.2.4 Delete an Iterator: *DeleteIterator*

```
INT_type DeleteIterator(iterator)
   ITERATOR_type iterator;
```

Real Name

delgen

Synopsis

This routine is used to delete an iterator.

Parameters

`iterator`	The iterator to delete.

Return Value

`ERROR_INT`	An error has occurred. `TKError` will contain an indication of the exact error (see the **Errors** part below).
otherwise	The iterator has been deleted.

Side Effects

Any dynamic storage used by this iterator will be released.

Semantics

The application programmer must remember to delete an iterator when it is no longer to be used. There are actions that are performed upon deletion that affect the number of iterators that can exist simultaneously for a structure.

Errors

`NOGENS_err`	No more iterators may be created for this structure.

Implementation Status

Implemented.

20
External Representations

In Part 5 of ISO IS 8613, the ODA standard defines two datastream formats (external representations) that may be used to interchange ODA documents between systems. ODIF (Office Document Interchange Format) is an interchange format that is defined in accordance with the ASN.1 standards [31, 32]. ODL (Office Document Language) is a format that allows an ODA document to be represented in SGML [33]. Routines for dealing with ODIF are described in section 20.1 and ODL, which is unsupported, is described in section 20.2.

Before an external representation of a document may be written, all of the constituents in the document (except for the document profile) must have appropriate *identifiers*. Identifiers are used in ODA to uniquely designate constituents. There are identifier attributes for each kind of constituent:

- "object identifier" for object descriptions,
- "object class identifier" for object class descriptions,
- "layout style identifier" for layout styles,
- "presentation style identifier" for presentation styles,
- "content identifier - logical" for content portions,
- "content identifier - layout" for content portions.

The programmer may choose to let the tool kit label automatically all constituents before the document is written. Alternately, the application program may perform some or all of the labeling and explicitly specify which structures the tool kit should label. These possibilities are described in section 20.3.

20.1 ODIF

ODIF uses the ASN.1 definition techniques and transfer syntax [31, 32] to provide a binary datastream format for the interchange of ODA documents. ASN.1 defines methods for defining and encoding arbitrary, typed values. A typed value is represented as a triple. The first element of the triple is a binary encoding of the type of the value. This type may be a universal type, defined by ASN.1, or a type that is private to an international standard or to this particular application. The second element of the triple is a binary encoding of the length of the value. This encoding is quite flexible and allows both fixed-length values and variable-length values. The last element of the triple is the binary representation of the value. This representation may contain additional ASN.1 encodings of values. ASN.1, thus, provides a general mechanism for representing data in a machine-independent manner.

There are two interchange format classes defined by ODA for representing a document in ODIF. Interchange format class 'A' is a format that can be used to represent any legal ODA document. Interchange format class 'B' is restricted to representing formatted documents that obey a specific set of constraints. These constraints are described in section 4.2, Part 5 of ISO IS 8613.

The tool kit provides a routine for reading a document represented in ODIF and a routine for writing an ODIF representation of a document. Two implementations of ODIF support exist. The first, a manually coded version, is accessed as the `ReadODIF` and `WriteODIF` routines (see sections 20.1.1 and 20.1.2 respectively). The second is based on the pepy utility of the Isode Development Environment [56]. These are found in section 20.1.3 for the isode-based ReadODIF and 20.1.4 for the isode-based WriteODIF.

20.1.1 Read an ODIF Representation: *ReadODIF*

```
DOCUMENT_type ReadODIF(id)
   INT_type id;
```

Real Name

rdodif

Synopsis

This routine is used to read an ODIF representation of an ODA document into a new document.

Parameters

`id`	The input descriptor to read from (section 10.2.4).

Return Value

`ERROR_DOCUMENT`

An error has occurred. `TKError` will contain an indication of the exact error (see the **Errors** part below).

otherwise	The document has been read and the corresponding data structures have been constructed and the return value is the document specification value.

Side Effects

Before returning, this routine will indicate termination on the input descriptor, `id`, by calling `LOCAL_ReadBytes` with a `length` parameter of -1 (section 10.2.4.1).

Semantics

A new document is created for each call to this routine.

Errors

`BADPARM_err`	One of the parameters is invalid.
`INTERNAL_err`	An error internal to the structure or operation of the tool kits has occurred.
`NOMEM_err`	No more memory was available for dynamic allocation.
`ODIF_err`	ODIF datastream is invalid.

Implementation Status

The following attributes are supported: "alignment", "alternative representation", "application comments", "block alignment", "border", "character fonts", "character orientation", "character path", "character spacing", "clipping", "code extension announcers", "colour", "compression", "concatenation", "content architecture class", "content identifier - layout", "content identifier - logical", "content portions", "content type", "default value lists" (see below for exceptions), "fill order", "first line offset", "fonts list" (see below for restrictions), "formatting indicator", "generator for subordinates", "generic layout structure", "generic logical structure", "graphic character sets", "graphic rendition", "image dimensions", "indivisibility", "initial offset", "interchange format class", "itemisation", "kerning offset", "layout category", "layout object class", "layout path", "layout style", "layout style identifier", "layout styles", "line progression", "line spacing", "medium type", "new layout object", "number of discarded pels", "number of lines", "number of pels per line", "object class", "object class identifier", "object identifier", "object type", "offset", "pairwise kerning", "pel path", "pel spacing", "pel transmission density", "permitted categories", "position", "presentation style", "presentation style identifier", "presentation styles", "proportional line spacing", "protection", "resource", "separation", "spacing ratio", "specific layout structure", "specific logical structure", "subordinates", "transparency", "unit scaling", "user-readable comments", "user-visible name".

The "default value lists" attribute is implemented with the exception of the following. For page components, the following attributes are not implemented: "dimensions", "transparency", "page position", "medium type", "colour". For frame components, the following attributes are not implemented: "position", "dimensions", "transparency", "layout path", "colour", "border". For block components, the following attributes are not implemented: "position", "dimensions", "transparency", "colour", "border".

Only the following fields of the entries for the attribute "fonts list" are supported: ISO/FONTNAME, ISO/FAMILY, ISO/POSTURE, ISO/WEIGHT, ISO/PROPWDTH, ISO/DSNSIZE, ISO/DSCLASS, ISO/STRUCTUR, ISO/ESCLASS, ISO/VARSCRIPT, ISO/LSCRPT, ISO/RSCRPT, ISO/VXOFFSET, ISO/VYOFFSET, ISO/VXSCALE, ISO/VYSCALE.

Due to a restriction in the implementation, datastreams will not be read correctly unless all content portions contain an appropriate value for the "type of coding" attribute.

20.1.2 Write an ODIF Representation: *WriteODIF*

```
INT_type WriteODIF(od, document, format)
    INT_type od;
    DOCUMENT_type document;
    VALUE_type format;
```

Real Name

wrodif

Synopsis

This routine is used to write an ODIF representation of a specified document.

Parameters

`od`	The output descriptor to write to (section 10.2.4).
`document`	The internal document to write.
`format`	The interchange format class to write. This parameter must have one of the legal values for the attribute "interchange format class" (section 17.1.8.14).

Return Value

`ERROR_INT`	An error has occurred. `TKError` will contain an indication of the exact error (see the **Errors** part below).
otherwise	The specified document has been written.

Side Effects

Before returning, this routine will indicate termination on the output descriptor, `od`, by calling `LOCAL_WriteBytes` with a `length` parameter of -1 (see section 10.2.4.2).

Semantics

If the parameter `format` is `dp_INTERCHANGE_FORMAT_CLASS_B`, then the document must obey the restrictions described in section 4.2, Part 5 of ISO IS 8613.

Errors

`BADPARM_err`	One of the parameters is invalid.
`INTERNAL_err`	An error internal to the structure or operation of the tool kits has occurred.

Implementation Status

This routine has the same limitation as `ReadODIF` (section 20.1.1), except that it also supports output of the following attributes: "balance", "page position", "imaging order", "same layout object".

Interchange format class 'B' is not supported.

20.1.3 Read an ODIF Representation on Unix: *isode_ReadODIF*

```
DOCUMENT_type isode_ReadODIF(infile)
   FILE *infile;
```

Real Name

isred

Synopsis

This routine is used to read an ODIF representation of an ODA document into a new document.

Parameters

`infile`	The Unix `FILE *` stream to read from.

Return Value

`ERROR_DOCUMENT`
: An error has occurred. `TKError` will contain an indication of the exact error (see the **Errors** part below).

otherwise
: The document has been read and the corresponding data structures have been constructed and the return value is the document specification value.

Side Effects

This routine does not close the file descriptor when done (unlike `ReadODIF`).

Semantics

A new document is created for each call to this routine.

Errors

`BADPARM_err`	One of the parameters is invalid.
`INTERNAL_err`	An error internal to the structure or operation of the tool kits has occurred.
`NOMEM_err`	No more memory was available for dynamic allocation.
`ODIF_err`	ODIF datastream is invalid.

Implementation Status

Only the following attributes are supported by this routine: "alternative representation", "application comments", "content identifier - layout", "content identifier - logical", "content information", "content portions", "default value lists", "generator for subordinates", "layout style", "layout style identifier", "object class", "object class identifier", "object identifier", "object type", "presentation style", "presentation style identifier", "protection", "resource", "subordinates", "type of coding", "user-readable comments", "user-visible name".

20.1.4 Write an ODIF Representation on Unix: *isode_WriteODIF*

```
INT_type isode_WriteODIF(outfile, document,
                         format)
   FILE *outfile;
   DOCUMENT_type document;
   VALUE_type format;
```

Real Name

iswrt

Synopsis

This routine is used to write an ODIF representation of a specified document on Unix.

Parameters

`outfile`	The Unix `FILE *` output descriptor to write to.
`document`	The internal document to write.
`format`	The interchange format class to write. This parameter must have one of the legal values for the attribute "interchange format class" (section 17.1.8.14).

Return Value

`ERROR_INT`	An error has occurred. `TKError` will contain an indication of the exact error (see the **Errors** part below).
otherwise	The specified document has been written.

Side Effects

Unlike `WriteODIF`, this routine does not close its output descriptor.

Semantics

If the parameter `format` is `dp_INTERCHANGE_FORMAT_CLASS_B`, then the document must obey the restrictions described in section 4.2, Part 5 of ISO IS 8613.

Errors

`BADPARM_err`	One of the parameters is invalid.
`INTERNAL_err`	An error internal to the structure or operation of the tool kits has occurred.

Implementation Status

This routine has the same restrictions as `isode_ReadODIF` (section 20.1.3). The interchange format class 'B' is not supported.

20.2 ODL

The CMU ODA Tool Kit currently contains no support for the ODL datastream.

20.3 Constituent Identifier Attributes

There are attributes whose values are identifiers for constituents: "object identifier", "object class identifier", "layout style identifier", "presentation style identifier", "content identifier - logical" and "content identifier - layout". Each identifier is a sequence of non-negative integers.

Within ODA there are attributes whose values reference constituents. Each reference takes the form of an integer, which specifies some unambiguous representation for the constituent's identifier. For example, the attribute "subordinates" has as its value a sequence of non-negative integers, each representing a subordinate constituent. Each integer corresponds to the final integer in the "object identifier" attribute of a subordinate constituent.

Within the tool kit, these references take the form of objects of type `CONSTITUENT`. When writing an external representation of the document, however, an appropriate identifier must be assigned to each identifier attribute and references to constituents must be represented by non-negative integers.

The tool kit provides mechanisms to allow the application program to ignore the complexities of identifiers: appropriate identifiers will be assigned to constituents before an external representation is generated for the document. This is accomplished by specifying *automatic labeling* in the initialization call (section 14.1).

Alternatively, the application program may desire to assign these identifiers explicitly. This may be the case if, for example, the application profile being used assigns a specific semantics to forms of identifiers. In this case, *automatic labeling* should not be specified during initialization. The application program is then responsible for assigning appropriate values to each identifier. If any identifier is unassigned, or if an illegal identifier is assigned then, unpredictable results may occur.

20.3.1 Assign Identifiers for an Entire Document: *LabelDocument*

```
INT_type LabelDocument(document)
   DOCUMENT_type document;
```

Real Name

lbldoc

Synopsis

This routine is used to assign appropriate identifiers to an entire document.

Parameters

`document`	The document to label.

Return Value

`ERROR_INT`	An error has occurred. `TKError` will contain an indication of the exact error (see the **Errors** part below).
otherwise	The document has been labeled.

Side Effects

Values will be assigned to the appropriate identifier attributes for each constituent of the document.

Any identifiers existing for a constituent will be replaced.

Semantics

Errors

`BADPARM_err`	One of the parameters is invalid.
`INTERNAL_err`	An error internal to the structure or operation of the tool kits has occurred.

Implementation Status

Existing identifiers will not be deleted, but will be replaced (this results in a core leak).

20.3.2 Assign Identifiers for a Logical or Layout Structure: *LabelStructure*

```
INT_type LabelStructure(root)
   CONSTITUENT root;
```

Real Name

lblstr

Synopsis

This routine is used to assign appropriate identifiers to the constituents of a logical or layout structure.

Parameters

`root`	The document logical or document layout root of the structure to label.

Return Value

`ERROR_INT`	An error has occurred. `TKError` will contain an indication of the exact error (see the **Errors** part below).
otherwise	The structure has been labeled.

Side Effects

All of the constituents in the designated structure have been labeled.

Semantics

Any identifiers existing for a constituent will be replaced.

Errors

`BADPARM_err`	One of the parameters is invalid.
`INTERNAL_err`	An error internal to the structure or operation of the tool kits has occurred.

Implementation Status

Existing identifiers will not be deleted, but will be replaced (this results in a core leak).

20.3.3 Assign Identifiers to the Layout or Presentation Styles: *LabelStyles*

```
INT_type LabelStyles(document, islayout)
   DOCUMENT_type document;
   BOOL_type islayout;
```

Real Name

lblsty

Synopsis

This routine is used to assign identifiers to the layout styles or presentation styles.

Parameters

`document`	The document whose styles are to be labeled.
`type`	If `BOOL_true`, then the layout styles will be labeled. Otherwise, the presentation styles will be labeled.

Return Value

`ERROR_INT`	An error has occurred. `TKError` will contain an indication of the exact error (see the **Errors** part below).
otherwise	The designated styles have been labeled.

Side Effects

All of the constituents in the designated structure have been labeled.

Semantics

Any identifiers existing for a constituent will be replaced.

Errors

`BADPARM_err`	One of the parameters is invalid.
`INTERNAL_err`	An error internal to the structure or operation of the tool kits has occurred.

Implementation Status

Existing identifiers will not be deleted, but will be replaced (this results in a core leak).

21
Affecting Run-time Behavior

This chapter describes the use of routines that can affect the run-time behavior of the tool kit. In particular, these routines can be used to

- ensure that designated data structures are kept in primary memory during the execution of certain time-critical sections of code;
- improve the efficiency of access to frequently-used attributes of some constituents.

Use of these routines is not mandatory—the tool kit will perform correctly using default actions. The application programmer is advised to be careful in the use of these routines as it is easy to worsen the performance of an application by attempting to optimize code that is not well understood.

The descriptions in this chapter are to be considered as informational only. None of the routines are implemented and calling any of them will result in an error return.

21.1 Managing Memory

The tool kit has been designed to allow parts of the internal representation of an ODA document to be kept in secondary memory. This is to allow the tool kit to be used in an environment where the amount of primary memory is severely restricted. The tool kit contains code that manages the allocation of parts of the document by invoking site-defined routines for choosing and moving objects between primary and secondary memories. The levels of granularity for this allocation are an attribute (and its value) and objects of type `STRING_type`.

The default action taken by the tool kit, however, may turn out to be inefficient for particular applications. For example, an application may make extensive use of a long `STRING_type` object that is frequently moved between primary and secondary storage. In such cases, it may be useful to ensure that the string in question is kept in memory during crucial sections of code. The routines in this section may be used to affect the allocation of objects to primary memory.

21.1.1 Keep a Constituent in Primary Memory: *HoldConstituent*

```
INT_type HoldConstituent(constituent)
   CONSTITUENT constituent;
```

Real Name

hldcns

Synopsis

This routine is used to ensure that a constituent (that is, all of its attributes and their values) is kept in primary memory until a corresponding `ReleaseConstituent` call is made.

Parameters

`constituent`	The constituent to hold in primary memory.

Return Value

`ERROR_INT`	An error has occurred. `TKError` will contain an indication of the exact error (see the **Errors** part below).
otherwise	The constituent is in primary memory and will remain there until a corresponding call is made to `ReleaseConstituent`.

Side Effects

None.

Errors

`BADPARM_err`	One of the parameters is invalid.
`INTERNAL_err`	An error internal to the structure or operation of the tool kits has occurred.
`NOMEM_err`	No more memory was available for dynamic allocation.

Implementation Status

Unimplemented.

21.1.2 Release Constituent from Memory: *ReleaseConstituent*

```
INT_type ReleaseConstituent(constituent)
   CONSTITUENT constituent;
```

Real Name

rlscns

Synopsis

This routine is used to allow a constituent to be released from primary memory. In other words, the designated constituent need not be kept in primary memory when not in use.

Parameters

`constituent`	The constituent to be released.

Return Value

`ERROR_INT`	An error has occurred. `TKError` will contain an indication of the exact error (see the **Errors** part below).
otherwise	The constituent has been marked as releasable from primary memory.

Side Effects

None.

Errors

`BADPARM_err`	One of the parameters is invalid.
`INTERNAL_err`	An error internal to the structure or operation of the tool kits has occurred.
`NOMEM_err`	No more memory was available for dynamic allocation.

Implementation Status

Unimplemented.

21.1.3 Keep an Attribute in Primary Memory: *HoldAttr*

```
INT_type HoldAttr(attribute)
   INT_type attribute;
```

Real Name

hldatt

Synopsis

This routine is used to ensure that an attribute is kept in primary memory until a corresponding `ReleaseAttr` call is made.

Parameters

`attribute`	The identifier for the attribute to hold in primary memory.

Return Value

`ERROR_INT`	An error has occurred. `TKError` will contain an indication of the exact error (see the **Errors** part below).
otherwise	The attribute and its value is in primary memory and will remain there until the corresponding `ReleaseAttr` call.

Side Effects

None.

Errors

`BADPARM_err`	One of the parameters is invalid.
`INTERNAL_err`	An error internal to the structure or operation of the tool kits has occurred.
`NOMEM_err`	No more memory was available for dynamic allocation.

Implementation Status

Unimplemented.

21.1.4 Release Attribute from Memory: *ReleaseAttr*

```
INT_type ReleaseAttr(attribute)
   INT_type attribute;
```

Real Name

rlsatt

Synopsis

This routine is used to allow an attribute and its value to be released from primary memory. In other words, the designated attribute need not be kept in primary memory when not in use.

Parameters

`attribute`	The identifier of the attribute to be released.

Return Value

`ERROR_INT` An error has occurred. `TKError` will contain an indication of the exact error (see the **Errors** part below).

otherwise The attribute has been marked as releasable from primary memory.

Side Effects

None.

Errors

`BADPARM_err` One of the parameters is invalid.

`INTERNAL_err` An error internal to the structure or operation of the tool kits has occurred.

`NOMEM_err` No more memory was available for dynamic allocation.

Implementation Status

Unimplemented.

21.1.5 Keep a String in Primary Memory: *HoldString*

```
INT_type HoldString(string)
   STRING_type string;
```

Real Name

hldstr

Synopsis

This routine is used to ensure that a `STRING_type` object is kept in primary memory until a corresponding `ReleaseString` call is made.

Parameters

`string` The `STRING_type` object to keep in memory.

Return Value

`ERROR_INT` An error has occurred. `TKError` will contain an indication of the exact error (see the **Errors** part below).

otherwise The `STRING_type` object is in primary memory and will remain there until the corresponding `ReleaseString` call.

Side Effects

None.

Errors

`BADPARM_err`	One of the parameters is invalid.
`INTERNAL_err`	An error internal to the structure or operation of the tool kits has occurred.
`NOMEM_err`	No more memory was available for dynamic allocation.

Implementation Status

Unimplemented.

21.1.6 Release String from Memory: *ReleaseString*

```
INT_type ReleaseString(string)
   STRING_type string;
```

Real Name

rlsstr

Synopsis

This routine is used to allow a `STRING_type` object to be released from primary memory. In other words, the designated string need not be kept in primary memory when not in use.

Parameters

`string`	The string to be marked as releasable from primary memory.

Return Value

`ERROR_INT`	An error has occurred. `TKError` will contain an indication of the exact error (see the **Errors** part below).
otherwise	The `STRING_type` object is in primary memory and will remain there until the corresponding `ReleaseString` call.

Side Effects

None.

Errors

BADPARM_err	One of the parameters is invalid.
INTERNAL_err	An error internal to the structure or operation of the tool kits has occurred.
NOMEM_err	No more memory was available for dynamic allocation.

Implementation Status

Unimplemented.

21.2 Affecting Search Time for Attributes

The tool kit uses an on-demand allocation scheme for attributes. That is, the storage for an attribute and its value is allocated when the attribute is first assigned a value and released when the attribute is deleted. This means that only enough storage for the attributes actually applied to a constituent need be allocated. The scheme we have implemented, however, requires that a search be made to locate each attribute.

In general, the attributes are ordered and the search scheme will locate some attributes of a constituent with fewer steps than others. Some applications may find that there are certain attributes that are used frequently and others that are used infrequently. In this case, the default ordering of the attributes may prove inefficient.

The routines described in this section allow the application programmer to specify priorities for attributes for classes of constituents. The priority of an attribute is a non-negative integer that indicates the relative likelihood of the attribute being used. The higher the priority, the more frequently the attribute is expected to be referenced.

Priorities are divided into three categories: 0, greater than 0 but less than 100, and 100. The meanings of these categories are as follows:

0	The attribute is expected not to be used at all. The tool kit is encouraged to improve the efficiency of other (non-zero) attributes at the expense of this attribute.
0<*priority*≤100	A priority in this category is a relative priority and indicates that this attribute is expected to be used more frequently than attributes with lower priorities.
100	This attribute is expected to be present always and frequently used during execution. The tool kit is encouraged to improve the efficiency of access to this attribute as much

as possible. In particular, the tool kit is free to ensure that this attribute is always present for constituents of the specified class, whether or not the attribute is actually in use.

Despite the priority assigned, it is legal to use any attribute, although some attributes may be inefficient to use. If no priority is assigned, an attribute is assumed to have priority 50.

21.3 Access the Priority of Components: *ComponentAttrPriority*

```
INT_type ComponentAttrPriority(set, type,
                               class, attr,
                               priority)
   BOOL_type set;
   VALUE_type type;
   BOOL_type class;
   INT_type attr;
   INT_type *priority;
```

Real Name

compri

Synopsis

This routine is used to set or get the priority assigned to an attribute for a category of components. The category is specified by the `type` and `class` parameters.

Parameters

`set`	If this parameter is `BOOL_true`, then this is an assignment of a priority to an attribute of the specified category. Otherwise, this is a request to obtain the current priority of the attribute.
`type`	This indicates the type of the component. It must be one of the legal values for the attribute "object type", defined in section 17.1.1.1.
`class`	If `BOOL_true`, then this category of components will be object class descriptions. Otherwise, it will be an object description.
`attr`	The attribute to be prioritized.
`priority`	The address of a variable that will contain the priority for this attribute.

Return Value

`ERROR_INT`	An error has occurred. `TKError` will contain an indication of the exact error.
otherwise	If `set` is `BOOL_false`, the priority for this attribute is returned in `*priority`. Otherwise the priority has been assigned.

Side Effects

None.

Semantics

See section 21.2 for a description of the meaning of priorities.

Implementation Status

Unimplemented.

21.3.1 Access the Priority for Styles: *StyleAttrPriority*

```
INT_type StyleAttrPriority(set, type,
                           attribute,
                           priority)
   BOOL_type set;
   VALUE_type type;
   INT_type attribute;
   INT_type *priority;
```

Real Name

stypri

Synopsis

This routine is used to set or get the priority of an attribute for a category of styles: either layout or presentation styles.

Parameters

`set`	If this parameter is `BOOL_true`, then this is an assignment of a priority to an attribute of the specified category. Otherwise, this is a request to obtain the current priority of the attribute.
`type`	An integer denoting the style desired. It must be one of the values `LAYOUT_value` or `PRESENTATION_value`.

`attribute`	The attribute to be prioritized.
`priority`	The address of a variable that will contain the priority for this attribute.

Return Value

`ERROR_INT`	An error has occurred. `TKError` will contain an indication of the exact error.
otherwise	If `set` is `BOOL_false`, the priority for this attribute is returned in `*priority`, else the priority has been assigned.

Side Effects

None.

Implementation Status

Unimplemented.

21.3.2 Access the Priority for Contents: *ContentAttrPriority*

```
INT_type ContentAttrPriority(set, type,
                             attribute,
                             priority)
   BOOL_type set;
   INT_type type;
   INT_type attribute;
   INT_type *priority;
```

Real Name

conpri

Synopsis

This routine is used to get or set a priority for a presentation attribute that applies to a basic component description.

Parameters

`set`	If this parameter is `BOOL_true`, then this is an assignment of a priority to an attribute of the specified category. Otherwise, this is a request to obtain the current priority of the attribute.
`type`	The legal values of this parameter are currently unspecified.

`attribute` The attribute to be prioritized.

`priority` The address of a variable that contains the priority for this attribute.

Return Value

`ERROR_INT` An error has occurred. `TKError` will contain an indication of the exact error.

otherwise If `set` is `BOOL_false`, the priority for this attribute is returned in `*priority`, else the priority is assigned.

Side Effects

None.

Semantics

This routine sets the priority for a presentation attribute that applies to a basic component description. Note, however, that the type of content architecture is specified in the call.

Implementation Status

Unimplemented.

21.3.3 Access the Attribute Priority for the Document Profile: *AttrDocProfPriority*

```
INT_type AttrDocProfPriority(set, attribute,
                             priority)
   BOOL_type set;
   INT_type attribute;
   INT_type *priority;
```

Real Name

dppri

Synopsis

This routine is used to get or set the priority of an attribute for the document profile.

Parameters

`set` If this parameter is `BOOL_true`, then this is an assignment of a priority to an attribute of the specified category. Otherwise, this is a request to obtain the current priority of the attribute.

`attribute`	The attribute to be prioritized.
`priority`	The address of a variable that contains the priority for this attribute.

Return Value

`ERROR_INT`	An error has occurred. `TKError` will contain an indication of the exact error.
otherwise	If `set` is `BOOL_false`, the priority for this attribute is returned in `*priority`, else the priority has been assigned.

Side Effects

None.

Implementation Status

Unimplemented.

22
Debugging Support

The CMU ODA Tool Kit provides a number of routines for use in debugging an application program that manipulates an ODA document. These routines provide what we consider to be a minimal set of useful debugging functions. These routines allow the application programmer to write a human-readable form of the current state of parts of the ODA document structure. The programmer may write the state of the entire document, a specified subgraph or individual constituents.

22.1 Notation for Debugging Output

The output notation used by the tool kit debugging routines is intended to be readable by humans and not to serve as unambiguous input for a parser. As such, we will not give a formal definition of the output format but merely describe its salient points and leave it to the reader's common sense and the self-describing nature of the format to fill in the missing information.

The primary unit of output is, of course, the constituent, which is represented as:

```
<document #>/<node #>: <constituent kind>
   ("<attribute_1>" <value_1>)
   ("<attribute_2>" <value_2>)
   . . .
   ("<attribute_n>" <value_n>)
```

For example,

```
1/50: GENERIC COMPOSITE LOGICAL OBJECT
    ("Application Comments" `Figure`)
    ("Generator for Subordinates"
           OPT REP CHO[OPT REP 1/244:
                      OPT REP 1/245:
                      OPT REP 1/246:])
    ("Object Class Identifier" "2 0 1 2 3 4 5")
    ("Object Type" 'composite logical object')
    ("User-readable Comments"
               "Generic figure for document")
```

The combination of <document #> and <constituent #> provides a unique identifier for each constituent that is used to refer to the constituent from other constituents (note the uses of such identifiers in the attribute "generator for subordinates", above).

Here are some other points that may not be obvious about the debugging format:

- Spaces and new lines outside of bracketed quotes are meaningless.
- If the value of an attribute is a list, then the values comprising the list are separated by spaces.
- Attribute values that are character strings are enclosed in double quote marks (").
- Attribute values that are octet strings are enclosed in single reverse quote marks (`).
- Non-printable characters and bytes are represented as

  ```
  \ooo
  ```

  where 'ooo' is the value of the character (byte) represented in octal notation.
- The values of some attributes are followed by comments enclosed in square brackets ([]). These comments are explanatory and are not part of the document. For example, in the following

  ```
  ("Type of Coding" 2 8 3 6 0
     ['character content encoding'])
  ```

  the comment indicates that the ASN.1 identifier that is the value of the "type of coding" attribute specifies 'character content encoding'.
- Attribute values that are drawn from a set of designated values specified by the ODA standard are enclosed in single quotes (`').

22.2 Debugging Routines

There are debugging routines for writing the state of

- a specified constituent,
- a subgraph reachable from a specified constituent,
- the entire document.

These routines may be used to achieve selective debugging output for various document parts. For example, the state of any of the specific or generic logical or layout structures may be written out by calling the `WriteSubgraphState` routine with the root of the appropriate structure. The state of all of the layout or presentation styles may be written out by calling `WriteStylesState` as described in section 22.2.3.

22.2.1 Write the State of the Entire Document: *WriteDocumentState*

```
INT_type WriteDocumentState(od, document)
   INT_type od;
   DOCUMENT_type document;
```

Real Name

dbgdoc

Synopsis

This routine is used to output the state of the entire document for debugging purposes.

Parameters

`od`	The output descriptor (section 10.2.4) to write the state to.
`document`	The document to write.

Return Value

`ERROR_INT`	An error has occurred. `TKError` will contain an indication of the exact error (see the **Errors** part below).
otherwise	The state of the document has been output.

Side Effects

None.

Errors

`BADPARM_err`	One of the parameters is invalid.
`INTERNAL_err`	An error internal to the structure or operation of the tool kits has occurred.
`LOCAL_err`	A locally-defined error, not covered by any other available error, has occurred.

Implementation Status

This routine has limitations on the attributes that it will output. See the entry for `WriteConstituentState` (section 22.2.4) for details.

22.2.2 Write the State of a Subgraph: *WriteSubgraphState*

```
INT_type WriteSubgraphState(od, constituent)
   INT_type od;
   CONSTITUENT constituent;
```

Real Name

dbgsbg

Synopsis

This routine is used to output the state of a subgraph of the document for debugging.

Parameters

`od`	The output descriptor (section 10.2.4) to write the state to.
`constituent`	A constituent representing the root of the subgraph.

Return Value

`ERROR_INT`	An error has occurred. `TKError` will contain an indication of the exact error (see the **Errors** part below).
otherwise	The state of the subgraph has been output.

Side Effects

None.

Semantics

This routine will output the state of the constituents that are *descendants* of the specified constituent. (See section 19.2 for a definition of *descendant*.)

Errors

`BADPARM_err`	One of the parameters is invalid.
`INTERNAL_err`	An error internal to the structure or operation of the tool kits has occurred.
`LOCAL_err`	A locally-defined error, not covered by any other available error, has occurred.

Implementation Status

This routine has limitations on the attributes that it will output. See the entry for `WriteConstituentState` (section 22.2.4) for details.

22.2.3 Write the State of Layout or Presentation Styles: *WriteStylesState*

```
INT_type WriteStylesState(od, document,
                          islayout)
    INT_type od;
    DOCUMENT_type document;
    BOOL_type islayout;
```

Real Name

dbgsty

Synopsis

This routine is used to output the states of the layout or presentation styles of a document for debugging.

Parameters

`od`	The output descriptor (section 10.2.4) to write the states to.
`document`	The document whose styles are to be output.
`islayout`	If this parameter is `BOOL_true`, then either the layout styles are output, or the presentation styles are output.

Return Value

`ERROR_INT`	An error has occurred. `TKError` will contain an indication of the exact error (see the **Errors** part below).
otherwise	The states of the indicated styles have been output.

Side Effects

None.

Errors

`BADPARM_err`	One of the parameters is invalid.
`INTERNAL_err`	An error internal to the structure or operation of the tool kits has occurred.
`LOCAL_err`	A locally-defined error, not covered by any other available error, has occurred.

Implementation Status

This routine has limitations on the attributes that it will output. See the entry for `WriteConstituentState` (section 22.2.4) for details.

22.2.4 Write the State of a Constituent: *WriteConstituentState*

```
INT_type WriteConstituentState(od,
                               constituent)
   INT_type od;
   CONSTITUENT constituent;
```

Real Name

dbgcns

Synopsis

This routine is used to write out the state of a specified constituent for debugging.

Parameters

`od`	The output descriptor (section 10.2.4) to write the state to.
`constituent`	The constituent to be debugged.

Return Value

`ERROR_INT`	An error has occurred. `TKError` will contain an indication of the exact error (see the **Errors** part below).

otherwise The state of the constituent has been output.

Side Effects

None.

Errors

`BADPARM_err`	One of the parameters is invalid.
`INTERNAL_err`	An error internal to the structure or operation of the tool kits has occurred.
`LOCAL_err`	A locally-defined error, not covered by any other available error, has occurred.

Implementation Status

This routine does not support the following attributes (it will not, however, return an error—the attribute name will simply appear without a value): "authors", "bindings", "border", "colour", representations "content architecture classes", "content generator", "copyright", "dimensions", "distribution list", "document application profile defaults", "document reference", "edge rendition", "encoding announcer", "filled area rendition", "itemisation", "line layout table", "line rendition", "local file reference", "marker rendition", "medium type", "medium types", "ODA version", "owners", "picture dimensions", "position", "preparers", "references to other documents", "region of interest specification", "resource document", "resources", "revision history", "same layout object", "superseded documents", "text rendition", "transformation specification", "transparency specification", "types of coding".

22.2.5 Write a Textual Description of an ODIF File on Unix: *isode_PrintODIF*

```
INT_type isode_PrintODIF(infile)
   FILE *infile;
```

Real Name

isprdf

Synopsis

This routine is used to write out an ODIF file in a form that describes its ASN.1 structure.

Parameters

`od`	The Unix `FILE *` output descriptor to write the ODIF description to.

Return Value

`ERROR_INT`	An error has occurred. `TKError` will contain an indication of the exact error (see the **Errors** part below).
otherwise	The ODIF has been described.

Side Effects

None.

Errors

`BADPARM_err`	One of the parameters is invalid.
`INTERNAL_err`	An error internal to the structure or operation of the tool kits has occurred.
`LOCAL_err`	A locally-defined error, not covered by any other available error, has occurred.

Implementation Status

Implemented.

4 The Raster Graphics Tool Kit

23
Introduction to the Raster Graphics Tool Kit

The Raster Graphics Tool Kit is a portable set of subroutines and data structures that makes possible the manipulation of bit-mapped graphics images. The tool kit may be used within applications programs to support many widely-used graphics storage formats. The tool kit provides graphics editing primitives to allow for the changing, clipping and copying of graphics images. This package may be used independently to provide graphics support, or in conjunction with the CMU ODA Tool Kit (see part 3 of this book) to provide raster graphics manipulation for documents stored in the Office Documentation Architecture (ODA) format.

23.1 Rasters

The graphics images that the tool kit manipulates are called *rasters*. A raster is defined to be a rectangular matrix of *picture elements*, or *pels*. At times, a raster will also be referred to as a *bitmap*. This will be the case when there are only two states for the pels in the raster, such as black or white, and each can, thus, be represented by a single bit.

A raster image manipulated by the tool kit is of type `RASTER_type`. This structure stores relevant information such as the dimensions of the raster, the type of pels being used, and pointers to the pels. The `RASTER_type` structure may be created by several routines, such as `MakeRaster`, `ReadRaster`, `ReadMemRaster` and `ExtractRaster`. Rasters may be deleted by the `DeleteRaster` function call. All of these routines are described in chapter 25.

23.2 Coordinate System

The pels within a raster, as mentioned above, form a 2-dimensional matrix. Each pel may be addressed by a unique set of coordinates *(x, y)*. The coordinate (0, 0) refers to the pel in the uppermost horizontal row and the leftmost vertical column in the matrix. The x coordinate increases as pels are located further towards the right and the y coordinate increases as pels are located further towards the bottom of the raster. A raster of size

$$width \times height$$

has pels which range from (0, 0) in the top-left corner to (*width*-1, *height*-1) in the bottom-right corner.

23.3 Pels

A pel is the smallest portion of a raster that may be individually manipulated; it is the picture element. There are several different ways that pels may be represented. The Raster Graphics Tool Kit currently supports only two-colored rasters: a pel may be either *BLACK* or *WHITE*. These states correspond to a pel being *SET* (black) or *UNSET* (white). The UNSET, or WHITE, state is the background state.

In a bitmap, pels may be set to black or white by using either the `AssignBit` or `AssignPel` routines. As `AssignBit` is specifically designed for bitmaps, it is the optimal choice when the desired raster is known to be a bitmap.[65]

23.4 Graphics Formats

There are many different ways that raster images may be stored. The techniques differ in compression schemes, header information, bit packing order and row padding. The Raster Graphics Tool Kit makes support for a diversity of formats simple and straightforward. The tool kit allows rasters to be read or written in

---

[65]The `AssignPel` routine is currently the same as the `AssignBit` routine, since the Raster Graphics Tool Kit does not support multicolored pels. However, `AssignPel` would be used to modify the color of a pel. Programs that will be dealing exclusively with bitmaps should always use `AssignBit`, as this routine is optimized for bitmaps.

several widely-used graphics formats. Each supported graphics format is listed below (and fully described in section 24.5):

- Amiga Interchange File Format (IFF)
- Andrew (ATK) Raster Format
- CompuServe's Graphics Interchange Format (GIF)
- Group 3 Facsimile Encoding (CCITT Recommendation T.4)
- Group 4 Facsimile Encoding (CCITT Recommendation T.6)
- MacPaint Document Format
- ODA Bitmap Format
- Portable BitMap Format (PBM)
- Postscript (output format only)
- Sun Raster Format
- X Bitmap Format

23.5 Memory Management

One of the goals of the Raster Graphics Tool Kit is to provide a graphics system that may be used on small machines, those without large amounts of primary memory or without built-in virtual memory. The tool kit performs its own memory-mapping. If the rasters that the tool kit is handling are too large to fit in primary memory, the tool kit will perform automatic swapping to a secondary storage device. An entire raster or portions of a raster may be swapped. The swapping takes place at a row-by-row level and is transparent to the user of the tool kit—all raster and pel manipulation functions will work on rasters that are in either primary or secondary storage.

The user does, however, have the ability to improve the performance of the swapping mechanism. When a large amount of manipulation is to be performed on a raster, it might be desirable to retain the entire raster in memory so that the support routines execute quickly without any swapping overhead. Two routines, `HoldRaster` and `ReleaseRaster`, allow the programmer to specify that a raster should, if at all possible, be held in primary memory, or to specify that a raster may be swapped to a backing store if needed (the default for rasters).

23.6 Error Handling

All of the routines defined by the tool kit indicate an error in invocation by returning a distinguished value. Once an error has been detected, the application program may obtain further information about the nature of the error by inspecting the global variable `RGError`.

After a routine has returned an error indication, and before the next Tool Kit routine has been invoked, the global `INT_type` variable `RGError` contains an indication of the kind of error that occurred. Each routine description in this manual contains an Errors part that lists the error codes that may be found in `RGError` if an error occurs during the execution of the routine.

23.7 Site Dependencies

Like the CMU ODA Tool Kit, the Raster Graphics Tool Kit depends on a number of types, definitions and routines that are defined by the local installer. Chapter 10 describes these entities in detail. In addition, there is an additional site-dependent type that must be defined to use the Raster Graphics Tool Kit. This type, `FLOAT_type`, is defined in the next section.

The Raster Graphics Tool Kit uses portions of code from existing graphics packages. These packages were not written with the same strict portability guidelines as the rest of the tool kit. These packages have been modified to increase their portability, and they have been tested on a number of systems. If the C compiler at the installation site supports the most common C routines, then there should be no problem compiling the tool kit.

The site-dependent type `SWAP_INFO_type` is normally undefined unless the site has decided to implement swapping to secondary memory. The Raster Graphics Tool Kit, however, provides a dummy definition for this type, because of the manner in which the tool kit is implemented. If you are going to implement swapping, remember to remove this definition.

23.7.1 FLOAT_type

The type `FLOAT_type` is used by the tool kit for handling scaling factors, such as determining the size of PostScript output on a page. The type used to define `FLOAT_type` must be a type that can be assigned floating point values either by explicit assignment or by the results of an arithmetic operation such as division. The usual definition for this type is:

```
typedef float FLOAT_type;
```

24
Tool Kit Data Structures

This chapter describes the user-visible types defined by the Raster Graphics Tool Kit, in the first three sections. The raster formats supported by the tool kit are defined in the remainder of this chapter.

24.1 The Raster Structure: *RASTER_type*

An application program may manipulate multiple rasters. To distinguish among the rasters, each is identified by a value of type `RASTER_type`. The structure of this type is hidden from the programmer and values are created, inspected and modified by tool kit routines.

The programmer should never directly access or modify any of the fields in this structure, but rather should use appropriate functions to achieve the desired affect. `RasterWidth` and `RasterHeight` are used to determine the size of the raster, `RasterStatus` is used to determine whether or not a raster is held in memory, `RasterPelFormat` is used to determine the pel format for the raster and the access functions for pels or bits are used for individual pel manipulation.

There is a predefined object of type `RASTER_type`: `RG_ERROR_RASTER`. The object `RG_ERROR_RASTER` is guaranteed to be different from any other value of type `RASTER_type`. `RG_ERROR_RASTER` is used as an error return value from some of the routines described in chapter 25.

24.2 The Type PEL_type

This `PEL_type` structure is used for passing information between the program and the tool kit. `PEL_type` is defined as follows:

```
typedef struct {
    TAG_type            pel_tag;
    union {
        struct {
            INT_type    red;
            INT_type    green;
            INT_type    blue;
        } RGB;
        struct {
            INT_type    hue;
            INT_type    saturation;
            INT_type    intensity;
        } HSI;
        VALUE_type      bit;
        INT_type        colormap;
    } pel;
} PEL_type;
```

Currently, the only legal value for the field `pel_tag` is `PEL_BITS`, in which case the field `bit` is present. The field `bit` must have one of the values `BIT_BLACK` or `BIT_WHITE`.

24.3 Initialization Parameters: *INIT_RGPARMS_type*

The `INIT_RGPARMS_type` structure specifies initialization parameters for the Raster Graphics Tool Kit. A pointer to this structure with the appropriate information filled in must be passed to the `InitRGToolKit` routine (section 25.1) prior to any tool kit use.

```
typedef struct {
    INT_type        memsize;
    VALUE_type      default_format;
} INIT_RGPARMS_type;
```

`memsize`
: The number of bytes of primary memory to use for storing rasters. The tool kit will begin swapping rasters when new memory allocations would exceed this limit. If this number is -1, then the tool kit will use all available primary memory.

`default_format`
: This is the format that will be used to output rasters whenever the `FORMAT_DETERMINE` tag is used and the raster has been created without a default image. This field must be set to one of the distinguished values. See section 24.4 on the `FORMAT_type` for more details.

24.4 Bitmap Format Specifications: *FORMAT_type*

Bitmaps may be stored in many different formats using the Raster Graphics tool kit. Therefore, a means of specifying what format a given routine should use is required. The `FORMAT_type` structure provides a simple mechanism for specifying what graphics format to use, and other parameters to that format. The data structure is:

```
typedef struct {
    TAG_type            format_tag;
    union {
        FORMAT_ANDREW_type   andrew;
        FORMAT_GIF_type      gif;
        FORMAT_GROUP3_type   g3;
        FORMAT_GROUP4_type   g4;
        FORMAT_IFF_type      iff;
        FORMAT_MACPAINT_type macpaint;
        FORMAT_ODABM_type    oda_bitmap;
        FORMAT_PBM_type      pbm;
        FORMAT_PS_type       ps;
        FORMAT_SUN_type      sun;
        FORMAT_XBM_type      xbm;
    } format_body;
} FORMAT_type;
```

The `format_tag` field is a field that specifies the name of the format. The legal values for this field are constants whose names are determined by removing the "`_type`" characters from the union's field names. For example, to specify the GIF format, the `format_tag` field would be assigned the value `FORMAT_GIF`. If this field is given the value `FORMAT_DETERMINE`, then the tool kit will attempt to determine the format. In the case of a raster read, the tool kit will attempt to identify the format by information stored in the header of most graphics formats. For raster writes, the tool kit will use the format used to read in the raster (by inspecting the `format` field of the `RASTER_type` structure). If no format has been specified (such as if a raster has been created by `MakeRaster` or `ExtractRaster` and not read in), then the default format as specified in the `INIT_RGPARMS_type` structure will be used to initialize the tool kit. A complete list of valid tag values is given in table 24-1.

The tool kit value is unable to detect the types of certain rasters due to the nature of the representation. Group 4 Facsimile and ODA Bitmap Encoding Schemes do not have a magic number at the header, nor do they have any kind of header that tells how wide the rasters are. Without this information, detecting and reading these bitstreams is impossible. The programmer should thus use care when using `FORMAT_DETERMINE` if there is any possibility of confusion.

Tag	Format
`FORMAT_GIF`	CompuServe GIF format
`FORMAT_GROUP3`	Group 3 facsimile format
`FORMAT_GROUP4`	Group 4 facsimile format
`FORMAT_MACPAINT`	MacPaint format
`FORMAT_ODABM`	ODA bitmap encoding
`FORMAT_PBM`	Portable Bitmap Format
`FORMAT_POSTSCRIPT`	PostScript output format
`FORMAT_SUN`	Sun raster format
`FORMAT_XBM`	X bitmap format

Table 24-1. Supported Raster Formats

The various fields in the union contain information that applies to specific graphics formats. Such information might contain additional parameters to algorithms or bit packing information. Each graphics format has its own structure defining any information.

24.5 Supported Raster Graphics Formats

Several raster formats are supported by the Raster Graphics Tool Kit. The tool kit input and output routines may read or write a raster stored in any one of these formats (except for PostScript, for which only output is supported). Each format has its own data structure that holds information that tells the tool kit exactly how the raster is stored.

To specify the format information, the program must define a `FORMAT_type` structure. Next, it must set the tag field, `format_tag`, of the structure to identify the raster format. Before calling the desired tool kit routine, the specific format information for the raster must be specified. A sample sequence of instructions for setting up a format structure follows:

```
FORMAT_type    format;
RASTER_type    *raster;
INT_type       id;

/* Point id to the desired datastream */
format.format_tag = FORMAT_ANDREW;
format.format_body.andrew.object_id = 0;
raster = ReadRaster(id, &format);
```

The specific parameter structures for each graphics format are described in the remainder of this chapter.

24.5.1 Amiga IFF Format (*FORMAT_IFF*)

The graphics interchange file format used on the Amiga personal computer, developed by Electronic Arts, is called "EA IFF 1985," or IFF for short. It is a general formatting scheme used to represent various types of data, such as graphics images, digitized sound or formatted text. The Raster Graphics Tool Kit will only handle IFF datastreams for raster images. Although IFF images may be multi-colored, the tool kit will correctly handle only two-colored rasters. The tool kit may successfully parse a multi-colored IFF datastream; however, all color information will be lost and all non-black pels will be converted to white pels for manipulation by the tool kit.

To specify the Amiga IFF format, the tag field of the `FORMAT_type` structure must be set to the value `FORMAT_IFF`. There are no other parameters to set.

24.5.2 Andrew Raster Format (*FORMAT_ANDREW*)

The Andrew raster format developed for the Andrew Toolkit at Carnegie Mellon University is an ASCII-encoded, byte-wise compression scheme.

In order to specify the Andrew raster format, the tag field of the `FORMAT_type` structure must be set to the value `FORMAT_ANDREW`. There is one other parameter that must be set:

`object_id` A raster represents one type of object in the Andrew Toolkit. To distinguish different objects of the same type within an application, an object identifier is assigned to each object. Before writing a raster, this field may be set to a value of type `INT_type` that represents the object identifier desired for any raster images to be written. If the data object will be output in its own separate stream or file, then the value of `object_id` is immaterial. Upon reading a raster, the `object_id` will be set to the object identifier of the input raster, and this value may be used within a tool kit application as desired.

24.5.3 CompuServe GIF Format (*FORMAT_GIF*)

CompuServe has defined a graphics interchange format, called GIF, for storing and transmitting color raster images. The Raster Graphics Tool Kit will correctly handle only those rasters that have two colors. The tool kit may succeed in parsing a raster with multi-colored pels; however, all non-black pels will be transformed into white pels.

To specify the GIF raster format, the tag field of the `FORMAT_type` structure must be set to the value `FORMAT_GIF`. There are no other parameters to set.

24.5.4 Group 3 Facsimile Format (*FORMAT_GROUP3*)

The Group 3 Format is a graphics standard proposed by the CCITT in Recommendation T.4 [13] as a way to transmit two-colored bitmaps on facsimile apparatus.

In order to specify the Group 3 format, the tag field of the `FORMAT_type` structure must be set to the value `FORMAT_GROUP3`. There are several other parameters to this structure that must be set:

`msb` This field specifies the order in which the bits should be stored in bytes upon input or output. The value assigned to this field declares the bit that gets stored as the most significant bit of the byte. The legal values are:

- `LEFTBIT`: The leftmost pel (or bit) in a group of eight successive bits is the most significant bit in a byte.
- `RIGHTBIT`: The rightmost pel (or bit) in a group of eight successive bits is the most significant bit in a byte.
- `UNKNOWN`: This value is only valid when a read operation is to be performed. The tool kit will attempt to determine the ordering of the bits in the raster being read. (Not currently implemented).

`pad_width` — The Group 3 format allows for each horizontal line of pels to be padded with a fill signal to ensure that lines will not be sent in less than the minimum transmission time for the line. This parameter must be assigned a value of type `INT_type` that specifies the number of bits to which each coded scan line will be extended by a fill signal, if necessary. If this value is zero, then no padding will take place. This field is only valid for write operations, and is ignored when a read takes place.

`k` — There are two different types of coding schemes in the Group 3 format. The one-dimensional scheme consists of horizontally coding white and black run lengths while the two-dimensional scheme is a mixture of one-dimensionally coded lines and vertically-encoded lines. Each line is coded based on differences from the previous line. In this format, the value k-1 specifies the maximum number of successive lines that may be coded two-dimensionally before a line of one-dimensionally coded pels shall occur. The parameter that must be specified is this `k` parameter. Valid `INT_type` values for this field are 1, 2 and 4. The value 1 specifies that the entire raster is encoded one-dimensionally. The values 2 and 4 specify two-dimensional coding for standard vertical resolution and higher vertical resolution, respectively. Note that the field name `k` is in lower-case.

The Group 3 parameter data structure is as follows:

```
typedef struct {
    VALUE_type     msb;
    INT_type       k;
    INT_type       pad_width;
} FORMAT_GROUP3_type;
```

24.5.5 Group 4 Facsimile Format (*FORMAT_GROUP4*)

The Group 4 Format is a graphics standard proposed by the CCITT in Recommendation T.6 [15] as a way to transmit two-colored bitmaps on a facsimile apparatus.

To specify the Group 4 format, the tag field of the `FORMAT_type` structure must be set to the value `FORMAT_GROUP4`. There are several other parameters to this structure that must be set:

`msb` — This field specifies the order in which the bits should be stored in bytes upon input or output. The value assigned to this field declares which bit gets stored as the most significant bit of the byte. The legal values are:

- `LEFTBIT`: The leftmost pel (or bit) in a group of eight successive bits is the most significant bit.
- `RIGHTBIT`: The rightmost pel (or bit) in a group of eight successive bits is the most significant bit.
- `UNKNOWN`: This value is only valid when a read operation is to be performed. The tool kit will attempt to determine the ordering of the bits in the raster being read. (This is currently implemented).

`width` A Group 4 raster does not contain any end-of-line information, so the reader *must* know the width of the image. This parameter must be assigned a value of type `INT_type` that specifies the width (in pels) of the raster to be read. This field is only valid for read operations, and is ignored when a write takes place.

`pad_size` The Group 4 format specifications allow for a raster to be padded to a certain size in order that it may be transmitted in a specified time. This field must be assigned a value of type `INT_type` that specifies the total number of bits that the fax image should occupy. If this number is less than the total number of bits required for storage of the raster, then no padding occurs. Thus, a value of zero guarantees no padding will take place. Padding is performed by adding zero bits to the end of the raster data.

The Group 4 parameter data structure is as follows:

```
typedef struct {
    VALUE_type     msb;
    INT_type       pad_size;
    INT_type       width;
} FORMAT_GROUP4_type;
```

24.5.6 MacPaint Format (*FORMAT_MACPAINT*)

This is a popular graphics format used on Macintosh personal computers for the interchange of full-page images. All rasters stored in this format have dimensions of 576 bits wide by 720 bits tall, or one 8 by 10 inch page at 72 bits per inch.

To specify the MacPaint graphics format, the tag field of the `FORMAT_type` structure must be set to the value `FORMAT_MACPAINT`. There is one other parameter to this structure that must be set:

`macbin` Often, a MacPaint document will be embedded in another formatting scheme called MacBinary. When reading a MacBinary-formatted raster, this field should be assigned a value of type INT_type that specifies whether or not MacBinary formatting will be interpreted. A value of 0 indicates that the datastream the tool kit will read or write should only contain the data portion of the MacPaint image. A value of 1 indicates that the old MacBinary format is used, and a value of 2 means MacBinary II formatting. In practice, these two formats are very similar, and specifying either 1 or 2 for the macbin field will allow the tool kit to read either format. The tool kit will not write rasters using the MacBinary protocol, so this field is ignored on write operations.

The MacPaint parameter data structure is as follows:

```
typedef struct {
    INT_type            left;
    INT_type            right;
    INT_type            top;
    INT_type            bottom;
    INT_type            macbin;
}   FORMAT_MACPAINT_type;
```

Note that the first four fields of the structure are ignored.

24.5.7 ODA Bitmap Encoding Scheme (*FORMAT_ODABM*)

This bitmap format is specified as one of the allowable ways to store a pel array in an ODA document.

To specify the ODA Bitmap Encoding Scheme, the tag field of the FORMAT_type structure must be set to the value FORMAT_ODABM. There is one other parameter to this structure that must be set:

width — An ODA Bitmap raster does not contain any end-of-line information within it. Thus, the reader of such a raster must know the width of the image before reading. This parameter should be assigned a value of type INT_type that specifies the width (in pels) of the raster to read. This field is only valid for read operations, and is ignored when a write takes place.

The ODA Bitmap Encoding parameter data structure is as follows:

```
typedef struct {
    INT_type width;
} FORMAT_ODABM_type;
```

24.5.8 Portable Bitmap Format (*FORMAT_PBM*)

The Portable Bitmap Format (PBM) is a straightforward way of storing two-colored bitmaps: it stores each bit in a raster as either an ASCII '1' or an ASCII '0'. PBM was originally designed as a lowest common denominator in a family of graphics conversion programs.

To use the PBM format, the tag field of the `FORMAT_type` structure must be set to the value `FORMAT_PBM`. There are no other parameters that need to be set to use PBM.

24.5.9 PostScript Output (*FORMAT_POSTSCRIPT*)

The Raster Graphics Tool Kit can produce rasters in PostScript form. It will not, however, read in a PostScript raster. Attempting to read a PostScript raster will produce a `RG_BADPARM_err` from `ReadRaster` (section 25.14). To specify the PostScript format, the tag field of the `FORMAT_type` structure must be set to the value `FORMAT_POSTSCRIPT`. There is one other parameter to this structure that must be set:

`scale` This field should be assigned a value of type `FLOAT_type` that specifies a scaling factor for the image. A factor of 1.0 will produce a PostScript raster that will be at 72 pels per inch resolution when printed. Scaling factors lower than 1.0 will produce smaller images, and factors larger than 1.0 will produce larger images. If an image is scaled so large that it would not be able to be imaged on a 8.5" × 11" piece of paper, then an error occurs on writing. Obviously, this parameter is only valid for write operations.

The ODA Bitmap Encoding parameter data structure is as follows:

```
typedef struct {
   FLOAT_type          scale;
   CHAR_type           *name;
} FORMAT_POSTSCRIPT_type;
```

Note that last field of the structure is ignored.

24.5.10 Sun Raster Format (*FORMAT_SUN*)

Sun computers store rasters in a format called the Sun Raster format. These rasters may have more than one color. The Raster Graphics Tool Kit, however, will only correctly handle those rasters that have two colors. The tool kit may succeed in parsing a raster with multi-colored pels; however, all color information other than black and white will be lost.

To specify the Sun Raster format, the tag field of the `FORMAT_type` structure must be set to the value `FORMAT_SUN`. There are no other parameters that must be set.

24.5.11 X Bitmap Format (*FORMAT_XBM*)

This is the format used in the X Window System for storing uniplanar bitmaps. To specify the X Bitmap format, the tag field of the `FORMAT_type` structure must be set to the value `FORMAT_XBM`. There are no other parameters that must be set.

25
Raster Graphics Tool Kit Routines

This chapter describes the user-visible routines in the Raster Graphics Tool Kit. These routines can be classified broadly into five categories (described in the order in which they are presented): initialization and finalization routines (`InitRGToolKit` and `FinalRGToolKit`), creation and destruction routines (`MakeRaster`, `ExtractRaster` and `DeleteRaster`), assignment and access routines (`AssignBit`, `AssignPel`, `GetBit`, `GetPel`, `InsertRaster`, `RasterPelFormat`, `RasterHeight` and `RasterWidth`), input and output routines (`ReadMemRaster`, `ReadRaster`, `WriteMemRaster` and `WriteRaster`) and miscellaneous routines (`HoldRaster`, `ReleaseRaster` and `RasterStatus`).

25.1 Initialize the Raster Graphics Tool Kit: *InitRGToolKit*

```
INT_type InitRGToolKit(parms)
   INIT_RGPARMS_type *parms;
```

Real Name

rgintk

Synopsis

Initialize the Raster Graphics Tool Kit before any tool kit functions are used.

Parameters

`parms`	A pointer to a structure that contains fields that are used to configure the tool kit.

Return Value

`RG_ERROR_INT`	An error has occurred; `RGError` will indicate the error.

otherwise The tool kit has been initialized successfully and tool kit functions may be called.

Semantics

This routine will initialize the Raster Graphics Tool Kit. A copy of the parameter will be retained by the tool kit, and the values of the fields will be checked by the tool kit routines. This means that changing the values of fields may cause changes in the run-time behavior of the tool kit. The application program must be sure that the structure whose address is passed in will be available until the `FinalRGToolKit` routine is called.

If the CMU ODA Tool Kit is also being used, this routine must be called after the CMU ODA Tool Kit initialization routine (section 14.1).

Errors

`RG_BADPARM_err`
One of the parameters is invalid.

`RG_INTERNAL_err`
An error internal to the tool kit has occurred.

`RG_NOFINAL_err`
The Raster Graphics Tool Kit was already initialized and has not been finalized (by calling `FinalRGToolKit`)

`RG_NOMEM_err` No more memory was available for dynamic allocation.

`RG_LOCAL_err` A locally defined error has occurred.

Implementation Status

Implemented.

25.2 Finalize the Raster Graphics Tool Kit: *FinalRGToolKit*

```
INT_type FinalRGToolKit()
```

Real Name

rgfitk

Synopsis

This routine is used to indicate that the application will no longer call routines in the Raster Graphics Tool Kit.

Return Value

`RG_ERROR_INT`	An error has occurred; `RGError` will indicate the error.
otherwise	The tool kit has been successfully finalized.

Semantics

This function cleans up after the tool kit has been used.

It will release all dynamically allocated memory used by the Raster Graphics Tool Kit. This function will also close all input and output channels that have not already been terminated, by sending a length of -1 to the input and output routines.

The tool kit may not be finalized unless a prior call to `InitRGToolKit` has been made.

If the CMU ODA Tool Kit is also being used, this routine must be called before the ODA Tool Kit finalization routine (section 14.2).

Errors

`RG_INTERNAL_err`
: An error internal to the tool kit has occurred.

`RG_NOFINAL_err`
: The tool kit was never initialized (via a call to `InitRGToolKit`) or has not been initialized since the last call to this routine.

Implementation Status

Implemented.

25.3 Create a Raster Graphics Portion: *MakeRaster*

```
RASTER_type *MakeRaster(Width, Height,
                        PelFormat)
   INT_type Width;
   INT_type Height;
   VALUE_type PelFormat;
```

Real Name

rgramr

Synopsis

This function creates a blank raster structure of size `Width` × `Height`.

Parameters

`Width`	The width of the raster to create.
`Height`	The height of the raster to create.
`PelFormat`	Determines the type of pels to use in this raster.

Return Value

`RASTER_ERROR`	An error has occurred; `RGError` indicates the error.
otherwise	Returns a pointer to the raster structure for the newly created raster.

Semantics

This function creates a new, blank raster of dimensions `Width` × `Height`. The raster initially has no pels turned on. The created raster is not initially held in primary memory; that is, it may be swapped out of primary memory.

A raster created by `MakeRaster` may be deleted by calling `DeleteRaster`.

The minimum size for a raster is 1 × 1 pels.

Errors

`RG_BADPARM_err`
: One of the parameters is invalid.

`RG_INTERNAL_err`
: An error internal to the tool kit has occurred.

`RG_NOMEM_err`
: The tool kit was unable to allocate enough memory to provide support for the extracted bitmap.

Implementation Status

The tool kit currently assumes that all rasters will be created with the `PelFormat` parameter set to `PEL_BITS`. This means that all rasters will be bitmaps. Thus, this parameter is currently ignored. In future versions of the tool kit, this value may be one of the values described under the function `RasterPelFormat`.

25.4 Extract a Subraster from a Raster: *ExtractRaster*

```
RASTER_type *ExtractRaster(raster, left, top,
                           right, bottom)
   RASTER_type *raster;
   INT_type left;
   INT_type top;
   INT_type right;
   INT_type bottom;
```

Real Name

rgraex

Synopsis

Creates a new raster from a portion of an existing raster.

Parameters

`raster`	A pointer to the raster from which the new raster will be extracted.
`left, top`	The upper-left coordinate of the rectangular area to extract.
`right, bottom`	The lower-right coordinate of the rectangular area to extract.

Return Value

`RG_ERROR_RASTER`	An error has occurred; `RGError` indicates the error.
otherwise	A pointer to the newly created raster structure whose image is taken as the source raster.

Semantics

Given a source raster, this function creates (and returns) a new raster whose bitmap was copied from the specified portion of the source raster. The upper-left corner of the rectangular area that is copied is specified by the offset coordinate (`left`, `top`); the lower-right corner is specified by (`right`,`bottom`).

The extraction is inclusive of the upper-left and lower-right corners. For example, the pel at coordinate (`left`,`top`) in the original raster will become the pel at (0,0) in the new raster. The new raster will have dimensions

```
(right - left + 1) × (bottom - top + 1)
```

If the region specified does not lie wholly within the source raster, then an error will occur.

The new raster has its data copied from the original raster. Any change to either raster after the extraction will have no effect on the other raster.

The new raster will have pels of the same type and will have the same default raster format as the original raster.

The new raster will not be held in memory upon its creation. Thus, if not enough memory exists to create the extracted region in memory, it will be swapped out to secondary storage as it is being created.

The raster created by this function may be deleted by using the `DeleteRaster` routine.

Errors

`RG_BADPARM_err`
: One of the parameters is invalid.

`RG_INTERNAL_err`
: An error internal to the tool kit has occurred.

`RG_NOMEM_err` The tool kit was unable to allocate enough memory to provide support for the extracted bitmap.

`RG_LOCAL_err` A locally defined error has occurred.

Implementation Status

Implemented.

25.5 Delete a Raster Graphics Portion: *DeleteRaster*

```
INT_type DeleteRaster(raster)
   RASTER_type *raster;
```

Real Name

rgradr

Synopsis

Delete a raster and deallocate all associated memory.

Parameters

`raster` The raster structure to delete.

Return Value

`RG_ERROR_INT`	An error has occurred; `RGError` will indicate the error.
otherwise	The entire raster has been deallocated; all swapped out portions have been removed from the secondary storage.

Semantics

This function deletes a raster. It removes all of the raster image data from secondary storage. It deallocates all memory associated with the raster data and the raster structure. The program must not use the raster once it has been deleted.

Errors

`RG_BADPARM_err`	One of the parameters is invalid.
`RG_INTERNAL_err`	An internal tool kit error has occurred.
`RG_LOCAL_err`	A locally-defined error has occurred.

Implementation Status

Implemented.

25.6 Assign a Value to a Pel: *AssignBit*

```
INT_type AssignBit(raster, x, y, value)
   RASTER_type *raster;
   INT_type x;
   INT_type y;
   VALUE_type value;
```

Real Name

rgpeab

Synopsis

In the case of 2-dimensional rasters, this makes a pel either BLACK or WHITE, depending upon `value`.

Parameters

`raster`	Pointer to the raster in which the desired pel exists.
`x`	The X coordinate of the pel to change.
`y`	The Y coordinate of the pel to change.
`value`	Either `BIT_BLACK` or `BIT_WHITE`.

Return Value

`RG_ERROR_INT`	An error has occurred; `RGError` indicates the error.
otherwise	The desired pel has been set.

Semantics

If the specified raster is a one-dimensional bitmap, then this function will set the pel located at coordinate `(x,y)` to the BLACK or WHITE state. The parameter to be passed as `value` must be either `BIT_BLACK` or `BIT_WHITE`, depending on the desired color for the pel.

If the `y`th row in the raster is swapped out of memory at the time of the function call, then at least that row will be loaded back into primary memory. Swapping of other rasters may be necessary.

If the desired pel is not located within the raster, then an error occurs.

Errors

`RG_BADPARM_err`
: One of the parameters was invalid.

`RG_NOMEM_err`
: A portion of the raster that was swapped out could not be loaded into primary memory due to a lack of free space.

`RG_PELRANGE_err`
: The position specified by `(x,y)` does not lie within the raster.

Implementation Status

In a multi-planar raster, this function will set the pel located at the coordinates (x,y) to either the current "black" or "white" color for pels.

25.7 Assign a Value (Color) to a Pel: *AssignPel*

```
INT_type AssignPel(raster, x, y, value)
   RASTER_type *raster;
   INT_type x;
   INT_type y;
   PEL_type *value;
```

Real Name

rgpeap

Synopsis

Sets the pel `(x,y)` in the specified `raster` to either BLACK or WHITE.

Parameters

`raster`	The raster in which the desired pel exists.
`x`	The X coordinate of the pel to change.
`y`	The Y coordinate of the pel to change.
`value`	Pointer to a pel value specifier structure.

Return Value

`RG_ERROR_INT`	An error has occurred; `RGError` indicates the error.
otherwise	The desired pel has been set to the appropriate value.

Semantics

Refer to the `AssignBit` routine for a description of this routine.

Errors

`RG_BADPARM_err`
One of the parameters is invalid.

`RG_INTERNAL_err`
An error internal to the tool kit has occurred.

`RG_NOMEM_err` A portion of the raster that was swapped out could not be loaded into primary memory due to a lack of free space.

Implementation Status

In the current implementation of the tool kit, multi-colored pels are not supported. This routine functions identically to the `AssignBit` routine. In the future, this routine may provide for assigning various colors to pels.

25.8 Determine the Value of a Bit: *GetBit*

```
VALUE_type GetBit(raster, x, y)
   RASTER_type *raster;
   INT_type x;
   INT_type y;
```

Real Name

rgpegb

Synopsis

Determines the value of the bit at coordinates `(x,y)` in `raster`.

Parameters

`raster`	The raster in which the desired pel exists.
`x`	The X coordinate of the pel.
`y`	The Y coordinate of the pel.

Return Value

`VALUE_err`	An error has occurred; `RGError` indicates the error.
`BIT_BLACK`	The specified bit is set (on).
`BIT_SET`	The specified bit is cleared (off).

Semantics

This routine returns the value of the pel (which must be a bit) located at position `(x,y)` in the raster pointed to by `raster`. If the pel is not within the range of the raster, then an error occurs. If the raster is not a bitmap, then an error occurs (note that this error is not possible in the current version of the tool kit); `GetPel` should be used instead.

If the y'th row in the raster is swapped out of memory at the time of the function call, then at least that row will be loaded back into primary memory. Swapping of other rasters may be necessary.

Errors

`RG_BADPARM_err`
: One of the parameters is invalid.

`RG_INTERNAL_err`
: An error internal to the tool kit has occurred.

`RG_PELRANGE_err`
: The coordinates for the pel do not lie within the raster.

`RG_PELFORMAT_err`
: The raster in question is not a bitmap and may not be used in bit-only functions.

`RG_NOMEM_err`
: A portion of the raster that was swapped out could not be loaded into primary memory due to a lack of free space.

Implementation Status

Implemented.

25.9 Determine the Value (Color) of a Pel: *GetPel*

```
INT_type GetPel(raster, x, y, pel)
   RASTER_type *raster;
   INT_type x;
   INT_type y;
   PEL_type *pel;
```

Real Name

rgpegp

Synopsis

Returns the value of the specified pel `(x,y)` in the raster.

Parameters

`raster`	The raster in which the desired pel exists.
`x`	The X coordinate of the pel.
`y`	The Y coordinate of the pel.
`pel`	This is a user-allocated structure that is unspecified upon entry and holds the pel information upon return.

Return Value

`RG_ERROR_INT`	An error has occurred; `RGError` indicates the error.
otherwise	The structure pointed to by pel has been set to the appropriate value, which depends upon what type of pels are used in the raster.

Semantics

This function loads the structure pointed to by `pel` with the value of the pel at location `(x,y)` within the specified `raster`.

If the y'th row in the raster is swapped out of memory at the time of the function call, then at least that row will be loaded back into primary memory. Swapping of other raster rows may be necessary.

It is the programmer's responsibility to allocate and deallocate the memory for `pel`.

If the pel described by `(x,y)` is not within the specified raster, then an error occurs.

Errors

`RG_BADPARM_err`
: One of the parameters is invalid.

`RG_INTERNAL_err`
: An error internal to the tool kit has occurred.

`RG_PELRANGE_err`
: The coordinates for the pel do not lie within the raster.

`RG_NOMEM_err`
: A portion of the raster that was swapped out could not be loaded into primary memory due to a lack of free space.

`RG_LOCAL_err`
: A locally defined error has occurred.

Implementation Status

In the current version of the Raster Graphics Tool Kit, multicolored pels are not supported. This routine will load the bit field of the `PEL_type` structure with either `BIT_BLACK` or `BIT_WHITE`, depending on the status of the desired pel.

25.10 Overwrite a Portion of a Raster: *InsertRaster*

```
INT_type InsertRaster(SourceRaster,
                      DestRaster, x,y)
   RASTER_type *SourceRaster;
   RASTER_type *DestRaster;
   INT_type x;
   INT_type y;
```

Real Name

rgrais

Synopsis

Places one raster into another raster, beginning at coordinates `(x,y)`.

Parameters

`SourceRaster`	Pointer to the raster that is to be inserted.
`DestRaster`	Pointer to the raster onto which the `SourceRaster` will be placed.
`x,y`	The (x, y) coordinates in `DestRaster` where the upper-left corner of the `SourceRaster` will start.

Return Value

`RG_ERROR_INT`	An error has occurred; `RGError` will indicate the error.
otherwise	The source raster has been inserted into the destination raster (see **Semantics**, below).

Semantics

This function takes one raster and pastes it onto another. The `SourceRaster` will be copied into the `DestRaster`. `SourceRaster` will be unaffected by this command. The location of `SourceRaster` within `DestRaster` is determined by the offset parameters `x` and `y`, where (`x`, `y`) represents the offset from the top-left corner of `DestRaster`. For example, if `x` and `y` were 10 and 20, then the top-left corner of `SourceRaster` would be located at position (10,20) within `DestRaster`.

Only the portion of `SourceRaster` that fits wholly within `DestRaster` will be copied. Thus, the size of `DestRaster` remains unchanged by this function. Any portion of the source raster that does not fit within the destination raster will be clipped (the size of `SourceRaster` is left unchanged).

Errors

`RG_BADPARM_err`
One of the parameters is invalid.

`RG_INTERNAL_err`
An error internal to the tool kit has occurred.

`RG_PELFORMAT_err`
The pel formats in the two rasters do not match.

Implementation Status

Implemented.

25.11 Determine the Type of Pels in a Raster: *RasterPelFormat*

```
VALUE_type RasterPelFormat(raster)
   RASTER_type *raster;
```

Real Name

rgrapf

Synopsis

This routine returns the type of pel used in the raster.

Parameters

`raster`	A pointer to the raster whose pels are being examined.

Return Value

`PEL_BITS`	Pels are bits, the raster is a bitmap.
`PEL_RGB`	Pels are represented by RGB (Red, Green, Blue) values.
`PEL_HSI`	Pels are represented by HSI (Hue, Saturation, Intensity) values.
`PEL_CMAP`	Pels are represented by an index into a color map.

Semantics

This routine checks the raster structure for the specified raster and returns the type of pel used in that raster.

Errors

`RG_BADPARM_err`
One of the parameters is invalid.

`RG_INTERNAL_err`
An error internal to the tool kit has occurred.

Implementation Status

In the current version of the tool kit, rasters are not multi-colored and this routine always returns the value `PEL_BITS`.

25.12 Determine the Size of a Raster: *RasterHeight*, *RasterWidth*

```
INT_type RasterHeight(raster)

INT_type RasterWidth(raster)

   RASTER_type *raster;
```

Real Name

rgrarh, rgrarw

Synopsis

These functions return either the height or width of the specified raster.

Parameters

`raster`	A pointer to the raster for which sizing information is desired.

Return Value

`RG_ERROR_INT`	An error has occurred. `RGError` will contain an indication of the exact error.
otherwise	Returns the height or width of the raster.

Semantics

These two routines return either the height or width of the specified raster.

Errors

`RG_BADPARM_err`
: One of the parameters is invalid.

`RG_INTERNAL_err`
: An error internal to the tool kit has occurred.

Implementation Status

Implemented.

25.13 Read a Bitmap from Memory into a Raster: *ReadMemRaster*

```
RASTER_type *ReadMemRaster(memory, format)
   POINTER_type *memory;
   FORMAT_type *format;
```

Real Name

rgrirm

Synopsis

This routine is used to read a bitmap from a section of memory into a raster structure.

Parameters

`memory`	The address of a pointer variable that points at the starting memory location to read from.
`format`	The format in which the bitmap is stored.

Return Value

`ERROR_RASTER`	An error has occurred. `RGError` will contain an indication of the exact error.
otherwise	This routine returns a raster structure that is used to refer to the given raster.

Semantics

`ReadMemRaster` reads a bitmap from primary memory and stores it in an internal format so that it may be manipulated by other tool kit functions. The tool kit will interpret the data it reads according to the `format` parameter. if the data is not in the format specified, then an error occurs.

The parameter `memory` needs to be the address of a user-allocated variable that in turn points to the starting memory location to read from. The tool kit will modify this parameter as it is reading the raster—after the raster has been read this variable will point to the end of the raster text in memory. A typical calling sequence skeleton would look like this:

```
BYTE_type        *buffer;
BYTE_type        *bufptr;
RASTER_type      *raster;
FORMAT_type      format;

buffer = ReadFileToMemory(filename);
   /* Not a TK Routine */
bufptr = buffer;
raster = ReadMemRaster(&bufptr, &format);
```

The reason for the line `bufptr = buffer` is so that a pointer to the beginning of the buffer is maintained, since `bufptr` will be changed by the call to `ReadMemRaster`. (The code to initialize the `format` structure has been omitted.)

This function first allocates memory for the data structure and then proceeds to read in the bitmap. If there is not enough space left in primary memory for the raster to be loaded, then the tool kit will try to swap out any other raster in primary memory that is currently *not being held*. If there still would not be enough memory to read in the bitmap, then the tool kit will attempt to both read and swap out the raster at the same time (as each row of the bitmap is read in, it will be swapped out to secondary storage). This will allow the tool kit to manipulate bitmaps that are larger than primary memory can hold. If there is not enough primary memory free to hold the individual raster rows, then an error occurs and the raster is not loaded.

Errors

`RG_BADPARM_err`
: One of the parameters is invalid.

`RG_INTERNAL_err`
: An error internal to structure of tool kit.

`RG_NOFORM_err`
: The format of the input memory does not match the format specified.

`RG_FORMBAD_err`
: The input memory was identified as the specified format, but it contains a data error.

`RG_LOCAL_err`
: A locally defined error has occurred.

Implementation Status

Implemented.

25.14 Read a Bitmap From a Descriptor Into a Raster: *ReadRaster*

```
RASTER_type *ReadRaster(id, format)
   INT_type id;
   VALUE_type *format;
```

Real Name

rgrird

Synopsis

This routine is used to read a bitmap from an input descriptor into a raster structure.

Parameters

`id`	The input descriptor to read from.
`format`	The format in which the bitmap is stored.

Return Value

`RG_ERROR_RASTER`
: An error has occurred; `RGError` will contain an indication of the exact error.

otherwise
: This routine returns a raster structure that is used to refer to the given raster.

Semantics

`ReadRaster` reads a bitmap from the datastream specified by `id` and stores it in an internal format so that it may be manipulated by other tool kit functions. The tool kit will interpret the data it reads according to the format parameter; if the data is not in the format specified, then an error occurs.

This function first allocates memory for the data structure and then proceeds to read in the bitmap. If there is not enough space left in primary memory for the raster to be loaded, then the tool kit will try to swap out any other raster in primary memory that is currently *not being held.* If there still would not be enough memory to read in the bitmap, then the tool kit will attempt to both read and swap out the raster at the same time (as each row of the bitmap is read in, it will be swapped out to secondary storage). This will allow the tool kit to manipulate bitmaps that are larger than primary memory can hold. If there is not enough primary memory free to hold the individual raster rows, then an error occurs and the raster is not loaded.

Errors

`RG_BADPARM_err`
: One of the parameters is invalid.

`RG_INTERNAL_err`
: An error internal to structure of tool kit.

`RG_NOFORM_err` The format of the input stream does not match the format specified.

`RG_FORMBAD_err`
The input stream was identified as the specified format, but it contains a data error.

`RG_LOCAL_err` A locally defined error has occurred.

Implementation Status

Implemented.

25.15 Write a Raster to Memory in a Specified Format:*WriteMemRaster*

```
BYTE_type *WriteMemRaster(raster, format,
                          start, size)
   RASTER_type *raster;
   FORMAT_type *format;
   INT_type start;
   INT_type *size;
```

Real Name

rgriwm

Synopsis

This routine is used to write a bitmap to primary memory in a specified format.

Parameters

`raster` The internal raster to write.

`format` The format in which the bitmap is stored.

`start` The number of the starting byte within the format to write to memory.

`size` A pointer to an `INT_type` that holds the number of bytes to write to memory. If this parameter is zero, the entire raster is written. This parameter always returns with the actual number of bytes written.

Return Value

`RG_ERROR_POINTER`
An error has occurred; `RGError` will indicate the error.

otherwise The address of the starting memory location where the specified raster has been written successfully.

Semantics

This function takes a bitmap stored in an internal raster structure and writes it to main memory in an output format specified by `format`. The function returns a pointer to the first byte of the data written. Note that size is a *pointer* to an `INT_type` number. When this number is greater than zero, only the bytes that fall within the range

```
[start, start + *size - 1]
```

are written to memory (thus, `*size` bytes are written). If `*size` is zero upon entry, all bytes from `start` to the end of the raster are written. Thus, this routine allows for portions of the raster to be written to memory as well as the entire raster. The value `*size` always returns with the actual number of bytes written.

Errors

`RG_BADPARM_err`
One of the parameters is invalid.

`RG_INTERNAL_err`
An error internal to the tool kit has occurred.

`RG_NOMEM_err` No memory available for dynamic allocation.

`RG_LOCAL_err` A locally defined error has occurred.

Implementation Status

Implemented.

25.16 Write a Raster in a Specified Bitmap Format: *WriteRaster*

```
INT_type WriteRaster(od, raster, format)
   INT_type od;
   RASTER_type *raster;
   FORMAT_type *format;
```

Real Name

rgriwr

Synopsis

This routine is used to write a bitmap to an output stream in the specified format.

Parameters

`od`	The output descriptor.
`raster`	The internal raster to write.
`format`	The format in which the bitmap is stored.

Return Value

`RG_ERROR_INT`	An error has occurred; `RGError` will indicate the error.
otherwise	The specified raster has been written successfully.

Semantics

This function takes a raster stored in an internal raster structure and writes it to the output descriptor `od` as specified by format. After the raster is written, the output channel is not closed. This allows a raster to be written in the middle of a file.

Errors

`RG_BADPARM_err`
One of the parameters is invalid.

`RG_INTERNAL_err`
An error internal to the tool kit has occurred.

`RG_LOCAL_err` An error in the writing of the file has occurred.

Implementation Status

Implemented.

25.17 Keep a Raster in Primary Memory: *HoldRaster*

```
INT_type HoldRaster(raster)
   RASTER_type *raster;
```

Real Name

rgraho

Synopsis

Attempt to place an entire raster in primary memory.

Parameters

`raster`	The raster to hold in primary memory.

Return Value

`RG_ERROR_INT`	An error has occurred; `RGError` will indicate the error. The raster is not guaranteed to be in primary memory.
otherwise	The raster is in primary memory and will remain there at least until the corresponding ReleaseRaster call is made.

Semantics

At times it will be desirable to have an entire raster in primary memory. This will allow for the rapid manipulation of a raster without delays from memory swapping. Calling this function causes the tool kit to examine the specified raster. If any portion of the raster is swapped out of memory, then those portions will be reloaded, if possible.

When a raster is being held, it will not be swapped out of primary memory. The actual flag for holding a bitmap is a counter within the raster structure for the bitmap to be held. Each time a raster is passed to `HoldRaster`, this counter is increased by one. Each time the raster is released by `ReleaseRaster`, this counter gets decreased by one. Whenever this counter is non-zero and the tool kit needs to swap out primary memory, it will skip over the bitmap being held.

Errors

`RG_BADPARM_err`	One of the parameters is invalid.
`RG_INTERNAL_err`	An error internal to the tool kit has occurred.
`RG_NOMEM_err`	No more memory was available for dynamic allocation.
`RG_LOCAL_err`	A locally defined error has occurred.

Implementation Status

Implemented.

25.18 Allow a Raster to be Released from Memory: *ReleaseRaster*

```
INT_type ReleaseRaster(raster)
   RASTER_type *raster;
```

Real Name

rgrare

Synopsis

Allows a raster to be released from primary memory: the raster may be swapped out to secondary storage while not in use.

Parameters

`raster`	The raster to be released.

Return Value

`RG_ERROR_INT`	An error has occurred; `RGError` indicates the error.
otherwise	The raster has been marked as releasable from primary memory.

Semantics

This function decreases the counter for the specified raster by one. If the counter becomes zero, the raster may be swapped out of primary memory. It does not force the raster to secondary storage, but allows it to be swapped.

This function has no effect on a raster if the raster is not being held.

Errors

`RG_BADPARM_err`
One of the parameters is invalid.

`RG_INTERNAL_err`
An error internal to the tool kit has occurred.

Implementation Status

Implemented.

25.19 Determine the Hold/Release Status of a Raster: *RasterStatus*

```
VALUE_type RasterStatus(raster)
   RASTER_type *raster;
```

Real Name

rgrast

Synopsis

This routine determines whether a raster is held in memory.

Parameters

`raster` A pointer to the raster.

Return Value

`RG_STATUS_HELD`
The specified raster is being held.

`RG_STATUS_RELEASED`
The specified raster is not being held.

`RG_ERROR_VALUE`
An error has occurred. `RGError` will contain an indication of the exact error.

Semantics

This routine returns the status of a raster: whether it is currently being held in primary memory by the function `HoldRaster`. If the raster is being held, the function returns `RG_STATUS_HELD`. If the raster was once held but is now released (by the function `ReleaseRaster`), or if the raster has never been held, then `RG_STATUS_RELEASED` is returned.

Errors

`RG_BADPARM_err`
One of the parameters is invalid.

`RG_INTERNAL_err`
An error internal to the tool kit has occurred.

Implementation Status

Implemented.

5 References and Appendixes

References

[1] Association for Computing Machinery (editor).
Proceedings of the ACM Conference on Document Processing Systems.
Association for Computing Machinery, 1988.

[2] Adobe Systems Incorporated.
PostScript Language: Tutorial and Cookbook.
Addison-Wesley, 1985.

[3] Adobe Systems Incorporated.
PostScript Language: Reference Manual.
Addison-Wesley, 1985.

[4] Jaap Akkerhuis.
Spacing Out on Troff.
;login: 13(1):9-12, January/February, 1988.

[5] *Standard for Electronic Manuscript Preparation and Markup*
American Association of Publishers, Washington, D.C., 1986.
Electronic Manuscript Series.

[6] Nancy Andrews.
Rich Text Format Standard Makes Transferring Text Easier.
Microsoft Systems Journal :63-70, March, 1987.

[7] Ann Arbor Softworks.
FULLWRITE PROFESSIONAL
Ann Arbor Softworks, 2393 Teller Rd., Suite 106, Newbury Park, CA 91320, 1987.

[8] *(ANSI X3.122-1986) Metafile for the Storage and Transfer of Picture Description Information*
American National Standards Institute, 1986.

[9] APDA.
PICT File Format Notes and Disk Set
APDA, APDA, 20525 Mariani Ave., MS: 33G, Cupertino, CA 95014, 1988.

[10] E. Bertino, F. Rabitti and C. Thanos.
MULTOS - a Filing Server for Multimedia Documents.
Internal document, Istituto Elaborazione, Consiglio Nazionale delle Ticherche, Pisa (Italy).

[11] Nathaniel Borenstein, Craig Everhart, Jonathan Rosenberg and Adam Stoller.
A Multi-media Message System for Andrew.
In *Proceedings of the USENIX Winter Conference*, pages 37-42. February, 1988.

[12] U. Bormann, C. Bormann and C. Bathe.
ISOTEXT—A WYSIWYG Editing and Formatting System for ODA and SGML Documents.
In *Proceedings of the 5th Annual EXPRIT Conference*. North-Holland, November, 1988.

[13] *Standardisation of Group 3 Facsimile Apparatus for Document Transmission*
CCITT, 1984.
CCITT Recommendation T.4.

[14] *Open Document Architecture*
CCITT, 1988.
CCITT Recommendations T.411—T.418.

[15] *Facsimile Coding Schemes and Coding Control Functions for Group 4 Facsimile Apparatus*
CCITT, 1988.
CCITT Recommendation T.6.

[16] P. Constantopoulos, Y. Yeorgaroudakis, M. Theodoridou, D. Konstantas, K. Kreplin, H. Eirund, A. Fitas, P. Savino, A. Converti, L. Martino, F. Rabitti, E. Bertino and C. Thanos.
Office Document Retrieval in MULTOS.
In *Proceedings of the 3rd ESPIRIT Technical Week.* 1986.

[17] Digital Equipment Corporation.
DEC Standard Runoff (DSR) User's Guide
Digital Equipment Corporation, Maynard, Mass, 1979.

[18] Digital Equipment Corporation.
Digital Technical Journal.
Technical Report Volume 2, Number 1, Digital Equipment Corporation, 1990.
This issue of the journal is dedicated to CDA, DEC's Compound Document Architecture, of which DDIF is a component.

[19] ESPRIT.
Esprit: European Strategic Programme for Research and Development in Information Technology (1987 Annual Report).
Office for Official Publications of the European Communities, 1987.
Commission of the European Communities, D.G.XIII: Telecommunications, Information Industries and Innovation (Catalogue number: CD-NA-11352-EN-C).

[20] S. I. Feldman.
Make—a Program for Maintaining Programs.
Software — Practice and Experience 9(4):255-265, 1979.

[21] James Gosling.
EMACS—screen editor
UniPress Software, Inc., Suite 312, 2025 Lincoln, NJ, 08817, 1983.

[22] Project INCA.
DAP H'88: Functionality and Fall-back Specification.
Technical Report, Project INCA, October, 1988.
Deliverable 3.2-Appendix B, INCA Partners: GEC, Modcomp, Nixdorf, Olivetti, UCL.

[23] A. N. Habermann, Robert Ellison, Raul Medina Mora, Peter H. Feiler, David S. Notkin, Gail E. Kaiser, David B. Garlan and Steven Popovich.
The Second Compendium of Gandalf Documentation.
Research Report, Carnegie-Mellon University, 1982.

[24] John H. Howard.
An Overview of the Andrew File System.
In *Proceedings of the USENIX Winter Conference*, pages 23-26. February, 1988.

[25] *Document Content Architecture: Revisable-Form-Text Reference*
IBM, 1983.

[26] *(ISO 10179) Information processing — Text composition — Document style, semantics and specification language*
International Organization for Standardization (ISO), 1989.

[27] *(ISO 2022) Information processing—ISO 7-bit and 8-bit coded character sets—Code extension techniques*
International Organization for Standardization (ISO), 1986.

[28] *(ISO 6429) Information processing — ISO 7-bit and 8-bit coded character sets — Additional control functions for character-imaging devices*
International Organization for Standardization (ISO), 1983.
Reference number: ISO 6429-1983 (E).

[29] *(ISO 646) Information processing — ISO 7-bit encoded character set for information interchange*
International Organization for Standardization (ISO), 1983.
Reference number: ISO 646-1983 (E).

[30] *(ISO 8613) Information processing—Text and Office Systems—Office Document Architecture (ODA)*
International Organization for Standardization (ISO), 1988.
Parts 1-8.

[31] *(ISO 8824) Information processing — Open Systems Interconnection — Specification of Abstract Syntax Notation One (ASN.1)*
International Organization for Standardization (ISO), 1987.
Reference number: ISO 8824: 1987 (E).

[32] *(ISO 8825) Information processing—Open Systems Interconnection—Specification of basic encoding rules for Abstract Syntax Notation One (ASN.1)*
International Organization for Standardization (ISO), 1986.

[33] *(ISO 8879) Information processing—Text and office systems—Standard Generalized Markup Language (SGML)*
International Organization for Standardization (ISO), 1986.

[34] *(ISO 9541-1) Information processing — Font and character information interchange*
International Organization for Standardization (ISO), 1987.
Reference number: ISO 8825: 1987 (E), parts 1-7.

[35] Lynnea Johnson.
MacWrite
Apple Computer, 20525 Mariani Ave., Cupertino, CA 95014, 1983.

[36] B. W. Kernighan.
PIC — A Language for Typesetting Graphics.
Software Practice & Experience 12(1):1-20, January, 1982.

[37] B. W. Kernighan and L. L. Cherry.
A System for Typesetting Mathematics.
Communications of the ACM 18(3):151-157, 1975.

[38] Donald E. Knuth.
TEX and METAFONT: New Directions in Typesetting.
Digital Press, 1979.

[39] M. E. Lesk.
Typing documents on the UNIX System: using the -ms macros with troff and nroff
AT&T Bell Laboratories, Muray Hill, New Jersey, January 1979.
Unix Programmer's Manual, Section 8.

[40] M. E. Lesk.
TBL — A Program to format tables
AT&T Bell Laboratories, Murray Hill, New Jersey, 1979.
UNIX Programmer's Manual, Section 10.

[41] Macworld.
SoftQuad Author/Editor 1.1.
Macworld :162-164, December, 1988.

[42] Sandra A. Mamrak, Michael J. Kaelbling, Charles K. Nicholas and Michael Share.
Chameleon: A system for Solving the Data-Translation Problem.
IEEE Transactions on Software Engineering 15(9):1090-1108, September, 1989.

[43] Microsoft Corporation.
Microsoft Excel
1.50 edition, Microsoft, Bellevue, WA, 1988.

[44] Microsoft Corporation.
Microsoft Word, User's Guide
Microsoft Corporation, 1988.

[45] *Markup Requirements and Generic Style Specification for Electronic Printed Output and Exchange of Text*
Military Specification, 1989.

[46] Robert A. Morris.
Is What You See Enough to Get?: a Description of the Interleaf Publishing System.
In *PROTEXT II, Proceedings of the Second International Conference on Text Processing Systems*, pages 56-81. October, 1985.

[47] National Science Foundation.
EXPRES Project, Solicitation for Research Groups.

[48] *Stable Implementation Agreements for Open Systems Interconnection Protocols, Version 2 Edition 3*
National Institute for Standards and Technology, 1989.
Chapter 16 is devoted to ODA.

[49] Adrian Nye.
Xlib Programming Manual for Version 11
O'Reilly and Associates, 1988.
The Definitive Guides to the X Window System, Volume One.

[50] Adrian Nye (editor).
Xlib Reference Manual for Version 11
O'Reilly and Associates, 1988.
The Definitive Guides to the X Window System, Volume Two.

[51] J.F. Ossanna.
NROFF/TROFF User's Manual
AT&T Bell Laboratories, Murray Hill, New Jersey, January 1979.
Unix Programmer's Manual, Section 22.

[52] Andrew J. Palay, Wilfred J. Hansen, Mark Sherman, Maria G. Wadlow, Thomas P. Neuendorffer, Zalman Stern, Miles Bader and Thom Peters.
The Andrew Toolkit - An Overview.
In *Proceedings of the USENIX Winter Conference*, pages 9-21. February, 1988.

[53] R. F. Pennell and S. F. Feather.
Stored ODA (SODA) Interface
The ESPRIT Project, 1987.
ESPRIT Project 1024, Piloting of the Office Document Architecture.

[54] Thomas William Reps.
Generating Language-based Environments.
PhD thesis, Cornell University, 1982.

[55] Marshall T. Rose.
The Open Book: A Practical Perspective on OSI.
Prentice Hall, 1990.

[56] Marshall T. Rose.
The ISO Development Environment: User's Manual
Performance Systems International, Inc., Mountain View, CA, 1990.
Volumes 1-5.

[57] Jonathan Rosenberg, Paul Crumley, Ann Marks, Mark Sherman and Maria Wadlow.
The CMU ODA Tool Kit, Site Installation Guide & Application Programmer's Interface
Information Technology Center, Carnegie Mellon University, 1988.
Out of print.

[58] Frankie E. Spielman, Editor.
Standards for the Interchange of Large Format Tiled Raster Documents
National Institute for Standards and Technology, 1988.
NISTIR 88-4017.

[59] Sun Microsystems.
User's Manual for the Sun Workstation
Sun Microsystems, 1984.
Part Number 800-1107-01, Release 1.1.

[60] Robert H. Thomas, Harry C. Forsdick, Terrence R. Crowley, Richard W. Schaaf, Raymond S. Tomlinson and Virginia M. Travers.
Diamond: A Multimedia Message System Built on a Distributed Architecture.
IEEE Computer 18(12):65-78, December, 1985.

[61] *Scribe Document Production Software (User Manual)*
Unilogic, Ltd., 1985.

[62] C. J. Van Wyk.
A High-Level Language for Specifying Pictures.
ACM Transactions on Graphics 1(2):163-182, 1982.

[63] Xerox Corporation.
Interpress Electronic Printing Standard.
In *Xerox System Integration Guide*. Xerox Corporation, Stamford, Connecticut, 1983.

A
Example Programs

In this appendix, we present example programs illustrating the use of both the CMU ODA Tool Kit and the Raster Graphics Tool Kit. The first section contains examples of use of the CMU ODA Tool Kit, and the second section has examples using the Raster Graphics Tool Kit.

A.1. The CMU ODA Tool Kit

In this section, we present four programming examples illustrating the use of the CMU ODA Tool Kit. These examples come in two pairs. Each pair includes a document creator, which creates a document and then writes out its ODIF representation. This ODIF representation is to be read by the corresponding document reader. All example programs have been run successfully on the IBM RT-PC machine running IBM's AOS operating system[66].

We provide two different aids for reference in these examples. First, the sample programs contain annotations that designate particular lines in the code that are referenced in the supporting text. These annotations appear in the code in parentheses at the right margin in the code. Second, to facilitate reference to parts 3 and 4 of this book, we provide four tables, in section A.5, that cross reference features used in our examples. Table A-1 lists attributes, table A-2 lists procedures, table A-3 lists data types and tables A-4 and A-5 list constants.

A.1.1. Example 1

In the first pair of examples, a simple logical structure for an ODA document is built by the document creator, and then interpreted by the document reader. The sample document is shown in figure A-1.

---

[66]AOS is essentially Berkeley 4.3BSD Unix.

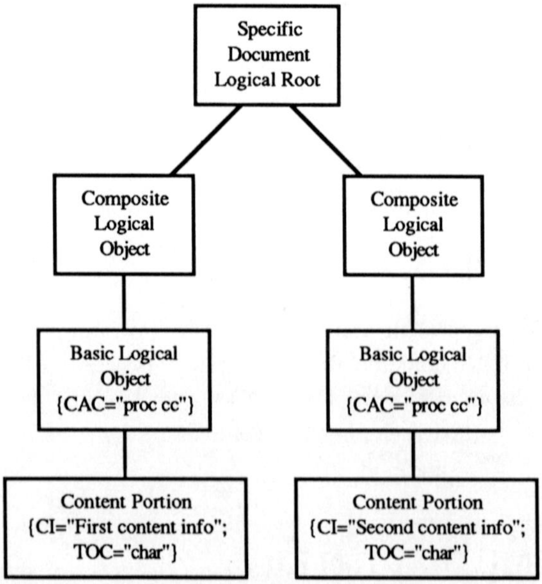

Figure A-1. Document for the First Pair of Example Programs

The document consists of a specific document logical root that has two subordinates, both composite logical objects. Each composite logical object has one subordinate that is a basic logical object. Each basic logical object has one content portion, which has associated content . The "type of coding" attribute is associated with each content portion to indicate that the content is character. Each basic logical object has the "content architecture class" attribute to indicate that the content below the basic logical object is processable character. We now proceed to discuss the document creator. Each discussion will include a commentary on the code and tool kit used by the example.

A.1.1.1. Document Creator: Example 1

The code for this document creator begins on page 585. The program passes through five distinct phases. These are initialization, creating the document and constituents, wiring the document together with the appropriate attributes and attaching content, writing the document out in its ODIF form as well as in the human readable form and, lastly, finalization.

The initialization begins at line (A.1), where the `INIT_PARMS_type` data structure, `InitParms`, is assigned initial values and then the `InitToolKit` routine is called. The `InitParms` indicates that the tool kit is to automatically label the document with ASN.1 object identifiers when writing the ODIF interchange form of the document, that tool kit error output is to go the the Unix standard error output, and that errors are to be formatted and printed by the tool kit. The `LOCAL_Init` routine is called to initialize the site-dependent routines in the tool kit. Note that the call to `LOCAL_Init` must be made first.

Next the document and constituents are created by calls to `MakeDocument`, `MakeComponent` and `MakeContent`. The code for this begins at (A.2) and ends at (A.3). Note that all five components are created by calls to `MakeComponent`, and that `MakeContent` is called for the content portions.

Subsequently, the document's constituents are wired together by attaching the appropriate attributes. The components are connected by the "subordinates" attribute, and the basic logical components are attached to the content portions by the "content portions" attribute.

The code for connecting constituents together starts at line (A.4) and ends at line line (A.7). It proceeds by making a sequence using the `MakeSequence` procedure. The body of the sequence is loaded with the `constituent`'s for the composite logical objects. The `SetAttr` tool kit procedure is called to associate the "subordinates" attribute with the document logical root. Lastly, the sequence is deleted by a call to `DeleteSequence` to free the local copy of the sequence, since the tool kit makes its own copy during the call of `SetAttr`. Similar comments apply to wiring the basic logical objects to the composite logical objects and the content portions to the basics logical objects.

Two other attributes are added in this example. The "content information" attribute is attached to each content portion in lines (A.8) through (A.9). This is accomplished by first creating a sequence of bytes to hold the content by a call to the `MakeSequence` procedure. Next, the content stored in a character array in the program is copied over to the byte sequence, and the characters are coerced to `BYTE_type`. The `SetAttr` procedure is used to attach content information to each content portion. At this point, the local sequence for the content information is deleted by a call to the procedure `DeleteSequence`. The "type of coding" attribute is added to each content portion by a call to `SetAttr`. This is not required by ODA, since the standard provides a default for determining the coding used in a content portion. However, our `ReadODIF` routine needs the "type of coding" attribute to parse an ODIF stream, so we include the attribute explicitly on content portions. The value for the "type of coding" attribute is the tool kit defined constant `at_TYPE_OF_CODING_cc`. This code is found at lines (A.10) through (A.11). Finally, the "content architecture class" attribute is attached to each basic logical component.

At this point, the document has been built and is ready to be output. Two versions of the document are output to files, beginning at (A.12) and continuing through (A.13). The first file is to hold the ODIF form of the document. A Unix file is created, then, the `WriteODIF` procedure is invoked to write the document. Note that our site-dependent routines interpret file descriptors as Unix file numbers. Closing the file requires that procedure

`LOCAL_WriteBytes` is called with a length argument of `-1`. This indicates that the application is done with the file. The second file is a human-readable version of the document. The `WriteDocumentState` routine is called to do the output.

The program ends at (A.14) by calling the finalization routine `FinalToolKit` and then calling the local module finalization routine `LOCAL_Final`. Note that the tool kit finalization routine must be called before the local routine.

We now comment on different tool kit features presented in this example. First, note that the `InitToolKit` routine is passed the address of the initialization parameter structure. This permits the program to make changes to the parameters. The tool kit obtains the value of the parameters by reference, using the address that it retains from the `InitToolKit` call.

Accessing the body of a sequence is somewhat messy. For example, consider lines (A.5) and (A.6), where the body of the sequence used for the document logical root's subordinates is loaded. The sequence is a pointer to a structure that contains generic sequence value information that the user must not modify. This information includes the sequence's length, tag and a pointer to the value of the sequence body that is a union with one variant for each element type. Thus, addressing an element of the sequence is clumsy as the code shows.

There is an interesting point with respect to `SetAttr`. `SetAttr` is a general service routine that can be used to set the value of any attribute in ODA. The value to assign to the attribute is passed by reference. But, given that some tool kit objects are already pointers of one flavor or another, the value passed is not necessarily written using the C address operator (`&`). For example, sequences are addresses, and so, when passed to `SetAttr`, they need only be cast to `POINTER_type`. For attributes where the application declares the structure and does not dynamically allocate storage, then the address of the structure must be passed and coerced to `POINTER_type`.

Lastly, note that constants must be carefully cast when passing them to the tool kit. For example, the `length` argument passed to `MakeSequence` is cast to `INT_type`. Another constant requiring casting is the file descriptor passed to the document write routines.

```
/*
 *
 * This example illustrates the construction of a simple
 *  document containing specific logical structure only.
 *  The content portions contain processable character content.
 *
 * A call to initialize the CMU ODA Tool Kit is made.
 *
 * A document is created. Then the constituents of the document
 *      are created: the specific logical root, two composite
 *      logical objects, two basic logical objects and two
 *      content portions.
 * These constituents are wired together with the
 *      "subordinates" attribute (between the components), and
 *      the "content portions" attribute (to connect basic
 *      logical objects to the content portions).
 * Some simple content is associated with the content portions,
 *      and the "content architecture class" attribute is
 *      attached to the basic components.
 *
 * Finally the document is printed using the debug routine,
 *      WriteDocumentState, and the document's ODIF is written
 *      by a call to WriteODIF.
 *
 * FinalToolKit is done last.
 *
 * To facilitate readability, no checking of CMU ODA Tool Kit
 *      returns is performed.
 */

#include <odatk.h>

/* for AOS UNIX */
#include <sys/file.h>
#include <stdio.h>

#define IGNORE_MASK  (PARM_MASK_type) 0
    /* for attributes without parameters */

main()
{
    INT_type ReturnCode;          /* for Tool Kit returns */

    DOCUMENT_type Document;       /* the document */

    CONSTITUENT SpecificRoot;     /* root of the specific
                                     logical structure */

    CONSTITUENT Composite[2];     /* composite logical objects */
    CONSTITUENT Basic[2];         /* basic logical objects */

    CONSTITUENT ContentPortions[2]; /* content portions */

    SEQUENCE_CONSTITUENT_type atSubordinates;
                    /* value for the "subordinates" */
    SEQUENCE_CONSTITUENT_type atContentPortions;
                    /* value for "content portions" */
    SEQUENCE_BYTE_type atContentInformation;
                    /* value for "content information" */

    char        *Content[2] = {
                "First Content Information",
                "Second Content Information"
    };

    INIT_PARMS_type InitParms;  /* to initialize tool kit */
```

```
    int od;       /* output descriptor for writing document */

    int i,j;      /* indices for looping */

/*                                                              (A.1)
*/
    /* set up init parms to initialize the tool kit */
    InitParms.automatic_label = BOOL_true;
    /* UNIX only - use of stderr for error output */
    InitParms.error_od = (INT_type) fileno( stderr );
    InitParms.format_errors = BOOL_true;
    InitParms.print_warnings = BOOL_true;
    InitParms.output_filter = NULL_FILTER_ROUTINE;
    ReturnCode = LOCAL_Init();
    ReturnCode = InitToolKit( &InitParms );

/*                                                              (A.2)
*/
    /* create the document */
    Document = MakeDocument();

    /* create the document logical root */
    SpecificRoot = MakeComponent(Document,
                            at_OBJECT_TYPE_doc_logical_root,
                            SPECIFIC_COMPONENT);

   /* create composite & basic components & content portions */
    for( i = 0; i < 2; i++){

        Composite[i] = MakeComponent(Document,
                        at_OBJECT_TYPE_comp_logical_obj,
                        SPECIFIC_COMPONENT);

        Basic[i] = MakeComponent(Document,
                            at_OBJECT_TYPE_bas_logical_obj,
                            SPECIFIC_COMPONENT);

        ContentPortions[i] = MakeContent( Document,
                                    CHARACTER_CONTENT );

    }
/*                                                              (A.3)
*/
/*                                                              (A.4)
*/
    /* wire the document together */
```

```
    /* wire root to the composites */
    /* create the sequence for the subordinates attribute */
    atSubordinates = MakeSequence( SEQUENCE_CONSTITUENT_tag,
                                   (INT_type) 2 );
/*                                                          (A.5)
*/
    atSubordinates->sequence_value.constituents[0] =
                                            Composite[0];
    atSubordinates->sequence_value.constituents[1] =
                                            Composite[1];
/*                                                          (A.6)
*/
    ReturnCode = SetAttr( SpecificRoot, at_SUBORDINATES,
                          (POINTER_type) atSubordinates,
                          IGNORE_MASK );
    ReturnCode = DeleteSequence( atSubordinates );
/*                                                          (A.7)
*/
    /*
     * Wire basics to the composites.
     * Attach at_CONTENT_ARCH_CLASS to the basic components
     * (they are processable character content).
     */
    atSubordinates = MakeSequence( SEQUENCE_CONSTITUENT_tag,
                                   (INT_type) 1);
    for( i = 0; i < 2; i++){

        atSubordinates->sequence_value.constituents[0] =
                                            Basic[i];
        ReturnCode = SetAttr(Composite[i], at_SUBORDINATES,
                             (POINTER_type) atSubordinates,
                             IGNORE_MASK);
        /* this makes copy of value so we can reuse sequence */
        ReturnCode = SetAttr(Basic[i], at_CONTENT_ARCH_CLASS,
                  (POINTER_type) at_CONTENT_ARCH_CLASS_proc_cc,
                  IGNORE_MASK);

    }
    ReturnCode = DeleteSequence( atSubordinates );

    /*
     * Wire the content portions to the basic components.
     * Attach "content information" and "type of content"
     *  to the content portions.
     */
    atContentPortions = MakeSequence( SEQUENCE_CONSTITUENT_tag,
                                      (INT_type) 1 );
    for( i = 0; i < 2; i++){

        atContentPortions->sequence_value.constituents[0] =
                                ContentPortions[i];
        ReturnCode = SetAttr( Basic[i], at_CONTENT_PORTIONS,
                              (POINTER_type) atContentPortions,
                              IGNORE_MASK );
/*                                                          (A.8)
*/
        atContentInformation = MakeSequence(SEQUENCE_BYTE_tag,
                                       strlen(Content[i]));
        for( j = 0; j < strlen( Content[i] ); j++){
            atContentInformation->sequence_value.bytes[j] =
                (BYTE_type) Content[i][j];
        }
```

```
        ReturnCode = SetAttr(ContentPortions[i],
                              at_CONTENT_INFORMATION,
                              (POINTER_type) atContentInformation,
                              IGNORE_MASK);
        ReturnCode = DeleteSequence( atContentInformation );
/*                                                              (A.9)
*/
/*                                                              (A.10)
*/
        ReturnCode = SetAttr(ContentPortions[i],
                              at_TYPE_OF_CODING,
                              (POINTER_type) at_TYPE_OF_CODING_cc,
                              IGNORE_MASK);
/*                                                              (A.11)
*/
    }
    ReturnCode = DeleteSequence( atContentPortions );
/*                                                              (A.12)
*/
    /* Write the ODIF form of  the document */
    od = open( "./OutEx1.odif", O_WRONLY|O_CREAT, 0600 );
            /* UNIX OPEN */
    ReturnCode = WriteODIF( (INT_type) od, Document,
                            dp_INTERCHANGE_FMT_CLASS_a );
    /* next statement CLOSES the file */
    ReturnCode = LOCAL_WriteBytes(od, (BYTE_type *) 0,
                                  (INT_type) -1);

    /* write a human readable version of the document */
    od = open( "./OutEx1.dump", O_WRONLY|O_CREAT, 0600 );
            /* UNIX OPEN */
    ReturnCode = WriteDocumentState( (INT_type) od, Document );
    /* next statement closes the file */
    ReturnCode = LOCAL_WriteChars(od, (CHAR_type *) 0,
                                  (INT_type) -1);

/*                                                              (A.13)
*/
/*                                                              (A.14)
*/
    /* finalize the tool kit */
    ReturnCode = FinalToolKit();
    ReturnCode = LOCAL_Final();

}
```

A.1.1.2. Document Reader: Example 1

The code for this example begins on page 591. This program reads and processes the document written by the previous example program. The program reads the document and then walks it. As basic components are encountered, the content type and content are obtained. Processing of content depends on the content type. Although processing for raster graphics and geometric graphics content is unimplemented, the processing for character content prints out the content. Comments on the details of how all this happens are given below. Note that this program handles errors by printing an error message on the Unix standard error output and then aborting by a call to the Unix exit system call.

Beginning at (A.15) and ending at (A.16), the program initializes itself. This requires setting up an `INIT_PARMS_type` structure. The fields of this structure are set to indicate that the tool kit should format error messages when they occur, should use the Unix standard error output for printing error messages and that no filtering of output characters need be done. The actual initialization is done by calling the `LOCAL_Init` routine, followed by a call to `InitToolKit`.

After initializing the tool kit, the program then proceeds to read in the ODIF document. This is accomplished between (A.17) and (A.18). The Unix file for the ODIF datastream is first opened by a call to the Unix open system call. Next, the `ReadODIF` routine is called. If the read is successful, the tool kit returns to the program a `DOCUMENT_type` created as a result of the read. Lastly, the Unix file is closed. Note that the Unix file descriptor (an `int` in C) must be cast to `INT_type` when calling `ReadODIF`.

Next the program begins to traverse the document. This requires locating the document logical root, by calling the `FindDocumentRoot` tool kit routine. This returns the `CONSTITUENT` for the document logical root. With the document logical root in hand, the traversal can now begin. The traversal is done by the `WalkComponent` routine, between lines (A.19) and (A.20).

Before discussing `WalkComponent`, we comment on how the main line terminates beginning at line (A.21) and ending at line (A.22). Basically, it finalizes the tool kit by calling `FinalToolKit` and then `LOCAL_Final`.

The `WalkComponent` routine is responsible for traversing the children of the component that is passed as an argument. This component must not be a basic logical object or a content portion. It must be the document logical root or a composite logical object. Therefore, the component's children are given by the "subordinates" attribute.

`WalkComponent`, which starts at line (A.23), proceeds by performing a check to determine the type of the component passed as an argument. This is done beginning at (A.24). First, the value of the "object type" attribute is obtained. If the type of the component is 'basic logical object', then the walker for basic components is called (see the code between (A.25) and (A.26)). After this call completes, a return from `WalkComponent` is made to short circuit subsequent processing by `WalkComponent`. Next, the component's type is checked again, this time to ensure that it is either the document logical root or a logical composite object—see lines (A.27) through (A.28). If this test fails, the program aborts. By line (A.28), `WalkComponent` is guaranteed to have a component which is either the document logical root or a composite logical object. At this point, the traversal of the component's children can begin.

To traverse the children, `WalkComponent` first checks to see if the component has the "subordinates" attribute. It does this by calling the tool kit `HasAttr` procedure, at line (A.29) through line (A.31). If the component does not have the "subordinates" attribute, which is detected at line (A.30), then the call to `WalkComponent` returns, because the traversal of the component's children has been trivially completed. If the component does have the "subordinates" attribute, then the value of the "subordinates" attribute is obtained by a call to the `GetAttr` subroutine at line (A.32). If the value was successfully obtained, the value returned is of type `SEQUENCE_CONSTITUENT_type`. The loop at line (A.33) examines each child in turn making a recursive call to `WalkComponent`. After the traversal of all children has been completed, the value of the "subordinates" attribute for the component is no longer needed, and so the storage allocated by the call to `GetAttr` for the `SEQUENCE_CONSTITUENT_type` is deallocated by a call to the tool kit procedure `DeleteSequence` at line (A.34).

Traversal of the basic logical components is done by the `WalkBasic` procedure, which begins at line (A.35). This routine can assume that the component it was passed is a basic logical component (because of the testing done by the `WalkComponent`). Therefore, it can proceed to traverse the component without any initial checking. The processing begins by obtaining the content kind associated with the basic logical component by calling the `GetContentKind` routine at line (A.36). This routine will use the defaulting rules of ODA for obtaining the content kind. Note that the defaulting rules defined by ODA for this are not the same as the usual ones. The search selectively examines the document, looking for either the "content architecture class" or "content type" attribute. The `GetContentKind` routine correctly performs this search in accordance with ODA's defaulting rules for obtaining content kind.

After successfully obtaining the content kind, the content is obtained at line (A.37) by a call to `GetContent`. The ODA defaulting rules for obtaining content are complex, involve an abbreviated search of the document, and must account for the presence of "content information" and "content generator" attributes. The `GetContent` routine implements all of these rules. The content is returned as a `SEQUENCE_BYTE_type`.

Once the content has been successfully obtained, `WalkBasic` then switches at line (A.38) depending on the content kind. Note that cases for raster and geometric graphics are provided, but that no processing is done. If the content kind is character, then the `CharacterContent` routine is called to process the content. `WalkBasic` ends by deallocating the space for the content that was allocated by the `GetContent` call. This is accomplished at line (A.39) by a call to the `DeleteSequence` procedure.

The routine `CharacterContent` processes character content. Its one argument is a `SEQUENCE_BYTE_type` for the content. The processing done prints out the content on the Unix standard output. To do this requires converting the ODA `BYTE_type` content to the C <u>`char`</u> type. This is done by a loop shown at line (A.39) that performs the output by calling the Unix `putchar` library utility. A new line is printed when all content characters have been printed, as the ODA content may not include one.

Some general points on this example are worth noting. First, the means for traversing the document is to have the program use the value of the "subordinates" attribute, and to iterate over the subordinates. The second reader example given in section A.1.2.2 accomplishes the traversal very differently by using an `ITERATOR_type` object and support for iterators. Also note that the tool kit procedures `GetContentKind` and `GetContent` are very powerful as they implement complex ODA semantics. Thus, the ODA application programmer is spared the task of rummaging around in the document for content kind and content.

```
/*
 * This example shows how to read and traverse the document
 *  constructed in example A.1.1.1.
 *
 * A call to initialize the CMU ODA Tool Kit is made.
 *
 * The document is read in by a call to ReadODIF and the
 *  document logical root is located.
 *
 * A call to the recursive document walker is made to initiate
 *  the traversal of the document. Comments on the walker and
 *  other routines are given in those subroutines.
 *
 * main ends with a call to finalize the tool kit.
 */

#include <odatk.h>

/* for AOS UNIX */
#include <sys/file.h>
#include <stdio.h>

/* forward declarations for procedures */
void WalkComponent();
void WalkBasic();
void CharacterContent();

/* values for the defaulting parameter to GetAttr */
#define     DEFAULT     BOOL_true
#define     NO_DEFAULT  BOOL_false

/* ignore parameter masks for attributes without them */
#define     IGNORE_MASK (PARM_MASK_type *) 0

main()
{
    INT_type ReturnCode;          /* Tool Kit returns */
    DOCUMENT_type Document;       /* the document */
    CONSTITUENT SpecificRoot;     /* root of specific logical */
    INIT_PARMS_type InitParms;    /* initialize the tool kit */
```

```
    int id;     /* file descriptor for writing document */
/*                                                          (A.15)
*/
    /* set up init parms before initializing the tool kit */
    InitParms.automatic_label = BOOL_true;
    /* UNIX only - use of stderr for error output */
    InitParms.error_od = (INT_type) fileno( stderr );
    InitParms.format_errors = BOOL_true;
    InitParms.print_warnings = BOOL_true;
    InitParms.output_filter = NULL_FILTER_ROUTINE;
    ReturnCode = LOCAL_Init();
    ReturnCode = InitToolKit( &InitParms );
/*                                                          (A.16)
*/
/*
  */
/*                                                          (A.17)
*/
    /* read in the ODIF form of the document */
    id = open( "./OutEx1.odif", O_RDONLY );
    Document = ReadODIF( (INT_type) id );
    if( Document == ERROR_DOCUMENT ){
        fprintf( stderr, "Can not Read the ODIF\n" );
        exit ( 01 );
    }

    /* Tell TK done with descriptor and to close file */
    ReturnCode = LOCAL_ReadBytes((INT_type) id,
                                 (BYTE_type *) 0,
                                 (INT_type) -1);

/*                                                          (A.18)
*/
/*                                                          (A.19)
*/
    /* locate the document logical root */
    SpecificRoot = FindDocumentRoot( Document,
                        SPECIFIC_DOC_LOGICAL_ROOT );
    if( SpecificRoot == ERROR_CONSTITUENT ){
        fprintf(stderr, "Can't find specific logical root\n");
        exit( 01 );
    }

    /* start the recursive walk of the document */
    WalkComponent( SpecificRoot );
/*                                                          (A.20)
*/
/*                                                          (A.21)
*/
```

```
    /* finalize the tool kit */
    ReturnCode = FinalToolKit();
    ReturnCode = LOCAL_Final();
/*                                                          (A.22)
*/
}
/*                                                          (A.23)
*/
void WalkComponent( component )
    CONSTITUENT component;
{
    /*
     * Find the type of the component.
     *
     * If document root or composite logical then find the
     *  "subordinates" attribute and iterate over the
     *  subordinates making recursive calls.
     *
     * If a basic logical component call the basic component
     *  walker.
     *
     * Check to ensure that the component is either the
     *  document logical root or a composite logical object.
     *
     * Get the value of the "subordinates" attribute.
     *
     * Loop through the subordinates making recursive calls to
     *  this routine.
     */

    INT_type ReturnCode;
    VALUE_type ComponentType;
    SEQUENCE_CONSTITUENT_type Subordinates;
    int i;
/*
  */
/*                                                          (A.24)
*/
    /* get the at_OBJECT_TYPE */
    ReturnCode = GetAttr( component, at_OBJECT_TYPE,
                          (POINTER_type) &ComponentType,
                          NO_DEFAULT, IGNORE_MASK );
    if( ReturnCode == ERROR_INT ){
        fprintf(stderr,
            "Can't get component type in WalkComponent\n");
        exit( 01 );
    }
/*
  */
/*                                                          (A.25)
*/
    /* if basic component then deal with walking content */
    if( ComponentType == at_OBJECT_TYPE_bas_logical_obj ){
        WalkBasic( component );
        return;
    }
/*
```

```
  */
/*                                                                   (A.26)
*/
    /*
     * just to be safe make sure that the component is doc log
     * root or a composite logical.
     */
/*
  */
/*                                                                   (A.27)
*/
    if( ComponentType != at_OBJECT_TYPE_doc_logical_root &&
        ComponentType != at_OBJECT_TYPE_comp_logical_obj ){
        fprintf(stderr, "Layout component in WalkComponent\n");
        exit( 01 );
    }
/*
  */
/*                                                                   (A.28)
*/
/*                                                                   (A.29)
*/
    /* find the subordinates */
    ReturnCode = HasAttr( component, at_SUBORDINATES );
    if (ReturnCode == ERROR_INT) {
      fprintf(stderr,
             "Error HasAttr: subordinates in WalkComponent\n");
        exit( 01 );
    }
/*
  */
/*                                                                   (A.30)
*/
    if( ReturnCode == NO_ATTR){
        /* stop the recursion */
        return;
    }
/*
  */
/*                                                                   (A.31)
*/
/*                                                                   (A.32)
*/
    ReturnCode = GetAttr( component, at_SUBORDINATES,
                          (POINTER_type) &Subordinates,
                          NO_DEFAULT, IGNORE_MASK );
    if( ReturnCode == ERROR_INT ){
        fprintf( stderr,
                 "Can't get subordinates in WalkComponent\n" );
        exit( 01 );
    }
/*
  */
/*                                                                   (A.33)
*/
    for( i = 0; i < Subordinates->length; i++){
   WalkComponent(Subordinates->sequence_value.constituents[i]);
    }
/*
```

```
   */
/*                                                          (A.34)
*/
    ReturnCode = DeleteSequence( Subordinates );
}
/*
   */
/*                                                          (A.35)
*/
void WalkBasic( component )
    CONSTITUENT component;
{
    /*
     * ASSERT: the component must be a basic logical
     *  object.
     *
     * First get the content type.
     *
     * Then get the content.
     *
     * Call the appropriate content specific routine.
     */

    VALUE_type ContentKind;

    SEQUENCE_BYTE_type Content;

    INT_type ReturnCode;
/*                                                          (A.36)
*/
    ContentKind = GetContentKind( component );
    if( ContentKind == ERROR_kind ){
        fprintf(stderr,
                "Can't get content kind in WalkBasic\n");
        exit( 01 );
    }
/*                                                          (A.37)
*/
    Content = GetContent( component );
    if( Content == ERROR_SEQUENCE ){
        fprintf( stderr, "Can't get content in WalkBasic\n" );
        exit( 01 );
    }
```

```
    /* Switch on ContentKind to process content */
    /*
     * Note that there is a spot for each kind of content
     *  even though we only handle character content.
     */
/*                                                        (A.38)
*/
    switch( ContentKind ){
    case FORM_CC_kind:
    case FP_CC_kind:
    case PROC_CC_kind:
        CharacterContent( Content );
        break;
    case FORM_RG_kind:
    case FP_RG_kind:
    case FORM_GG_kind:
        break;
    default:
        fprintf( stderr, "Bad ContentKind in WalkBasic\n" );
        exit( 01 );
    }
/*
  */
/*                                                        (A.39)
*/
    ReturnCode = DeleteSequence( Content );
}

void CharacterContent( Content )
    SEQUENCE_BYTE_type Content;
{
    /*
     * Print out the content.
     *
     * This will only pretty print short content (i.e. fits on
     *  one line ).
     */

    INT_type i;

    printf( "\nContent for one of the basic components:\n" );

    for( i = (INT_type) 0; i < Content->length; i++){

        putchar( (char) Content->sequence_value.bytes[i] );

    }

    printf( "\n" );
}
```

A.1.2. Example 2

The second set of sample programs work with the document whose specific and generic logical structures are shown in figure A-2. In both the generic and specific structures, there is a document root that has one child. This child is a composite logical object or object class, and it has one child, which is a basic logical object or object class. In the specific logical structure, the basic logical object has one content portion, which has associated character content.

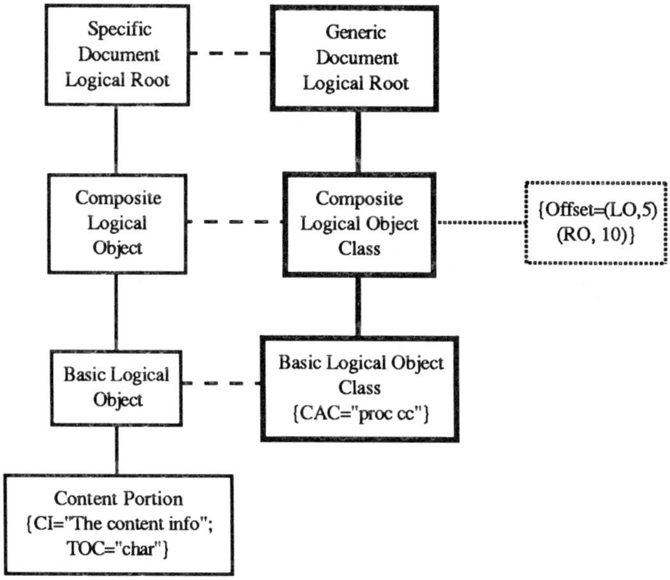

Figure A-2. Document for the Second Pair of Example Programs

The specific structure is wired together with the "subordinates" attribute and the "content portions" attribute. The "subordinates" attribute is used to connect the document logical root to the composite logical object, and to connect the composite logical object to the basic logical object. The "content portions" attribute is used to connect the basic logical object to the content portion.

The wiring in the generic structure uses the "generator for subordinates" attribute. The form of the construction expression used is an optionally repeated construction expression. This recursive construction expression allows for a sequence of required object classes. The document logical root object class permits optional repetitions of a sequence of the generic logical composite object class, and the composite logical object class permits optional repetitions of a sequence of the basic logical object class.

Associations between the logical objects in the specific structure and the object classes in the generic structure are made with the "object class" attribute.

To specify the content type, the "content architecture class" attribute is associated with the generic basic object class, and the "type of coding" attribute is attached to the content portion in the specific logical structure.

One layout directive attribute, "offset", is placed in the document. This attribute applies to the basic logical object. The attribute is not associated with the basic logical object, but rather, is on a layout style associated with a default value list. The default value list is associated with the composite object class in the generic structure.

A.1.2.1. Document Creator: Example 2

In this section, we discuss the program that creates the second document. The program begins on page 602. The program is straight line in structure and passes through a number of phases. In the first phase, the tool kit is initialized. Next, the document and all the components are created. Subsequently, the specific logical structure is constructed with the "subordinates" and "content portions" attributes. Then, the construction expression, used as the value for the "generator for subordinates" attribute, is created and built, and the generic logical structure structure is wired together with the "generator for subordinates" attribute. In the next phase, the correspondence between components in the specific logical structure and those in the generic logical structure is made using the "object class" attribute. Next, content typing information is added to the document. Finally, the structure to hold the value of the "offset" attribute is built. This entails initializing a default value list, making a layout style, making the default value list's basic logical component point at the layout style, and attaching the properly initialized value of the "offset" attribute to the layout style. The program then writes out the document in both the binary ODIF form and in the human-readable debugging form. Lastly, finalization is performed.

The first phase, initialization of the tool kit, occurs at lines (A.40) through (A.41). An `INIT_PARMS_type` structure is initialized to indicate that the tool kit should format and print error messages when errors occur, that the tool kit should not do any filtering of characters when outputting error messages, and should automatically label the document with ASN.1 object identifiers when writing the ODIF representation. The tool kit is initialized by calling `LOCAL_Init` and then `InitToolKit`. Note that `LOCAL_Init` must be called before `InitToolKit`.

Once the tool kit has been initialized, the document is created at line (A.42) by a call to `MakeDocument`, which returns a `DOCUMENT_type` object. Subsequently, the components are created. The `MakeComponent` routine is used to create all the components in both the specific logical and generic logical structures. The one content portion is created by calling `MakeContent`.

After creating the document and constituents, the program proceeds to wire the specific structure together with the "subordinates" and "content portions" attributes. This activity begins on line (A.43) and continues through (A.46). The "subordinates" attribute is used to wire the document logical root to the composite logical object. To do this, a sequence is first made at line (A.44) by a call to `MakeSequence`, as the value of the "subordinates" attribute is a sequence of constituents. The value of the sequence is set to that of the composite logical object, and then the "subordinates" attribute is associated with the document logical root by a call to `SetAttr`. Next, the value of the sequence is changed to that of the basic logical object, and `SetAttr` is called to associate this value with the composite logical object for its "subordinates" attribute. This sequence is now deallocated by a call to the `DeleteSequence` routine at line (A.45). Now, the content portion is wired to the basic logical object. A sequence is created by a call to `MakeSequence`, as the value of the "content portions" attribute is a sequence of constituents. The value of the sequence is set to the content portion, and `SetAttr` is called to associate this value with the basic logical object. Note that we can reuse the sequence in multiple calls to `SetAttr` since the tool kit makes its own copy of the sequence when an attribute is defined. Finally, this sequence is deallocated by a call to `DeleteSequence`.

Next, the generic structure is wired together using the "generator for subordinates" attribute. First, memory space for the top level construction expression is allocated by a call to the `LOCAL_Malloc` routine. We use this tool kit routine because space will later be deallocated by a call to the tool kit. Next, the construction expression, which is the value for the "generator for subordinates" attribute, must be built. The common fields of the construction expression are first set. This is accomplished beginning at line (A.47). The top level of the construction expression indicates that it is an optional and repeated construction expression. Memory for this recursive construction expression is allocated at line (A.48). Note that the `LOCAL_Malloc` procedure is used, as we will later use a tool kit routine to deallocate space for the construction expression. The value of the factor in the top-level construction is now set to be the recursive construction that was just allocated—this is done at line (A.48). Now the recursive construction expression can be loaded. Its top level indicates that the value is a choice of required objects, which will be set later on in the actual wiring process.

To complete the wiring of the generic logical structure, the factor in the recursive construction is set to the `CONSTITUENT` for the composite generic logical object. Then, `SetAttr` is called to associate the full construction expression value with the document logical generic root for the "generator for subordinates" attribute, at line (A.49). Next the factor in the recursive construction expression is set to the `CONSTITUENT` for the basic generic logical object. Now, `SetAttr` is called to associate the "generator for subordinates" attribute with the composite logical object using the modified construction expression for the value. Lastly, memory for the entire construction expression is reclaimed by a call to the `DeleteConstExpr` routine at line (A.50).

In the next phase, the correspondence between the specific and generic structures is established by using the "object class" attribute. Lines (A.51) through (A.52) show three `SetAttr` calls that make this correspondence. Note that this attribute is attached in the specific structure between one specific component and the one in the generic structure with the same type.

Now the program proceeds to add content to the document, and attributes that indicate the type of content. To add the content, a sequence of bytes is created by a call to `MakeSequence`. This is done at line (A.53). Next, the content is copied into the sequence's body; note that this requires a coercion to `BYTE_type` from the C <u>char</u> type. The "content information" attribute is now set for the content portion by a call to `SetAttr` at line (A.54), and the content sequence is deallocated at line (A.55) by a call to `DeleteSequence`. Now, two attributes must be added to indicate the content type. First, the "type of coding" attribute is attached to the content portion to indicate that the content is character. This is accomplished by the `SetAttr` call at line (A.56). Next, the "content architecture class" attribute is attached to the basic logical object class, by calling `SetAttr`, to indicate that the content is processable character. This is done by the `SetAttr` call at line (A.57).

The next part of the program adds the "offset" attribute to the document on a layout style associated with a default value list placed on the composite generic logical object class. This procedure requires several steps. First, a default value list data structure is created by a call to `InitializeDefaultValueLists` at line (A.58). This data structure consists of five constituents, one for each type of constituent that can appear in a default value list. This particular default value list is initialized to contain only a basic logical constituent. The layout style is made at line (A.59) by a call to `MakeStyle`. Next, the style is associated with the basic logical component in the default value list by a call to `SetAttr` at line (A.60). Note that the "layout style" attribute is used for this. Now the value of the "offset" attribute is set to indicate that the leading offset is 5, and that the right hand offset is 10. Then, this value is associated with the layout style by a call to `SetAttr` at line (A.61). Note that the parameter mask argument of the `SetAttr` call indicates the presence of the leading and right hand offsets. Thus, the trailing and left hand offsets are not set. Finally, the default value list is attached to the composite generic logical object class by a call to `SetAttr` at line (A.62). The "default value lists" attribute is used for this.

Although not shown in this example, the tool kit provides a short cut for some of these operations. The tool kit routine `SetLayoutAttr` takes an attribute destined for a layout style and a constituent that can specify a layout style, and places the specified attribute on an appropriate layout style. If the constituent already specifies a layout style, it is reused; otherwise a layout style is created and attached to the constituent.

Now, the program proceeds to write the document out in both the ODIF and human-readable debugging form. First, to write the ODIF representation, a Unix file is opened and created at line (A.63). Now, the document is written by a call to `WriteODIF`. Note that the Unix file descriptor is cast to `INT_type` when making this call. To close the file, the tool kit and Unix must be called.

First, the file is closed with the tool kit by calling LOCAL_WriteBytes passing the length argument as (INT_type) -1. Writing the human-readable debugging output is done in similar fashion. There are only two differences. First, WriteDocumentState is called instead of WriteODIF. As the human-readable debugging output is characters and not bytes, closing the file with the tool kit is accomplished by calling LOCAL_WriteChars instead of LOCAL_WriteBytes. Note that the close is indicated to the tool kit by passing a length of (INT_type) -1.

The program ends with finalization. The tool kit is first finalized by calling FinalToolKit. Then, the local module is finalized by calling LOCAL_Final.

A few comments on this example are appropriate. We have seen the use of some of ODA's more complex aspects. The most important point to note is that the tool kit attempts to support these messy aspects. For example, there is the construction expression that is used for the value of the "generator for subordinates" attribute. The construction expression built in this example is the same as the one that we used in our EXPRES interchanges. Building up such a construction expression is a fair amount of work. But once the basic structure has been established, small changes can be made to create different construction expression values. In this example, the only change required in the value of the construction expression used for the "generator for subordinates" attribute for both the generic logical root and the generic logical composite object class was in the factor in the recursive construction expression. Although building the construction expression was messy, deallocating it required only one tool kit call. To be able to use this facility required the use of the LOCAL_Malloc procedure as the tool kit can only reclaim storage that it has allocated.

It is interesting to consider how the value to SetAttr is passed. Note that how the value of the attribute is passed depends on the type of the attribute. For CONSTITUENTs and sequences, the value can be passed directly (but must be cast to POINTER_type). But for the "offset" attribute, whose value is a structure, the address of the structure must be passed (the address must be cast to POINTER_type). Similar comments apply to a default value list passed to SetAttr.

In the last example, starting on page 602, we show how to read the document created by this program.

```
/*
 * This example constructs a simple document
 *  including specific and generic logical structure.
 *
 * Initialize the CMU ODA Tool Kit.
 *
 * Create the components needed: specific logical root,
 *  specific composite logical object,
 *  specific basic logical object,
 *  content portion,
 *  generic logical root,
 *  generic composite object,
 *  generic basic object.
 *
 * Wire the specific structure together with
 *  "subordinates" and "content portions" attributes.
 *
 * Wire the generic structure together with
 *  "generator for subordinates".
 *
 * Connect corresponding pieces of the specific and generic
 *  structures with the "object class" attribute.
 *
 * Write out the document in odif and human readable form.
 *
 * Call FinalToolKit.
 *
 * To facilitate readability, no checking of CMU ODA Tool Kit
 *  return codes is done.
 */
#include <odatk.h>

/* for AOS UNIX */
#include <sys/file.h>
#include <stdio.h>

/* Parmeter mask for attributes with no parameters */
#define IGNORE_MASK  (PARM_MASK_type) 0

main()
{

    INT_type ReturnCode;    /* for Tool Kit returns */

    DOCUMENT_type Document; /* the document */

    /* Constituents for the specific structure */
    CONSTITUENT SpecificRoot;
    CONSTITUENT SpecificComposite;
    CONSTITUENT SpecificBasic;
    CONSTITUENT ContentPortion;

    /* Constituents for the generic structure */
    CONSTITUENT GenericRoot;
    CONSTITUENT GenericComposite;
    CONSTITUENT GenericBasic;

    at_SUBORDINATES_type at_Subordinates;
    at_CONTENT_PORTIONS_type at_ContentPortions;

    at_GEN_FOR_SUBORDINATES_type *at_GeneratorForSubordinates;
    at_GEN_FOR_SUBORDINATES_type *RecursiveExpr;
    SEQUENCE_CONSTRUCTION_TERM_type ConstructionTerms;

    at_CONTENT_INFORMATION_type ContentInfo;
```

```
    at_DEFAULT_VALUE_LISTS_type at_DefaultValueList;
    CONSTITUENT LayoutStyle;
    at_OFFSET_type at_Offset;

    char *Content = { "The Content Information"};
    INIT_PARMS_type InitParms;  /* init the tool kit */
    int od;      /* output descriptor for writing document */
    int i;
/*                                                          (A.40)
*/
    /* set up init parms to initialize tool kit */
    InitParms.automatic_label = BOOL_true;
    /* AOS UNIX only - use of stderr for error output */
    InitParms.error_od = (INT_type) fileno( stderr );
    InitParms.format_errors = BOOL_true;
    InitParms.print_warnings = BOOL_true;
    InitParms.output_filter = NULL_FILTER_ROUTINE;
    ReturnCode = LOCAL_Init();
    ReturnCode = InitToolKit( &InitParms );
/*                                                          (A.41)
*/
/*                                                          (A.42)
*/
    /* create the document */
    Document = MakeDocument();

    /* Create the constituents for the specific structure */
    SpecificRoot = MakeComponent(Document,
                                at_OBJECT_TYPE_doc_logical_root,
                                SPECIFIC_COMPONENT);
    SpecificComposite = MakeComponent(Document,
                                at_OBJECT_TYPE_comp_logical_obj,
                                SPECIFIC_COMPONENT);
    SpecificBasic = MakeComponent(Document,
                            at_OBJECT_TYPE_bas_logical_obj,
                            SPECIFIC_COMPONENT);
    ContentPortion = MakeContent(Document, CHARACTER_CONTENT);

    /* Create the constituents for the generic structure */
    GenericRoot = MakeComponent(Document,
                                at_OBJECT_TYPE_doc_logical_root,
                                GENERIC_COMPONENT);
    GenericComposite = MakeComponent(Document,
                                at_OBJECT_TYPE_comp_logical_obj,
                                GENERIC_COMPONENT);
    GenericBasic = MakeComponent(Document,
                                 at_OBJECT_TYPE_bas_logical_obj,
                                 GENERIC_COMPONENT);
```

```
    /*
     * Wire the specific structure together with "subordinates"
     *  & the "content portions" attributes.
     */
/*                                                         (A.43)
*/
/*                                                         (A.44)
*/
    /* specific root to the specific composite */
    at_Subordinates = MakeSequence( SEQUENCE_CONSTITUENT_tag,
                                    (INT_type) 1 );
    at_Subordinates->sequence_value.constituents[0] =
                                        SpecificComposite;
    ReturnCode = SetAttr( SpecificRoot, at_SUBORDINATES,
                          (POINTER_type) at_Subordinates,
                          IGNORE_MASK );
    at_Subordinates->sequence_value.constituents[0] =
                                        SpecificBasic;
    ReturnCode = SetAttr( SpecificComposite, at_SUBORDINATES,
                          (POINTER_type) at_Subordinates,
                          IGNORE_MASK );
    ReturnCode = DeleteSequence( at_Subordinates );

    at_ContentPortions = MakeSequence(SEQUENCE_CONSTITUENT_tag,
                                      (INT_type) 1);
    at_ContentPortions->sequence_value.constituents[0] =
                                        ContentPortion;
    ReturnCode = SetAttr( SpecificBasic, at_CONTENT_PORTIONS,
                          (POINTER_type) at_ContentPortions,
                          IGNORE_MASK );
/*                                                         (A.45)
*/
    ReturnCode = DeleteSequence( at_ContentPortions );
/*                                                         (A.46)
*/
    /* wire the generic structure together
       with "generator for subordinates" */

    at_GeneratorForSubordinates =
        (at_GEN_FOR_SUBORDINATES_type *)
           LOCAL_Malloc((INT_type)
                        sizeof(at_GEN_FOR_SUBORDINATES_type));
/*                                                         (A.47)
*/
    /* Fill in the 'common' fields of the generator */
    at_GeneratorForSubordinates->construction_tag =
                                        TERM_tag;
    at_GeneratorForSubordinates->
                construction_value.term.term_tag = OPT_REP_tag;
    at_GeneratorForSubordinates->
                    construction_value.term.factor.factor_tag =
                                                    EXPR_tag;
/*                                                         (A.48)
*/
    /* fill in common fields of recursive construction expr */
    RecursiveExpr = (CONSTRUCTION_EXPR_type *)
       LOCAL_Malloc((INT_type) sizeof(CONSTRUCTION_EXPR_type));

        /*
         * We use LOCAL_Malloc as the generator will
         * be freed as a unit by calling DeleteConstExp
         */
```

```
    at_GeneratorForSubordinates->
          construction_value.term.factor.factor_value.expr =
                                                   RecursiveExpr;
    RecursiveExpr->construction_tag = CHOICE_tag;
    ConstructionTerms =
            MakeSequence(SEQUENCE_CONSTRUCTION_TERM_tag,
                        (INT_type) 1);
    RecursiveExpr->construction_value.construction =
                                              ConstructionTerms;
    ConstructionTerms->
            sequence_value.construction_terms[0].term_tag =
                                                        REQ_tag;
    ConstructionTerms->
       sequence_value.construction_terms[0].factor.factor_tag =
                                             OBJECT_CLASS_ID_tag;

    /* wire the generic root & generic composite together */
    ConstructionTerms->sequence_value.construction_terms[0].
                        factor.factor_value.object_class_id =
                                                GenericComposite;
/*                                                        (A.49)
*/
    ReturnCode = SetAttr( GenericRoot, at_GEN_FOR_SUBORDINATES,
                          (POINTER_type)
                               at_GeneratorForSubordinates,
                          IGNORE_MASK );

    /* wire generic composite & generic basic together */
    ConstructionTerms->sequence_value.construction_terms[0].
                       factor.factor_value.object_class_id =
                                                    GenericBasic;
    ReturnCode = SetAttr(GenericComposite,
                     at_GEN_FOR_SUBORDINATES,
                     (POINTER_type) at_GeneratorForSubordinates,
                     IGNORE_MASK);

/*                                                        (A.50)
*/
    ReturnCode = DeleteConstExpr(at_GeneratorForSubordinates);
/*                                                        (A.51)
*/
    /* connect specific & generic structures with
       "object class" */
    ReturnCode = SetAttr( SpecificRoot, at_OBJECT_CLASS,
                          (POINTER_type) GenericRoot,
                          IGNORE_MASK );
    ReturnCode = SetAttr( SpecificComposite, at_OBJECT_CLASS,
                          (POINTER_type) GenericComposite,
                          IGNORE_MASK );
    ReturnCode = SetAttr( SpecificBasic, at_OBJECT_CLASS,
                          (POINTER_type) GenericBasic,
                          IGNORE_MASK );
```

```
/*                                                         (A.52)
*/
   /* Ornament the structure */
/*                                                         (A.53)
*/
   /* Ornament the specific. content portion */
   ContentInfo = MakeSequence( SEQUENCE_BYTE_tag,
                               strlen( Content ) );
   for( i = (INT_type) 0; i < strlen( Content ); i++){
       ContentInfo->sequence_value.bytes[i] =
                              (BYTE_type) Content[i];
   }
/*                                                         (A.54)
*/
   ReturnCode = SetAttr( ContentPortion,
                         at_CONTENT_INFORMATION,
                         (POINTER_type) ContentInfo,
                         IGNORE_MASK );
/*                                                         (A.55)
*/
   ReturnCode = DeleteSequence( ContentInfo );
/*                                                         (A.56)
*/
   ReturnCode = SetAttr( ContentPortion, at_TYPE_OF_CODING,
                         (POINTER_type) at_TYPE_OF_CODING_cc,
                         IGNORE_MASK );
/*                                                         (A.57)
*/
   /* Add content architecture class to generic basic */
   ReturnCode = SetAttr(GenericBasic, at_CONTENT_ARCH_CLASS,
               (POINTER_type) at_CONTENT_ARCH_CLASS_proc_cc,
               IGNORE_MASK);

   /*
    * Build the default value list structure to attach to the
    *  generic composite object.
    */
/*                                                         (A.58)
*/
   ReturnCode =
     InitializeDefaultValueLists(Document,
                                 &at_DefaultValueList,
                                 BAS_LOGICAL_ATTRIBUTES_parm);
/*                                                         (A.59)
*/
   LayoutStyle = MakeStyle( Document, LAYOUT_STYLE );
/*                                                         (A.60)
*/

   ReturnCode =
       SetAttr( at_DefaultValueList.basic_logical_attributes,
                at_LAYOUT_STYLE, (POINTER_type) LayoutStyle,
                IGNORE_MASK );
   at_Offset.leading_offset = (INT_type) 5;
   at_Offset.right_hand_offset = (INT_type) 10;
/*                                                         (A.61)
*/
   ReturnCode = SetAttr(LayoutStyle, at_OFFSET,
                        (POINTER_type) &at_Offset,
                        (PARM_MASK_type)
                        LEADING_OFFSET_parm |
                           RIGHT_HAND_OFFSET_parm);
```

```
    /* The previous code could have been collapsed into:
    ReturnCode = SetLayoutAttr(
                    at_DefaultValueList.basic_logical_attributes,
                    at_OFFSET,
                    (POINTER_type) &at_Offset,
                    (PARM_MASK_type)
                    LEADING_OFFSET_parm |
                            RIGHT_HAND_OFFSET_parm);
    */

/*                                                              (A.62)
*/
    ReturnCode = SetAttr( GenericComposite,
                          at_DEFAULT_VALUE_LISTS,
                          (POINTER_type) &at_DefaultValueList,
                          IGNORE_MASK );
/*                                                              (A.63)
*/
    od = open( "./OutEx2.odif", O_WRONLY|O_CREAT, 0600 );
            /* AOS UNIX OPEN */
    ReturnCode = WriteODIF( (INT_type) od, Document,
                            dp_INTERCHANGE_FMT_CLASS_a );
    /* next statement closes the file */
    ReturnCode = LOCAL_WriteBytes((INT_type) od,
                                  (BYTE_type *) 0,
                                  (INT_type) -1);

    /* write a human readable version of the document */
    od = open( "./OutEx2.dump", O_WRONLY|O_CREAT, 0600 );
            /* AOS UNIX OPEN */
    ReturnCode = WriteDocumentState( (INT_type) od, Document );

    /* next statement closes the file */
    ReturnCode = LOCAL_WriteChars((INT_type) od,
                                  (CHAR_type *) 0,
                                  (INT_type) -1);

    /* finalize the tool kit */
    ReturnCode = FinalToolKit();
    ReturnCode = LOCAL_Final();

}
```

A.1.2.2. Document Reader: Example 2

This sample program reads and processes the document built by the previous program. The program traverses the document, then processes it to obtain the value of the "offset" attribute.

First, the program initializes the tool kit. Then it reads in the ODIF document. To traverse the document, the program uses an iterator tool kit object. Next, the traversal of the specific logical structure begins. In visiting a constituent in the specific logical structure, the object type of the constituent is obtained first, since it is only for basic logical objects that any processing is done. For basic logical objects, the value of the "offset" attribute is obtained and printed. Note that when an unrecoverable error occurs, the program prints an error message on the Unix standard error output and then aborts by calling the Unix exit system routine.

Beginning at (A.64), the program initializes the tool kit. To do this, an INIT_PARMS_type structure is loaded to indicate that the tool kit should format error messages and print them on the Unix standard error output. No filtering of output characters is to be done on output. When writing a document in ODIF, the tool kit should automatically label the document with ASN.1 object identifiers. The actual initialization involves making two tool kit calls. The first is to LOCAL_Init and the second is to InitToolKit. Note that the call to LOCAL_Init must be made first.

Beginning at line (A.65), the ODIF document is read. First, the Unix file is opened for reading. Next, the ReadODIF routine is called to read the document. Note that the Unix file descriptor must be cast to INT_type when calling the tool kit. When ReadODIF completes successfully, it returns a DOCUMENT_type object that was just read. Finally, the file is closed by calling LOCAL_ReadBytes with a length argument of -1.

Next the program proceeds to create the iterator that will be used to traverse the document. To do this, it must first locate the root of the tree. As the traversal will be of the specific logical structure, the root of the specific logical structure is obtained by a call to the FindDocumentRoot. Now the program can create the iterator for the traversal by calling the MakeSubgraphIterator procedure. This call, shown at line (A.66), indicates that the traversal order is to be parents before children and depth first. This will return an ITERATOR_type object that will be used in subsequent calls.

With the iterator in hand, the routine WalkComponent is called to carry out the traversal. Before discussing the details of the traversal, note that the program ends by doing two things. First, it deallocates the iterator by a call to DeleteIterator. Then, it finalizes the tool kit by calling FinalToolKit and LOCAL_Final. Note that the call to FinalToolKit must be made first.

The WalkComponent routine is responsible for performing an iterative traversal of the document. Using this iterator, all the components in the specific logical structure will be visited in the order specified to the MakeSubgraphIterator call—parents first, depth first in this case. Thus, there is a C for loop that controls the visiting of components. The NextConstituent routine is used to obtain the next constituent to visit. Iteration continues until NextConstituent returns either ERROR_CONSTITUENT (some error has occurred) or NULL_CONSTITUENT (the traversal has completed normally).

Each component is processed as follows. The value of the "object type" attribute is obtained at line (A.67), and is used to determine how to process the component. The switch statement at line (A.68) controls this processing. Note that no processing is done if the component is the document logical root or a composite logical object; components from the layout structure are also ignored. Illegal values of object type are treated as non-recoverable errors. Only basic logical objects are processed, by calling the `WalkBasic2` routine.

The `WalkBasic2` routine is responsible for recovering the value of the "offset" attribute associated with the constituent passed as its argument. First, the `GetAttr` routine is called. Note that the address of a pointer to the `at_OFFSET_type` structure is passed to `GetAttr`. Note also that the tool kit is being instructed to perform ODA defaulting, and to return parameter information. Subsequently, the value of offset is printed out. The content is processed by calling the `WalkBasic` routine discussed in section A.1.1.2. Lastly, the offset structure allocated by the call to `GetAttr` is deallocated by a call to `LOCAL_Free` at line (A.69). Given that the tool kit allocated this space, the tool kit must deallocate it.

There are three interesting things to note about this example. First, the power of `GetAttr` is demonstrated. In contrast to the previous program, which performed a lot of work to build up the document, this one requires surprisingly little code. Most of the messy work, specifically the defaulting, is done by the tool kit. Second, the way the value of the attribute is returned from `GetAttr` when the value is a structure (as in the case of the "offset" attribute) deserves mention. The semantics of `GetAttr` state that a copy of the value is returned to the application. The tool kit will create a copy of the value of the "offset" attribute and will return its address—hence, the use of a pointer to a pointer for the `value` argument to `GetAttr`. The call to `GetAttr` used to obtain the value of the "offset" attribute uses defaulting. Therefore, the value of the "offset" attribute located on the default value list associated with the composite logical object class will be found, and will specify values for the leading and right hand offsets. The value returned for the values of the trailing and left hand offsets will be the ODA default values. The third interesting feature to note is the iterator used to traverse the document. A simple iterative program structure results when using the iterator. In contrast, the previous reader example, discussed in section A.1.1.2, used a recursive structure having a loop that could visit the subordinates of a given component.

```
/*
 * Example to illustrate reading and traversing the document
 *  of A.1.2.1.
 *
 * Initialize the tool kit then read in the document.
 *
 * Locate the document specific logical root, make an
 *  iterator and then start the recursive traversal.
 */

#include <odatk.h>

/* for AOS UNIX */
#include <sys/file.h>
#include <stdio.h>
```

```
/* for attributes without parameters */
#define IGNORE_MASK (PARM_MASK_type *) 0

/* for defaulting arguement ot GetAttr */
#define DEFAULT BOOL_true
#define NO_DEFAULT BOOL_false

/* Forward declarations for procedures */
void WalkComponent();
void WalkBasic2();

/* Reuse subroutine from Example 1 */
void WalkBasic();

main()
{

    CONSTITUENT SpecificRoot;

    INT_type ReturnCode;         /* for Tool Kit returns */

    DOCUMENT_type Document;      /* the document */

    INIT_PARMS_type InitParms;   /* to initialize tool kit */

    int id;                      /* input file descriptor */

    ITERATOR_type Iterator;
/*                                                          (A.64)
*/
    /* set up init parms before initializing tool kit */
    InitParms.automatic_label = BOOL_true;
    /* AOS UNIX only - use of stderr for error output */
    InitParms.error_od = (INT_type) fileno( stderr );
    InitParms.format_errors = BOOL_true;
    InitParms.print_warnings = BOOL_true;
    InitParms.output_filter = NULL_FILTER_ROUTINE;
    ReturnCode = LOCAL_Init();
    ReturnCode = InitToolKit( &InitParms );
/*                                                          (A.65)
*/
    /* read in the ODIF form of the document */
    id = open( "./OutEx2.odif", O_RDONLY );
    Document = ReadODIF( (INT_type) id );
    if( Document == ERROR_DOCUMENT ){
        fprintf( stderr, "Can not Read the ODIF\n" );
        exit ( 01 );
    }

    /* Tell TK done with descriptor and close file */
    ReturnCode = LOCAL_ReadBytes((INT_type) id,
                                 (BYTE_type *) 0,
                                 (INT_type) -1);

    /* locate the document logical root */
    SpecificRoot = FindDocumentRoot( Document,
                        SPECIFIC_DOC_LOGICAL_ROOT );
    if( SpecificRoot == ERROR_CONSTITUENT ){
        fprintf(stderr, "Can't find specific logical root\n");
        exit( 01 );
    }
/*                                                          (A.66)
*/
    /* make iterator for Parents-first, depth-first traverse */
    Iterator = MakeSubgraphIterator( SpecificRoot,
                                     PARENTS_FIRST,
                                     DEPTH_FIRST );
```

```
    /* start the recursive walk of the document */
    WalkComponent( Iterator );

    ReturnCode = DeleteIterator( Iterator );

    /* finalize */
    ReturnCode = FinalToolKit();
    ReturnCode = LOCAL_Final();

}

void WalkComponent( Iterator )
ITERATOR_type Iterator;
{

    /*
     * Use the iterator to traverse the specific structure.
     *
     * During the visit of a constituent, get its object type
     *  and process accordingly.
     */

    CONSTITUENT Constituent;

    at_OBJECT_TYPE_type ObjectType;

    INT_type ReturnCode;

    for( Constituent = NextConstituent( Iterator );
         Constituent != NULL_CONSTITUENT &&
            Constituent != ERROR_CONSTITUENT;
         Constituent = NextConstituent( Iterator ) ){
/*                                                          (A.67)
*/
        ReturnCode = GetAttr( Constituent, at_OBJECT_TYPE,
                              (POINTER_type) &ObjectType,
                              NO_DEFAULT, IGNORE_MASK );
        if( ReturnCode == ERROR_INT ){
            fprintf(stderr,
                    "Can't get object type in WalkComponent\n");
            exit( 01 );
        }
/*                                                          (A.68)
*/
        switch( ObjectType ){
        case at_OBJECT_TYPE_doc_logical_root:
        case at_OBJECT_TYPE_comp_logical_obj:
            break;
        case at_OBJECT_TYPE_bas_logical_obj:
            WalkBasic2( Constituent );
            break;
        case at_OBJECT_TYPE_doc_layout_root:
        case at_OBJECT_TYPE_page_set:
        case at_OBJECT_TYPE_page:
        case at_OBJECT_TYPE_frame:
        case at_OBJECT_TYPE_block:
            break;
        default:
            fprintf( stderr,
                     "Bad object type in WalkComponent\n" );
            exit( 01 );
        }

    }

}
```

```
void WalkBasic2( Constituent )
    CONSTITUENT Constituent;
{

    /*
     * A specific basic logical component has been
     *  encountered.
     * Obtain its offset.
     * Print out any content associated with it.
     */

    at_OFFSET_type *at_Offset;
    PARM_MASK_type OffsetParmMask;

    INT_type ReturnCode;

    ReturnCode = GetAttr( Constituent, at_OFFSET,
                    (POINTER_type) &at_Offset,
                    DEFAULT, &OffsetParmMask );
    if( ReturnCode == ERROR_INT ){
        fprintf( stderr, "Can't obtain offset of a basic\n" );
        exit( 01 );
    }

    printf("For a basic component the offset is:\n" );
    printf("\tLeading Offset %d, Trailing Offset %d\n",
           at_Offset->leading_offset,
           at_Offset->trailing_offset );
    printf("\tLeft Hand Offset %d, Right Hand Offset %d\n",
           at_Offset->left_hand_offset,
           at_Offset->right_hand_offset );
/*                                                          (A.69)
*/
    ReturnCode = LOCAL_Free( (POINTER_type) at_Offset );

    WalkBasic( Constituent );
        /* from example 1 */

}
```

A.1.3. Conclusions

In the first half of the appendix, we have presented four sample programs—two pairs, each consisting of a document creator and a document reader. Although the documents created were relatively simple, the examples illustrate much of the tool kit's functionality. They show how several of the important tool kit objects, `DOCUMENT_type`, `CONSTITUENT`, `SEQUENCE_type` and `ITERATOR_type`, are created and manipulated. Also illustrated are the `SetAttr` and `GetAttr` routines. The second example illustrates the power of the defaulting mechanism implemented by `GetAttr`. The support for the `ITERATOR_type` is displayed in the second reader example. Contrasting the two reader examples, the first not using an iterator and the second one using an iterator, illustrates the power of the tool kit's support for iteration. Dealing with reading and writing the binary ODIF interchange format is extremely messy. This is completely hidden from the application by the `WriteODIF` and

ReadODIF routines. Finally, the two document creator examples illustrate the support for debugging that creates a human-readable representation of the document.

A.2. Raster Graphics Tool Kit Example Programs

This section contains three example programs that show typical uses of the Raster Graphics Tool Kit. The first program demonstrates how to read and write rasters from disk files; in effect, it is a simple conversion program. The second program reads a raster image, performs some simple cut, paste and editing functions on it, and writes the modified raster to a new file. The final program demonstrates how to use the Raster Graphics Tool Kit with the CMU ODA Tool Kit.

A.2.1. Raster Input and Output with Format Conversion

This program, which begins on page 614, reads a bitmap image stored in Group 3 Fax format from disk and converts it into a PostScript file.

Before any Raster Graphics Tool Kit functions can be called, the tool kit must be initialized. This begins at line (A.70) and ends at line (A.71). First, the site-dependent support must be initialized by a call to `LOCAL_Init`. Then certain parameters must be specified for the `InitRGToolKit` function. In the program, the structure `RGInit` is used to hold these parameters. Currently, the only two parameters specified are `default_format` and `memsize`. Since the program will be writing rasters in PostScript, we specify it as the default bitmap format. The default format is used when writing rasters that have no output format explicitly declared. The `memsize` parameter is set to -1 to indicate that all available memory will be allocated for holding rasters. Once the call to `InitRGToolKit` has been made, the tool kit is ready to be used.

Next, the program prepares to read the input raster, beginning at line (A.72). The input file is a raster image stored in Group 3 format. To specify this, the `format_tag` field of a `FORMAT_type` structure must be set to `FORMAT_GROUP3`. Then, the specific Group 3 parameters may be set. In this case, the program is expecting a one-dimensionally encoded bitmap with the most significant bit in a byte representing the rightmost pel in an octet. We also specify the padding width for reads, although it is not required.

After the format structure has been set up, the `ReadRaster` function reads the image in from the specified input descriptor. The code for this starts on line

(A.73). Should an error occur due to a damaged or incorrectly formatted file, `ReadRaster` will return `RG_ERROR_RASTER`.

Once an image has been successfully loaded, it may be manipulated by the tool kit routines. In this example, however, the image will simply be written to a different file in PostScript format. The code for this manipulation begins at line (A.74). The format parameters for PostScript must be specified. The scaling factor is set to 1.0 so that no scaling will take place. The `name` parameter is not used in the current version of the tool kit. However, to ensure compatibility with future versions of the Raster Graphics Tool Kit, one should set this field to a string naming the image.

As with reading rasters, the `WriteRaster` function may be called once the required formatting parameters have been specified. Should an error occur in writing the image, `WriteRaster` will return `RG_ERROR_INT`. Note that the error handler in the sample program does not merely exit should an error occur, but it goes through the cleanup step of calling `FinalRGToolKit` to make sure everything is shut down properly.

Once a file has been written using `WriteRaster` (at line (A.75)), it must be closed. This is accomplished by calling `LOCAL_WriteBytes` with a `-1` for the length to properly flush the I/O buffer and close the file.

Finally, once all tool kit routines are complete, `FinalRGToolKit` and `LOCAL_Final` should be called. The code for this is at line (A.76).

```
/*\
 *  File:    RasEx1.c
 *
 *  This program reads a raster image stored in Group 3 Fax
 *  format from a file and converts it into a PostScript file.
 *  The program uses the Raster Graphics Toolkit and is
 *  intended to be used as an example of the tool kit's use.
\*/

#include <rgtk.h>
#include <stdio.h>
#include <sys/file.h>

extern INT_type      RGError;      /* Toolkit errors here */

main(argc, argv)
    int      argc;
    char     **argv;
{
    INIT_RGPARMS_type   RGInit;  /* RGTK init parameters */
    RASTER_type         *Raster; /* The raster */
    FORMAT_type         Format;  /* Specifies raster format */
    int                 rval;
    int                 id,od;   /* UNIX file descriptors */
    int                 error;

    /* Check command line arguments */

    if (argc < 3) {
        fprintf(stderr,"Usage:  gr3tops <infile> <outfile>");
        goto LEAVE;
    }
```

```
    /* Initialize the RG Toolkit, specifying Group 3 as the
     * default raster format and use of all available primary
     * memory.
     */
/*                                                      (A.70)
*/
    RGInit.default_format = FORMAT_POSTSCRIPT;
    RGInit.memsize = -1;

    if (LCL_ERROR_INT == LOCAL_Init()){
        fprintf(stderr,"Can't Initialize Local support\n");
        rval = -1;
        goto LEAVE;
    }

    if (InitRGToolKit(&RGInit) == RG_ERROR_INT) {
        fprintf(stderr,"Can't Initialize Tool Kit\n");
        rval = -1;
        goto LEAVE;
    }
/*                                                      (A.71)
*/
    /* Specify the Group 3 parameters for the input file */
/*                                                      (A.72)
*/
    Format.format_tag                  = FORMAT_GROUP3;
    Format.format_body.g3.msb          = RIGHTBIT;
    Format.format_body.g3.k            = 1;
    Format.format_body.g3.pad_width    = 0;    /* opt on reads */

    if ((id = open(argv[1], O_RDONLY)) < 0) {
        fprintf(stderr,"Could not open input file.\n\n");
        rval = -2;
        goto CLEANUP;
    }
/*                                                      (A.73)
*/
    /* Perform the read */
    Raster = ReadRaster((INT_type)id, &Format);

    if (Raster == RG_ERROR_RASTER) {
        fprintf(stderr, "Input file damaged or incorrect\n");
        close(id);
        rval = -3;
        goto CLEANUP;
    }
    LOCAL_ReadBytes((INT_type) id,
                    (BYTE_type *) 0,
                    (INT_type) -1);

    od = open(argv[2], O_WRONLY|O_CREAT|O_TRUNC, 0666);
    if (od < 0) {
        fprintf(stderr,"Could not open output file.\n\n");
        rval = -4;
        goto CLEANUP;
    }

    /* Specify output parms for PostScript with 1.0x scaling */
/*                                                      (A.74)
*/
    Format.format_tag             = FORMAT_POSTSCRIPT;
    Format.format_body.ps.scale = 1.0;
    Format.format_body.ps.name  = "PS_Output";
```

```
    /* Write raster from memory to file in PostScript */
/*                                                        (A.75)
*/
    error = WriteRaster((INT_type) od, Raster, &Format);
    if (error == RG_ERROR_INT) {
        fprintf(stderr,
                "Error %d in writing raster to file.\n",
                RGError);
        rval = -5;
    }
    /* Needed to flush I/O and close file */
    LOCAL_WriteBytes( (INT_type) od,
                      (BYTE_type *) 0,
                      (INT_type) -1);

    rval = 0;

CLEANUP:
/*                                                        (A.76)
*/
    FinalRGToolKit();    /* Make sure everything's tidy */
    LOCAL_Final();

LEAVE:

    exit(rval);
}

/* End of file gr3tops.c */
```

A.3. Manipulating Rasters

This program reads a raster image stored in X Bitmap format from a file and performs some straightforward editing operations on the raster. It extracts a portion of the original raster, inverts the bits in the new image, inserts the picture into another raster and draws a border around the picture.

As in the previous example, the first thing to do is initialize the site-specific support and the Raster Graphics Tool Kit. The code for this begins at line (A.77) and ends at line (A.78). Since the program will read and write X Bitmaps, `FORMAT_XBM` is specified as the default raster format. Next, the image is read from the specified file, at line (A.79), using `ReadRaster`.

To extract a portion of a raster, one needs to know the size of the region; this program uses some simple math to determine a rectangular region in the center of the original raster. The `ExtractRaster` function then returns a new raster, which was copied from the specified part of the original raster. This code begins at line (A.80) and ends at line (A.81).

Each bit in the new image is individually toggled by reading the value of the bit (`GetBit`), toggling it and re-writing it in place (`AssignBit`). This is done by the loop at line (A.82).

A new raster is created by calling `MakeRaster`. This new raster is slightly larger than the edited raster so that it can hold a border around the image. The `InsertRaster` function is used to paste the extracted, inverted raster into the new one. This is done at line (A.83).

To draw the border, `AssignBit` is used on the bits surrounding the inserted image. The loop at line (A.84) controls this.

Finally, the new raster is written to the specified file and the program cleans up the tool kit. This is done between lines (A.85) and (A.86).

```
/*\
 *  File:   RasEx2.c
 *
 *  This program reads a raster image stored in X Bitmap format
 *  from a file, extracts a rectangular region from the center
 *  of the bitmap, inverts the bits in this region, and then
 *  pastes the region back in place into the original raster
 *  and writes out the results.  Not very practical, but it
 *  shows how you can use the RGTK to manipulate bitmaps.
\*/

#include <rgtk.h>
#include <stdio.h>
#include <sys/file.h>

extern INT_type     RGError;        /* Toolkit errors here */

main(argc, argv)
    int     argc;
    char    **argv;
{
    INIT_RGPARMS_type   Initialize;
    RASTER_type         *Raster, *Region, *NewRaster;
    FORMAT_type         Format;
    VALUE_type          Bit;
    int                 rval;
    int                 fd;
    int                 Width, Height, Left, Top, row, col;

    /* Check command line arguments */

    if (argc < 3) {
        fprintf(stderr, "Usage: flipbit <infile> <outfile>\n");
        goto LEAVE;
    }

    /* Initialize the RG Toolkit, specifying X Bitmap as the
     * default raster format and use of all available primary
     * memory.
     */
/*                                                          (A.77)
*/
    Initialize.default_format = FORMAT_XBM;
    Initialize.memsize = -1;

    if (LCL_ERROR_INT == LOCAL_Init()){
        fprintf(stderr,"Can't Initialize Local support\n");
        rval = -1;
        goto LEAVE;
    }
```

```
    if (InitRGToolKit(&Initialize) == RG_ERROR_INT) {
        fprintf(stderr,"Can't Initialize Tool Kit\n");
        rval = -1;
        goto LEAVE;
    }
/*                                                          (A.78)
*/
    /* Specify that the image is an X Bitmap */

    Format.format_tag                  = FORMAT_XBM;

    if ((fd = open(argv[1], O_RDONLY)) < 0) {
        fprintf(stderr,"Could not open input file.\n\n");
        rval = -2;
        goto CLEANUP;
    }
/*                                                          (A.79)
*/
    printf("Reading file...\n");
    Raster = ReadRaster((INT_type)fd, &Format);

    if (Raster == RG_ERROR_RASTER) {
        fprintf(stderr, "Input file damaged or incorrect\n");
        close(fd);
        rval = -3;
        goto CLEANUP;
    }
    LOCAL_ReadBytes((INT_type) fd,
                    (BYTE_type *) 0,
                    (INT_type) -1);

/*                                                          (A.80)
*/
    Width = RasterWidth(Raster);     /* Get the size of the */
    Height = RasterHeight(Raster);   /* raster and calculate */
    Left = Width / 4;                /* the left/right and */
    Top = Height / 4;                /* top/bottom margins */
    Width = Width / 2;               /* & the new width */
    Height = Height / 2;             /* height. */

    /* Grab a region from the middle of the original raster */
    Region = ExtractRaster(Raster,
                           Left, Top, Width + Left,
                           Height + Top);
/*                                                          (A.81)
*/
/*                                                          (A.82)
*/
    /* Toggle all the bits in the new raster */
    for (row = 0; row < Height; row++) {
        for (col = 0; col < Width; col++) {
            Bit = GetBit(Region, col, row);
            Bit = 1 - Bit;
            AssignBit(Region, col, row, Bit);
        }
    }

    /* Create a new raster with the inverted region from
     * original raster located in the middle.
     */
/*                                                          (A.83)
*/
    NewRaster = MakeRaster(Width + 20, Height + 20, PEL_BITS);
    InsertRaster(Region, NewRaster, 10, 10);
```

```
    /* Now use Height and Width to refer to the border size */
    Height += 15; Width += 15;
/*                                                              (A.84)
*/
    /* Draw the two vertical border lines */
    for (row = 5; row < Height; row++) {
        AssignBit(NewRaster, 5, row, BIT_BLACK);
        AssignBit(NewRaster, Width, row, BIT_BLACK);
    }

    /* Draw the two horizontal border lines */
    for (col = 5; col < Width; col++) {
        AssignBit(NewRaster, col, 5, BIT_BLACK);
        AssignBit(NewRaster, col, Height, BIT_BLACK);
    }
/*                                                              (A.85)
*/
    fd = open(argv[2], O_WRONLY|O_CREAT|O_TRUNC, 0666);
    if (fd < 0) {
        fprintf(stderr,"Could not open output file.\n\n");
        rval = -4;
        goto CLEANUP;
    }

    /* Write manipulated raster to file in XBM Format */

    if (WriteRaster(fd, NewRaster, &Format) == RG_ERROR_INT) {
        fprintf(stderr,
                "Error %d in writing raster to file.\n",
                RGError);
        rval = -5;
        close(fd);
        goto CLEANUP;
    }
    /* Needed to flush I/O and close file */
    LOCAL_WriteBytes((INT_type) fd,
                     (BYTE_type *) 0,
                     (INT_type) -1);

    rval = 0;

CLEANUP:

    FinalRGToolKit();           /* Make sure everything's tidy */
    LOCAL_Final();

/*                                                              (A.86)
*/
LEAVE:

    exit(rval);
}

/* End of file gr3tops.c */
```

A.4. Example Program 3

This program, shown starting on page 621, is an example of how to use both the CMU ODA Tool Kit and the Raster Graphics Tool Kit. It walks through an

ODIF document, searching for raster content portions. Each time it finds one, it will write that raster to disk in PostScript format.

To use both tool kits, you must include both header files ***odatk.h*** and ***rgtk.h*** as shown at line (A.87). The local I/O module and the two tool kits must be separately initialized by the appropriate function calls (`LOCAL_Init` first and then `InitToolKit` and `InitRGToolKit` in either order). This will allow all functions from either tool kit to be used. The initializations are done between lines (A.88) and (A.89).

The traversal of an ODIF document is fully described in section A.1.1.2, and the main line code beginning at (A.90) is essentially the same as this earlier example. This example is structured similarly; the only exception is that it handles raster graphics content as well as character content. The program first reads in an ODIF stream from the file specified on the command line using the `ReadODIF` function. Assuming there are no errors, the program then locates the specific logical root of the document and begins to walk recursively through the components.

Once in the function `WalkComponent`, the type of object is determined via an appropriate `GetAttr` call. If the object is a basic logical object, then `WalkBasic` traverses the object, and the recursion continues.

`WalkBasic` takes the basic logical object and determines the type of content it is dealing with (by calling `GetContentKind`) at line (A.92). It gets the content portion containing the data by a call to `GetContent` at line (A.93). Then, the program branches off into a subroutine for handling the appropriate content kind. For character content, the program simply prints out the character string.

The function `RasterGraphicsContent` gets the content portion from the basic logical component using the `GetAttr` procedure as shown at line (A.94). At this point, the program needs to know how the image data is stored. The subroutine `GetContentRasterFormat` will automatically decide what format is being used and set up a `FORMAT_type` data structure suitable for `ReadRaster`. This is called at line (A.95). Note that this subroutine is not part of either tool kit. Its function needs to be provided by an application developer. As the routine is written, it assumes a complete implementation of the `ReadODIF`. However, as noted in section A.1.1.1, the current release of the CMU ODA Tool Kit requires that "type of coding" be present on a content portion. Therefore, the various defaulting rules that the code implements will never be used. Further, as documented in sections 20.1.1 and 20.1.2, there is no support for reading or writing the "document application profile defaults" attribute. Thus, even if `ReadODIF` could read content portions that lack the "type of coding" attribute, the value of the "document application profile defaults" would not be available.

GetContentRasterFormat first needs to know what format is used in the internal representation of the image. It determines this by checking to see if the content portion has a "type of coding" attribute. If it does, it obtains the value using the GetAttr procedure. If not, the function checks to see if there are any default settings for the coding type. If there are, it uses the default. If there is no default explicitly specified in the content portion, Group 4 (type T6) is automatically used. Next, the function tries to determine how many pels are on a scan line. Again, if no number is explicitly specified, defaulting rules are used. Finally, GetContentRasterFormat loads up the FORMAT_type structure with the appropriate parameters for use by the Raster Graphics Tool Kit.

Once all the formatting information has been decoded, the raster may be read from the coded form into a RASTER_type object by a call to ReadMemRaster at line (A.96). Then the image is written to a disk file in PostScript format by using WriteRaster as shown between lines (A.97) and (A.98). The program will give each raster in a document a different name, starting at RasterA.ps.

Once the document has been traversed, all three of the main modules must be cleaned up by calling FinalRGToolKit, FinalToolKit and LOCAL_Final—the inverse order the packages were initialized in. The cleanup starts at line (A.91).

```
/*
 * File:  RasEx3.c
 *
 * This program is an example of how to use both the ODATK and
 * the RGTK at the same time.  It walks through an ODIF
 * document, searching for raster content portions.  Each time
 * it finds one, it will write that raster out to a disk file
 * in PostScript format.
 */

#include <odatk.h>
#include <rgtk.h>
/*                                                          (A.87)
*/
/* for AOS UNIX */
#include <sys/file.h>
#include <stdio.h>

/* forward declarations for procedures */
void WalkComponent();
void WalkBasic();
void CharacterContent();
void RasterGraphicsContent();
void GetContentRasterFormat();

DOCUMENT_type GlobalDocument;

/* values for the defaulting parameter to GetAttr */
#define     DEFAULT     BOOL_true
#define     NO_DEFAULT  BOOL_false

/* ignore parameter masks for attributes without them */
#define     IGNORE_MASK (PARM_MASK_type *) 0
```

```
main(argc, argv)
    int     argc;
    char    **argv;
{
    INT_type            ReturnCode;   /* Tool Kit returns */
    DOCUMENT_type       Document;     /* Doc to traverse */
    CONSTITUENT         SpecificRoot; /* Root specific log */
    INIT_PARMS_type     InitParms;    /* Init the ODATK */
    INIT_RGPARMS_type   InitRGParms;  /* Init the RGTK */

    int id;                           /* UNIX file desc */

    /* Set up the ODA Tool Kit */
/*                                                        (A.88)
*/
    InitParms.automatic_label = BOOL_true;
    InitParms.error_od = (INT_type) fileno( stderr );
    InitParms.format_errors = BOOL_true;
    InitParms.print_warnings = BOOL_true;
    InitParms.output_filter = NULL_FILTER_ROUTINE;
    ReturnCode = LOCAL_Init();
    ReturnCode = InitToolKit( &InitParms );

    /* Set up the RG Tool Kit */

    InitRGParms.default_format = FORMAT_GROUP4;
    InitRGParms.memsize = -1;
    ReturnCode = InitRGToolKit(&InitRGParms);
/*                                                        (A.89)
*/
/*                                                        (A.90)
*/
    /* Read in the ODIF form of the document */
    id = open(argv[1], O_RDONLY );
    Document = (DOCUMENT_type) ReadODIF( (INT_type) id );
    if (Document == ERROR_DOCUMENT){
        fprintf(stderr, "Can not Read the ODIF\n");
        exit (01);
    }
    LOCAL_ReadBytes((INT_type) id,
                    (BYTE_type *) 0,
                    (INT_type) -1);

    GlobalDocument = Document;      /* We may need this */

    /* Locate the document logical root */
    SpecificRoot = FindDocumentRoot(Document,
                        SPECIFIC_DOC_LOGICAL_ROOT);
    if( SpecificRoot == ERROR_CONSTITUENT ){
        fprintf(stderr, "Can't find specific logical root\n");
        exit( 01 );
    }

    printf("\n\n******** Walking over document ********\n\n");

    /* Start the recursive walk of the document */
    WalkComponent( SpecificRoot );
/*                                                        (A.91)
*/
    /* Finalize the tool kits and local support */
    ReturnCode = FinalRGToolKit();
    ReturnCode = FinalToolKit();
    ReturnCode = LOCAL_Final();
}
```

```
void WalkComponent(component)
    CONSTITUENT component;
{
    /*
     * Find the type of the component.
     *
     * If document root or composite logical then find the
     *  "subordinates" attribute and iterate over the
     *  subordinates making recursive calls.
     *
     * If a basic logical component call the basic component
     *  walker.
     *
     * Check to ensure that the component is either the
     *  document logical root or a composite logical object.
     *
     * Get the value of the "subordinates" attribute.
     *
     * Loop through the subordinates making recursive calls to
     *  this routine.
     */

    INT_type                    ReturnCode;
    VALUE_type                  ComponentType;
    SEQUENCE_CONSTITUENT_type   Subordinates;
    int                         i;

    /* Get the at_OBJECT_TYPE */
    ReturnCode = GetAttr( component, at_OBJECT_TYPE,
                          (POINTER_type) &ComponentType,
                          NO_DEFAULT, IGNORE_MASK );
    if( ReturnCode == ERROR_INT ){
        fprintf(stderr,
            "Can't get component type in WalkComponent\n");
        exit( 01 );
    }

    /* If basic component then deal with walking content */
    if( ComponentType == at_OBJECT_TYPE_bas_logical_obj ){
        WalkBasic( component );
        return;
    }

    /*
     * Just to be safe make sure that the component is doc log
     * root or a composite logical.
     */

    if( ComponentType != at_OBJECT_TYPE_doc_logical_root &&
        ComponentType != at_OBJECT_TYPE_comp_logical_obj ){
        fprintf(stderr, "Layout component in WalkComponent\n");
        exit( 01 );
    }
```

```
    /* Find the subordinates */
    ReturnCode = HasAttr( component, at_SUBORDINATES );
    if (ReturnCode == ERROR_INT) {
      fprintf(stderr,
            "Error HasAttr: subordinates in WalkComponent\n");
        exit( 01 );
    }
    if( ReturnCode == NO_ATTR){
        /* stop the recursion */
        return;
    }
    ReturnCode = GetAttr( component, at_SUBORDINATES,
                           (POINTER_type) &Subordinates,
                           NO_DEFAULT, IGNORE_MASK );
    if( ReturnCode == ERROR_INT ){
        fprintf( stderr,
                 "Can't get subordinates in WalkComponent\n" );
        exit( 01 );
    }

    for( i = 0; i < Subordinates->length; i++){
        WalkComponent(
            Subordinates->sequence_value.constituents[i]
        );
    }

    ReturnCode = DeleteSequence( Subordinates );
}
void WalkBasic( component )
    CONSTITUENT component;
{
    /*
     * ASSERT: the component must be a basic logical
     *  object.
     *
     * First get the content type.
     *
     * Then get the content.
     *
     * Call the appropriate content specific routine.
     */

    VALUE_type ContentKind;

    SEQUENCE_BYTE_type Content;

    INT_type ReturnCode;
/*                                                          (A.92)
*/
    ContentKind = GetContentKind( component );
    if( ContentKind == ERROR_kind ){
        fprintf(stderr,
                "Can't get content kind in WalkBasic\n");
        exit( 01 );
    }
/*                                                          (A.93)
*/
    Content = GetContent( component );
    if( Content == ERROR_SEQUENCE ){
        fprintf( stderr, "Can't get content in WalkBasic\n" );
        exit( 01 );
    }
```

```
    /* Switch on ContentKind to process content */
    /*
     * Note that there is a spot for each kind of content
     *  even though we only handle character content.
     */
    switch( ContentKind ){
    case FORM_CC_kind:
    case FP_CC_kind:
    case PROC_CC_kind:
        CharacterContent( Content );
        break;
    case FORM_RG_kind:
    case FP_RG_kind:
        RasterGraphicsContent( component, Content );
        break;
    case FORM_GG_kind:
        printf("Can't process geometric graphic content\n");
        break;
    default:
        fprintf( stderr, "Bad ContentKind in WalkBasic\n" );
        exit( 01 );
    }

    ReturnCode = DeleteSequence( Content );
}

/*
 * RASTERGRAPHICSCONTENT:  Handle raster graphics portion.
 */

void RasterGraphicsContent( Component, ContentInformation )
    CONSTITUENT        Component;
    SEQUENCE_BYTE_type ContentInformation;
{
    SEQUENCE_CONSTITUENT_type       atContentPortions;
    FORMAT_type                     Format;
    RASTER_type                     *Raster;
    int                             od;
    CONSTITUENT                     Content;
    POINTER_type                    CurrentByte;

    static char                     FileName[] = "RasterA.ps";
#define UPDATE_NAME (6)

    /* Get the content portion of the current raster so that
     * all of the attributes can be determined.
     */
/*                                                     (A.94)
*/
    GetAttr(Component,
            at_CONTENT_PORTIONS,
            (POINTER_type) &atContentPortions,
            NO_DEFAULT, IGNORE_MASK );

    Content =
        atContentPortions->sequence_value.constituents[0];

    printf("\nRaster content found.\n");

    /* Load the Format structure with the necessary parameters
     * depending upon the content's attributes.
     */
/*                                                     (A.95)
*/
    GetContentRasterFormat(Component, Content, &Format);
    CurrentByte = ContentInformation->sequence_value.bytes;
```

```
    /* Read the raster from the content information into a
     * RASTER_type structure for further processing
     */
/*                                                          (A.96)
*/
    Raster = ReadMemRaster(&CurrentByte, &Format);
    if (Raster == RG_ERROR_RASTER) {
        printf("Error reading raster.\n");
    }

    printf("Raster read OK (%d x %d)\n",RasterWidth(Raster),
                                    RasterHeight(Raster));
/*                                                          (A.97)
*/
    /* Write the raster out to disk in PostScript format */

    if ((od = open(FileName, O_WRONLY | O_CREAT, 0666)) < 0) {
        printf("Error opening file %s\n",FileName);
    }
    Format.format_tag = FORMAT_POSTSCRIPT;
    Format.format_body.ps.scale = 1.0;

    WriteRaster((INT_type) od, Raster, &Format);
    LOCAL_WriteBytes((INT_type) od,
                     (BYTE_type *) 0,
                     (INT_type) -1);

    printf("Wrote raster content as file %s\n",FileName);
/*                                                          (A.98)
*/
    FileName[UPDATE_NAME]++;     /* New name for next raster */
}

void GetContentRasterFormat(Component, Content, Format)
    CONSTITUENT     Component;
    CONSTITUENT     Content;
    FORMAT_type     *Format;
{
    rg_PEL_TRANS_DENSITY_type       Density;
    at_TYPE_OF_CODING_type          TypeCoding;
    INT_type                        Coding;
    dp_DOC_APP_PROF_DEFAULTS_type   *dp;
    INT_type                        PelsPerLine;

    /* Check to see if content has Type of Coding attribute;
     * if it does, get it.  If not, get the document's default
     * settings.  Then see if there is a default for type of
     * coding; if so, get it.  Otherwise just use the default
     * coding type.
     *
     * Note: We could have used LookUpObjectId to decode the
     * ASN.1 identifier present in TypeCoding
     */
```

```
if (HAS_ATTR == HasAttr(Content, at_TYPE_OF_CODING)) {
    GetAttr(Content, at_TYPE_OF_CODING,
            (POINTER_type) &TypeCoding,
            BOOL_false, (PARM_MASK_type *) 0);
    Coding = TypeCoding->sequence_value.ints[4];
    DeleteSequence(TypeCoding);
}
else if (HAS_ATTR == HasAttr(GlobalDocument,
                             dp_DOC_APP_PROF_DEFAULTS)) {
    GetAttr(GlobalDocument, dp_DOC_APP_PROF_DEFAULTS,
            (POINTER_type) &dp,
            BOOL_false, (PARM_MASK_type *) 0);
    if (dp->rg_coding != NULL_CONSTITUENT &&
        HAS_ATTR == HasAttr(dp->rg_coding,
                            at_TYPE_OF_CODING)) {
        GetAttr(dp->rg_coding, at_TYPE_OF_CODING,
                (POINTER_type) &TypeCoding,
                BOOL_false, (PARM_MASK_type *) 0);
        Coding = TypeCoding->sequence_value.ints[4];
        DeleteSequence(TypeCoding);
    }
    LOCAL_Free(dp);
}
else {
    Coding = 0;
}

/* Check to see if the content has Number of Pels Per Line
 * attribute; if it does, get it.  If not, get the
 * Transmission Pel Density attribute (must be present
 * if Pels Per Line absent) and use the appropriate value.
 */

if (HAS_ATTR == HasAttr(Content,
                        rg_NUMBER_OF_PELS_PER_LINE)) {
    GetAttr(Content, rg_NUMBER_OF_PELS_PER_LINE,
            (POINTER_type) &PelsPerLine,
            BOOL_false, (PARM_MASK_type *) 0);
    printf("Got the # of pels per line\n");
}
else {
    GetAttr(Component, rg_PEL_TRANS_DENSITY,
            (POINTER_type) &Density,
            BOOL_true, (PARM_MASK_type *) 0);

    switch (Density) {
        case rg_PEL_TRANS_DENSITY_1: PelsPerLine = 10368;
                                     break;
        case rg_PEL_TRANS_DENSITY_2: PelsPerLine =  5184;
                                     break;
        case rg_PEL_TRANS_DENSITY_3: PelsPerLine =  3456;
                                     break;
        case rg_PEL_TRANS_DENSITY_4: PelsPerLine =  2592;
                                     break;
        case rg_PEL_TRANS_DENSITY_5: PelsPerLine =  2074;
                                     break;
        case rg_PEL_TRANS_DENSITY_6: PelsPerLine =  1728;
                                     break;
        default:  printf("Default density error.\n");
    }
}

printf("PelsPerLine: %d, Coding: %d\n",PelsPerLine,Coding);
```

```
    /* Load the Format structure's parameters according to
     * coding type.  The values used for coding are taken from
     * the ASN.1 object identifier, if present.  The Coding
     * variable is the 5th element in the sequence of integers
     * which makes up an ASN.1 object, and is the one which
     * determines the Type of Coding.
     */

    switch (Coding) {
        case 0:         /* Group 4, T6 Encoding */
            Format->format_tag = FORMAT_GROUP4;
            Format->format_body.g4.msb = LEFTBIT;
            Format->format_body.g4.width = PelsPerLine;
            break;
        case 1:         /* Group 3, T4 1-dim    */
        case 2:         /* Group 3, T4 2-dim    */
            Format->format_tag = FORMAT_GROUP3;
            Format->format_body.g3.msb = LEFTBIT;
            break;
        case 3:         /* ODA Bitmap Encoding  */
            Format->format_tag = FORMAT_ODABM;
            break;
    }
}

void CharacterContent( Content )
    SEQUENCE_BYTE_type Content;
{
    /*
     * Print out the content.
     *
     * This will only pretty print short content (i.e. fits on
     *  one line ).
     */

    INT_type i;

    printf( "\nContent for one of the basic components:\n" );

    for( i = (INT_type) 0; i < Content->length; i++){

        putchar( (char) Content->sequence_value.bytes[i] );

    }

    printf( "\n" );
}
```

A.4.1. Summary

In this section, we have presented three examples illustrating the use of the raster graphics tool kit. These examples show how to convert between different raster formats, how to manipulate rasters and how to use the Raster Graphics Tool Kit with the CMU ODA Tool Kit.

A.5. Cross-reference Tables

Attribute	Section
"content architecture class"	17.1.1.11
"content information"	17.1.7.4
"content portions"	17.1.1.8
"default value lists"	17.1.1.17
"generator for subordinates"	17.1.1.4
"layout style"	17.1.3.2
"object class"	17.1.1.6
"object type"	17.1.1.1
"subordinates"	17.1.1.7
"type of coding"	17.1.7.3

Table A-1. Attributes Used in Examples

Procedure name	Section
DeleteSequence	13.13.5
FinalToolKit	14.2
FindDocumentRoot	19.1.1
GetAttr	16.1.3
GetContent	16.1.13
GetContentKind	16.1.12
HasAttr	16.2
InitializeDefaultValueLists	13.12.10
InitToolKit	14.1
LOCAL_Final	10.2.3
LOCAL_Free	10.2.5.3
LOCAL_Init	10.2.2
LOCAL_Malloc	10.2.5.1
LOCAL_WriteBytes	10.2.4.2
LOCAL_WriteChars	10.2.4.4
MakeComponent	15.4
MakeContent	15.6
MakeDocument	15.1
MakeSequence	13.13.4
MakeStyle	15.5
MakeSubgraphIterator	19.2.1
NextConstituent	19.2.3
ReadODIF	20.1.1
SetAttr	16.1.2
WriteDocumentState	22.2.1
WriteODIF	20.1.2

Table A-2. Procedures Used in Examples

Type	Section
at_DEFAULT_VALUE_LISTS_type	17.1.1.17
at_GEN_FOR_SUBORDINATES_type	17.1.1.4
at_OFFSET_type	17.1.5.8
BOOL_type	10.1.4
BYTE_type	10.1.3
CHAR_type	10.1.2
CONSTITUENT	13.2
CONSTRUCTION_EXPR_type	18.4
DOCUMENT_type	13.3
INIT_PARMS_type	14.1.1
INT_type	10.1.1
ITERATOR_type	13.4
PARM_MASK_type	10.1.7
POINTER_type	10.1.8
SEQUENCE_BYTE_type	13.13.1
SEQUENCE_CONSTITUENT_type	13.13.1
SEQUENCE_CONSTRUCTION_TERM_type	13.13.1
SEQUENCE_type	13.13.1
VALUE_type	10.1.5

Table A-3. Types Used in Examples

Constant	Section
at_OBJECT_TYPE_bas_logical_obj	17.1.1.1
at_OBJECT_TYPE_block	17.1.1.1
at_OBJECT_TYPE_comp_logical_obj	17.1.1.1
at_OBJECT_TYPE_doc_layout_root	17.1.1.1
at_OBJECT_TYPE_doc_logical_root	17.1.1.1
at_OBJECT_TYPE_frame	17.1.1.1
at_OBJECT_TYPE_page	17.1.1.1
at_OBJECT_TYPE_page_set	17.1.1.1
at_CONTENT_ARCH_CLASS	17.1.1.11
at_CONTENT_ARCH_CLASS_proc_cc	17.1.1.11
at_CONTENT_INFORMATION	17.1.7.4
at_CONTENT_PORTIONS	17.1.1.8
at_DEFAULT_VALUE_LISTS	17.1.1.17
at_GEN_FOR_SUBORDINATES	17.1.1.4
at_LAYOUT_STYLE	17.1.3.2
at_OBJECT_CLASS	17.1.1.6
at_SUBORDINATES	17.1.1.7
at_TYPE_OF_CODING	17.1.7.3
at_TYPE_OF_CODING_cc	17.1.7.3
dp_INTERCHANGE_FMT_CLASS_a	17.1.8.14
BAS_LOGICAL_ATTRIBUTES_parm	13.12.10
BOOL_false	10.1.4
BOOL_true	10.1.4
CHARACTER_CONTENT	15.6
CHOICE_tag	18.4
DEPTH_FIRST	19.2.1
ERROR_kind	16.1.12
ERROR_CONSTITUENT	13.2
ERROR_DOCUMENT	13.3
ERROR_INT	10.1.1
ERROR_SEQUENCE	13.13.1

Table A-4. Constants Used in Examples, I

Constant	Section
EXPR_tag	18.4
FORM_CC_kind	16.1.12
FORM_GG_kind	16.1.12
FORM_RG_kind	16.1.12
FP_CC_kind	16.1.12
FP_RG_kind	16.1.12
GENERIC_COMPONENT	15.4
LAYOUT_STYLE	17.1.3.2
LEADING_OFFSET_parm	17.1.5.8
NO_ATTR	16.2
NULL_CONSTITUENT	13.2
PROC_CC_kind	16.1.12
OBJECT_CLASS_ID_tag	18.4
PARENTS_FIRST	19.2.1
REQ_tag	18.4
RIGHT_HAND_OFFSET_parm	17.1.5.8
SEQUENCE_BYTE_tag	13.13.1
SEQUENCE_CONSTITUENT_tag	13.13.1
SEQUENCE_CONSTRUCTION_TERM_tag	13.13.1
SPECIFIC_COMPONENT	15.4
SPECIFIC_DOC_LOGICAL_ROOT	19.1.1
TERM_tag	18.4

Table A-5. Constants Used in Examples, II

B
Building and Installing the Tool Kits

This chapter provides details on how to build and install the tool kit on several machine types and operating systems: BSD Unix, System V Unix, MS-DOS, Macintosh MPW and VAX/VMS.

B.1. Contents of the *oda* Directory

The CMU ODA Tool Kit contribution to the X.V11R4 Distribution is located at path ***contrib/andrew/oda***. The ***oda*** directory contains seven top-level subdirectories, as follows:

doc The only files of any concern in this directory are the file ***COPYRITE*** (which contains the full text of the copyright for most of the software in the ***oda*** directory) and this document.

site This is the root directory for all of the site-dependent sources for the tool kits. Under this directory is a directory for each configuration that is currently supported.

src This directory contains the source for the main part of the CMU ODA Tool Kit.

valid This directory contains files necessary for constructing and executing a program to validate the installation of the CMU ODA Tool Kit. This is not a validation of the correctness of the CMU ODA Tool Kit with respect to the ODA standard, but simply a means for checking that the local dependencies specified during installation (chapter 10) are correct.

raster This directory contains the source code for building the Raster Graphics Tool Kit library (described in part 4 of this book). This library contains routines for translating among a number of common raster formats. The raster library may be used independently of the CMU ODA Tool Kit.

isode This directory contains the sources for constructing an alternate ODIF parser/unparser for the CMU ODA Tool Kit. The ODIF parser/unparser that is integral to the CMU ODA Tool Kit has the advantage of being portable to non-Unix machines. Unfortunately, the integral ODIF support is not at all complete. The ODIF support in this directory, although it can be built only on Unix machines, is more complete and robust.

prog This directory contains several programs that we have found useful in using the CMU ODA Tool Kit and the Raster Graphics Tool Kit.

B.1.1. Local Directory Structures

If you will be compiling the tool kits on a Unix machine, you may simply build directly in the directories as they are unpacked from the tape. On a non-Unix machine, however, you must copy the contents of the ODA directories to a directory structure on the machine that mimics that on the tape.

Even though the contents of the ***isode*** directory can not be built on a Macintosh configuration, you must still create this director, although you may leave them both empty.

B.2. Building

Instructions for building the contents of each top-level directory are provided in the next sections. Each section contains a description of the contents of the directory and a discussion of whether you may want to build that directory. In addition, for each supported configuration, the commands for building are specified.

B.2.1. Disk Space and Memory Requirements

The purpose of this section is to provide an idea of the amount of resources that will be required to build parts of the ODA contribution.

The exact hardware configuration that you will require to build things depends on a number of factors: what compile-time switches you use (section B.7), whether you turn on compiler debugging, what is running on the machine, etc.

In order to give you an idea of the configuration required, the following lists the hardware and software configurations we used in building this release during our last round of testing:

BSD Unix IBM RT-PC, BSD 4.3, 8MB main memory, 40MB hard disk (connection to central file system), HC compiler version 1.2.

System V Unix IBM RT-PC, AIX 2.2.1, 8MB main memory, 40MB hard disk (connection to central file system).

Macintosh MPW Macintosh II, 4MB main memory, 20MB hard disk, System 6.03, MPW 3.0

MS-DOS MS-DOS version 3.30, Make version 4.07, C compiler version 5.10

VAX/VMS VAX/VMS version V5.1, C compiler version V3.0-031

Table B-1 specifies the amount of disk space required for each directory on each configuration. The figure reported includes all source files, object files and resultant libraries or executables[67].

configuration	site	src	valid	raster	isode	prog
BSD Unix	0.04	5.6	1.7	0.8	6.1	2.1
SysV Unix	0.05	8.1	3.2	0.8	8.9	7.2
Mac[68]	0.03	3.0	0.3	0.5	2.8	0.7
MS-DOS	0.03	6.1	0.6	0.4	n/a	0.9
VMS	0.06	3.1	0.5	0.5	2.5	0.5

Table B-1. Disk Space Requirements, MB

B.2.2. The *site* Directory

The site directory contains a directory for each supported configuration. There are currently five directories:

site/bsdunix for BSD Unix

site/sysvunix for System V Unix (includes IBM AIX)

[67]The ***isode*** sources can not be compiled on non-Unix configurations, so the figure given is for sources only. The programs `prodif` and `convert` can not be built on non-Unix configurations, so the figure given does not include these programs on non-Unix machines.

[68]The Macintosh C compiler was run so that no debugging information was generated.

site/mac	for Macintosh MPW
site/pcdos	for IBM PC-compatible MS-DOS
site/vms	for VAX VMS

B.2.2.1. Building in the *site* Directory

If you want to build the CMU ODA Tool Kit on a configuration that is currently unsupported, read section B.6 first and create the necessary files. You may also want to read section B.7 at this time, which describes a number of actions that may be modified at compile time and must be done within the appropriate subdirectory of ***site***. We expect, however, that the default actions will be found satisfactory for most situations.

In order to build the library for your configuration, you must first position yourself in the appropriate subdirectory of ***site***. The commands to issue then depend on your configuration, as follows:

BSD Unix

```
make -f Make.bsd_unix liblocal.a
```

System V Unix

```
make -f Make.sysv_unix liblocal.a
```

VAX VMS

```
mms liblocal.olb
```

MS-DOS

```
make make.dos
```

Macintosh MPW

```
duplicate Make.mac makefile
BuildProgram liblocal.a
```

B.2.3. The *src* Directory

The ***src*** directory contains the source files for the primary library comprising the CMU ODA Tool Kit. Unless you are only interested in the stand-alone Raster Graphics Tool Kit (part 4 of this book) you should proceed to build the library in this directory.

B.2.3.1. Building in the *src* Directory

Once you are positioned in the ***src*** directory, issue the following commands, dependent on configuration:

BSD Unix

```
make -f Make.bsd_unix liboda.a
```

System V Unix

```
make -f Make.sysv_unix liboda.a
```

VAX VMS

```
mms liboda.olb
```

MS-DOS

```
make make1.dos objects
make make2.dos liboda.lib
```

Macintosh MPW

```
duplicate Make.mac makefile
BuildProgram liboda.a
```

B.2.4. The *valid* Directory

See section B.4 for details on running the validation program.

B.2.4.1. Building in the *valid* Directory

Once you are positioned in the ***valid*** directory, issue the following commands, dependent on configuration:

BSD Unix

```
make -f Make.bsd_unix valid
```

System V Unix

```
make -f Make.sysv_unix valid
```

VAX VMS

```
mms valid
```

MS-DOS

```
make make.dos
```

Macintosh MPW

```
duplicate Make.mac makefile
BuildProgram valid
```

B.2.5. The *raster* Directory

The Raster Graphics Tool Kit, described in part 4 of this book, is a set of subroutines that are useful for converting between a number of raster formats. The CMU ODA Tool Kit and the Raster Graphics Tool Kit are independent of each other (although both are dependent on the library built in the ***local*** directory). You may, in fact, find the Raster Graphics Tool Kit to be of use in a program manipulating an ODA document. In that case, you would include the Raster Graphics Tool Kit library during the link phase (section 12.4).

B.2.5.1. Building in the *raster* Directory

After positioning yourself in the ***raster*** directory, the appropriate commands to issue depend on the configuration, as follows:

BSD Unix

```
make -f Make.bsd_unix libraster.a
```

System V Unix

```
make -f Make.sysv_unix libraster.a
```

VAX VMS

```
mms libraster.olb
```

MS-DOS

```
make make.dos
```

Macintosh MPW

```
duplicate Make.mac makefile
BuildProgram libraster.a
```

B.2.6. The *isode* Directory

The ***isode*** directory contains source code for building a set of routines for parsing/unparsing ODIF. These routines are an alternative to the standard routines provided with the CMU ODA Tool Kit (chapter 20). The CMU ODA Tool Kit is independent of this library, but this library can only be used in conjunction with the CMU ODA Tool Kit.

B.2.6.1. Building in the *isode* Directory

The contents of this library may only be compiled on Unix configurations. After positioning yourself in the ***isode*** directory, the appropriate commands to issue depend on the configuration, as follows:

BSD Unix

```
make -f Make.bsd_unix libisode.a
```

System V Unix

```
make -f Make.sysv_unix libisode.a
```

VAX VMS

Not applicable.

MS-DOS

Not applicable.

Macintosh MPW

Not applicable.

B.2.7. The *prog* Directory

The ***prog*** directory contains four programs that we have found to be useful:

- ***proda***: given an ODIF stream (on standard input), ***proda*** will write (on standard output) a human-readable representation of the document denoted by the stream[69]. This program uses the standard ODIF parser/unparser.
- ***prtext***: given an ODIF stream (on standard input), ***prtext*** will write (on standard output) the printable characters from the character content portions.
- ***prodif***: given an ODIF stream (on standard input), ***proda*** will write (on standard output) a human-readable representation of the stream. This differs from ***proda*** in that this representation is a trace of the ASN.1 stream and displays the components in the order of their appearance in the stream. ***prodif*** uses the alternate parser/unparser included in the ***isode*** directory and, thus, can be built only on Unix configurations.
- ***convert***: converts between raster formats (***convert*** will prompt for its inputs and outputs).

Be aware that ***proda*** and ***prodif*** do not accept the same set of datastreams. In particular, ***proda*** will accept only those streams produced by the standard ODIF parser and ***prodif*** will accept only those produced by the ***isode*** parser.

The use of these programs is further restricted in that ***prodif*** and ***convert*** can only be built on Unix configurations. In addition, the ***convert*** program will not compile as is on VAX VMS, Macintosh or MS-DOS configurations[70].

B.2.7.1. Building in the *prog* Directory

After positioning yourself in the ***prog*** directory, the appropriate commands to issue depend on the configuration, as follows:

---

[69]The representation is described in chapter 22.

[70]The program can be modified to compile (we believe) on these configurations by changing the line

```
#include <fcntl.h>
```

to be

```
#include <file.h>
```

For the Macintosh, it is only necessary to remove the third parameter on calls to `open`.

BSD Unix

```
make -f Make.bsd_unix proda prtext prodif \
   convert
```

System V Unix

```
make -f Make.sysv_unix proda prtext prodif \
   convert
```

VAX VMS

```
mms proda, prtext
```

MS-DOS

```
make make.dos
```

Macintosh MPW

```
duplicate Make.mac makefile
BuildProgram proda prtext
```

B.3. Installing the Tool Kits

After building as desired you may wish to install the tools and include files in a location other than the directories in which they were built. The following is a description for each directory of the files that should be installed elsewhere for use by application programs.

site: The subdirectory of the ***site*** directory that corresponds to your configuration (section B.2.2) contains both include files and an object library. You should always install these files regardless of the particular parts of the ODA suite that you wish to use.

Copy the files ****.h*** to an appropriate include directory.

Copy the file ***liblocal.**** to an appropriate library directory.

src: This directory contains a large number of include files and a single object library. You should install these files if you wish to make use of the CMU ODA Tool Kit.

Copy the files ****.h*** to an appropriate include directory.

Copy the file ***liboda.**** to an appropriate library directory.

isode: This directory contains a number of include files and a single object library. You should install these files if you wish to make use of the alternate ODIF parser/unparser with the CMU ODA Tool Kit.

Copy the files ***.h*** to an appropriate include directory.

Copy the file ***libisode.**** to an appropriate library directory.

raster: This directory contains a number of include files and a single object library. You should install these files if you wish to make use of the Raster Graphics Tool Kit.

Copy the files ***.h*** to an appropriate include directory.

Copy the file ***libraster.**** to an appropriate library directory.

B.4. Validating the Installation

The distribution comes with a program that may be used to validate the installation of the CMU ODA Tool Kit at your site. Note that this program does not attempt to validate the correctness of the tool kit (that is, its conformance to ODA).

To build the validation suite, there is a set of site-dependent routines that must be defined, all of which are in the file ***vlocal.c***. These routines, which are only necessary for building the validator and not for the tool kit, are described in section B.5.

Once these routines have been created (they already exist for the supported sites) and the validation suite has been built (section B.2), the program may be run by typing its name with no arguments:

```
valid
```

The validation suite checks the installed tool kit by building a set of documents using a subset of tool kit routines considered essential to the functionality of the tool kit. The suite then writes them out in ODIF, and compares the resulting files to standard ones shipped with the tool kit.

The manner in which the validation program is executed depends on configuration, as follows:

BSD Unix

```
valid
```

System V Unix

```
valid
```

VAX VMS

```
run valid
```

MS-DOS

```
valid
```

Macintosh MPW

```
valid
```

The output from running ***valid*** should be as follows:

```
**Starting Validation Suite**

Running test "test0": simple TK link test
Test succeeded

Running test "test1": empty document test
Creating output for comparison...
Storing output...
Comparing files...
Test succeeded

Running test "test2": document profile test
Creating output for comparison...
>>> Tool Kit warning <<<
[INT_PAIR_CkVal]: Attribute dp_UNIT_SCALING -
   can not check value[71]
Storing output...
Comparing files...
Test succeeded

Running test "test3": presentation-style test
Creating output for comparison...
Storing output...
Comparing files...
Test succeeded
```

[71]This warning message will not appear if you have defined the DONT_FORMAT_ERRORS switch (section B.7).

```
Running test "test4": character-content test
Creating output for comparison...
Storing output...
Comparing files...
Test succeeded

Running test "test5": SEQUENCE test
Creating output for comparison...
Storing output...
Comparing files...
Test succeeded

Running test "test6": ASN.1 object ident.
   test
Creating output for comparison...
Storing output...
Comparing files...
Test succeeded

Running test "test7": undefined
Creating output for comparison...
Storing output...
Comparing files...
Test succeeded

All tests successful.
Exited validation suite.
```

If any errors are reported by the validation suite, the tool kit has not been built correctly on your system.

B.5. Site-dependent Validation Routines

There are three site-dependent routines that must be implemented in order to build and execute the validation program. Whenever possible, it is recommended that these routines be written and that the validation program be run whenever a new tool kit is built.

These routines, which are described in the next three sections, are separated from the rest of the site-dependent routines because these routines make assumptions about file names and file i/o. For this reason, there may be installations at which these routines can not be defined without making changes to the validation program. At these sites, the validation program need not be built.

B.5.1. Validation Routine to Return Output Descriptor: *VALID_GetStdout*

```
INT_type VALID_GetStdout()
```

Real Name

vlstd

Synopsis

This routine should return an output descriptor that corresponds to the desired location for output from the validation program.

Return Value

`ERROR_INT`	An error has ocurred.
otherwise	An output descriptor to which all informational output from the validation progam will be sent.

Side Effects

The side effects to this routine are site dependent.

Semantics

All human-readable output from the validation program will go to this descriptor. It should probably refer to the terminal device on which the validation program is being run.

Errors

`LOCAL_err`	A locally-defined error, not covered by any other available error, has occurred.

B.5.2. Validation Routine to Open a File for Reading: *VALID_ReadOpen*

```
INT_type VALID_ReadOpen(name)
   CHAR_type *name;
```

Real Name

vlrdo

Synopsis

This routine is used to open a file name for reading and associate the open file with an input descriptor.

Parameters

`name`	The name of the file to open. The file is assumed to be in the current directory.

Return Value

`ERROR_INT`	An error has ocurred.
otherwise	An input descriptor associated with the specified file.

Side Effects

The side effects to this routine are site dependent.

Semantics

The validation program assumes that the file names used are legal at all sites. This may be a bad assumption for some environments.

The file must be positioned at its beginning.

Errors

`LOCAL_err`	A locally-defined error, not covered by any other available error, has occurred.

B.5.3. Validation Routine to Open a File for Writing: *VALID_WriteOpen*

```
INT_type VALID_WriteOpen(name)
   CHAR_type *name;
```

Real Name

vlwro

Synopsis

This routine is used to open a file name for writing and associate the open file with an input descriptor.

Parameters

`name`	The name of the file to open. The file is assumed to be in the current directory.

Return Value

`ERROR_INT`	An error has ocurred.
otherwise	An input descriptor associated with the specified file.

Side Effects

A file with the appropriate name will be created if one does not exist.
Other side effects to this routine are site dependent.

Semantics

The validation program assumes that the file names used are legal at all sites. This may be a bad assumption for some environments.
The file must be opened so that any previous contents of the file will be discarded.

Errors

`LOCAL_err`	A locally-defined error, not covered by any other available error, has occurred.

B.6. Supporting a New Site

The minimum requirements for supporting a new site (one without a valid site subdirectory) are that appropriate versions of the files ***lcsite.h***, ***local.c*** and ***local.h*** must be created. Descriptions of the requirements of these files may be found in chapter 10.

It is also recommended that the site-dependent validation routines be implemented so that the build can be validated (section B.4).

The site installer may use the files in ***site/bsdunix*** or ***site/pcdos*** as examples.

B.7. Compile-time Switches

B.7.1. The DEBUG Switch

This switch is currently unused.

B.7.2. The DONT_FORMAT_ERRORS Switch

The default action of the routines in the tool kit in the presence of an error is to format a descriptive error message, which is then written to a designated output descriptor if specified (section 14.1). Although useful and descriptive, the construction of these error messages requires a non-trivial amount of code and data space. Should you wish to eliminate this space, you should define the symbol `DONT_FORMAT_ERRORS` in ***lcsite.h***. Be aware that, in this case, not only will no error messages be constructed, but that no message at all will be displayed (although the appropriate error value will be returned).

B.7.3. The SWAPPING Switch

This switch is currently unused.

B.7.4. The STRUCT_ASSIGN Switch

This switch is currently unused.

B.7.5. The USE_BCOPY Switch

This switch is currently unused.

B.7.6. The USE_LONG_NAMES Switch

To be portable to many environments the CMU ODA Tool Kit ensures that all names that are visible to the linker consist of six or fewer characters from a restricted character set. The longer names that appear in the source code are mapped to these shorter names using macros. While this mapping aids portability, it also makes interactive debugging more difficult because the actual name of a routine is a cryptic string that must be found in the manual.

On machines with relaxed linker restrictions (specifically, where mixed-case character strings including the character '_' up to length 31 are allowed) it is desirable to define the switch `USE_LONG_NAMES`. If this symbol is defined, the tool kit will not perform the name mapping and this may make debugging easier.

C
Error Codes

This appendix contains a list of all error codes that may be placed in the global `TKError` (section 12.3) by the tool kits' routines and a short description of the generic meaning of each code. In many cases, this generic description will be quite unspecific and not of much help in determining the cause of the error. The contents of the buffer `TKErrorMesg` (section 12.3) will be found to be extremely helpful in such circumstances.

C.1. CMU ODA Tool Kit Error Codes

`BADATTR_err` An attribute has been used incorrectly.

`BADPARM_err` A parameter to the routine is invalid.

`DESCRIPTOR_err`
An input or output descriptor has been used incorrectly.

`EXISTS_err` An attempt was made to create another instantiation of an object that may exist only once.

`INTERNAL_err` An error internal to the tool kit has occurred (this indicates an error within the tool kit itself and not in its use).

`LOCAL_err` An error in one of the site-dependent routines has occurred.

`NOATTR_err` When attempting to obtain the value of an attribute, this indicates that the attribute is not associated with the constituent (this may, in fact, not indicate an error at all).

`NODOC_err` *not used*

`NOFINAL_err` The tool kit was not finalized (by calling `FinalToolKit`) before an attempt to reinitialize it.

`NOGENS_err` No more iterators may be created for the specified structure.

`NOMEM_err` Insufficient memory is available for allocating an object.

`NOPROF_err` The document has no document profile.

`NOREPLACE_err` An attempt has been made to modify the value of a read-only attribute.

`NOROOT_err` No root for the specified structure could be found.

`ODIF_err` The ODIF datastream is invalid.

`UNIMPL_err` A routine or feature is unimplemented.

`WALK_err` Tree is not structured properly for walking (for example, a loop exists in the specific structure).

C.2. Raster Graphics Tool Kit Error Codes

`RG_BADPARM_err`
A parameter to the routine is invalid.

`RG_NOMEM_err` Insufficient memory is available for allocating an object or swapping.

`RG_PELRANGE_err`
The coordinates for a pel do not lie within the raster.

`RG_INTERNAL_err`
An error internal to the tool kit has occurred.

`RG_LOCAL_err` A locally-defined error has occurred.

`RG_NOFINAL_err`
The tool kit was never initialized (via a call to `InitRGToolKit`) or has not been initialized since the last call to this routine.

`RG_PELFORMAT_err`
The raster in question is not a bitmap and may not be used in bit-only functions.

`RG_NOFORM_err` The format of the input memory does not match the format specified.

`RG_FORMBAD_err`
The input memory was identified as the specified format, but it contains a data error.

D
Current Restrictions and Bugs

This appendix contains a description of the current known bugs and limitations of the CMU ODA Tool Kit and Raster Graphics Tool Kit. We have classified something as a bug or limitation if the implementation of a routine differs from its description in this book.

D.1. The CMU ODA Tool Kit

The bugs and limitations of the CMU ODA Tool Kit are organized in two sections. Section D.1.1 describes bugs that are not specific to particular attributes. Section D.1.2 lists problems with tool kit support for specific attributes. We have separated the information about specific attributes because it describes the vast majority of limitations and is described easily by tables (section D.1.2).

D.1.1. Miscellaneous Bugs and Limitations

Following is a list of known bugs and limitations that are not specific to any single attribute:

- The CMU ODA Tool Kit routines that evaluate expressions are all unimplemented and will return an error (and set `TKError` to `UNIMPL_err`). These routines are `EvalStringExpr` (section 18.1.2), `EvalNumExpr` (section 18.2.2), `EvalObjIdExpr` (section 18.3.2) and `EvalRefBind` (section 18.5.2).

  This implies that when the value of an attribute is obtained (with `GetAttr` or one of its relatives), expressions in the value can not be evaluated.

- The determination of default values for attributes may be incorrect in some cases: resource documents are not supported, and an improper value may be returned if the default must be obtained from the document profile attribute "document application profile defaults".

- The routines for deleting sequences, `DeleteSequenceDeeply` (section 13.13.6), and deleting construction expressions, `DeleteConstExpr` (section 18.4.2), will behave incorrectly if any object within the structures is referenced by multiple pointers.

- The routine `GetAttr` (section 16.1.3) and its relatives do not check that the requested attribute is actually applicable to the specified constituent.

- Use of the `output_filter` field of the tool kit initialization structure is unimplemented (section 14.1). Any value specified for this field will be ignored.

- tool kit finalization (routine `FinalToolKit`, section 14.2) returns successfully, but performs no actions.

- The font routines `Dsclass2Ints` (section 13.12.2) and `Ints2Dsclass` (section 13.12.3) support all font values listed in section 13.11.2.2. The listing in that section, however, is not complete.

- The routines used to label parts of a document (section 20.3) do not remove any existing values for identifiers. The existing values will simply be replaced, resulting in lost storage.

D.1.2. Specific Attributes

This section consists of eight tables that provide details about the implementation support for each ODA attribute. The attributes are listed in alphabetical order by their ODA name. The columns of the tables have the following meanings:

Name the ODA name of the attribute.

Check the entries in this column indicate the status of semantic checking (by the attribute assignment routines) for the attribute in the tool kit, as follows:

●: attribute checking is implemented

○: attribute checking is unimplemented

◑: attribute checking is unimplemented, except that if the attribute value is a sequence, the type of the sequence is checked[72].

R ODIF The routine for reading an ODIF datastream (`ReadODIF`, section 20.1.1) supports this attribute.

W ODIF The routine for writing an ODIF datastream (`WriteODIF`, section 20.1.2) supports this attribute.

ISODE R The ISODE routine for reading an ODIF datastream (`isode_ReadODIF`, section 20.1.3) supports this attribute.

ISODE W The ISODE routine for writing an ODIF datastream (`isode_WriteODIF`, section 20.1.4) supports this attribute.

Debug The tool kit debugging routines (chapter 22) will output a human-readable form of this attribute.

D.2. The Raster Graphics Tool Kit

- The uncompressed mode read on Group 3 is untested.
- The `memsize` parameter in the `INIT_RGPARMS_type` probably does not function correctly in determining when to swap out memory, and a value of -3 should always be assigned to this field. This means that swapping will occur whenever the system runs out of memory (as determined by calls to `LOCAL_Malloc`).

---

[72]Only the following fields of the entries for the attribute "fonts list" are supported: ISO/FONTNAME, ISO/FAMILY, ISO/POSTURE, ISO/WEIGHT, ISO/PROPWDTH, ISO/DSNSIZE, ISO/DSCLASS, ISO/STRUCTUR, ISO/ESCLASS,ISO/VARSCRIPT, ISO/LSCRPT, ISO/RSCRPT, ISO/VXOFFSET, ISO/VYOFFSET, ISO/VXSCALE, ISO/VYSCALE.

Name	Check	R ODIF	W ODIF	ISODE R	ISODE W	Debug
"abstract"	◑	○	○	○	○	●
"access rights"	◑	○	○	○	○	●
"additional information"	◑	○	○	○	○	●
"alignment"	◑	○	○	○	○	●
"alternative representation"	◑	●	●	●	○	●
"alternative representation character sets"	◑	○	○	○	○	●
"application comments"	◑	●	●	●	●	●
"authorization"	◑	○	○	○	○	●
"authors"	●	○	○	○	○	○
"balance"	◑	○	●	○	○	●
"bindings"	○	○	○	○	○	○
"block alignment"	●	○	○	○	○	●
"block alignments"	●	○	○	○	○	●
"border"	●	●	●	○	○	○
"character fonts"	●	●	●	○	○	●
"character orientation"	●	○	○	○	○	●
"character path"	●	●	●	○	○	●
"character spacing"	●	●	●	○	○	●
"clipping"	●	●	●	○	○	●
"code extension announcers"	◑	●	●	○	○	●
"coding attributes"	●	○	○	○	○	●

Table D-1. Attribute Limitations, I

Name	Check	R ODIF	W ODIF	ISODE R	ISODE W	Debug
"colour representations"	○	○	○	○	○	○
"colour"	●	●	●	○	○	○
"colours"	●	○	○	○	○	○
"comments character sets"	◑	○	○	○	○	●
"compression"	●	●	●	○	○	●
"concatenation"	●	●	●	○	○	●
"content architecture class"	◑	●	●	○	○	●
"content architecture classes"	●	○	○	○	○	○
"content generator"	●	○	○	○	○	○
"content identifier - layout"	◑	●	●	●	●	●
"content identifier - logical"	◑	●	●	●	●	●
"content information"	◑	●	●	●	●	●
"content portions"	◑	●	●	●	●	●
"content type"	◑	●	●	○	○	●
"copyright"	◑	○	○	○	○	○
"creation date and time"	◑	○	○	○	○	●
"default value lists"	○	●	●	●	●	●
"dimensions"	●	●	●	○	○	○
"distribution list"	◑	○	○	○	○	○
"document application profile defaults"	○	○	○	○	○	○

Table D-2. Attribute Limitations, II

Name	Check	R ODIF	W ODIF	ISODE R	ISODE W	Debug
"document application profile"	●	○	○	○	○	●
"document architecture class"	●	○	○	○	○	●
"document date and time"	◑	○	○	○	○	●
"document reference"	◑	○	○	○	○	○
"document size"	○	○	○	○	○	●
"document type"	◑	○	○	○	○	●
"edge rendition"	○	○	○	○	○	○
"expiry date and time"	◑	○	○	○	○	●
"external document class"	◑	○	○	○	○	●
"fill order"	●	●	●	○	○	●
"fill orders"	●	○	○	○	○	●
"filled area rendition"	○	○	○	○	○	○
"first line offset"	●	●	●	○	○	●
"fonts list"	◑	●	●	○	○	●
"formatting indicator"	●	●	●	○	○	●

Table D-3. Attribute Limitations, III

Name	Check	R ODIF	W ODIF	ISODE R	ISODE W	Debug
"generator for subordinates"	●	●	●	●	●	●
"generic layout structure"	●	●	●	○	○	●
"generic logical structure"	●	●	●	○	○	●
"geometric graphics encoding announcer"	○	○	○	○	○	○
"graphic character sets"	◑	●	●	○	○	●
"graphic character subrepertoire"	◑	○	○	○	○	●
"graphic rendition"	◑	●	●	○	○	●
"image dimensions"	●	●	●	○	○	●
"imaging order"	◑	○	●	○	○	●
"indentation"	●	●	●	○	○	●
"indivisibility"	●	●	●	○	○	●
"initial offset" (character)	●	●	●	○	○	●
"initial offset" (raster)	●	○	○	○	○	●
"interchange format class"	●	●	●	○	○	●
"itemisation"	●	●	●	○	○	●
"kerning offset"	●	●	●	○	○	●
"languages"	○	○	○	○	○	●
"layout category"	◑	●	●	○	○	●
"layout object class"	○	●	●	○	○	●
"layout path"	●	●	●	○	○	●
"layout paths"	●	○	○	○	○	●

Table D-4. Attribute Limitations, IV

Name	Check	R ODIF	W ODIF	ISODE R	ISODE W	Debug
"layout style identifier"	◑	●	●	●	●	●
"layout style"	○	●	●	●	●	●
"layout styles"	◑	●	●	○	○	●
"line layout table"	●	○	○	○	○	○
"line progression" (character)	●	●	●	○	○	●
"line progression" (raster)	●	○	○	○	○	●
"line rendition"	●	○	○	○	○	●
"line spacing"	●	●	●	○	○	●
"local file reference"	○	○	○	○	○	○
"local filing date and time"	◑	○	○	○	○	●
"logical source"	◑	○	○	○	○	●
"marker rendition"	○	○	○	○	○	○
"medium type"	●	●	●	○	○	○
"medium types"	●	○	○	○	○	○
"new layout object"	◑	●	●	○	○	●
"number of discarded pels"	●	●	●	○	○	●
"number of lines"	●	●	●	○	○	●
"number of objects per page"	●	○	○	○	○	●
"number of pages"	●	○	○	○	○	●
"number of pels per line"	●	●	●	○	○	●

Table D-5. Attribute Limitations, V

Name	Check	R ODIF	W ODIF	ISODE R	ISODE W	Debug
"object class identifier"	◑	●	●	●	●	●
"object class"	◑	●	●	●	●	●
"object identifier"	●	●	●	●	●	●
"object type"	●	●	●	●	●	●
"ODA version"	○	○	○	○	○	●
"offset"	●	●	●	○	○	●
"organizations"	○	○	○	○	○	●
"orphan size"	●	○	○	○	○	●
"owners"	◑	○	○	○	○	○
"page dimensions"	◑	○	○	○	○	●
"page position"	◑	○	●	○	○	●
"page positions"	◑	○	●	○	○	●
"pairwise kerning"	●	●	●	○	○	●
"pel path"	●	●	●	○	○	●
"pel spacing"	●	●	●	○	○	●
"pel transmission density"	●	●	●	○	○	●
"permitted categories"	○	●	●	○	○	●
"picture dimensions"	○	○	○	○	○	○
"picture orientation"	○	○	○	○	○	○
"position"	●	●	●	○	○	○

Table D-6. Attribute Limitations, VI

Name	Check	R ODIF	W ODIF	ISODE R	ISODE W	Debug
"preparers"	◑	○	○	○	○	○
"presentation features"	◑	○	○	○	○	○
"presentation style identifier"	◑	●	●	●	●	●
"presentation style"	●	●	●	●	●	●
"presentation styles"	◑	●	●	○	○	●
"profile character sets"	◑	○	○	○	○	●
"proportional line spacing"	●	●	●	○	○	●
"protection"	●	●	●	●	●	●
"protections"	◑	○	○	○	○	●
"purge date and time"	◑	○	○	○	○	●
"references to other documents"	◑	○	○	○	○	○
"region of interest specification"	○	○	○	○	○	○
"release date and time"	◑	○	○	○	○	●
"resource document"	○	○	○	○	○	○
"resource"	○	●	●	●	●	●
"resources"	◑	○	○	○	○	○
"revision history"	○	○	○	○	○	○
"security classification"	◑	○	○	○	○	●
"same layout object"	●	○	●	○	○	○

Table D-7. Attribute Limitations, VII

Name	Check	R ODIF	W ODIF	ISODE R	ISODE W	Debug
"separation"	●	●	●	○	○	●
"spacing ratio"	●	●	●	○	○	●
"specific layout structure"	●	●	●	○	○	●
"specific logical structure"	●	●	●	○	○	●
"start date and time"	◑	○	○	○	○	●
"status"	◑	○	○	○	○	●
"subject"	◑	○	○	○	○	●
"subordinates"	◑	●	●	●	●	●
"superseded documents"	◑	○	○	○	○	○
"synchronization"	◑	○	○	○	○	●
"text rendition"	○	○	○	○	○	○
"title"	◑	○	○	○	○	●
"transformation specification"	○	○	○	○	○	○
"transparencies"	◑	○	○	○	○	●
"transparency specification"	○	○	○	○	○	○
"transparency"	●	○	○	○	○	●
"type of coding"	◑	●	●	●	●	●
"types of coding"	◑	○	○	○	○	○
"unit scaling"	○	●	●	○	○	●
"user-readable comments"	●	●	●	●	●	●
"user specific codes"	○	○	○	○	○	○
"user-visible name"	●	●	●	●	●	●
"widow size"	●	○	○	○	○	●

Table D-8. Attribute Limitations, VIII

E
C Portability Guidelines

- Keep lines ≤ 72 characters long.
- Do not use L constants (e.g., `1L`)
- Do not use structure assignments.
- Do not use enumerated types.
- Do not rely on execution order, especially with ++ and --, for example:

  ```
  b[i] = a[i++];
  ```

  has no guarantees about i on the left side. This same guideline applies to ++ and -- in procedure calls.
- Do not use bit fields.
- Use structure definitions instead of bit sequences in long integers for holding many values that are extracted and set.
- No data declarations are permitted in blocks (some symbolic debuggers have trouble with this).
- Do not use a variable number of parameters to procedures.
- Do not use the `setjmp` facility.
- Do not use signal.
- Limit identifier length to less than 32 characters (≤ 31 characters).
- Do not use control-L for page breaks.
- Maximum length of a string constant (includes the terminating null character, '\0') is 512.
- Maximum level of nesting for struct/union definitions is 10 levels.
- Maximum size of a macro expansion is 512 bytes.

- Maximum number of actual arguments to a macro definition is 8.
- Maximum length of an actual macro argument is 256 bytes.
- Maximum level of nesting for `#if`, `#ifdef` and `#ifndef` directives is 32.
- Maximum level of nesting for include files is 10.
- Do not have more than 32 kilobytes of global data.
- Always cast constants used as actual parameters.
- Do not use `void`.
- Do not initialize dynamic variables in routines.
- Do not assume that macro expansion within string literals will work.
- Do not rely on `union` types for overlaying fields in memory.
- Do not assume either text or lexeme substitution will occur during macro expansion.

F
Alphabetical List of Routines

F.1. Routines Sorted by Alias

Alias	Real Name	Section
AppendSequence	appseq	13.13.9
AppendSubordinate	appsub	16.1.11
Array2StrctName	ary2sn	13.12.4
AssignBit	rgpeab	25.6
AssignPel	rgpeap	25.7
AttrDocProfPriority	dppri	21.3.3
Chars2String	chrstr	13.14.3
ComponentAttrPriority	compri	21.3
ConcatString	catstr	13.14.5
ContentAttrPriority	conpri	21.3.2
ContractSequence	conseq	13.13.8
CopyConstituent	cpcons	15.8
CopySequence	cpyseq	13.13.10
DeleteConstExpr	dlcnex	18.4.2
DeleteConstituent	delcns	15.9
DeleteDocument	deldoc	15.3
DeleteIterator	delgen	19.2.4
DeleteRaster	rgradr	25.5
DeleteSequence	delseq	13.13.5
DeleteSequenceDeeply	dlsqd	13.13.6
DeleteString	delstr	13.14.4
Dsclass2Ints	ds2int	13.12.2
EvalNumExpr	evlnum	18.2.2
EvalObjIdExpr	evlobi	18.3.2

Table F-1. Routines Sorted by Alias, I

Alias	Real Name	Section
EvalRefBind	evlrbv	18.5.2
EvalStringExpr	evlstr	18.1.2
ExpandSequence	expseq	13.13.7
ExtractRaster	rgraex	25.4
FinalToolKit	tkfinl	14.2
FinalRGToolKit	rgfitk	25.2
FindDocumentProfile	fnddp	19.1.2
FindDocumentRoot	fndrt	19.1.1
GetAttr	getatt	16.1.3
GetAttrName	gattnm	16.6
GetBit	rgpegb	25.8
GetContent	gcontt	16.1.13
GetContentKind	gcntkn	16.1.12
GetDocument	getdoc	15.2
GetIntAttr	ginatt	16.1.5
GetParmMask	getprm	16.5
GetValueAttr	gvlatt	16.1.7
GetValueName	gvalnm	16.7
HasAttr	hasatt	16.2
HoldAttr	hldatt	21.1.3
HoldConstituent	hldcns	21.1.1
HoldString	hldstr	21.1.5
InitializeDefaultValueLists	intdvl	13.12.10
InitRGToolKit	rgintk	25.1
InitToolKit	tkinit	14.1
InsertRaster	rgrais	25.10
Ints2Dsclass	int2ds	13.12.3
isode_PrintODIF	isprdf	22.2.5
isode_ReadODIF	isred	20.1.3
isode_WriteODIF	iswrt	20.1.4
LabelDocument	lbldoc	20.3.1
LabelStructure	lblstr	20.3.2
LabelStyles	lblsty	20.3.3
LOCAL_Bcopy	lclbcp	10.2.5.4
LOCAL_Final	lclfin	10.2.3
LOCAL_Free	lclfre	10.2.5.3

Table F-2. Routines Sorted by Alias, II

Alias	Real Name	Section
LOCAL_Init	lclini	10.2.2
LOCAL_Malloc	lclmal	10.2.5.1
LOCAL_ReadBytes	lrdbyt	10.2.4.1
LOCAL_ReadChars	lrdchr	10.2.4.3
LOCAL_Realloc	lclrlc	10.2.5.2
LOCAL_SwapIn	lclswi	10.3.2
LOCAL_SwapOut	lclswo	10.3.1
LOCAL_WriteBytes	lwrbyt	10.2.4.2
LOCAL_WriteChars	lwrchr	10.2.4.4
LookUpObjectId	lkobid	13.12.8
MakeComponent	mkcomp	15.4
MakeContent	mkcont	15.6
MakeDocument	mkdoc	15.1
MakeDocumentProfile	mkdocp	15.7
MakeRaster	rgramr	25.3
MakeSequence	mkseq	13.13.4
MakeString	mkstr	13.14.2
MakeStyle	mkstyl	15.5
MakeStylesIterator	mkstgn	19.2.2
MakeSubgraphIterator	mksbgn	19.2.1
NextConstituent	nxtcns	19.2.3
output_filter	*not applicable.*	14.1.1.1
RasterHeight	rgrarh	25.12
RasterPelFormat	rgrapf	25.11
RasterStatus	rgrast	25.19
RasterWidth	rgrard	25.12
ReadMemRaster	rgrirm	25.13
ReadODIF	rdodif	20.1.1
ReadRaster	rgrird	25.14
ReleaseAttr	rlsatt	21.1.4
ReleaseConstituent	rlscns	21.1.2
ReleaseRaster	rgrare	25.18
ReleaseString	rlsstr	21.1.6
RemAttr	rematt	16.3
SetAttr	setatt	16.1.2
SetFontFamily	stffst	13.12.7
SetFontFamilyArray	stffar	13.12.6

Table F-3. Routines Sorted by Alias, III

Alias	Real Name	Section
SetIntAttr	sinatt	16.1.4
SetLayoutAttr	setlat	16.1.8
SetPresAttr	setrat	16.1.9
SetStringAttr	ststat	16.1.10
SetValueAttr	svlatt	16.1.6
String2StrctName	str2sn	13.12.5
StyleAttrPriority	stypri	21.3.1
UnspecifyDefaultValueLists	unsdvl	13.12.9
UnspecifyFont	unsfnt	13.12.1
UnspecParms	unsprm	16.4
VALID_GetStdout	vlstd	B.5.1
VALID_ReadOpen	vlrdo	B.5.2
VALID_WriteOpen	vlwro	B.5.3
WriteConstituentState	dbgcns	22.2.4
WriteDocumentState	dbgdoc	22.2.1
WriteMemRaster	rgriwm	25.15
WriteODIF	wrodif	20.1.2
WriteRaster	rgriwr	25.16
WriteStylesState	dbgsty	22.2.3
WriteSubgraphState	dbgsbg	22.2.2

Table F-4. Routines Sorted by Alias, IV

F.2. Routines Sorted by Real Name

Real Name	Alias	Section
not applicable.	`output_filter`	14.1.1.1
appseq	`AppendSequence`	13.13.9
appsub	`AppendSubordinate`	16.1.11
ary2sn	`Array2StrctName`	13.12.4
catstr	`ConcatString`	13.14.5
chrstr	`Chars2String`	13.14.3
compri	`ComponentAttrPriority`	21.3
conpri	`ContentAttrPriority`	21.3.2
conseq	`ContractSequence`	13.13.8
cpcons	`CopyConstituent`	15.8
cpyseq	`CopySequence`	13.13.10
dbgcns	`WriteConstituentState`	22.2.4
dbgdoc	`WriteDocumentState`	22.2.1
dbgsbg	`WriteSubgraphState`	22.2.2
dbgsty	`WriteStylesState`	22.2.3
delcns	`DeleteConstituent`	15.9
deldoc	`DeleteDocument`	15.3
delgen	`DeleteIterator`	19.2.4
delseq	`DeleteSequence`	13.13.5
delstr	`DeleteString`	13.14.4
dlcnex	`DeleteConstExpr`	18.4.2
dlsqd	`DeleteSequenceDeeply`	13.13.6
dppri	`AttrDocProfPriority`	21.3.3
ds2int	`Dsclass2Ints`	13.12.2
evlnum	`EvalNumExpr`	18.2.2

Table F-5. Routines Sorted by Real Name, I

Real Name	Alias	Section
evlobi	EvalObjIdExpr	18.3.2
evlrbv	EvalRefBind	18.5.2
evlstr	EvalStringExpr	18.1.2
expseq	ExpandSequence	13.13.7
fnddp	FindDocumentProfile	19.1.2
fndrt	FindDocumentRoot	19.1.1
gattnm	GetAttrName	16.6
gcntkn	GetContentKind	16.1.12
gcontt	GetContent	16.1.13
getatt	GetAttr	16.1.3
getdoc	GetDocument	15.2
getprm	GetParmMask	16.5
ginatt	GetIntAttr	16.1.5
gvalnm	GetValueName	16.7
gvlatt	GetValueAttr	16.1.7
hasatt	HasAttr	16.2
hldatt	HoldAttr	21.1.3
hldcns	HoldConstituent	21.1.1
hldstr	HoldString	21.1.5
int2ds	Ints2Dsclass	13.12.3
intdvl	InitializeDefaultValueLists	13.12.10
isprdf	isode_PrintODIF	22.2.5
isred	isode_ReadODIF	20.1.3
iswrt	isode_WriteODIF	20.1.4
lbldoc	LabelDocument	20.3.1
lblstr	LabelStructure	20.3.2
lblsty	LabelStyles	20.3.3
lclbcp	LOCAL_Bcopy	10.2.5.4
lclfin	LOCAL_Final	10.2.3
lclfre	LOCAL_Free	10.2.5.3
lclini	LOCAL_Init	10.2.2
lclmal	LOCAL_Malloc	10.2.5.1
lclrlc	LOCAL_Realloc	10.2.5.2
lclswi	LOCAL_SwapIn	10.3.2
lclswo	LOCAL_SwapOut	10.3.1

Table F-6. Routines Sorted by Real Name, II

Real Name	Alias	Section
lkobid	LookUpObjectId	13.12.8
lrdbyt	LOCAL_ReadBytes	10.2.4.1
lrdchr	LOCAL_ReadChars	10.2.4.3
lwrbyt	LOCAL_WriteBytes	10.2.4.2
lwrchr	LOCAL_WriteChars	10.2.4.4
mkcomp	MakeComponent	15.4
mkcont	MakeContent	15.6
mkdoc	MakeDocument	15.1
mkdocp	MakeDocumentProfile	15.7
mksbgn	MakeSubgraphIterator	19.2.1
mkseq	MakeSequence	13.13.4
mkstgn	MakeStylesIterator	19.2.2
mkstr	MakeString	13.14.2
mkstyl	MakeStyle	15.5
nxtcns	NextConstituent	19.2.3
rdodif	ReadODIF	20.1.1
rematt	RemAttr	16.3
rgpeab	AssignBit	25.6
rgpeap	AssignPel	25.7
rgpegb	GetBit	25.8
rgfitk	FinalRGToolKit	25.2
rgintk	InitRGToolKit	25.1
rgradr	DeleteRaster	25.5
rgraex	ExtractRaster	25.4
rgrais	InsertRaster	25.10
rgramr	MakeRaster	25.3
rgrapf	RasterPelFormat	25.11
rgrard	RasterWidth	25.12
rgrare	ReleaseRaster	25.18
rgrarh	RasterHeight	25.12
rgrast	RasterStatus	25.19
rgrird	ReadRaster	25.14
rgrirm	ReadMemRaster	25.13
rgriwm	WriteMemRaster	25.15
rgriwr	WriteRaster	25.16

Table F-7. Routines Sorted by Real Name, III

Real Name	Alias	Section
rlsatt	`ReleaseAttr`	21.1.4
rlscns	`ReleaseConstituent`	21.1.2
rlsstr	`ReleaseString`	21.1.6
setatt	`SetAttr`	16.1.2
setlat	`SetLayoutAttr`	16.1.8
setrat	`SetPresAttr`	16.1.9
sinatt	`SetIntAttr`	16.1.4
stffar	`SetFontFamilyArray`	13.12.6
stffst	`SetFontFamily`	13.12.7
str2sn	`String2StrctName`	13.12.5
ststat	`SetStringAttr`	16.1.10
stypri	`StyleAttrPriority`	21.3.1
svlatt	`SetValueAttr`	16.1.6
tkfinl	`FinalToolKit`	14.2
tkinit	`InitToolKit`	14.1
unsdvl	`UnspecifyDefaultValueLists`	13.12.9
unsfnt	`UnspecifyFont`	13.12.1
unsprm	`UnspecParms`	16.4
vlrdo	`VALID_ReadOpen`	B.5.2
vlstd	`VALID_GetStdout`	B.5.1
vlwro	`VALID_WriteOpen`	B.5.3
wrodif	`WriteODIF`	20.1.2

Table F-8. Routines Sorted by Real Name, IV

G
Alphabetical List of Attributes

G.1. Attributes Sorted by Name

Name	Identifier	Section
"abstract"	dp_ABSTRACT	17.1.8.39
"access rights"	dp_ACCESS_RIGHTS	17.1.8.65
"additional information"	dp_ADDITIONAL_INFORMATION	17.1.8.56
"alignment"	cc_ALIGNMENT	17.1.9.1
"alternative representation character sets"	dp_ALT_REP_CHARACTER_SETS	17.1.8.18
"alternative representation"	at_ALT_REP	17.1.7.5
"application comments"	at_APPLICATION_COMMENTS	17.1.1.14
"authorization"	dp_AUTHORIZATION	17.1.8.63
"authors"	dp_AUTHORS	17.1.8.51
"balance"	at_BALANCE	17.1.2.4
"bindings"	at_BINDINGS	17.1.1.16
"block alignment"	at_BLK_ALGNM	17.1.5.1
"block alignments"	dp_BLOCK_ALIGNMENTS	17.1.8.23
"border"	at_BORDER	17.1.2.3
"character fonts"	cc_CHARACTER_FONTS	17.1.9.2
"character orientation"	cc_CHAR_ORIENTATION	17.1.9.3
"character path"	cc_CHARACTER_PATH	17.1.9.4
"character spacing"	cc_CHARACTER_SPACING	17.1.9.5
"clipping"	rg_CLIPPING	17.1.10.1
"code extension announcers"	cc_CODE_EXT_ANNOUNCERS	17.1.9.6

Table G-1. Attributes Sorted by Name, I

Name	Identifier	Section
"coding attributes"	dp_CODING_ATTRIBUTES	17.1.8.30
"colour representations"	gg_COLOUR_REPS	17.1.12.7
"colour"	at_COLOUR	17.1.2.10
"colours"	dp_COLOURS	17.1.8.26
"comments character sets"	dp_COMMENTS_CHARACTER_SETS	17.1.8.17
"compression"	rg_COMPRESSION	17.1.11.1
"concatenation"	at_CONCATENATION	17.1.5.2
"content architecture class"	at_CONTENT_ARCH_CLASS	17.1.1.11
"content architecture classes"	dp_CONTENT_ARCH_CLASSES	17.1.8.13
"content generator"	at_CONTENT_GENERATOR	17.1.1.5
"content identifier - layout"	at_CONTENT_ID_LAYOUT	17.1.7.2
"content identifier - logical"	at_CONTENT_ID_LOGICAL	17.1.7.1
"content information"	at_CONTENT_INFORMATION	17.1.7.4
"content portions"	at_CONTENT_PORTIONS	17.1.1.8
"content type"	at_CONTENT_TYPE	17.1.1.12
"copyright"	dp_COPYRIGHT	17.1.8.52
"creation date and time"	dp_CREATION_DATE_AND_TIME	17.1.8.41
"default value lists"	at_DEFAULT_VALUE_LISTS	17.1.1.17
"dimensions"	at_DIMENSIONS	17.1.2.2
"distribution list"	dp_DISTRIBUTION_LIST	17.1.8.55
"document application profile defaults"	dp_DOC_APP_PROF_DEFAULTS	17.1.8.11
"document application profile"	dp_DOC_APP_PROFILE	17.1.8.10
"document architecture class"	dp_DOC_ARCH_CLASS	17.1.8.12
"document date and time"	dp_DOCUMENT_DATE_AND_TIME	17.1.8.40
"document reference"	dp_DOCUMENT_REFERENCE	17.1.8.37
"document size"	dp_DOCUMENT_SIZE	17.1.8.60
"document type"	dp_DOCUMENT_TYPE	17.1.8.38
"edge rendition"	gg_EDGE_RENDITION	17.1.12.6
"expiry date and time"	dp_EXPIRY_DATE_AND_TIME	17.1.8.43
"external document class"	dp_EXTERNAL_DOCUMENT_CLASS	17.1.8.7
"fill order"	at_FILL_ORDER	17.1.5.3
"fill orders"	dp_FILL_ORDERS	17.1.8.24
"filled area rendition"	gg_FILLED_AREA_REND	17.1.12.5
"first line offset"	cc_1ST_LINE_OFFSET	17.1.9.7

Table G-2. Attributes Sorted by Name, II

Name	Identifier	Section
"fonts list"	dp_FONTS_LIST	17.1.8.34
"formatting indicator"	cc_FORMATTING_INDICATOR	17.1.9.17
"generator for subordinates"	at_GEN_FOR_SUBORDINATES	17.1.1.4
"generic layout structure"	dp_GENERIC_LAY_STRUCT	17.1.8.1
"generic logical structure"	dp_GENERIC_LOG_STRUCT	17.1.8.3
"geometric graphics encoding announcer"	gg_ENCODING_ANNOUNCER	17.1.12.1
"graphic character sets"	cc_GRAPHIC_CHARACTER_SETS	17.1.9.8
"graphic character subrepertoire"	cc_GRAPHIC_CHAR_SUBREP	17.1.9.9
"graphic rendition"	cc_GRAPHIC_RENDITION	17.1.9.10
"image dimensions"	rg_IMAGE_DIMENSIONS	17.1.10.6
"imaging order"	at_IMAGING_ORDER	17.1.2.8
"indentation"	cc_INDENTATION	17.1.9.19
"indivisibility"	at_INDIVISIBILITY	17.1.5.4
"initial offset"	cc_INITIAL_OFFSET	17.1.9.18
"initial offset"	rg_INITIAL_OFFSET	17.1.10.4
"interchange format class"	dp_INTERCHANGE_FMT_CLASS	17.1.8.14
"itemisation"	cc_ITEMISATION	17.1.9.11
"kerning offset"	cc_KERNING_OFFSET	17.1.9.12
"languages"	dp_LANGUAGES	17.1.8.62
"layout category"	at_LAYOUT_CATEGORY	17.1.5.5
"layout object class"	at_LAYOUT_OBJECT_CLASS	17.1.5.6
"layout path"	at_LAYOUT_PATH	17.1.2.5
"layout paths"	dp_LAYOUT_PATHS	17.1.8.21
"layout style identifier"	at_LAYOUT_STYLE_IDENTIFIER	17.1.4.1
"layout style"	at_LAYOUT_STYLE	17.1.3.2
"layout styles"	dp_LAYOUT_STYLES	17.1.8.5
"line layout table"	cc_LINE_LAYOUT_TABLE	17.1.9.13
"line progression"	cc_LINE_PROGRESSION	17.1.9.14
"line progression"	rg_LINE_PROGRESSION	17.1.10.2
"line rendition"	gg_LINE_RENDITION	17.1.12.2

Table G-3. Attributes Sorted by Name, III

Name	Identifier	Section
"line spacing"	cc_LINE_SPACING	17.1.9.15
"local file reference"	dp_LOCAL_FILE_REFERENCE	17.1.8.59
"local filing date and time"	dp_LOCAL_FILING_DATE_TIME	17.1.8.42
"logical source"	at_LOGICAL_SOURCE	17.1.2.6
"marker rendition"	gg_MARKER_RENDITION	17.1.12.3
"medium type"	at_MEDIUM_TYPE	17.1.2.12
"medium types"	dp_MEDIUM_TYPES	17.1.8.20
"new layout object"	at_NEW_LAYOUT_OBJECT	17.1.5.7
"number of discarded pels"	rg_NUMBER_DISCARDED_PELS	17.1.11.4
"number of lines"	rg_NUMBER_OF_LINES	17.1.11.2
"number of objects per page"	dp_NUMBER_OBJECTS_PER_PAGE	17.1.8.32
"number of pages"	dp_NUMBER_OF_PAGES	17.1.8.61
"number of pels per line"	rg_NUMBER_OF_PELS_PER_LINE	17.1.11.3
"object class identifier"	at_OBJECT_CLASS_IDENTIFIER	17.1.1.3
"object class"	at_OBJECT_CLASS	17.1.1.6
"object identifier"	at_OBJECT_IDENTIFIER	17.1.1.2
"object type"	at_OBJECT_TYPE	17.1.1.1
"ODA version"	dp_ODA_VERSION	17.1.8.15
"offset"	at_OFFSET	17.1.5.8
"organizations"	dp_ORGANIZATIONS	17.1.8.48
"orphan size"	cc_ORPHAN_SIZE	17.1.9.20
"owners"	dp_OWNERS	17.1.8.50
"page dimensions"	dp_PAGE_DIMENSIONS	17.1.8.19
"page position"	at_PAGE_POSITION	17.1.2.11
"page positions"	dp_PAGE_POSITIONS	17.1.8.28
"pairwise kerning"	cc_PAIRWISE_KERNING	17.1.9.16
"pel path"	rg_PEL_PATH	17.1.10.3
"pel spacing"	rg_PEL_SPACING	17.1.10.7

Table G-4. Attributes Sorted by Name, IV

Name	Identifier	Section
"pel transmission density"	rg_PEL_TRANS_DENSITY	17.1.10.5
"permitted categories"	at_PERMITTED_CATEGORIES	17.1.2.7
"picture dimensions"	gg_PICTURE_DIMENSIONS	17.1.12.12
"picture orientation"	gg_PICTURE_ORIENTATION	17.1.12.11
"position"	at_POSITION	17.1.2.1
"preparers"	dp_PREPARERS	17.1.8.49
"presentation features"	dp_PRESENTATION_FEATURES	17.1.8.31
"presentation style identifier"	at_PRESENTATION_STYLE_ID	17.1.6.1
"presentation style"	at_PRESENTATION_STYLE	17.1.1.10
"presentation styles"	dp_PRESENTATION_STYLES	17.1.8.6
"profile character sets"	dp_PROFILE_CHARACTER_SETS	17.1.8.16
"proportional line spacing"	cc_PROP_LINE_SPACING	17.1.9.21
"protection"	at_PROTECTION	17.1.3.1
"protections"	dp_PROTECTIONS	17.1.8.22
"purge date and time"	dp_PURGE_DATE_AND_TIME	17.1.8.45
"references to other documents"	dp_REFS_TO_OTHER_DOCS	17.1.8.57
"region of interest specification"	gg_REGION_INTEREST_SPEC	17.1.12.10
"release date and time"	dp_RELEASE_DATE_AND_TIME	17.1.8.46
"resource document"	dp_RESOURCE_DOCUMENT	17.1.8.8
"resource"	at_RESOURCE	17.1.1.9
"resources"	dp_RESOURCES	17.1.8.9
"revision history"	dp_REVISION_HISTORY	17.1.8.47
"same layout object"	at_SAME_LAYOUT_OBJECT	17.1.5.9
"security classification"	dp_SECURITY_CLASSIFICATION	17.1.8.64
"separation"	at_SEPARATION	17.1.5.10
"spacing ratio"	rg_SPACING_RATIO	17.1.10.8
"specific layout structure"	dp_SPECIFIC_LAY_STRUCT	17.1.8.2
"specific logical structure"	dp_SPECIFIC_LOG_STRUCT	17.1.8.4

Table G-5. Attributes Sorted by Name, V

Name	Identifier	Section
"start date and time"	dp_START_DATE_AND_TIME	17.1.8.44
"status"	dp_STATUS	17.1.8.53
"subject"	dp_SUBJECT	17.1.8.36
"subordinates"	at_SUBORDINATES	17.1.1.7
"superseded documents"	dp_SUPERSEDED_DOCUMENTS	17.1.8.58
"synchronization"	at_SYNCHRONIZATION	17.1.5.11
"text rendition"	gg_TEXT_RENDITION	17.1.12.4
"title"	dp_TITLE	17.1.8.35
"transformation specification"	gg_TRANSFORMATION_SPEC	17.1.12.9
"transparencies"	dp_TRANSPARENCIES	17.1.8.25
"transparency specification"	gg_TRANSPARENCY_SPEC	17.1.12.8
"transparency"	at_TRANSPARENCY	17.1.2.9
"type of coding"	at_TYPE_OF_CODING	17.1.7.3
"types of coding"	dp_TYPES_OF_CODING	17.1.8.29
"unit scaling"	dp_UNIT_SCALING	17.1.8.33
"user specific codes"	dp_USER_SPECIFIC_CODES	17.1.8.54
"user-readable comments"	at_USER_READABLE_COMMENTS	17.1.1.13
"user-visible name"	at_USER_VISIBLE_NAME	17.1.1.15
"widow size"	cc_WIDOW_SIZE	17.1.9.22

Table G-6. Attributes Sorted by Name, VI

G.2. Attributes Sorted by Identifier

Identifier	Name	Section
at_ALT_REP	"alternative representation"	17.1.7.5
at_APPLICATION_COMMENTS	"application comments"	17.1.1.14
at_BALANCE	"balance"	17.1.2.4
at_BINDINGS	"bindings"	17.1.1.16
at_BLK_ALGNM	"block alignment"	17.1.5.1
at_BORDER	"border"	17.1.2.3
at_COLOUR	"colour"	17.1.2.10
at_CONCATENATION	"concatenation"	17.1.5.2
at_CONTENT_ARCH_CLASS	"content architecture class"	17.1.1.11
at_CONTENT_GENERATOR	"content generator"	17.1.1.5
at_CONTENT_ID_LAYOUT	"content identifier - layout"	17.1.7.2
at_CONTENT_ID_LOGICAL	"content identifier - logical"	17.1.7.1
at_CONTENT_INFORMATION	"content information"	17.1.7.4
at_CONTENT_PORTIONS	"content portions"	17.1.1.8
at_CONTENT_TYPE	"content type"	17.1.1.12
at_DEFAULT_VALUE_LISTS	"default value lists"	17.1.1.17
at_DIMENSIONS	"dimensions"	17.1.2.2
at_FILL_ORDER	"fill order"	17.1.5.3
at_GEN_FOR_SUBORDINATES	"generator for subordinates"	17.1.1.4
at_IMAGING_ORDER	"imaging order"	17.1.2.8
at_INDIVISIBILITY	"indivisibility"	17.1.5.4
at_LAYOUT_CATEGORY	"layout category"	17.1.5.5
at_LAYOUT_OBJECT_CLASS	"layout object class"	17.1.5.6
at_LAYOUT_PATH	"layout path"	17.1.2.5
at_LAYOUT_STYLE	"layout style"	17.1.3.2
at_LAYOUT_STYLE_IDENTIFIER	"layout style identifier"	17.1.4.1
at_LOGICAL_SOURCE	"logical source"	17.1.2.6
at_MEDIUM_TYPE	"medium type"	17.1.2.12

Table G-7. Attributes Sorted by Identifier, I

Identifier	Name	Section
at_NEW_LAYOUT_OBJECT	"new layout object"	17.1.5.7
at_OBJECT_CLASS	"object class"	17.1.1.6
at_OBJECT_CLASS_IDENTIFIER	"object class identifier"	17.1.1.3
at_OBJECT_IDENTIFIER	"object identifier"	17.1.1.2
at_OBJECT_TYPE	"object type"	17.1.1.1
at_OFFSET	"offset"	17.1.5.8
at_PAGE_POSITION	"page position"	17.1.2.11
at_PERMITTED_CATEGORIES	"permitted categories"	17.1.2.7
at_POSITION	"position"	17.1.2.1
at_PRESENTATION_STYLE	"presentation style"	17.1.1.10
at_PRESENTATION_STYLE_ID	"presentation style identifier"	17.1.6.1
at_PROTECTION	"protection"	17.1.3.1
at_RESOURCE	"resource"	17.1.1.9
at_SAME_LAYOUT_OBJECT	"same layout object"	17.1.5.9
at_SEPARATION	"separation"	17.1.5.10
at_SUBORDINATES	"subordinates"	17.1.1.7
at_SYNCHRONIZATION	"synchronization"	17.1.5.11
at_TRANSPARENCY	"transparency"	17.1.2.9
at_TYPE_OF_CODING	"type of coding"	17.1.7.3
at_USER_READABLE_COMMENTS	"user-readable comments"	17.1.1.13
at_USER_VISIBLE_NAME	"user-visible name"	17.1.1.15
cc_1ST_LINE_OFFSET	"first line offset"	17.1.9.7
cc_ALIGNMENT	"alignment"	17.1.9.1
cc_CHAR_ORIENTATION	"character orientation"	17.1.9.3
cc_CHARACTER_FONTS	"character fonts"	17.1.9.2
cc_CHARACTER_PATH	"character path"	17.1.9.4
cc_CHARACTER_SPACING	"character spacing"	17.1.9.5
cc_CODE_EXT_ANNOUNCERS	"code extension announcers"	17.1.9.6

Table G-8. Attributes Sorted by Identifier, II

Identifier	Name	Section
cc_FORMATTING_INDICATOR	"formatting indicator"	17.1.9.17
cc_GRAPHIC_CHAR_SUBREP	"graphic character subrepertoire"	17.1.9.9
cc_GRAPHIC_CHARACTER_SETS	"graphic character sets"	17.1.9.8
cc_GRAPHIC_RENDITION	"graphic rendition"	17.1.9.10
cc_INDENTATION	"indentation"	17.1.9.19
cc_INITIAL_OFFSET	"initial offset"	17.1.9.18
cc_ITEMISATION	"itemisation"	17.1.9.11
cc_KERNING_OFFSET	"kerning offset"	17.1.9.12
cc_LINE_LAYOUT_TABLE	"line layout table"	17.1.9.13
cc_LINE_PROGRESSION	"line progression"	17.1.9.14
cc_LINE_SPACING	"line spacing"	17.1.9.15
cc_ORPHAN_SIZE	"orphan size"	17.1.9.20
cc_PAIRWISE_KERNING	"pairwise kerning"	17.1.9.16
cc_PROP_LINE_SPACING	"proportional line spacing"	17.1.9.21
cc_WIDOW_SIZE	"widow size"	17.1.9.22
dp_ABSTRACT	"abstract"	17.1.8.39
dp_ACCESS_RIGHTS	"access rights"	17.1.8.65
dp_ADDITIONAL_INFORMATION	"additional information"	17.1.8.56
dp_ALT_REP_CHARACTER_SETS	"alternative representation character sets"	17.1.8.18
dp_AUTHORIZATION	"authorization"	17.1.8.63
dp_AUTHORS	"authors"	17.1.8.51
dp_BLOCK_ALIGNMENTS	"block alignments"	17.1.8.23
dp_CODING_ATTRIBUTES	"coding attributes"	17.1.8.30
dp_COLOURS	"colours"	17.1.8.26
dp_COMMENTS_CHARACTER_SETS	"comments character sets"	17.1.8.17
dp_CONTENT_ARCH_CLASSES	"content architecture classes"	17.1.8.13
dp_COPYRIGHT	"copyright"	17.1.8.52
dp_CREATION_DATE_AND_TIME	"creation date and time"	17.1.8.41

Table G-9. Attributes Sorted by Identifier, III

Identifier	Name	Section
dp_DISTRIBUTION_LIST	"distribution list"	17.1.8.55
dp_DOC_APP_PROF_DEFAULTS	"document application profile defaults"	17.1.8.11
dp_DOC_APP_PROFILE	"document application profile"	17.1.8.10
dp_DOC_ARCH_CLASS	"document architecture class"	17.1.8.12
dp_DOCUMENT_DATE_AND_TIME	"document date and time"	17.1.8.40
dp_DOCUMENT_REFERENCE	"document reference"	17.1.8.37
dp_DOCUMENT_SIZE	"document size"	17.1.8.60
dp_DOCUMENT_TYPE	"document type"	17.1.8.38
dp_EXPIRY_DATE_AND_TIME	"expiry date and time"	17.1.8.43
dp_EXTERNAL_DOCUMENT_CLASS	"external document class"	17.1.8.7
dp_FILL_ORDERS	"fill orders"	17.1.8.24
dp_FONTS_LIST	"fonts list"	17.1.8.34
dp_GENERIC_LAY_STRUCT	"generic layout structure"	17.1.8.1
dp_GENERIC_LOG_STRUCT	"generic logical structure"	17.1.8.3
dp_INTERCHANGE_FMT_CLASS	"interchange format class"	17.1.8.14
dp_LANGUAGES	"languages"	17.1.8.62
dp_LAYOUT_PATHS	"layout paths"	17.1.8.21
dp_LAYOUT_STYLES	"layout styles"	17.1.8.5
dp_LOCAL_FILE_REFERENCE	"local file reference"	17.1.8.59
dp_LOCAL_FILING_DATE_TIME	"local filing date and time"	17.1.8.42
dp_MEDIUM_TYPES	"medium types"	17.1.8.20
dp_NUMBER_OBJECTS_PER_PAGE	"number of objects per page"	17.1.8.32
dp_NUMBER_OF_PAGES	"number of pages"	17.1.8.61
dp_ODA_VERSION	"ODA version"	17.1.8.15
dp_ORGANIZATIONS	"organizations"	17.1.8.48
dp_OWNERS	"owners"	17.1.8.50
dp_PAGE_DIMENSIONS	"page dimensions"	17.1.8.19
dp_PAGE_POSITIONS	"page positions"	17.1.8.28

Table G-10. Attributes Sorted by Identifier, IV

Identifier	Name	Section
dp_PREPARERS	"preparers"	17.1.8.49
dp_PRESENTATION_FEATURES	"presentation features"	17.1.8.31
dp_PRESENTATION_STYLES	"presentation styles"	17.1.8.6
dp_PROFILE_CHARACTER_SETS	"profile character sets"	17.1.8.16
dp_PROTECTIONS	"protections"	17.1.8.22
dp_PURGE_DATE_AND_TIME	"purge date and time"	17.1.8.45
dp_REFS_TO_OTHER_DOCS	"references to other documents"	17.1.8.57
dp_RELEASE_DATE_AND_TIME	"release date and time"	17.1.8.46
dp_RESOURCE_DOCUMENT	"resource document"	17.1.8.8
dp_RESOURCES	"resources"	17.1.8.9
dp_REVISION_HISTORY	"revision history"	17.1.8.47
dp_SECURITY_CLASSIFICATION	"security classification"	17.1.8.64
dp_SPECIFIC_LAY_STRUCT	"specific layout structure"	17.1.8.2
dp_SPECIFIC_LOG_STRUCT	"specific logical structure"	17.1.8.4
dp_START_DATE_AND_TIME	"start date and time"	17.1.8.44
dp_STATUS	"status"	17.1.8.53
dp_SUBJECT	"subject"	17.1.8.36
dp_SUPERSEDED_DOCUMENTS	"superseded documents"	17.1.8.58
dp_TITLE	"title"	17.1.8.35
dp_TRANSPARENCIES	"transparencies"	17.1.8.25
dp_TYPES_OF_CODING	"types of coding"	17.1.8.29
dp_UNIT_SCALING	"unit scaling"	17.1.8.33
dp_USER_SPECIFIC_CODES	"user specific codes"	17.1.8.54
gg_COLOUR_REPS	"colour representations"	17.1.12.7
gg_EDGE_RENDITION	"edge rendition"	17.1.12.6
gg_ENCODING_ANNOUNCER	"geometric graphics encoding announcer"	17.1.12.1
gg_FILLED_AREA_REND	"filled area rendition"	17.1.12.5
gg_LINE_RENDITION	"line rendition"	17.1.12.2

Table G-11. Attributes Sorted by Identifier, V

Identifier	Name	Section
gg_MARKER_RENDITION	"marker rendition"	17.1.12.3
gg_PICTURE_DIMENSIONS	"picture dimensions"	17.1.12.12
gg_PICTURE_ORIENTATION	"picture orientation"	17.1.12.11
gg_REGION_INTEREST_SPEC	"region of interest specification"	17.1.12.10
gg_TEXT_RENDITION	"text rendition"	17.1.12.4
gg_TRANSFORMATION_SPEC	"transformation specification"	17.1.12.9
gg_TRANSPARENCY_SPEC	"transparency specification"	17.1.12.8
rg_CLIPPING	"clipping"	17.1.10.1
rg_COMPRESSION	"compression"	17.1.11.1
rg_IMAGE_DIMENSIONS	"image dimensions"	17.1.10.6
rg_INITIAL_OFFSET	"initial offset"	17.1.10.4
rg_LINE_PROGRESSION	"line progression"	17.1.10.2
rg_NUMBER_DISCARDED_PELS	"number of discarded pels"	17.1.11.4
rg_NUMBER_OF_LINES	"number of lines"	17.1.11.2
rg_NUMBER_OF_PELS_PER_LINE	"number of pels per line"	17.1.11.3
rg_PEL_PATH	"pel path"	17.1.10.3
rg_PEL_SPACING	"pel spacing"	17.1.10.7
rg_PEL_TRANS_DENSITY	"pel transmission density"	17.1.10.5
rg_SPACING_RATIO	"spacing ratio"	17.1.10.8

Table G-12. Attributes Sorted by Identifier, VI

H
Alphabetical List of Types

Identifier	Section
ASN1_OBJECT_ID_type	13.8
at_ALT_REP_type	17.1.7.5
at_APPLICATION_COMMENTS_type	17.1.1.14
at_BALANCE_type	17.1.2.4
at_BINDINGS_type	17.1.1.16
at_BLK_ALGNM_type	17.1.5.1
at_BORDER_type	17.1.2.3
at_COLOUR_type	17.1.2.10
at_CONCATENATION_type	17.1.5.2
at_CONTENT_ARCH_CLASS_type	17.1.1.11
at_CONTENT_GENERATOR_type	17.1.1.5
at_CONTENT_ID_LAYOUT_type	17.1.7.2
at_CONTENT_ID_LOGICAL_type	17.1.7.1
at_CONTENT_INFORMATION_type	17.1.7.4
at_CONTENT_PORTIONS_type	17.1.1.8
at_CONTENT_TYPE_type	17.1.1.12
at_DEFAULT_VALUE_LISTS_type	17.1.1.17
at_DIMENSIONS_type	17.1.2.2
at_FILL_ORDER_type	17.1.5.3
at_GEN_FOR_SUBORDINATES_type	17.1.1.4

Table H-1. Types, I

Identifier	Section
at_IMAGING_ORDER_type	17.1.2.8
at_INDIVISIBILITY_type	17.1.5.4
at_LAYOUT_CATEGORY_type	17.1.5.5
at_LAYOUT_OBJECT_CLASS_type	17.1.5.6
at_LAYOUT_PATH_type	17.1.2.5
at_LAYOUT_STYLE_IDENTIFIER_type	17.1.4.1
at_LAYOUT_STYLE_type	17.1.3.2
at_LOGICAL_SOURCE_type	17.1.2.6
at_MEDIUM_TYPE_type	17.1.2.12
at_NEW_LAYOUT_OBJECT_type	17.1.5.7
at_OBJECT_CLASS_IDENTIFIER_type	17.1.1.3
at_OBJECT_CLASS_type	17.1.1.6
at_OBJECT_IDENTIFIER_type	17.1.1.2
at_OBJECT_TYPE_type	17.1.1.1
at_OFFSET_type	17.1.5.8
at_PAGE_POSITION_type	17.1.2.11
at_PERMITTED_CATEGORIES_type	17.1.2.7
at_POSITION_type	17.1.2.1
at_PRESENTATION_STYLE_ID_type	17.1.6.1
at_PRESENTATION_STYLE_type	17.1.1.10
at_PROTECTION_type	17.1.3.1
at_RESOURCE_type	17.1.1.9
at_SAME_LAYOUT_OBJECT_type	17.1.5.9
at_SEPARATION_type	17.1.5.10
at_SUBORDINATES_type	17.1.1.7
at_SYNCHRONIZATION_type	17.1.5.11
at_TRANSPARENCY_type	17.1.2.9
at_TYPE_OF_CODING_type	17.1.7.3

Table H-2. Types, II

Identifier	Section
at_USER_READABLE_COMMENTS_type	17.1.1.13
at_USER_VISIBLE_NAME_type	17.1.1.15
ATOMIC_STRING_EXPR_type	18.1
BINDING_PAIR_type	17.1.1.16
BINDING_VALUE_type	17.1.1.16
BOOL_type	10.1.4
BORDER_INFO_type	17.1.2.3
BYTE_type	10.1.3
cc_1ST_LINE_OFFSET_type	17.1.9.7
cc_ALIGNMENT_type	17.1.9.1
cc_CHAR_ORIENTATION_type	17.1.9.3
cc_CHARACTER_FONTS_type	17.1.9.2
cc_CHARACTER_PATH_type	17.1.9.4
cc_CHARACTER_SPACING_type	17.1.9.5
cc_CODE_EXT_ANNOUNCERS_type	17.1.9.6
cc_FORMATTING_INDICATOR_type	17.1.9.17
cc_GRAPHIC_CHAR_SUBREP_type	17.1.9.9
cc_GRAPHIC_CHARACTER_SETS_type	17.1.9.8
cc_GRAPHIC_RENDITION_type	17.1.9.10
cc_INDENTATION_type	17.1.9.19
cc_INITIAL_OFFSET_type	17.1.9.18
cc_ITEMISATION_type	17.1.9.11
cc_KERNING_OFFSET_type	17.1.9.12
cc_LINE_LAYOUT_TABLE_type	17.1.9.13
cc_LINE_PROGRESSION_type	17.1.9.14
cc_LINE_SPACING_type	17.1.9.15
cc_ORPHAN_SIZE_type	17.1.9.20
cc_PAIRWISE_KERNING_type	17.1.9.16

Table H-3. Types, III

Identifier	Section
cc_PROP_LINE_SPACING_type	17.1.9.21
cc_WIDOW_SIZE_type	17.1.9.22
CHAR_type	10.1.2
CONSTITUENT_type	13.2
CONSTRUCTION_EXPR_type	18.4
CONSTRUCTION_FACTOR_type	18.4
CONSTRUCTION_TERM_type	18.4
COPYRIGHT_INFO_type	17.1.8.52
CURR_INSTANCE_PARMS_type	18.3
DIMENSION_type	17.1.2.2
DOCUMENT_type	13.3
dp_ABSTRACT_type	17.1.8.39
dp_ACCESS_RIGHTS_type	17.1.8.65
dp_ADDITIONAL_INFORMATION_type	17.1.8.56
dp_ALT_REP_CHARACTER_SETS_type	17.1.8.18
dp_AUTHORIZATION_type	17.1.8.63
dp_AUTHORS_type	17.1.8.51
dp_BLOCK_ALIGNMENTS_type	17.1.8.23
dp_BORDERS_type	17.1.8.27
dp_CODING_ATTRIBUTES_type	17.1.8.30
dp_COLOURS_type	17.1.8.26
dp_COMMENTS_CHARACTER_SETS_type	17.1.8.17
dp_CONTENT_ARCH_CLASSES_type	17.1.8.13
dp_COPYRIGHT_type	17.1.8.52
dp_CREATION_DATE_AND_TIME_type	17.1.8.41
dp_DISTRIBUTION_LIST_type	17.1.8.55
dp_DOC_APP_PROF_DEFAULTS_type	17.1.8.11
dp_DOC_APP_PROFILE_type	17.1.8.10

Table H-4. Types, IV

Identifier	Section
dp_DOC_ARCH_CLASS_type	17.1.8.12
dp_DOCUMENT_DATE_AND_TIME_type	17.1.8.40
dp_DOCUMENT_REFERENCE_type	17.1.8.37
dp_DOCUMENT_SIZE_type	17.1.8.60
dp_DOCUMENT_TYPE_type	17.1.8.38
dp_EXPIRY_DATE_AND_TIME_type	17.1.8.43
dp_EXTERNAL_DOCUMENT_CLASS_type	17.1.8.7
dp_FILL_ORDERS_type	17.1.8.24
dp_FONTS_LIST_type	17.1.8.34
dp_GENERIC_LAY_STRUCT_type	17.1.8.1
dp_GENERIC_LOG_STRUCT_type	17.1.8.3
dp_INTERCHANGE_FMT_CLASS_type	17.1.8.14
dp_LANGUAGES_type	17.1.8.62
dp_LAYOUT_PATHS_type	17.1.8.21
dp_LAYOUT_STYLES_type	17.1.8.5
dp_LOCAL_FILE_REFERENCE_type	17.1.8.59
dp_LOCAL_FILING_DATE_TIME_type	17.1.8.42
dp_MEDIUM_TYPES_type	17.1.8.20
dp_NUMBER_OBJECTS_PER_PAGE_type	17.1.8.32
dp_NUMBER_OF_PAGES_type	17.1.8.61
dp_ODA_VERSION_type	17.1.8.15
dp_ORGANIZATIONS_type	17.1.8.48
dp_OWNERS_type	17.1.8.50
dp_PAGE_DIMENSIONS_type	17.1.8.19
dp_PAGE_POSITIONS_type	17.1.8.28
dp_PREPARERS_type	17.1.8.49
dp_PRESENTATION_FEATURES_type	17.1.8.31
dp_PRESENTATION_STYLES_type	17.1.8.6

Table H-5. Types, V

Identifier	Section
dp_PROFILE_CHARACTER_SETS_type	17.1.8.16
dp_PROTECTIONS_type	17.1.8.22
dp_PURGE_DATE_AND_TIME_type	17.1.8.45
dp_REFS_TO_OTHER_DOCS_type	17.1.8.57
dp_RELEASE_DATE_AND_TIME_type	17.1.8.46
dp_RESOURCE_DOCUMENT_type	17.1.8.8
dp_RESOURCES_type	17.1.8.9
dp_REVISION_HISTORY_type	17.1.8.47
dp_SECURITY_CLASSIFICATION_type	17.1.8.64
dp_SPECIFIC_LAY_STRUCT_type	17.1.8.2
dp_SPECIFIC_LOG_STRUCT_type	17.1.8.4
dp_START_DATE_AND_TIME_type	17.1.8.44
dp_STATUS_type	17.1.8.53
dp_SUBJECT_type	17.1.8.36
dp_SUPERSEDED_DOCUMENTS_type	17.1.8.58
dp_TITLE_type	17.1.8.35
dp_TRANSPARENCIES_type	17.1.8.25
dp_TYPES_OF_CODING_type	17.1.8.29
dp_UNIT_SCALING_type	17.1.8.33
dp_USER_SPECIFIC_CODES_type	17.1.8.54
FIXED_POSITION_type	17.1.2.1
FLOAT_type	23.7.1
FONT_PAIR_type	17.1.8.34
FONT_REFERENCE_type	13.11
FORMAT_type	24.4
FORMAT_GROUP3_type	24.5.4
FORMAT_GROUP4_type	24.5.5
FORMAT_ODABM_type	24.5.7
gg_COLOUR_REPS_type	17.1.12.7
gg_EDGE_RENDITION_type	17.1.12.6
gg_ENCODING_ANNOUNCER_type	17.1.12.1
gg_FILLED_AREA_REND_type	17.1.12.5
gg_LINE_RENDITION_type	17.1.12.2

Table H-6. Types, VI

Identifier	Section
gg_MARKER_RENDITION_type	17.1.12.3
gg_PICTURE_DIMENSIONS_type	17.1.12.12
gg_REGION_INTEREST_SPEC_type	17.1.12.10
gg_TEXT_RENDITION_type	17.1.12.4
gg_TRANSFORMATION_SPEC_type	17.1.12.9
gg_TRANSPARENCY_SPEC_type	17.1.12.8
IMAGE_SIZE_type	17.1.10.6
INIT_PARMS_type	14.1.1
INT_PAIR_type	13.5
INT_type	10.1.1
ISO9541_ID_type	13.11
ITERATOR_type	13.4
LAYOUT_OBJECT_type	13.9
MASK_type	10.1.7
MODAL_FONT_ATTR_type	13.11
NAME_ORG_type	13.7
NUMERIC_EXPR_type	18.2
NUMERIC_FUNCTION_type	18.2
OBJECT_ID_EXPR_type	18.3
OBJECT_REF_type	18.6
PARM_MASK_type	10.1.7
PEL_type	24.2
PERSONAL_NAME_type	13.6
POINTER_type	10.1.8
RASTER_type	24.1
REF_BINDING_VALUE_type	18.5
RESOURCE_MAP_type	17.1.8.9
REVISER_type	17.1.8.47
REVISION_HISTORY_type	17.1.8.47
rg_CLIPPING_type	17.1.10.1

Table H-7. Types, VII

Identifier	Section
rg_COMPRESSION_type	17.1.11.1
rg_IMAGE_DIMENSIONS_type	17.1.10.6
rg_INITIAL_OFFSET_type	17.1.10.4
rg_LINE_PROGRESSION_type	17.1.10.2
rg_NUMBER_DISCARDED_PELS_type	17.1.11.4
rg_NUMBER_OF_LINES_type	17.1.11.2
rg_NUMBER_OF_PELS_PER_LINE_type	17.1.11.3
rg_PEL_PATH_type	17.1.10.3
rg_PEL_SPACING_type	17.1.10.7
rg_PEL_TRANS_DENSITY_type	17.1.10.5
rg_SPACING_RATIO_type	17.1.10.8
RULE_type	17.1.2.2
SCORE_type	13.11
SCORES_type	13.11
SEQUENCE_ASN1_OBJECT_ID_type	13.13.1
SEQUENCE_at_MEDIUM_TYPE_type	13.13.1
SEQUENCE_ATOMIC_STR_EXPR_type	13.13.1
SEQUENCE_BINDING_PAIR_type	13.13.1
SEQUENCE_BYTE_type	13.13.1
SEQUENCE_CHAR_type	13.13.1
SEQUENCE_CONSTITUENT_type	13.13.1
SEQUENCE_CONSTRUCTION_TERM_type	13.13.1
SEQUENCE_COPYRIGHT_INFO_type	13.13.1
SEQUENCE_FONT_PAIR_type	13.13.1
SEQUENCE_INT_PAIR_type	13.13.1
SEQUENCE_INT_type	13.13.1
SEQUENCE_ISO9541_ID_type	13.13.1
SEQUENCE_NAME_ORG_type	13.13.1

Table H-8. Types, VIII

Identifier	Section
SEQUENCE_PERSONAL_NAME_type	13.13.1
SEQUENCE_RESOURCE_MAP_type	13.13.1
SEQUENCE_REVISER_type	13.13.1
SEQUENCE_REVISION_HISTORY_type	13.13.1
SEQUENCE_STRING_OR_ID_type	13.13.1
SEQUENCE_STRING_type	13.13.1
SEQUENCE_TAB_STOP_type	13.13.1
SEQUENCE_type	13.13.1
SEQUENCE_VALUE_type	13.13.1
STRING_EXPR_type	18.1
STRING_FUNCTION_type	18.1
STRING_OR_ID_type	13.10
STRING_type	13.14
STRUCTURED_NAME_type	13.11
SWAP_INFO_type	10.1.9
TAB_STOP_type	17.1.9.13
TAG_type	10.1.6
VALUE_type	10.1.5
VARIABLE_POSITION_type	17.1.2.1
VARIANT_SCRIPT_type	13.11
VARIANT_SCRIPTS_type	13.11
WRITING_MODES_type	13.11

Table H-9. Types, IX

I
List of Real and Alias Names

As mentioned in section 9.3, for reasons of portability, all link-time identifiers in the CMU ODA Tool Kit are restricted in length. For readability, however, such identifiers also have an *alias*, a longer readable name that is used throughout the source code. The alias is defined to map to the real name by a compile-time macro.

Although the long identifiers appear in the source code, only the shorter names are known to run-time debuggers. To aid in the debugging of applications using the CMU ODA Tool Kit or the Raster Graphics Tool Kit, this appendix contains listings of all alias-real name pairs, including those not visible to application programmers. The first listing is alphabetized by alias, the second by real name.

I.1. Alphabetized by Alias

Alias	Real name
ATTR_CheckDAPDCOnstituents	gatcdc
ATTR_FindConstituent	gatfco
ATTR_FindRightObjectType	gatfro
ATTR_GetConstituentKind	gatgck
ATTR_GetDocProfDefaults	gatgdp
ATTR_IsCCCoding	gaticc
ATTR_IsCCPresentation	gatccp
ATTR_IsGGCoding	gatigc
ATTR_IsGGPresentation	gatigp
ATTR_IsLayoutDirective	gatild
ATTR_IsPresentationAttribute	gatipa

Table I-1. Sorted by Alias, I

Alias	Real name
ATTR_IsRGCoding	gatirc
ATTR_IsRGPresentation	gatirp
ATTR_RummageThruDAPD	gatrtd
ANDY_ReadRaster	rganrd
ANDY_WriteRaster	rganwr
ATTR16_Dummy	ga16dm
ATTR17_Dummy	ga17dm
ATTR18_Dummy	ga18dm
ATTR_AccumulateContent	gatacc
ATTR_AddToConstituent	gatatc
ATTR_AllocASN1ObjectId	gataao
ATTR_AllocAtomicStrExpr	gataas
ATTR_AllocBindingPair	gatabp
ATTR_AllocCopyright	gatacr
ATTR_AllocCurrInstanceParms	gataci
ATTR_AllocFontPairBody	gatfpb
ATTR_AllocISO9541Id	gatiid
ATTR_AllocNameOrg	gatano
ATTR_AllocNumericExpr	gatane
ATTR_AllocNumericFunction	gatanf
ATTR_AllocObjectIdExpr	gataoi
ATTR_AllocObjectRef	gataor
ATTR_AllocPersonalName	gatapn
ATTR_AllocRefBindingValue	gatarb
ATTR_AllocResourceMap	gatarm
ATTR_AllocReviser	gatarv
ATTR_AllocRevisionHistory	gatarh
ATTR_AllocSequencePersonalNames	gatasp
ATTR_AllocSequenceRevisers	gatasr
ATTR_AllocStringExpr	gatase
ATTR_AllocStringFunction	gatasf
ATTR_AllocStructuredName	gatasn
ATTR_Alloc_String_or_id_value	gatasi
ATTR_AreWeDone	gatawd
ATTR_AttemptInheritFromDVL	gatadv
ATTR_AttrNameTbl	gatant

Table I-2. Sorted by Alias, II

Alias	Real name
ATTR_AttrTypeTbl	gatatt
ATTR_BORDER_INFO_CkVal	gatbik
ATTR_BORDER_INFO_Copy	gatbic
ATTR_BORDER_INFO_Deflt	gatbid
ATTR_CheckAndCleanMask	gatccm
ATTR_CheckDAPDConstituents	gatdpa
ATTR_CheckId	gatcid
ATTR_CheckStringExpr	gatkse
ATTR_CheckTabStop	gatcts
ATTR_Check_at_MEDIUM_TYPE	gatcmt
ATTR_CkValAtomicStrExpr	gatkas
ATTR_CkValBindingPair	gatkbp
ATTR_CkValCurrInstanceParms	gatkci
ATTR_CkValNumericExpr	gatkne
ATTR_CkValNumericFunction	gatknf
ATTR_CkValObjectIdExpr	gatkoi
ATTR_CkValObjectRef	gatkor
ATTR_CkValRefBindingValue	gatkrb
ATTR_CkValStringFunction	gatksf
ATTR_CkVal_Dfnpre	gfrkdf
ATTR_CkVal_ISO9541_Id_Type	gfrkis
ATTR_CkVal_Int_Pair_Type	gfrkip
ATTR_CkVal_Int_Type	gfrkit
ATTR_CkVal_Modal_Font_Attr_Type	gfrkmf
ATTR_CkVal_Score_Type	gfrksc
ATTR_CkVal_Scores_Type	gfrkst
ATTR_CkVal_Sequence_Int_Type	gfrksi
ATTR_CkVal_Structured_Name	gfrksn
ATTR_CkVal_Tabesc	gfrktb
ATTR_CkVal_Variant_Script_Type	gfrkvt
ATTR_CkVal_Variant_Scripts_Type	gfrkvs
ATTR_CkVal_Writing_Modes_Type	gfrkwm
ATTR_Ck_Fields	gfrkcf
ATTR_Ck_String_or_id_value	gatcsi
ATTR_CopyAtomicStrExpr	gatcas
ATTR_CopyBindingPair	gatcbp

Table I-3. Sorted by Alias, III

Alias	Real name
ATTR_CopyCopyRight	gatccr
ATTR_CopyCurrInstanceParms	gatcip
ATTR_CopyNameOrg	gatcno
ATTR_CopyNumericExpr	gatcne
ATTR_CopyNumericFunction	gatcnf
ATTR_CopyObjectIdExpr	gatoie
ATTR_CopyObjectRef	gatcor
ATTR_CopyPersonalName	gatcpn
ATTR_CopyRefBindingValue	gatrgv
ATTR_CopyResourceMap	gatcrm
ATTR_CopyReviser	gatcrv
ATTR_CopyRevisionHistory	gatcrh
ATTR_CopySequencePersonalNames	gatcsp
ATTR_CopySequenceRevisers	gatcsr
ATTR_CopyStringExpr	gatcse
ATTR_CopyStringFunction	gatcsf
ATTR_Copy_Dfnpre	gfrcdf
ATTR_Copy_Fields	gatcpf
ATTR_Copy_ISO9541_Id_Type	gfrcis
ATTR_Copy_Int_Pair_Type	gfrcip
ATTR_Copy_Int_Type	gfrcit
ATTR_Copy_Modal_Font_Attr_Type	gftcmf
ATTR_Copy_Score_Type	gfrcsc
ATTR_Copy_Scores_Type	gfrcst
ATTR_Copy_Sequence_Int_Type	gfrcsi
ATTR_Copy_Structured_Name	gfrcsn
ATTR_Copy_Tabesc	gfrctb
ATTR_Copy_Variant_Script_Type	gfrcvt
ATTR_Copy_Variant_Scripts_Type	gfrcvs
ATTR_Copy_Writing_Modes_Type	gfrcwm
ATTR_Final	gatfnl
ATTR_FindAndRemoveAttr	gatfar
ATTR_FindAttr	gfndat
ATTR_FindContentArchClass	gatfca
ATTR_FindLocal	gatflc
ATTR_FontPairCkVal	gfrkfp

Table I-4. Sorted by Alias, IV

Alias	Real name
ATTR_FontPairCopy	gatfpc
ATTR_FreeAsn1ObjectId	gatfa1
ATTR_FreeAtomicStrExpr	gatfas
ATTR_FreeBindingPair	gatfbp
ATTR_FreeCopyrightInfo	gatfcp
ATTR_FreeCurrInstanceParms	gatfci
ATTR_FreeNameOrg	gatfno
ATTR_FreeNumericExpr	gatfne
ATTR_FreeNumericFunction	gatfnf
ATTR_FreeObjectIdExpr	gatfoi
ATTR_FreeObjectRef	gatfor
ATTR_FreePersonalName	gatfpn
ATTR_FreeRefBindingValue	gatfrb
ATTR_FreeResourceMap	gatfrm
ATTR_FreeReviser	gatfrv
ATTR_FreeRevisionHistory	gatfrh
ATTR_FreeStringExpr	gatfse
ATTR_FreeStringFunction	gatfsf
ATTR_FreeStrings	gatfsr
ATTR_FreeTabStop	gatfts
ATTR_Free_Some_String	gatfst
ATTR_Free_String_or_id_value	gatfsi
ATTR_GetDefaults	gatgdf
ATTR_GetFontReferenceBase	gagfrb
ATTR_GetModalFontAttrBase	ggmfab
ATTR_GetObjClContentKind	gatgcl
ATTR_GetObjectContentKind	gatgoc
ATTR_GetScoreBase	gagscb
ATTR_GetScoresBase	ggscsb
ATTR_GetTabescBase	gagteb
ATTR_GetVariantScriptBase	gagvsb
ATTR_GetVariantScriptsBase	ggvssb
ATTR_GetWritingModesBase	gagwmb
ATTR_Hash	gathsh
ATTR_InheritForContentPortion	gaticp
ATTR_InheritFromConst	gatifc

Table I-5. Sorted by Alias, V

Alias	Real name
ATTR_InheritFromConstAndStyle	gatics
ATTR_InheritFromDVL	gatdvl
ATTR_InheritFromDVLConstituent	gatdvc
ATTR_InheritFromObjClass	gatioc
ATTR_InheritFromStyle	gatifs
ATTR_InhertForObject	gatifo
ATTR_Init	gatint
ATTR_KindOfAttr	gatkoa
ATTR_LoadDestination	gatldt
ATTR_LocateConstAttr	gatlca
ATTR_LookForContentGenerator	gatfcg
ATTR_LookInResourceDoc	gatlrd
ATTR_LookOnConstituent	gatloc
ATTR_LookOnObjClass	gatooc
ATTR_SetUpReturnToUser	gatsru
ATTR_SterilizeParmMask	gatsmk
ATTR_ThreadTree	gatttr
ATTR_UnThreadTree	gatutt
ATTS_FillInOneDefaultConst	gasfod
ATTS_Final	gasfnl
ATTS_Init	gasint
AddChars	gtkach
AppendSequence	appseq
AppendSubordinate	appsub
Array2StrctName	ary2sn
AssignBit	rgpeab
AssignIdAttr	glbaia
AssignPel	rgpeap
AttrDocProfPriority	dppri
BlackTreeTop	rgvbtt
BreakUpName	gasbun
CONSTITUENT_Alloc	gzabfa
CONSTITUENT_CkVal	gzabez
CONSTITUENT_Copy	gzabey
CONSTITUENT_Deflt	gzabfe
CONSTITUENT_Free	gzabfb

Table I-6. Sorted by Alias, VI

Alias	Real name
CONSTITUENT_Get	gzabfd
CONSTITUENT_Set	gzabfc
CONSTITUENT_WDbug	gzabex
CONSTRUCTION_EXPR_Alloc	gzabfi
CONSTRUCTION_EXPR_CkVal	gzabfh
CONSTRUCTION_EXPR_Copy	gzabfg
CONSTRUCTION_EXPR_Deflt	gzabfm
CONSTRUCTION_EXPR_Free	gzabfj
CONSTRUCTION_EXPR_Get	gzabfl
CONSTRUCTION_EXPR_Set	gzabfk
CONSTRUCTION_EXPR_WDbug	gzabff
CONSTRUCTION_FACTOR_Alloc	gcfall
CONSTRUCTION_FACTOR_CkVal	gcfckv
CONSTRUCTION_FACTOR_Copy	gcfcpy
CONSTRUCTION_FACTOR_Free	gcffre
CONSTRUCTION_TERM_Alloc	gctall
CONSTRUCTION_TERM_CkVal	gctckv
CONSTRUCTION_TERM_Copy	gctcpy
CONSTRUCTION_TERM_Free	gctfre
CONST_AddDoublyLinked	gcsadl
CONST_AddSpecialConstituent	gcsasc
CONST_Chameleon	gcscha
CONST_ChangeContentPortionRefs	gcsccr
CONST_CleanUpConstituent	gcscuc
CONST_CollaspeContentPortions	gcsccp
CONST_DoSetComponentWhatKind	gcdswk
CONST_FillInConstituent	gcsfic
CONST_Final	gcsfnl
CONST_Init	gcsint
CONST_InsertConstituent	gcsics
CONST_MakeConstituent	gcsmct
CONST_MoveConstituent	gcsmcs
CONST_RemoveConstituent	gcsrct
CONST_RemoveDoublyLinked	gcsrdl
CONST_RemoveSpecialConstituent	gcsrsc
CONST_SetComponentWhatKind	gcswk

Table I-7. Sorted by Alias, VII

Alias	Real name
CONST_SetDefComponentWhatKind	gcsdwk
CONST_SetDepWhatKind	gcspwk
CONST_SetIndepWhatKind	gcsiwk
CONST_SetOtherFields	gcsofs
Chars2String	chrstr
ComponentAttrPriority	compri
ConcatString	catstr
ContentAttrPriority	conpri
Contents	grocon
ContractSequence	conseq
CopyConstituent	cpcons
CopySequence	cpyseq
CreateAndAssignId	glbcid
DEBUG_Final	gdbfnl
DEBUG_Init	gdbint
DEBUG_WalkGenericObject	dbgwgo
DEBUG_WriteBytes	dbgwbs
DEBUG_WriteDebugConstituent	dbgwcn
DIMENSION_type_CkVal	gatdck
DIMENSION_type_Copy	gatdtc
DOC_CheckRoot	gitcrt
DOC_DocumentList	gdoclt
DOC_Final	gdcfnl
DOC_Init	gdcint
DOC_ZapConstituents	gdoczc
DeleteConstExpr	dlcnex
DeleteConstTerm	gasdct
DeleteConstituent	delcns
DeleteDocument	deldoc
DeleteIterator	delgen
DeleteRaster	rgradr
DeleteSequence	delseq
DeleteSequenceDeeply	dlsqd
DeleteString	delstr
Digits	gdigit

Table I-8. Sorted by Alias, VIII

Alias	Real name
Dsclass2Ints	ds2int
ERROR_ATTRIBUTE	geratt
ERROR_CONSTITUENT	gercon
ERROR_SEQUENCE	gerseq
ETC_GetNumber	rgetgn
ETC_MagicBuffer	rgvemb
ETC_MagicSize	rgvems
ETC_PutNumber	rgetpn
ETC_ReadRaster	rgetrd
EXPR_Final	gexfnl
EXPR_Init	gexint
Entering	rgentr
ErrorIndex	gtkein
EsclassTable	gdbetb
EvalNumExpr	evlnum
EvalObjIdExpr	evlobi
EvalRefBind	evlrbv
EvalStringExpr	evlstr
ExpandSequence	expseq
ExtractRaster	rgraex
FBM_ReadRaster	rgfbrd
FBM_WriteRaster	rgfbwr
FMT_Final	gfmfnl
FMT_FormatInt	gfmint
FMT_Init	gfmini
FONT_Final	fntfnl
FONT_Init	fntint
FinalRGToolKit	rgfitk
FinalToolKit	tkfinl
FindDocumentProfile	fnddp
FindDocumentRoot	fndrt
FontReferenceFields	gatfrf
FontReferenceInfo	gatfri

Table I-9. Sorted by Alias, IX

Alias	Real name
FormatInt	gfmfnt
FormattingErrors	gtkfer
G3_DecodeOne	rgg3do
G3_EncodeOne	rgg3eo
G3_GetBit	rgg3gb
G3_NextRun	rgg3nr
G3_PutBit	rgg3pb
G3_PutEOL	rgg3pe
G3_PutRunLength	rgg3pr
G3_ReadBitmap2	rgg3r2
G3_ReadRaster	rgg3rd
G3_Traverse	rgg3tr
G3_WriteBitmap2	rgg3w2
G3_WriteRaster	rgg3wr
G4_GetBit	rgg4gb
G4_NextRun	rgg4nr
G4_ReadRaster	rgg4rd
G4_Traverse	rgg4tr
G4_WriteBitmap	rgg4wb
G4_WriteRaster	rgg4wr
GIFEncode	rggfen
GIFNextPixel	rggfnp
GenForSub	grogsu
GenericError	generr
GetAttr	getatt
GetAttrName	gattnm
GetBit	rgpegb
GetContent	gcontt
GetContentKind	gcntkn
GetDocument	getdoc
GetIntAttr	ginatt
GetParmMask	getprm
GetPel	rgpegp
GetSeqValueField	gsqgsv
GetValueAttr	gvlatt
GetValueName	gvalnm

Table I-10. Sorted by Alias, X

Alias	Real name
HasAttr	hasatt
HoldAttr	hldatt
HoldConstituent	hldcns
HoldRaster	rgraho
HoldString	hldstr
INIT_Final	ginfnl
INIT_Init	ginint
INIT_RGParms	rgvrgp
INT_Alloc	gzabjy
INT_CkVal	gzabjx
INT_Copy	gzabjw
INT_Deflt	gzabkc
INT_Free	gzabjz
INT_Get	gzabkb
INT_PAIR_Alloc	gzabkg
INT_PAIR_CkVal	gzabkf
INT_PAIR_Copy	gzabke
INT_PAIR_Deflt	gzabkk
INT_PAIR_Free	gzabkh
INT_PAIR_Get	gzabkj
INT_PAIR_Set	gzabki
INT_PAIR_WDbug	gzabkd
INT_Set	gzabka
INT_WDbug	gzabjv
IO_Final	giofnl
IO_INT_Value	gwaioi
IO_Init	gioint
IO_PrintByte	iopbyt
IO_PrintChar	iopchr
IO_PrintInt	iopint
IO_PrintString	iopstr
IO_ReadByte	iorbyt
IO_ReadBytes	iorbts
IO_ReadChar	iorchr
IO_ReadChars	iorcrs
IO_WriteByte	iowbyt

Table I-11. Sorted by Alias, XI

Alias	Real name
IO_WriteBytes	iowbts
IO_WriteChar	iowchr
IO_WriteChars	iowcrs
ITER_AddGenericKids	gitagk
ITER_AddKidsToParent	gitakp
ITER_AddKidsToQueue	gitakq
ITER_AddSpecificKids	gitask
ITER_Final	gitfnl
ITER_GetObjectID	gitgid
ITER_Init	gitint
ITER_NextSubgraphConstituent	gitnsc
ITER_ShiftDownNodeArray	gitsda
ITER_ShiftUpNodeArray	gitsua
ITER_ValidateArraySize	gitvas
ITER_WalkConsExpr	gitwce
ITER_WalkConsFactor	gitwcf
ITER_WalkConsTerm	gitwct
ITER_WalkSeqConsTerm	gitwsq
IdChars	gidchr
Identical	graide
InitObjectIds	ginioi
InitParms	gintpm
InitRGToolKit	rgintk
InitRouts	ginirt
InitToolKit	tkinit
InitializeDefaultValueLists	intdvl
InsertRaster	rgrais
InternalSubgraphState	gdbiss
Ints2Dsclass	int2ds
IsLayoutDirective	gatild
IsPresentationAttribute	gatipa
LABEL_Final	glbfnl
LABEL_Init	glbint
LAYOUT_OBJECT_Alloc	gzabko
LAYOUT_OBJECT_CkVal	gzabkn

Table I-12. Sorted by Alias, XII

Alias	Real name
LAYOUT_OBJECT_Copy	gzabkm
LAYOUT_OBJECT_Deflt	gzabks
LAYOUT_OBJECT_Free	gzabkp
LAYOUT_OBJECT_Get	gzabkr
LAYOUT_OBJECT_Set	gzabkq
LAYOUT_OBJECT_WDbug	gzabkl
LST_AddElement	rglsad
LST_DeleteElement	rglsde
LST_FreeList	rglsfl
LST_MakeList	rglsml
LST_MoveHead	rglsmh
LST_NextElement	rglsne
LabelConstructionExpr	glblce
LabelConstructionFactor	glblcf
LabelConstructionTerm	glblct
LabelContentPortions	glbcpt
LabelDocument	lbldoc
LabelGenericStructure	glbgst
LabelSeqConstructionTerm	glbsct
LabelSpecificStructure	glblsp
LabelStructure	lblstr
LabelStyles	lblsty
LabelConstructionFactor	glblcf
Leaving	rgleav
LookUpObjectId	lkobid
MACP_ReadRaster	rgmcrd
MACP_WriteRaster	rgmcwr
MEM_Alloced	rgvmal
MEM_Final	gmefnl
MEM_Free	rgmmfr
MEM_Init	gmeint
MEM_Malloc	rgmmma
MEM_Realloc	rgmmre
MM_Final	gmmfin
MM_Free	gmmfre
MM_Init	gmmini

Table I-13. Sorted by Alias, XIII

Alias	Real name
MM_Malloc	gmmmal
MM_Realloc	gmmrel
MakeComponent	mkcomp
MakeContent	mkcont
MakeDocument	mkdoc
MakeDocumentProfile	mkdocp
MakeRaster	rgramr
MakeSequence	mkseq
MakeString	mkstr
MakeStyle	mkstyl
MakeStylesIterator	mkstgn
MakeSubgraphIterator	mksbgn
ModalFontAttrFields	gamfaf
ModalFontAttrInfo	gamfai
NULL_ATTRIBUTE	gnlatt
NULL_CONSTITUENT	gnlcon
NULL_SEQUENCE	gnlseq
Nbytes	gwanby
Ndigits	glbdig
NextConstituent	nxtcns
OBM_ReadRaster	rgobrd
OBM_WriteRaster	rgobwr
PBM_ReadRaster	rgpbrd
PBM_WriteRaster	rgpbwr
PEL_BitOffL	rgvbfl
PEL_BitOffR	rgvbfr
PEL_BitOnL	rgvbnl
PEL_BitOnR	rgvbnr
PRIOR_Final	gprfnl
PRIOR_Init	gprint
PS_WriteRaster	rgpswr
PositiveInteger	gwapos
PostureTable	gdbptb
PrintInt	giopin

Table I-14. Sorted by Alias, XIV

Alias	Real name
PropwdthTable	gdbpwt
RANS1_DefaultPresAttributes	gradpa
RASN1_Asn1ObjectId	graaoi
RASN1_BasicLogicalAttributes	grabla
RASN1_BlockAttributes	grabat
RASN1_CompLogicalAttributes	gracla
RASN1_CompleteStructName	gracsn
RASN1_ContentAttributes	gracat
RASN1_DefaultPresAttributes	gradpa
RASN1_FontType	grafty
RASN1_FrameAttributes	grafat
RASN1_GetIntPair	gragip
RASN1_GetString	gragst
RASN1_GetValue	gragva
RASN1_Identifier	graidn
RASN1_Length	gralng
RASN1_ObjectType	graoty
RASN1_Offset	graoff
RASN1_PageAttributes	grapat
RASN1_RasterAttributes	grarat
RASN1_RasterGrCodingAttrs	grarga
RASN1_ReadByte	grarbt
RASN1_ReadBytes	grarbs
RASN1_RgClipping	grargc
RASN1_RgImageDimensions	grarid
RASN1_RgPelSpacing	grarps
RASN1_RgSetAreaControlled	grarac
RASN1_RgSetSpacing	grarss
RASN1_SaveSeqStrings	grasss
RASN1_Separation	grasep
RASN1_SetValue	grasva
RASN1_String2Constituent	gras2c
RASN1_StructuredName	grasnm
RASN1_TabStop	gratst
RAST_AllocRaster	rgraal
RAST_FinalRasterList	rgrafi

Table I-15. Sorted by Alias, XV

Alias	Real name
RAST_InitRasterList	rgrain
RAST_SwapAlloc	rgrasa
RAST_SwapIn	rgrasi
RGError	rgverr
RIO_FromMemory	rgvfrm
RODIF_AddDocCharacteristics	groadc
RODIF_CharacterAttributes	grocat
RODIF_ConstrExpression	grocep
RODIF_ConstrFactor	grocft
RODIF_ConstrTerm	groctm
RODIF_ContentPortions	grocpt
RODIF_DefaultValueListsLayout	grodll
RODIF_DefaultValueListsLogical	grodvl
RODIF_DocumentCharacteristics	grodch
RODIF_DocumentProfile	grodpr
RODIF_Final	grdfnl
RODIF_FontList	groflt
RODIF_FontReference	grofrf
RODIF_FontsList	grofsl
RODIF_GenForSubordinates	grogfs
RODIF_GetSequenceByte	grogsb
RODIF_GetSequenceInt	grogsi
RODIF_Init	grdint
RODIF_LayoutDirectives	groldr
RODIF_LayoutObjClassBody	grolcb
RODIF_LayoutObject	grolbj
RODIF_LayoutObjectClass	groyoc
RODIF_LayoutStyle	groyst
RODIF_LogObjDescriptorBody	groldb
RODIF_LogicalObjClassBody	grolcl
RODIF_LogicalObject	grolgo
RODIF_LogicalObjectClass	groloc
RODIF_ModalFontAttrs	gromfa
RODIF_PresentationAttributes	gropat
RODIF_PresentationStyle	gropst
RODIF_SeqTableStop	grosts

Table I-16. Sorted by Alias, XVI

Alias	Real name
RODIF_SequenceConstituent	grosct
RODIF_SequenceIntPair	grosip
RODIF_VariantScript	grovsc
RODIF_VariantScripts	grovss
RODIF_WritingModes	growmd
RasterHeight	rgrarh
RasterPelFormat	rgrapf
RasterStatus	rgrast
RasterWidth	rgrarw
Read2sComplement	grar2c
ReadMemRaster	rgrirm
ReadODIF	rdodif
ReadRaster	rgrird
RecursiveFormatInt	gfmrfi
ReleaseAttr	rlsatt
ReleaseConstituent	rlscns
ReleaseRaster	rgrare
ReleaseString	rlsstr
RemAttr	rematt
SEQUENCE_ASN1_OBJECT_ID_Alloc	gzablu
SEQUENCE_ASN1_OBJECT_ID_CkVal	gzablt
SEQUENCE_ASN1_OBJECT_ID_Copy	gzabls
SEQUENCE_ASN1_OBJECT_ID_Deflt	gzably
SEQUENCE_ASN1_OBJECT_ID_Free	gzablv
SEQUENCE_ASN1_OBJECT_ID_Get	gzablx
SEQUENCE_ASN1_OBJECT_ID_Set	gzablw
SEQUENCE_ASN1_OBJECT_ID_WDbug	gzablr
SEQUENCE_BINDING_PAIR_Alloc	gzabmk
SEQUENCE_BINDING_PAIR_CkVal	gzabmj
SEQUENCE_BINDING_PAIR_Copy	gzabmi
SEQUENCE_BINDING_PAIR_Deflt	gzabmo
SEQUENCE_BINDING_PAIR_Free	gzabml
SEQUENCE_BINDING_PAIR_Get	gzabmn
SEQUENCE_BINDING_PAIR_Set	gzabmm
SEQUENCE_BINDING_PAIR_WDbug	gzabmh
SEQUENCE_BYTE_Alloc	gzabms

Table I-17. Sorted by Alias, XVII

Alias	Real name
SEQUENCE_BYTE_CkVal	gzabmr
SEQUENCE_BYTE_Copy	gzabmq
SEQUENCE_BYTE_Deflt	gzabmw
SEQUENCE_BYTE_Free	gzabmt
SEQUENCE_BYTE_Get	gzabmv
SEQUENCE_BYTE_Set	gzabmu
SEQUENCE_BYTE_WDbug	gzabmp
SEQUENCE_CONSTITUENT_Alloc	gzabna
SEQUENCE_CONSTITUENT_CkVal	gzabmz
SEQUENCE_CONSTITUENT_Copy	gzabmy
SEQUENCE_CONSTITUENT_Deflt	gzabne
SEQUENCE_CONSTITUENT_Free	gzabnb
SEQUENCE_CONSTITUENT_Get	gzabnd
SEQUENCE_CONSTITUENT_Set	gzabnc
SEQUENCE_CONSTITUENT_WDbug	gzabmx
SEQUENCE_COPYRIGHT_INFO_Alloc	gzabni
SEQUENCE_COPYRIGHT_INFO_CkVal	gzabnh
SEQUENCE_COPYRIGHT_INFO_Copy	gzabng
SEQUENCE_COPYRIGHT_INFO_Deflt	gzabnm
SEQUENCE_COPYRIGHT_INFO_Free	gzabnj
SEQUENCE_COPYRIGHT_INFO_Get	gzabnl
SEQUENCE_COPYRIGHT_INFO_Set	gzabnk
SEQUENCE_COPYRIGHT_INFO_WDbug	gzabnf
SEQUENCE_FONT_PAIR_Alloc	gzabnq
SEQUENCE_FONT_PAIR_CkVal	gzabnp
SEQUENCE_FONT_PAIR_Copy	gzabno
SEQUENCE_FONT_PAIR_Deflt	gzabnu
SEQUENCE_FONT_PAIR_Free	gzabnr
SEQUENCE_FONT_PAIR_Get	gzabnt
SEQUENCE_FONT_PAIR_Set	gzabns
SEQUENCE_FONT_PAIR_WDbug	gzabnn
SEQUENCE_INT_Alloc	gzabny
SEQUENCE_INT_CkVal	gzabnx
SEQUENCE_INT_Copy	gzabnw
SEQUENCE_INT_Deflt	gzaboc
SEQUENCE_INT_Free	gzabnz

Table I-18. Sorted by Alias, XVIII

Alias	Real name
SEQUENCE_INT_Get	gzabob
SEQUENCE_INT_PAIR_Alloc	gzabog
SEQUENCE_INT_PAIR_CkVal	gzabof
SEQUENCE_INT_PAIR_Copy	gzaboe
SEQUENCE_INT_PAIR_Deflt	gzabok
SEQUENCE_INT_PAIR_Free	gzaboh
SEQUENCE_INT_PAIR_Get	gzaboj
SEQUENCE_INT_PAIR_Set	gzaboi
SEQUENCE_INT_PAIR_WDbug	gzabod
SEQUENCE_INT_Set	gzaboa
SEQUENCE_INT_WDbug	gzabnv
SEQUENCE_NAME_ORG_Alloc	gzaboo
SEQUENCE_NAME_ORG_CkVal	gzabon
SEQUENCE_NAME_ORG_Copy	gzabom
SEQUENCE_NAME_ORG_Deflt	gzabos
SEQUENCE_NAME_ORG_Free	gzabop
SEQUENCE_NAME_ORG_Get	gzabor
SEQUENCE_NAME_ORG_Set	gzaboq
SEQUENCE_NAME_ORG_WDbug	gzabol
SEQUENCE_RESOURCE_MAP_Alloc	gzabow
SEQUENCE_RESOURCE_MAP_CkVal	gzabov
SEQUENCE_RESOURCE_MAP_Copy	gzabou
SEQUENCE_RESOURCE_MAP_Deflt	gzabpa
SEQUENCE_RESOURCE_MAP_Free	gzabox
SEQUENCE_RESOURCE_MAP_Get	gzaboz
SEQUENCE_RESOURCE_MAP_Set	gzaboy
SEQUENCE_RESOURCE_MAP_WDbug	gzabot
SEQUENCE_REVISION_HISTORY_Alloc	gzabpe
SEQUENCE_REVISION_HISTORY_CkVal	gzabpd
SEQUENCE_REVISION_HISTORY_Copy	gzabpc
SEQUENCE_REVISION_HISTORY_Deflt	gzabpi
SEQUENCE_REVISION_HISTORY_Free	gzabpf
SEQUENCE_REVISION_HISTORY_Get	gzabph
SEQUENCE_REVISION_HISTORY_Set	gzabpg
SEQUENCE_REVISION_HISTORY_WDbug	gzabpb
SEQUENCE_STRING_Alloc	gzabpm

Table I-19. Sorted by Alias, XIX

Alias	Real name
SEQUENCE_STRING_CkVal	gzabpl
SEQUENCE_STRING_Copy	gzabpk
SEQUENCE_STRING_Deflt	gzabpq
SEQUENCE_STRING_Free	gzabpn
SEQUENCE_STRING_Get	gzabpp
SEQUENCE_STRING_OR_ID_Alloc	gzabpu
SEQUENCE_STRING_OR_ID_CkVal	gzabpt
SEQUENCE_STRING_OR_ID_Copy	gzabps
SEQUENCE_STRING_OR_ID_Deflt	gzabpy
SEQUENCE_STRING_OR_ID_Free	gzabpv
SEQUENCE_STRING_OR_ID_Get	gzabpx
SEQUENCE_STRING_OR_ID_Set	gzabpw
SEQUENCE_STRING_OR_ID_WDbug	gzabpr
SEQUENCE_STRING_Set	gzabpo
SEQUENCE_STRING_WDbug	gzabpj
SEQUENCE_TAB_STOP_Alloc	gzabqc
SEQUENCE_TAB_STOP_CkVal	gzabqb
SEQUENCE_TAB_STOP_Copy	gzabqa
SEQUENCE_TAB_STOP_Deflt	gzabqg
SEQUENCE_TAB_STOP_Free	gzabqd
SEQUENCE_TAB_STOP_Get	gzabqf
SEQUENCE_TAB_STOP_Set	gzabqe
SEQUENCE_TAB_STOP_WDbug	gzabpz
SEQUENCE_VALUE_Alloc	gzabqk
SEQUENCE_VALUE_CkVal	gzabqj
SEQUENCE_VALUE_Copy	gzabqi
SEQUENCE_VALUE_Deflt	gzabqo
SEQUENCE_VALUE_Free	gzabql
SEQUENCE_VALUE_Get	gzabqn
SEQUENCE_VALUE_Set	gzabqm
SEQUENCE_VALUE_WDbug	gzabqh
SEQUENCE_at_MEDIUM_TYPE_Alloc	gzabmc
SEQUENCE_at_MEDIUM_TYPE_CkVal	gzabmb
SEQUENCE_at_MEDIUM_TYPE_Copy	gzabma
SEQUENCE_at_MEDIUM_TYPE_Deflt	gzabmg
SEQUENCE_at_MEDIUM_TYPE_Free	gzabmd

Table I-20. Sorted by Alias, XX

Alias	Real name
SEQUENCE_at_MEDIUM_TYPE_Get	gzabmf
SEQUENCE_at_MEDIUM_TYPE_Set	gzabme
SEQUENCE_at_MEDIUM_TYPE_WDbug	gzablz
SEQ_CopySeqElts2	gsqcp2
SEQ_CopySequenceElements	gsqcse
SEQ_CopyTabStop	gsqcts
SEQ_Final	gsqfnl
SEQ_Init	gsqint
SEQ_MakeSequence	gsqmks
STRING_Alloc	gzabqs
STRING_CkVal	gzabqr
STRING_Copy	gzabqq
STRING_Deflt	gzabqw
STRING_EXPR_Alloc	gzabra
STRING_EXPR_CkVal	gzabqz
STRING_EXPR_Copy	gzabqy
STRING_EXPR_Deflt	gzabre
STRING_EXPR_Free	gzabrb
STRING_EXPR_Get	gzabrd
STRING_EXPR_Set	gzabrc
STRING_EXPR_WDbug	gzabqx
STRING_Free	gzabqt
STRING_Get	gzabqv
STRING_OR_ID_Alloc	gzabri
STRING_OR_ID_CkVal	gzabrh
STRING_OR_ID_Copy	gzabrg
STRING_OR_ID_Deflt	gzabrm
STRING_OR_ID_Free	gzabrj
STRING_OR_ID_Get	gzabrl
STRING_OR_ID_Set	gzabrk
STRING_OR_ID_WDbug	gzabrf
STRING_Set	gzabqu
STRING_WDbug	gzabqp
STR_Final	gstfnl
STR_Init	gstint
ScoreFields	gatscf

Table I-21. Sorted by Alias, XXI

Alias	Real name
ScoreInfo	gatsci
ScoresFields	gascsf
ScoresInfo	gascsi
SequenceElementSize	gsqses
SequenceOfGenContents	gwosgc
SequenceOfLogClass	gwoslc
SequenceOfSpeContents	gwossc
SetAttr	setatt
SetFontFamily	stffst
SetFontFamilyArray	stffar
SetIntAttr	sinatt
SetLayoutAttr	setlat
SetPresAttr	setrat
SetSeqValueField	gsqsvf
SetStringAttr	ststat
SetStyleAttr	gasssa
SetValueAttr	svlatt
StrLength	giosln
String2StrctName	str2sn
StructurTable	gdbstb
StyleAttrPriority	stypri
Subordinates	grosub
TKERR_EndErrMsg	gtkene
TKERR_EndWarnMsg	gtkenw
TKERR_Final	gtkfnl
TKERR_Format	gtkfor
TKERR_FormatInt	gtkfmi
TKERR_FormatStringType	gtkfsy
TKERR_Init	gtkini
TKERR_StartErrMsg	gtkste
TKERR_StartWarnMsg	gtkstw
TKError	gtkerr
TKErrorMesg	gtkmsg
TabescFields	gattef
TabescInfo	gattei
TheContent	gratcn

Table I-22. Sorted by Alias, XXII

Alias	Real name
TruncMarker	gtktrm
Truncate	gtktru
TwosComplement	gwatwo
UnspecParms	unsprm
UnspecifyDefaultValueLists	unsdvl
UnspecifyFont	unsfnt
UnspecifyModalFont	gftumf
UnspecifyScore	gftusr
UnspecifyStructName	gftusn
UnspecifyVarScript	gftuvs
VALUE_Alloc	gzabrq
VALUE_CkVal	gzabrp
VALUE_CkValATValue	gatkat
VALUE_CkValCCValue	gatkcc
VALUE_CkValConstituent	gatkco
VALUE_CkValDPValue	gatkdp
VALUE_CkValRGValue	gatkrg
VALUE_CkValValue	gatkvl
VALUE_Copy	gzabro
VALUE_Deflt	gzabru
VALUE_Free	gzabrr
VALUE_Get	gzabrt
VALUE_Set	gzabrs
VALUE_WDbug	gzabrn
VariantScriptFields	gatvsf
VariantScriptInfo	gatvsi
VariantScriptsFields	gavssf
VariantScriptsInfo	gavssi
WASN1_9541IdType	gwsn95
WASN1_Asn1ObjectId	gwsnao
WASN1_BasicLogicalAttributes	gwsnbl
WASN1_Binding_Pair	gwsnbg
WASN1_BlockAttributes	gwsnba
WASN1_Border	gwsnbd
WASN1_CharacterFonts	gwsncf
WASN1_CompLogicalAttributes	gwsnca

Table I-23. Sorted by Alias, XXIII

Alias	Real name
WASN1_ConsitutentId	gwacid
WASN1_ConstituentId	gwsnci
WASN1_Constituent_LastId	gwsncl
WASN1_ContentPortionAttributes	gwsncp
WASN1_DefaultDimensions	gwsndd
WASN1_DefaultPosition	gwsndp
WASN1_Dimensions	gwsndm
WASN1_EOC	wsneoc
WASN1_FrameAttributes	gwsnfa
WASN1_GraphicRendition	gwsngr
WASN1_Header	gwsnhd
WASN1_IntPair	gwsnip
WASN1_IntType	gwsnit
WASN1_Integer	wsnint
WASN1_Itemisation	gwsnim
WASN1_LayoutObjectType	gwsnlo
WASN1_LayoutOjbectType	gwalot
WASN1_Length	wsnlen
WASN1_LineLayoutTable	gwsnll
WASN1_MediumType	gwsnmd
WASN1_ModalFontAttrs	gwamfa
WASN1_Modes	gwsnmo
WASN1_Offset	gwsnof
WASN1_PageAttributes	gwsnpa
WASN1_RasterCodingAttributes	gwsnrc
WASN1_Real	gwsnre
WASN1_RgClipping	gwsnrg
WASN1_RgImageDimensions	gwsnri
WASN1_RgPelSpacing	gwsnps
WASN1_Separation	gwsnsp
WASN1_Seq9541IdType	gwsns9
WASN1_SeqBytesInStruct	gwsnsb
WASN1_SeqIntsInStruct	gwsnsi
WASN1_SequenceBytes	gwasby
WASN1_SequenceStrings	gwasst
WASN1_Strings	gwsnsg

Table I-24. Sorted by Alias, XXIV

Alias	Real name
WASN1_StructuredName	gwsnst
WASN1_ValueType	wsnvlt
WASN1_VariantScript	gwavst
WASN1_VariantScripts	gwavss
WODIF_AdditionalDocChars	gwoadc
WODIF_CharacterAttributes	gwocha
WODIF_ContentPortions	gwocpt
WODIF_DefaultValueListsLayout	gwodll
WODIF_DefaultValueListsLogical	gwodvl
WODIF_DocumentCharacteristics	gwodch
WODIF_DocumentProDescriptor	gwodpd
WODIF_DocumentProfile	gwodpr
WODIF_Final	gwofnl
WODIF_FontPair	gwofpr
WODIF_FontReference	gwofrf
WODIF_FontsList	gwofls
WODIF_GenericContent	gwogct
WODIF_GenericLayout	gwoglo
WODIF_GenericLogical	gwoglg
WODIF_Init	gwoint
WODIF_LayoutDirectives	gwoldr
WODIF_LayoutObjClassDesc	gwoloc
WODIF_LayoutObjClassDescBody	gwolod
WODIF_LayoutObjDesc	gwolob
WODIF_LayoutObjDescBody	gwoldb
WODIF_LayoutStyle	gwolst
WODIF_LayoutStyleDescriptor	gwolsd
WODIF_LogicalObjClassDesc	gwoloc
WODIF_LogicalObjClassDescBody	gwolcb
WODIF_LogicalObjDesc	gwolod
WODIF_LogicalObjDescBody	gwoldb
WODIF_PresentationAttributes	gwopat
WODIF_PresentationStyle	gwopst
WODIF_PresentationStyleDesc	gwopsd
WODIF_RasterGraphicsAttributes	gworga
WODIF_SequenceConstituents	gwosct

Table I-25. Sorted by Alias, XXV

Alias	Real name
WODIF_SpecificContent	gwoscn
WODIF_SpecificLayout	gwosly
WODIF_SpecificLogical	gwoslo
WODIF_WalkConstructionExpr	gwowce
WODIF_WalkConstructionTerm	gwowct
WODIF_WalkGenericLayout	gwowkl
WODIF_WalkGenericLogical	gwowgl
WODIF_WalkSpecificLayout	gwowsy
WODIF_WalkSpecificLogical	gwogsl
WODIF_WriteConstructionExpr	gworce
WODIF_WriteConstructionFactor	gworcf
WODIF_WriteConstructionTerm	gworct
WODIF_WriteGenerator	gwowge
WalkAndLabelGenericStructure	glblgs
WalkAndLabelSpecificStructure	glblss
WalkConstructionExpr	gdbkce
WalkConstructionFactor	gdbkcf
WalkConstructionTerm	gdbkct
WalkSeqConstructionTerm	gdbksq
WalkSpecificObject	gdbwso
WeightTable	gdbwtb
WhiteTreeTop	rgvwtt
WriteAttributesState	gdbwas
WriteConstituentState	dbgcns
WriteConstructionExpr	gdbwce
WriteConstructionFactor	gdbwcf
WriteConstructionTerm	gdbwct
WriteDebugByte	gdbwdb
WriteDebugDsclass	gdbwds
WriteDebugFont	gdbwdf
WriteDebugISO9541Id	gdbw95
WriteDebugISO9541IdArray	gdbwar
WriteDebugModalFontAttr	gdbwfa
WriteDebugModes	gdbwms
WriteDebugReal	gdbwrl
WriteDebugScore	gdbwss

Table I-26. Sorted by Alias, XXVI

Alias	Real name
WriteDebugScores	gdbwsc
WriteDebugString	gdbwds
WriteDebugStructName	gdbwsn
WriteDebugVarScript	gdbwvs
WriteDebugVarScripts	gdbwvv
WriteDocumentState	dbgdoc
WriteFontEnum	gdbwfe
WriteMemRaster	rgriwm
WriteODIF	wrodif
WriteProcs	gwowpr
WriteRaster	rgriwr
WriteSeqConstructionTerm	gdbwsq
WriteStylesState	dbgsty
WriteSubgraphState	dbgsbg
WriteUnreachableConstituents	gdbwuc
WritingModesFields	gatwmf
WritingModesInfo	gatwmi
X_ReadRaster	rgxrd
X_WriteRaster	rgxwr
addnode	graand
alloc_fbm	rgfaal
at_BORDER_Alloc	gzabcg
at_BORDER_CkVal	gzabcf
at_BORDER_Copy	gzabce
at_BORDER_Deflt	gzabck
at_BORDER_Free	gzabch
at_BORDER_Get	gzabcj
at_BORDER_Set	gzabci
at_BORDER_WDbug	gzabcd
at_CONTENT_ARCH_CLASS_form_cc	cacfc
at_CONTENT_ARCH_CLASS_form_gg	cacfg
at_CONTENT_ARCH_CLASS_form_rg	cacfr
at_CONTENT_ARCH_CLASS_fp_cc	cacfpc
at_CONTENT_ARCH_CLASS_fp_rg	cacfpr
at_CONTENT_ARCH_CLASS_proc_cc	cacpc
at_DEFAULT_VALUE_LISTS_Alloc	gzabco

Table I-27. Sorted by Alias, XXVII

Alias	Real name
at_DEFAULT_VALUE_LISTS_CkVal	gzabcn
at_DEFAULT_VALUE_LISTS_Copy	gzabcm
at_DEFAULT_VALUE_LISTS_Deflt	gzabcs
at_DEFAULT_VALUE_LISTS_Free	gzabcp
at_DEFAULT_VALUE_LISTS_Get	gzabcr
at_DEFAULT_VALUE_LISTS_Set	gzabcq
at_DEFAULT_VALUE_LISTS_WDbug	gzabcl
at_DIMENSIONS_Alloc	gzabcw
at_DIMENSIONS_CkVal	gzabcv
at_DIMENSIONS_Copy	gzabcu
at_DIMENSIONS_Deflt	gzabda
at_DIMENSIONS_Free	gzabcx
at_DIMENSIONS_Get	gzabcz
at_DIMENSIONS_Set	gzabcy
at_DIMENSIONS_WDbug	gzabct
at_MEDIUM_TYPE_Alloc	gzabde
at_MEDIUM_TYPE_CkVal	gzabdd
at_MEDIUM_TYPE_Copy	gzabdc
at_MEDIUM_TYPE_Deflt	gzabdi
at_MEDIUM_TYPE_Free	gzabdf
at_MEDIUM_TYPE_Get	gzabdh
at_MEDIUM_TYPE_Set	gzabdg
at_MEDIUM_TYPE_WDbug	gzabdb
at_OFFSET_Alloc	gzabdm
at_OFFSET_CkVal	gzabdl
at_OFFSET_Copy	gzabdk
at_OFFSET_Deflt	gzabdq
at_OFFSET_Free	gzabdn
at_OFFSET_Get	gzabdp
at_OFFSET_Set	gzabdo
at_OFFSET_WDbug	gzabdj
at_POSITION_Alloc	gzabdu
at_POSITION_CkVal	gzabdt
at_POSITION_Copy	gzabds
at_POSITION_Deflt	gzabdy
at_POSITION_Free	gzabdv

Table I-28. Sorted by Alias, XXVIII

Alias	Real name
at_POSITION_Get	gzabdx
at_POSITION_Set	gzabdw
at_POSITION_WDbug	gzabdr
at_SAME_LAYOUT_OBJECT_Alloc	gzabec
at_SAME_LAYOUT_OBJECT_CkVal	gzabeb
at_SAME_LAYOUT_OBJECT_Copy	gzabea
at_SAME_LAYOUT_OBJECT_Deflt	gzabeg
at_SAME_LAYOUT_OBJECT_Free	gzabed
at_SAME_LAYOUT_OBJECT_Get	gzabef
at_SAME_LAYOUT_OBJECT_Set	gzabee
at_SAME_LAYOUT_OBJECT_WDbug	gzabdz
at_SEPARATION_Alloc	gzabek
at_SEPARATION_CkVal	gzabej
at_SEPARATION_Copy	gzabei
at_SEPARATION_Deflt	gzabeo
at_SEPARATION_Free	gzabel
at_SEPARATION_Get	gzaben
at_SEPARATION_Set	gzabem
at_SEPARATION_WDbug	gzabeh
at_TYPE_OF_CODING_T4_1	toct41
at_TYPE_OF_CODING_T4_2	toct42
at_TYPE_OF_CODING_T6	toct6
at_TYPE_OF_CODING_bitmap	tocbit
at_TYPE_OF_CODING_cc	toccc
at_TYPE_OF_CODING_gg	tocgg
bl_make	rgvblm
bl_term	rgvblt
bytecount	grabct
cc_ITEMISATION_Alloc	gzabes
cc_ITEMISATION_CkVal	gzaber
cc_ITEMISATION_Copy	gzabeq
cc_ITEMISATION_Deflt	gzabew
cc_ITEMISATION_Free	gzabet
cc_ITEMISATION_Get	gzabev
cc_ITEMISATION_Set	gzabeu
cc_ITEMISATION_WDbug	gzabep

Table I-29. Sorted by Alias, XXIX

Alias	Real name
compress	rgfgcm
contids	grocid
digits	giodig
doc	gradoc
dp_DOC_APP_PROF_DEFAULTS_Alloc	gzabfq
dp_DOC_APP_PROF_DEFAULTS_CkVal	gzabfp
dp_DOC_APP_PROF_DEFAULTS_Copy	gzabfo
dp_DOC_APP_PROF_DEFAULTS_Deflt	gzabfu
dp_DOC_APP_PROF_DEFAULTS_Free	gzabfr
dp_DOC_APP_PROF_DEFAULTS_Get	gzabft
dp_DOC_APP_PROF_DEFAULTS_Set	gzabfs
dp_DOC_APP_PROF_DEFAULTS_WDbug	gzabfn
dp_LOCAL_FILE_REFERENCE_Alloc	gzabfy
dp_LOCAL_FILE_REFERENCE_CkVal	gzabfx
dp_LOCAL_FILE_REFERENCE_Copy	gzabfw
dp_LOCAL_FILE_REFERENCE_Deflt	gzabgc
dp_LOCAL_FILE_REFERENCE_Free	gzabfz
dp_LOCAL_FILE_REFERENCE_Get	gzabgb
dp_LOCAL_FILE_REFERENCE_Set	gzabga
dp_LOCAL_FILE_REFERENCE_WDbug	gzabfv
dp_ODA_VERSION_Alloc	gzabgg
dp_ODA_VERSION_CkVal	gzabgf
dp_ODA_VERSION_Copy	gzabge
dp_ODA_VERSION_Deflt	gzabgk
dp_ODA_VERSION_Free	gzabgh
dp_ODA_VERSION_Get	gzabgj
dp_ODA_VERSION_Set	gzabgi
dp_ODA_VERSION_WDbug	gzabgd
dsclasses	gftdsc
error_sequence	gsqerr
fbm_GetByte	rgvfgb
fbm_Parameter	rgvfpa
fbm_PutByte	rgvfpb
fbmprintf	rgfbpf
fbmread	rgfbre
fbmwrite	rgfbwi

Table I-30. Sorted by Alias, XXX

Alias	Real name
free_fbm	rgfafr
get_long	rgfbgl
get_short	rgfbgs
gg_COLOUR_REPS_Alloc	gzabgo
gg_COLOUR_REPS_CkVal	gzabgn
gg_COLOUR_REPS_Copy	gzabgm
gg_COLOUR_REPS_Deflt	gzabgs
gg_COLOUR_REPS_Free	gzabgp
gg_COLOUR_REPS_Get	gzabgr
gg_COLOUR_REPS_Set	gzabgq
gg_COLOUR_REPS_WDbug	gzabgl
gg_EDGE_RENDITION_Alloc	gzabgw
gg_EDGE_RENDITION_CkVal	gzabgv
gg_EDGE_RENDITION_Copy	gzabgu
gg_EDGE_RENDITION_Deflt	gzabha
gg_EDGE_RENDITION_Free	gzabgx
gg_EDGE_RENDITION_Get	gzabgz
gg_EDGE_RENDITION_Set	gzabgy
gg_EDGE_RENDITION_WDbug	gzabgt
gg_ENCODING_ANNOUNCER_Alloc	gzabhe
gg_ENCODING_ANNOUNCER_CkVal	gzabhd
gg_ENCODING_ANNOUNCER_Copy	gzabhc
gg_ENCODING_ANNOUNCER_Deflt	gzabhi
gg_ENCODING_ANNOUNCER_Free	gzabhf
gg_ENCODING_ANNOUNCER_Get	gzabhh
gg_ENCODING_ANNOUNCER_Set	gzabhg
gg_ENCODING_ANNOUNCER_WDbug	gzabhb
gg_FILLED_AREA_REND_Alloc	gzabhm
gg_FILLED_AREA_REND_CkVal	gzabhl
gg_FILLED_AREA_REND_Copy	gzabhk
gg_FILLED_AREA_REND_Deflt	gzabhq
gg_FILLED_AREA_REND_Free	gzabhn
gg_FILLED_AREA_REND_Get	gzabhp
gg_FILLED_AREA_REND_Set	gzabho
gg_FILLED_AREA_REND_WDbug	gzabhj
gg_LINE_RENDITION_Alloc	gzabhu
gg_LINE_RENDITION_CkVal	gzabht

Table I-31. Sorted by Alias, XXXI

Alias	Real name
gg_LINE_RENDITION_Copy	gzabhs
gg_LINE_RENDITION_Deflt	gzabhy
gg_LINE_RENDITION_Free	gzabhv
gg_LINE_RENDITION_Get	gzabhx
gg_LINE_RENDITION_Set	gzabhw
gg_LINE_RENDITION_WDbug	gzabhr
gg_MARKER_RENDITION_Alloc	gzabic
gg_MARKER_RENDITION_CkVal	gzabib
gg_MARKER_RENDITION_Copy	gzabia
gg_MARKER_RENDITION_Deflt	gzabig
gg_MARKER_RENDITION_Free	gzabid
gg_MARKER_RENDITION_Get	gzabif
gg_MARKER_RENDITION_Set	gzabie
gg_MARKER_RENDITION_WDbug	gzabhz
gg_PICTURE_DIMENSIONS_Alloc	gzabik
gg_PICTURE_DIMENSIONS_CkVal	gzabij
gg_PICTURE_DIMENSIONS_Copy	gzabii
gg_PICTURE_DIMENSIONS_Deflt	gzabio
gg_PICTURE_DIMENSIONS_Free	gzabil
gg_PICTURE_DIMENSIONS_Get	gzabin
gg_PICTURE_DIMENSIONS_Set	gzabim
gg_PICTURE_DIMENSIONS_WDbug	gzabih
gg_REGION_INTEREST_SPEC_Alloc	gzabis
gg_REGION_INTEREST_SPEC_CkVal	gzabir
gg_REGION_INTEREST_SPEC_Copy	gzabiq
gg_REGION_INTEREST_SPEC_Deflt	gzabiw
gg_REGION_INTEREST_SPEC_Free	gzabit
gg_REGION_INTEREST_SPEC_Get	gzabiv
gg_REGION_INTEREST_SPEC_Set	gzabiu
gg_REGION_INTEREST_SPEC_WDbug	gzabip
gg_TEXT_RENDITION_Alloc	gzabja
gg_TEXT_RENDITION_CkVal	gzabiz
gg_TEXT_RENDITION_Copy	gzabiy
gg_TEXT_RENDITION_Deflt	gzabje
gg_TEXT_RENDITION_Free	gzabjb
gg_TEXT_RENDITION_Get	gzabjd

Table I-32. Sorted by Alias, XXXII

Alias	Real name
gg_TEXT_RENDITION_Set	gzabjc
gg_TEXT_RENDITION_WDbug	gzabix
gg_TRANSFORMATION_SPEC_Alloc	gzabji
gg_TRANSFORMATION_SPEC_CkVal	gzabjh
gg_TRANSFORMATION_SPEC_Copy	gzabjg
gg_TRANSFORMATION_SPEC_Deflt	gzabjm
gg_TRANSFORMATION_SPEC_Free	gzabjj
gg_TRANSFORMATION_SPEC_Get	gzabjl
gg_TRANSFORMATION_SPEC_Set	gzabjk
gg_TRANSFORMATION_SPEC_WDbug	gzabjf
gg_TRANSPARENCY_SPEC_Alloc	gzabjq
gg_TRANSPARENCY_SPEC_CkVal	gzabjp
gg_TRANSPARENCY_SPEC_Copy	gzabjo
gg_TRANSPARENCY_SPEC_Deflt	gzabju
gg_TRANSPARENCY_SPEC_Free	gzabjr
gg_TRANSPARENCY_SPEC_Get	gzabjt
gg_TRANSPARENCY_SPEC_Set	gzabjs
gg_TRANSPARENCY_SPEC_WDbug	gzabjn
identifier	groidn
isode_PrintODIF	isprdf
isode_ReadODIF	isred
isode_WriteODIF	iswrt
last_debug_counter	gdbctr
machine_byte_order	rgfbbo
next_document_id	gdocid
nextbyte	granby
null_sequence	gsqnul
object_ids	ginobj
put_long	rgfbpl
put_short	rgfbps
read_gif	rgfgrd
read_iff	rgfird
read_sun	rgfsrd
realconstituent	grorco
reference	gwaref
rg_CLIPPING_Alloc	gzabkw

Table I-33. Sorted by Alias, XXXIII

Alias	Real name
rg_CLIPPING_CkVal	gzabkv
rg_CLIPPING_Copy	gzabku
rg_CLIPPING_Deflt	gzabla
rg_CLIPPING_Free	gzabkx
rg_CLIPPING_Get	gzabkz
rg_CLIPPING_Set	gzabky
rg_CLIPPING_WDbug	gzabkt
rg_IMAGE_DIMENSIONS_Alloc	gzable
rg_IMAGE_DIMENSIONS_CkVal	gzabld
rg_IMAGE_DIMENSIONS_Copy	gzablc
rg_IMAGE_DIMENSIONS_Deflt	gzabli
rg_IMAGE_DIMENSIONS_Free	gzablf
rg_IMAGE_DIMENSIONS_Get	gzablh
rg_IMAGE_DIMENSIONS_Set	gzablg
rg_IMAGE_DIMENSIONS_WDbug	gzablb
rg_PEL_SPACING_Alloc	gzablm
rg_PEL_SPACING_CkVal	gzabll
rg_PEL_SPACING_Copy	gzablk
rg_PEL_SPACING_Deflt	gzablq
rg_PEL_SPACING_Free	gzabln
rg_PEL_SPACING_Get	gzablp
rg_PEL_SPACING_Set	gzablo
rg_PEL_SPACING_WDbug	gzablj
roots	gdbrot
write_gif	rgfgwr
write_iff	rgfiwr
write_sun	rgfswr
wt_make	rgvwhm
wt_term	rgvwht

Table I-34. Sorted by Alias, XXXIV

I.2. Alphabetized by Real Names

Real Name	Alias
appseq	AppendSequence
appsub	AppendSubordinate
ary2sn	Array2StrctName
cacfc	at_CONTENT_ARCH_CLASS_form_cc
cacfg	at_CONTENT_ARCH_CLASS_form_gg
cacfpc	at_CONTENT_ARCH_CLASS_fp_cc
cacfpr	at_CONTENT_ARCH_CLASS_fp_rg
cacfr	at_CONTENT_ARCH_CLASS_form_rg
cacpc	at_CONTENT_ARCH_CLASS_proc_cc
catstr	ConcatString
chrstr	Chars2String
compri	ComponentAttrPriority
conpri	ContentAttrPriority
conseq	ContractSequence
cpcons	CopyConstituent
cpyseq	CopySequence
dbgcns	WriteConstituentState
dbgdoc	WriteDocumentState
dbgsbg	WriteSubgraphState
dbgsty	WriteStylesState
dbgwbs	DEBUG_WriteBytes
dbgwcn	DEBUG_WriteDebugConstituent
dbgwgo	DEBUG_WalkGenericObject
delcns	DeleteConstituent
deldoc	DeleteDocument
delgen	DeleteIterator
delseq	DeleteSequence
delstr	DeleteString
dlcnex	DeleteConstExpr
dlsqd	DeleteSequenceDeeply
dppri	AttrDocProfPriority
ds2int	Dsclass2Ints
evlnum	EvalNumExpr

Table I-35. Sorted by Real Name, I

Real Name	Alias
evlobi	EvalObjIdExpr
evlrbv	EvalRefBind
evlstr	EvalStringExpr
expseq	ExpandSequence
fnddp	FindDocumentProfile
fndrt	FindDocumentRoot
fntfnl	FONT_Final
fntint	FONT_Init
ga16dm	ATTR16_Dummy
ga17dm	ATTR17_Dummy
ga18dm	ATTR18_Dummy
gagfrb	ATTR_GetFontReferenceBase
gagscb	ATTR_GetScoreBase
gagteb	ATTR_GetTabescBase
gagvsb	ATTR_GetVariantScriptBase
gagwmb	ATTR_GetWritingModesBase
gamfaf	ModalFontAttrFields
gamfai	ModalFontAttrInfo
gasbun	BreakUpName
gascsf	ScoresFields
gascsi	ScoresInfo
gasdct	DeleteConstTerm
gasfnl	ATTS_Final
gasfod	ATTS_FillInOneDefaultConst
gasint	ATTS_Init
gasssa	SetStyleAttr
gataao	ATTR_AllocASN1ObjectId
gataas	ATTR_AllocAtomicStrExpr
gatabp	ATTR_AllocBindingPair
gatacc	ATTR_AccumulateContent
gataci	ATTR_AllocCurrInstanceParms
gatacr	ATTR_AllocCopyright
gatadv	ATTR_AttemptInheritFromDVL
gatane	ATTR_AllocNumericExpr
gatanf	ATTR_AllocNumericFunction
gatano	ATTR_AllocNameOrg

Table I-36. Sorted by Real Name, II

Real Name	Alias
gatant	ATTR_AttrNameTbl
gataoi	ATTR_AllocObjectIdExpr
gataor	ATTR_AllocObjectRef
gatapn	ATTR_AllocPersonalName
gatarb	ATTR_AllocRefBindingValue
gatarh	ATTR_AllocRevisionHistory
gatarm	ATTR_AllocResourceMap
gatarv	ATTR_AllocReviser
gatase	ATTR_AllocStringExpr
gatasf	ATTR_AllocStringFunction
gatasi	ATTR_Alloc_String_or_id_value
gatasn	ATTR_AllocStructuredName
gatasp	ATTR_AllocSequencePersonalNames
gatasr	ATTR_AllocSequenceRevisers
gatatc	ATTR_AddToConstituent
gatatt	ATTR_AttrTypeTbl
gatawd	ATTR_AreWeDone
gatbic	ATTR_BORDER_INFO_Copy
gatbid	ATTR_BORDER_INFO_Deflt
gatbik	ATTR_BORDER_INFO_CkVal
gatcas	ATTR_CopyAtomicStrExpr
gatcbp	ATTR_CopyBindingPair
gatccm	ATTR_CheckAndCleanMask
gatccp	ATTR_IsCCPresentation
gatccr	ATTR_CopyCopyRight
gatcdc	ATTR_CheckDAPDCOnstituents
gatcid	ATTR_CheckId
gatcip	ATTR_CopyCurrInstanceParms
gatcmt	ATTR_Check_at_MEDIUM_TYPE
gatcne	ATTR_CopyNumericExpr
gatcnf	ATTR_CopyNumericFunction
gatcno	ATTR_CopyNameOrg
gatcor	ATTR_CopyObjectRef
gatcpf	ATTR_Copy_Fields
gatcpn	ATTR_CopyPersonalName
gatcrh	ATTR_CopyRevisionHistory

Table I-37. Sorted by Real Name, III

Real Name	Alias
gatcrm	ATTR_CopyResourceMap
gatcrv	ATTR_CopyReviser
gatcse	ATTR_CopyStringExpr
gatcsf	ATTR_CopyStringFunction
gatcsi	ATTR_Ck_String_or_id_value
gatcsp	ATTR_CopySequencePersonalNames
gatcsr	ATTR_CopySequenceRevisers
gatcts	ATTR_CheckTabStop
gatdck	DIMENSION_type_CkVal
gatdpa	ATTR_CheckDAPDConstituents
gatdtc	DIMENSION_type_Copy
gatdvc	ATTR_InheritFromDVLConstituent
gatdvl	ATTR_InheritFromDVL
gatfa1	ATTR_FreeAsn1ObjectId
gatfar	ATTR_FindAndRemoveAttr
gatfas	ATTR_FreeAtomicStrExpr
gatfbp	ATTR_FreeBindingPair
gatfca	ATTR_FindContentArchClass
gatfcg	ATTR_LookForContentGenerator
gatfci	ATTR_FreeCurrInstanceParms
gatfco	ATTR_FindConstituent
gatfcp	ATTR_FreeCopyrightInfo
gatflc	ATTR_FindLocal
gatfne	ATTR_FreeNumericExpr
gatfnf	ATTR_FreeNumericFunction
gatfnl	ATTR_Final
gatfno	ATTR_FreeNameOrg
gatfoi	ATTR_FreeObjectIdExpr
gatfor	ATTR_FreeObjectRef
gatfpb	ATTR_AllocFontPairBody
gatfpc	ATTR_FontPairCopy
gatfpn	ATTR_FreePersonalName
gatfrb	ATTR_FreeRefBindingValue
gatfrf	FontReferenceFields
gatfrh	ATTR_FreeRevisionHistory
gatfri	FontReferenceInfo

Table I-38. Sorted by Real Name, IV

Real Name	Alias
gatfrm	ATTR_FreeResourceMap
gatfro	ATTR_FindRightObjectType
gatfrv	ATTR_FreeReviser
gatfse	ATTR_FreeStringExpr
gatfsf	ATTR_FreeStringFunction
gatfsi	ATTR_Free_String_or_id_value
gatfsr	ATTR_FreeStrings
gatfst	ATTR_Free_Some_String
gatfts	ATTR_FreeTabStop
gatgck	ATTR_GetConstituentKind
gatgcl	ATTR_GetObjClContentKind
gatgdf	ATTR_GetDefaults
gatgdp	ATTR_GetDocProfDefaults
gatgoc	ATTR_GetObjectContentKind
gathsh	ATTR_Hash
gaticc	ATTR_IsCCCoding
gaticp	ATTR_InheritForContentPortion
gatics	ATTR_InheritFromConstAndStyle
gatifc	ATTR_InheritFromConst
gatifo	ATTR_InhertForObject
gatifs	ATTR_InheritFromStyle
gatigc	ATTR_IsGGCoding
gatigp	ATTR_IsGGPresentation
gatiid	ATTR_AllocISO9541Id
gatild	ATTR_IsLayoutDirective
gatild	IsLayoutDirective
gatint	ATTR_Init
gatioc	ATTR_InheritFromObjClass
gatipa	ATTR_IsPresentationAttribute
gatipa	IsPresentationAttribute
gatirc	ATTR_IsRGCoding
gatirp	ATTR_IsRGPresentation
gatkas	ATTR_CkValAtomicStrExpr
gatkat	VALUE_CkValATValue
gatkbp	ATTR_CkValBindingPair
gatkcc	VALUE_CkValCCValue

Table I-39. Sorted by Real Name, V

Real Name	Alias
gatkci	ATTR_CkValCurrInstanceParms
gatkco	VALUE_CkValConstituent
gatkdp	VALUE_CkValDPValue
gatkne	ATTR_CkValNumericExpr
gatknf	ATTR_CkValNumericFunction
gatkoa	ATTR_KindOfAttr
gatkoi	ATTR_CkValObjectIdExpr
gatkor	ATTR_CkValObjectRef
gatkrb	ATTR_CkValRefBindingValue
gatkrg	VALUE_CkValRGValue
gatkse	ATTR_CheckStringExpr
gatksf	ATTR_CkValStringFunction
gatkvl	VALUE_CkValValue
gatlca	ATTR_LocateConstAttr
gatldt	ATTR_LoadDestination
gatloc	ATTR_LookOnConstituent
gatlrd	ATTR_LookInResourceDoc
gatoie	ATTR_CopyObjectIdExpr
gatooc	ATTR_LookOnObjClass
gatrgv	ATTR_CopyRefBindingValue
gatrtd	ATTR_RummageThruDAPD
gatscf	ScoreFields
gatsci	ScoreInfo
gatsmk	ATTR_SterilizeParmMask
gatsru	ATTR_SetUpReturnToUser
gattef	TabescFields
gattei	TabescInfo
gattnm	GetAttrName
gatttr	ATTR_ThreadTree
gatutt	ATTR_UnThreadTree
gatvsf	VariantScriptFields
gatvsi	VariantScriptInfo
gatwmf	WritingModesFields
gatwmi	WritingModesInfo
gavssf	VariantScriptsFields
gavssi	VariantScriptsInfo

Table I-40. Sorted by Real Name, VI

Real Name	Alias
gcdswk	CONST_DoSetComponentWhatKind
gcfall	CONSTRUCTION_FACTOR_Alloc
gcfckv	CONSTRUCTION_FACTOR_CkVal
gcfcpy	CONSTRUCTION_FACTOR_Copy
gcffre	CONSTRUCTION_FACTOR_Free
gcntkn	GetContentKind
gcontt	GetContent
gcsadl	CONST_AddDoublyLinked
gcsasc	CONST_AddSpecialConstituent
gcsccp	CONST_CollaspeContentPortions
gcsccr	CONST_ChangeContentPortionRefs
gcscha	CONST_Chameleon
gcscuc	CONST_CleanUpConstituent
gcsdwk	CONST_SetDefComponentWhatKind
gcsfic	CONST_FillInConstituent
gcsfnl	CONST_Final
gcsics	CONST_InsertConstituent
gcsint	CONST_Init
gcsiwk	CONST_SetIndepWhatKind
gcsmcs	CONST_MoveConstituent
gcsmct	CONST_MakeConstituent
gcsofs	CONST_SetOtherFields
gcspwk	CONST_SetDepWhatKind
gcsrct	CONST_RemoveConstituent
gcsrdl	CONST_RemoveDoublyLinked
gcsrsc	CONST_RemoveSpecialConstituent
gcswk	CONST_SetComponentWhatKind
gctall	CONSTRUCTION_TERM_Alloc
gctckv	CONSTRUCTION_TERM_CkVal
gctcpy	CONSTRUCTION_TERM_Copy
gctfre	CONSTRUCTION_TERM_Free
gdbctr	last_debug_counter
gdbetb	EsclassTable
gdbfnl	DEBUG_Final
gdbint	DEBUG_Init
gdbiss	InternalSubgraphState

Table I-41. Sorted by Real Name, VII

Real Name	Alias
gdbkce	WalkConstructionExpr
gdbkcf	WalkConstructionFactor
gdbkct	WalkConstructionTerm
gdbksq	WalkSeqConstructionTerm
gdbptb	PostureTable
gdbpwt	PropwdthTable
gdbrot	roots
gdbstb	StructurTable
gdbw95	WriteDebugISO9541Id
gdbwar	WriteDebugISO9541IdArray
gdbwas	WriteAttributesState
gdbwce	WriteConstructionExpr
gdbwcf	WriteConstructionFactor
gdbwct	WriteConstructionTerm
gdbwdb	WriteDebugByte
gdbwdf	WriteDebugFont
gdbwds	WriteDebugDsclass
gdbwds	WriteDebugString
gdbwfa	WriteDebugModalFontAttr
gdbwfe	WriteFontEnum
gdbwms	WriteDebugModes
gdbwrl	WriteDebugReal
gdbwsc	WriteDebugScores
gdbwsn	WriteDebugStructName
gdbwso	WalkSpecificObject
gdbwsq	WriteSeqConstructionTerm
gdbwss	WriteDebugScore
gdbwtb	WeightTable
gdbwuc	WriteUnreachableConstituents
gdbwvs	WriteDebugVarScript
gdbwvv	WriteDebugVarScripts
gdcfnl	DOC_Final
gdcint	DOC_Init
gdigit	Digits
gdocid	next_document_id
gdoclt	DOC_DocumentList

Table I-42. Sorted by Real Name, VIII

Real Name	Alias
gdoczc	DOC_ZapConstituents
generr	GenericError
geratt	ERROR_ATTRIBUTE
gercon	ERROR_CONSTITUENT
gerseq	ERROR_SEQUENCE
getatt	GetAttr
getdoc	GetDocument
getprm	GetParmMask
gexfnl	EXPR_Final
gexint	EXPR_Init
gfmfnl	FMT_Final
gfmfnt	FormatInt
gfmini	FMT_Init
gfmint	FMT_FormatInt
gfmrfi	RecursiveFormatInt
gfndat	ATTR_FindAttr
gfrcdf	ATTR_Copy_Dfnpre
gfrcip	ATTR_Copy_Int_Pair_Type
gfrcis	ATTR_Copy_ISO9541_Id_Type
gfrcit	ATTR_Copy_Int_Type
gfrcsc	ATTR_Copy_Score_Type
gfrcsi	ATTR_Copy_Sequence_Int_Type
gfrcsn	ATTR_Copy_Structured_Name
gfrcst	ATTR_Copy_Scores_Type
gfrctb	ATTR_Copy_Tabesc
gfrcvs	ATTR_Copy_Variant_Scripts_Type
gfrcvt	ATTR_Copy_Variant_Script_Type
gfrcwm	ATTR_Copy_Writing_Modes_Type
gfrkcf	ATTR_Ck_Fields
gfrkdf	ATTR_CkVal_Dfnpre
gfrkfp	ATTR_FontPairCkVal
gfrkip	ATTR_CkVal_Int_Pair_Type
gfrkis	ATTR_CkVal_ISO9541_Id_Type
gfrkit	ATTR_CkVal_Int_Type
gfrkmf	ATTR_CkVal_Modal_Font_Attr_Type
gfrksc	ATTR_CkVal_Score_Type

Table I-43. Sorted by Real Name, IX

Real Name	Alias
gfrksi	ATTR_CkVal_Sequence_Int_Type
gfrksn	ATTR_CkVal_Structured_Name
gfrkst	ATTR_CkVal_Scores_Type
gfrktb	ATTR_CkVal_Tabesc
gfrkvs	ATTR_CkVal_Variant_Scripts_Type
gfrkvt	ATTR_CkVal_Variant_Script_Type
gfrkwm	ATTR_CkVal_Writing_Modes_Type
gftcmf	ATTR_Copy_Modal_Font_Attr_Type
gftdsc	dsclasses
gftumf	UnspecifyModalFont
gftusn	UnspecifyStructName
gftusr	UnspecifyScore
gftuvs	UnspecifyVarScript
ggmfab	ATTR_GetModalFontAttrBase
ggscsb	ATTR_GetScoresBase
ggvssb	ATTR_GetVariantScriptsBase
gidchr	IdChars
ginatt	GetIntAttr
ginfnl	INIT_Final
ginint	INIT_Init
ginioi	InitObjectIds
ginirt	InitRouts
ginobj	object_ids
gintpm	InitParms
giodig	digits
giofnl	IO_Final
gioint	IO_Init
giopin	PrintInt
giosln	StrLength
gitagk	ITER_AddGenericKids
gitakp	ITER_AddKidsToParent
gitakq	ITER_AddKidsToQueue
gitask	ITER_AddSpecificKids
gitcrt	DOC_CheckRoot
gitfnl	ITER_Final
gitgid	ITER_GetObjectID

Table I-44. Sorted by Real Name, X

Real Name	Alias
gitint	ITER_Init
gitnsc	ITER_NextSubgraphConstituent
gitsda	ITER_ShiftDownNodeArray
gitsua	ITER_ShiftUpNodeArray
gitvas	ITER_ValidateArraySize
gitwce	ITER_WalkConsExpr
gitwcf	ITER_WalkConsFactor
gitwct	ITER_WalkConsTerm
gitwsq	ITER_WalkSeqConsTerm
glbaia	AssignIdAttr
glbcid	CreateAndAssignId
glbcpt	LabelContentPortions
glbdig	Ndigits
glbfnl	LABEL_Final
glbgst	LabelGenericStructure
glbint	LABEL_Init
glblce	LabelConstructionExpr
glblcf	LabelConstructionFactor
glblcf	LabelConstructionFactor
glblct	LabelConstructionTerm
glblgs	WalkAndLabelGenericStructure
glblsp	LabelSpecificStructure
glblss	WalkAndLabelSpecificStructure
glbsct	LabelSeqConstructionTerm
gmefnl	MEM_Final
gmeint	MEM_Init
gmmfin	MM_Final
gmmfre	MM_Free
gmmini	MM_Init
gmmmal	MM_Malloc
gmmrel	MM_Realloc
gnlatt	NULL_ATTRIBUTE
gnlcon	NULL_CONSTITUENT
gnlseq	NULL_SEQUENCE
gprfnl	PRIOR_Final
gprint	PRIOR_Init

Table I-45. Sorted by Real Name, XI

Real Name	Alias
graand	addnode
graaoi	RASN1_Asn1ObjectId
grabat	RASN1_BlockAttributes
grabct	bytecount
grabla	RASN1_BasicLogicalAttributes
gracat	RASN1_ContentAttributes
gracla	RASN1_CompLogicalAttributes
gracsn	RASN1_CompleteStructName
gradoc	doc
gradpa	RANS1_DefaultPresAttributes
gradpa	RASN1_DefaultPresAttributes
grafat	RASN1_FrameAttributes
grafty	RASN1_FontType
gragip	RASN1_GetIntPair
gragst	RASN1_GetString
gragva	RASN1_GetValue
graide	Identical
graidn	RASN1_Identifier
gralng	RASN1_Length
granby	nextbyte
graoff	RASN1_Offset
graoty	RASN1_ObjectType
grapat	RASN1_PageAttributes
grar2c	Read2sComplement
grarac	RASN1_RgSetAreaControlled
grarat	RASN1_RasterAttributes
grarbs	RASN1_ReadBytes
grarbt	RASN1_ReadByte
grarga	RASN1_RasterGrCodingAttrs
grargc	RASN1_RgClipping
grarid	RASN1_RgImageDimensions
grarps	RASN1_RgPelSpacing
grarss	RASN1_RgSetSpacing
gras2c	RASN1_String2Constituent
grasep	RASN1_Separation
grasnm	RASN1_StructuredName

Table I-46. Sorted by Real Name, XII

Real Name	Alias
grasss	RASN1_SaveSeqStrings
grasva	RASN1_SetValue
gratcn	TheContent
gratst	RASN1_TabStop
grdfnl	RODIF_Final
grdint	RODIF_Init
groadc	RODIF_AddDocCharacteristics
grocat	RODIF_CharacterAttributes
grocep	RODIF_ConstrExpression
grocft	RODIF_ConstrFactor
grocid	contids
grocon	Contents
grocpt	RODIF_ContentPortions
groctm	RODIF_ConstrTerm
grodch	RODIF_DocumentCharacteristics
grodll	RODIF_DefaultValueListsLayout
grodpr	RODIF_DocumentProfile
grodvl	RODIF_DefaultValueListsLogical
groflt	RODIF_FontList
grofrf	RODIF_FontReference
grofsl	RODIF_FontsList
grogfs	RODIF_GenForSubordinates
grogsb	RODIF_GetSequenceByte
grogsi	RODIF_GetSequenceInt
grogsu	GenForSub
groidn	identifier
grolbj	RODIF_LayoutObject
grolcb	RODIF_LayoutObjClassBody
grolcl	RODIF_LogicalObjClassBody
groldb	RODIF_LogObjDescriptorBody
groldr	RODIF_LayoutDirectives
grolgo	RODIF_LogicalObject
groloc	RODIF_LogicalObjectClass
gromfa	RODIF_ModalFontAttrs
gropat	RODIF_PresentationAttributes
gropst	RODIF_PresentationStyle

Table I-47. Sorted by Real Name, XIII

Real Name	Alias
grorco	realconstituent
grosct	RODIF_SequenceConstituent
grosip	RODIF_SequenceIntPair
grosts	RODIF_SeqTableStop
grosub	Subordinates
grovsc	RODIF_VariantScript
grovss	RODIF_VariantScripts
growmd	RODIF_WritingModes
groyoc	RODIF_LayoutObjectClass
groyst	RODIF_LayoutStyle
gsqcp2	SEQ_CopySeqElts2
gsqcse	SEQ_CopySequenceElements
gsqcts	SEQ_CopyTabStop
gsqerr	error_sequence
gsqfnl	SEQ_Final
gsqgsv	GetSeqValueField
gsqint	SEQ_Init
gsqmks	SEQ_MakeSequence
gsqnul	null_sequence
gsqses	SequenceElementSize
gsqsvf	SetSeqValueField
gstfnl	STR_Final
gstint	STR_Init
gtkach	AddChars
gtkein	ErrorIndex
gtkene	TKERR_EndErrMsg
gtkenw	TKERR_EndWarnMsg
gtkerr	TKError
gtkfer	FormattingErrors
gtkfmi	TKERR_FormatInt
gtkfnl	TKERR_Final
gtkfor	TKERR_Format
gtkfsy	TKERR_FormatStringType
gtkini	TKERR_Init
gtkmsg	TKErrorMesg
gtkste	TKERR_StartErrMsg

Table I-48. Sorted by Real Name, XIV

Real Name	Alias
gtkstw	TKERR_StartWarnMsg
gtktrm	TruncMarker
gtktru	Truncate
gvalnm	GetValueName
gvlatt	GetValueAttr
gwacid	WASN1_ConsitutentId
gwaioi	IO_INT_Value
gwalot	WASN1_LayoutOjbectType
gwamfa	WASN1_ModalFontAttrs
gwanby	Nbytes
gwapos	PositiveInteger
gwaref	reference
gwasby	WASN1_SequenceBytes
gwasst	WASN1_SequenceStrings
gwatwo	TwosComplement
gwavss	WASN1_VariantScripts
gwavst	WASN1_VariantScript
gwoadc	WODIF_AdditionalDocChars
gwocha	WODIF_CharacterAttributes
gwocpt	WODIF_ContentPortions
gwodch	WODIF_DocumentCharacteristics
gwodll	WODIF_DefaultValueListsLayout
gwodpd	WODIF_DocumentProDescriptor
gwodpr	WODIF_DocumentProfile
gwodvl	WODIF_DefaultValueListsLogical
gwofls	WODIF_FontsList
gwofnl	WODIF_Final
gwofpr	WODIF_FontPair
gwofrf	WODIF_FontReference
gwogct	WODIF_GenericContent
gwoglg	WODIF_GenericLogical
gwoglo	WODIF_GenericLayout
gwogsl	WODIF_WalkSpecificLogical
gwoint	WODIF_Init
gwolcb	WODIF_LogicalObjClassDescBody
gwoldb	WODIF_LayoutObjDescBody

Table I-49. Sorted by Real Name, XV

Real Name	Alias
gwoldb	WODIF_LogicalObjDescBody
gwoldr	WODIF_LayoutDirectives
gwolob	WODIF_LayoutObjDesc
gwoloc	WODIF_LayoutObjClassDesc
gwoloc	WODIF_LogicalObjClassDesc
gwolod	WODIF_LayoutObjClassDescBody
gwolod	WODIF_LogicalObjDesc
gwolsd	WODIF_LayoutStyleDescriptor
gwolst	WODIF_LayoutStyle
gwopat	WODIF_PresentationAttributes
gwopsd	WODIF_PresentationStyleDesc
gwopst	WODIF_PresentationStyle
gworce	WODIF_WriteConstructionExpr
gworcf	WODIF_WriteConstructionFactor
gworct	WODIF_WriteConstructionTerm
gworga	WODIF_RasterGraphicsAttributes
gwoscn	WODIF_SpecificContent
gwosct	WODIF_SequenceConstituents
gwosgc	SequenceOfGenContents
gwoslc	SequenceOfLogClass
gwoslo	WODIF_SpecificLogical
gwosly	WODIF_SpecificLayout
gwossc	SequenceOfSpeContents
gwowce	WODIF_WalkConstructionExpr
gwowct	WODIF_WalkConstructionTerm
gwowge	WODIF_WriteGenerator
gwowgl	WODIF_WalkGenericLogical
gwowkl	WODIF_WalkGenericLayout
gwowpr	WriteProcs
gwowsy	WODIF_WalkSpecificLayout
gwsn95	WASN1_9541IdType
gwsnao	WASN1_Asn1ObjectId
gwsnba	WASN1_BlockAttributes
gwsnbd	WASN1_Border
gwsnbg	WASN1_Binding_Pair
gwsnbl	WASN1_BasicLogicalAttributes

Table I-50. Sorted by Real Name, XVI

Real Name	Alias
gwsnca	WASN1_CompLogicalAttributes
gwsncf	WASN1_CharacterFonts
gwsnci	WASN1_ConstituentId
gwsncl	WASN1_Constituent_LastId
gwsncp	WASN1_ContentPortionAttributes
gwsndd	WASN1_DefaultDimensions
gwsndm	WASN1_Dimensions
gwsndp	WASN1_DefaultPosition
gwsnfa	WASN1_FrameAttributes
gwsngr	WASN1_GraphicRendition
gwsnhd	WASN1_Header
gwsnim	WASN1_Itemisation
gwsnip	WASN1_IntPair
gwsnit	WASN1_IntType
gwsnll	WASN1_LineLayoutTable
gwsnlo	WASN1_LayoutObjectType
gwsnmd	WASN1_MediumType
gwsnmo	WASN1_Modes
gwsnof	WASN1_Offset
gwsnpa	WASN1_PageAttributes
gwsnps	WASN1_RgPelSpacing
gwsnrc	WASN1_RasterCodingAttributes
gwsnre	WASN1_Real
gwsnrg	WASN1_RgClipping
gwsnri	WASN1_RgImageDimensions
gwsns9	WASN1_Seq9541IdType
gwsnsb	WASN1_SeqBytesInStruct
gwsnsg	WASN1_Strings
gwsnsi	WASN1_SeqIntsInStruct
gwsnsp	WASN1_Separation
gwsnst	WASN1_StructuredName
gzabcd	at_BORDER_WDbug
gzabce	at_BORDER_Copy
gzabcf	at_BORDER_CkVal
gzabcg	at_BORDER_Alloc
gzabch	at_BORDER_Free

Table I-51. Sorted by Real Name, XVII

Real Name	Alias
gzabci	at_BORDER_Set
gzabcj	at_BORDER_Get
gzabck	at_BORDER_Deflt
gzabcl	at_DEFAULT_VALUE_LISTS_WDbug
gzabcm	at_DEFAULT_VALUE_LISTS_Copy
gzabcn	at_DEFAULT_VALUE_LISTS_CkVal
gzabco	at_DEFAULT_VALUE_LISTS_Alloc
gzabcp	at_DEFAULT_VALUE_LISTS_Free
gzabcq	at_DEFAULT_VALUE_LISTS_Set
gzabcr	at_DEFAULT_VALUE_LISTS_Get
gzabcs	at_DEFAULT_VALUE_LISTS_Deflt
gzabct	at_DIMENSIONS_WDbug
gzabcu	at_DIMENSIONS_Copy
gzabcv	at_DIMENSIONS_CkVal
gzabcw	at_DIMENSIONS_Alloc
gzabcx	at_DIMENSIONS_Free
gzabcy	at_DIMENSIONS_Set
gzabcz	at_DIMENSIONS_Get
gzabda	at_DIMENSIONS_Deflt
gzabdb	at_MEDIUM_TYPE_WDbug
gzabdc	at_MEDIUM_TYPE_Copy
gzabdd	at_MEDIUM_TYPE_CkVal
gzabde	at_MEDIUM_TYPE_Alloc
gzabdf	at_MEDIUM_TYPE_Free
gzabdg	at_MEDIUM_TYPE_Set
gzabdh	at_MEDIUM_TYPE_Get
gzabdi	at_MEDIUM_TYPE_Deflt
gzabdj	at_OFFSET_WDbug
gzabdk	at_OFFSET_Copy
gzabdl	at_OFFSET_CkVal
gzabdm	at_OFFSET_Alloc
gzabdn	at_OFFSET_Free
gzabdo	at_OFFSET_Set
gzabdp	at_OFFSET_Get
gzabdq	at_OFFSET_Deflt
gzabdr	at_POSITION_WDbug

Table I-52. Sorted by Real Name, XVIII

Real Name	Alias
gzabds	at_POSITION_Copy
gzabdt	at_POSITION_CkVal
gzabdu	at_POSITION_Alloc
gzabdv	at_POSITION_Free
gzabdw	at_POSITION_Set
gzabdx	at_POSITION_Get
gzabdy	at_POSITION_Deflt
gzabdz	at_SAME_LAYOUT_OBJECT_WDbug
gzabea	at_SAME_LAYOUT_OBJECT_Copy
gzabeb	at_SAME_LAYOUT_OBJECT_CkVal
gzabec	at_SAME_LAYOUT_OBJECT_Alloc
gzabed	at_SAME_LAYOUT_OBJECT_Free
gzabee	at_SAME_LAYOUT_OBJECT_Set
gzabef	at_SAME_LAYOUT_OBJECT_Get
gzabeg	at_SAME_LAYOUT_OBJECT_Deflt
gzabeh	at_SEPARATION_WDbug
gzabei	at_SEPARATION_Copy
gzabej	at_SEPARATION_CkVal
gzabek	at_SEPARATION_Alloc
gzabel	at_SEPARATION_Free
gzabem	at_SEPARATION_Set
gzaben	at_SEPARATION_Get
gzabeo	at_SEPARATION_Deflt
gzabep	cc_ITEMISATION_WDbug
gzabeq	cc_ITEMISATION_Copy
gzaber	cc_ITEMISATION_CkVal
gzabes	cc_ITEMISATION_Alloc
gzabet	cc_ITEMISATION_Free
gzabeu	cc_ITEMISATION_Set
gzabev	cc_ITEMISATION_Get
gzabew	cc_ITEMISATION_Deflt
gzabex	CONSTITUENT_WDbug
gzabey	CONSTITUENT_Copy
gzabez	CONSTITUENT_CkVal
gzabfa	CONSTITUENT_Alloc
gzabfb	CONSTITUENT_Free

Table I-53. Sorted by Real Name, XIX

Real Name	Alias
gzabfc	CONSTITUENT_Set
gzabfd	CONSTITUENT_Get
gzabfe	CONSTITUENT_Deflt
gzabff	CONSTRUCTION_EXPR_WDbug
gzabfg	CONSTRUCTION_EXPR_Copy
gzabfh	CONSTRUCTION_EXPR_CkVal
gzabfi	CONSTRUCTION_EXPR_Alloc
gzabfj	CONSTRUCTION_EXPR_Free
gzabfk	CONSTRUCTION_EXPR_Set
gzabfl	CONSTRUCTION_EXPR_Get
gzabfm	CONSTRUCTION_EXPR_Deflt
gzabfn	dp_DOC_APP_PROF_DEFAULTS_WDbug
gzabfo	dp_DOC_APP_PROF_DEFAULTS_Copy
gzabfp	dp_DOC_APP_PROF_DEFAULTS_CkVal
gzabfq	dp_DOC_APP_PROF_DEFAULTS_Alloc
gzabfr	dp_DOC_APP_PROF_DEFAULTS_Free
gzabfs	dp_DOC_APP_PROF_DEFAULTS_Set
gzabft	dp_DOC_APP_PROF_DEFAULTS_Get
gzabfu	dp_DOC_APP_PROF_DEFAULTS_Deflt
gzabfv	dp_LOCAL_FILE_REFERENCE_WDbug
gzabfw	dp_LOCAL_FILE_REFERENCE_Copy
gzabfx	dp_LOCAL_FILE_REFERENCE_CkVal
gzabfy	dp_LOCAL_FILE_REFERENCE_Alloc
gzabfz	dp_LOCAL_FILE_REFERENCE_Free
gzabga	dp_LOCAL_FILE_REFERENCE_Set
gzabgb	dp_LOCAL_FILE_REFERENCE_Get
gzabgc	dp_LOCAL_FILE_REFERENCE_Deflt
gzabgd	dp_ODA_VERSION_WDbug
gzabge	dp_ODA_VERSION_Copy
gzabgf	dp_ODA_VERSION_CkVal
gzabgg	dp_ODA_VERSION_Alloc
gzabgh	dp_ODA_VERSION_Free
gzabgi	dp_ODA_VERSION_Set
gzabgj	dp_ODA_VERSION_Get
gzabgk	dp_ODA_VERSION_Deflt
gzabgl	gg_COLOUR_REPS_WDbug

Table I-54. Sorted by Real Name, XX

Real Name	Alias
gzabgm	gg_COLOUR_REPS_Copy
gzabgn	gg_COLOUR_REPS_CkVal
gzabgo	gg_COLOUR_REPS_Alloc
gzabgp	gg_COLOUR_REPS_Free
gzabgq	gg_COLOUR_REPS_Set
gzabgr	gg_COLOUR_REPS_Get
gzabgs	gg_COLOUR_REPS_Deflt
gzabgt	gg_EDGE_RENDITION_WDbug
gzabgu	gg_EDGE_RENDITION_Copy
gzabgv	gg_EDGE_RENDITION_CkVal
gzabgw	gg_EDGE_RENDITION_Alloc
gzabgx	gg_EDGE_RENDITION_Free
gzabgy	gg_EDGE_RENDITION_Set
gzabgz	gg_EDGE_RENDITION_Get
gzabha	gg_EDGE_RENDITION_Deflt
gzabhb	gg_ENCODING_ANNOUNCER_WDbug
gzabhc	gg_ENCODING_ANNOUNCER_Copy
gzabhd	gg_ENCODING_ANNOUNCER_CkVal
gzabhe	gg_ENCODING_ANNOUNCER_Alloc
gzabhf	gg_ENCODING_ANNOUNCER_Free
gzabhg	gg_ENCODING_ANNOUNCER_Set
gzabhh	gg_ENCODING_ANNOUNCER_Get
gzabhi	gg_ENCODING_ANNOUNCER_Deflt
gzabhj	gg_FILLED_AREA_REND_WDbug
gzabhk	gg_FILLED_AREA_REND_Copy
gzabhl	gg_FILLED_AREA_REND_CkVal
gzabhm	gg_FILLED_AREA_REND_Alloc
gzabhn	gg_FILLED_AREA_REND_Free
gzabho	gg_FILLED_AREA_REND_Set
gzabhp	gg_FILLED_AREA_REND_Get
gzabhq	gg_FILLED_AREA_REND_Deflt
gzabhr	gg_LINE_RENDITION_WDbug
gzabhs	gg_LINE_RENDITION_Copy
gzabht	gg_LINE_RENDITION_CkVal
gzabhu	gg_LINE_RENDITION_Alloc
gzabhv	gg_LINE_RENDITION_Free

Table I-55. Sorted by Real Name, XXI

Real Name	Alias
gzabhw	gg_LINE_RENDITION_Set
gzabhx	gg_LINE_RENDITION_Get
gzabhy	gg_LINE_RENDITION_Deflt
gzabhz	gg_MARKER_RENDITION_WDbug
gzabia	gg_MARKER_RENDITION_Copy
gzabib	gg_MARKER_RENDITION_CkVal
gzabic	gg_MARKER_RENDITION_Alloc
gzabid	gg_MARKER_RENDITION_Free
gzabie	gg_MARKER_RENDITION_Set
gzabif	gg_MARKER_RENDITION_Get
gzabig	gg_MARKER_RENDITION_Deflt
gzabih	gg_PICTURE_DIMENSIONS_WDbug
gzabii	gg_PICTURE_DIMENSIONS_Copy
gzabij	gg_PICTURE_DIMENSIONS_CkVal
gzabik	gg_PICTURE_DIMENSIONS_Alloc
gzabil	gg_PICTURE_DIMENSIONS_Free
gzabim	gg_PICTURE_DIMENSIONS_Set
gzabin	gg_PICTURE_DIMENSIONS_Get
gzabio	gg_PICTURE_DIMENSIONS_Deflt
gzabip	gg_REGION_INTEREST_SPEC_WDbug
gzabiq	gg_REGION_INTEREST_SPEC_Copy
gzabir	gg_REGION_INTEREST_SPEC_CkVal
gzabis	gg_REGION_INTEREST_SPEC_Alloc
gzabit	gg_REGION_INTEREST_SPEC_Free
gzabiu	gg_REGION_INTEREST_SPEC_Set
gzabiv	gg_REGION_INTEREST_SPEC_Get
gzabiw	gg_REGION_INTEREST_SPEC_Deflt
gzabix	gg_TEXT_RENDITION_WDbug
gzabiy	gg_TEXT_RENDITION_Copy
gzabiz	gg_TEXT_RENDITION_CkVal
gzabja	gg_TEXT_RENDITION_Alloc
gzabjb	gg_TEXT_RENDITION_Free
gzabjc	gg_TEXT_RENDITION_Set
gzabjd	gg_TEXT_RENDITION_Get
gzabje	gg_TEXT_RENDITION_Deflt
gzabjf	gg_TRANSFORMATION_SPEC_WDbug

Table I-56. Sorted by Real Name, XXII

Real Name	Alias
gzabjg	gg_TRANSFORMATION_SPEC_Copy
gzabjh	gg_TRANSFORMATION_SPEC_CkVal
gzabji	gg_TRANSFORMATION_SPEC_Alloc
gzabjj	gg_TRANSFORMATION_SPEC_Free
gzabjk	gg_TRANSFORMATION_SPEC_Set
gzabjl	gg_TRANSFORMATION_SPEC_Get
gzabjm	gg_TRANSFORMATION_SPEC_Deflt
gzabjn	gg_TRANSPARENCY_SPEC_WDbug
gzabjo	gg_TRANSPARENCY_SPEC_Copy
gzabjp	gg_TRANSPARENCY_SPEC_CkVal
gzabjq	gg_TRANSPARENCY_SPEC_Alloc
gzabjr	gg_TRANSPARENCY_SPEC_Free
gzabjs	gg_TRANSPARENCY_SPEC_Set
gzabjt	gg_TRANSPARENCY_SPEC_Get
gzabju	gg_TRANSPARENCY_SPEC_Deflt
gzabjv	INT_WDbug
gzabjw	INT_Copy
gzabjx	INT_CkVal
gzabjy	INT_Alloc
gzabjz	INT_Free
gzabka	INT_Set
gzabkb	INT_Get
gzabkc	INT_Deflt
gzabkd	INT_PAIR_WDbug
gzabke	INT_PAIR_Copy
gzabkf	INT_PAIR_CkVal
gzabkg	INT_PAIR_Alloc
gzabkh	INT_PAIR_Free
gzabki	INT_PAIR_Set
gzabkj	INT_PAIR_Get
gzabkk	INT_PAIR_Deflt
gzabkl	LAYOUT_OBJECT_WDbug
gzabkm	LAYOUT_OBJECT_Copy
gzabkn	LAYOUT_OBJECT_CkVal
gzabko	LAYOUT_OBJECT_Alloc
gzabkp	LAYOUT_OBJECT_Free

Table I-57. Sorted by Real Name, XXIII

Real Name	Alias
gzabkq	LAYOUT_OBJECT_Set
gzabkr	LAYOUT_OBJECT_Get
gzabks	LAYOUT_OBJECT_Deflt
gzabkt	rg_CLIPPING_WDbug
gzabku	rg_CLIPPING_Copy
gzabkv	rg_CLIPPING_CkVal
gzabkw	rg_CLIPPING_Alloc
gzabkx	rg_CLIPPING_Free
gzabky	rg_CLIPPING_Set
gzabkz	rg_CLIPPING_Get
gzabla	rg_CLIPPING_Deflt
gzablb	rg_IMAGE_DIMENSIONS_WDbug
gzablc	rg_IMAGE_DIMENSIONS_Copy
gzabld	rg_IMAGE_DIMENSIONS_CkVal
gzable	rg_IMAGE_DIMENSIONS_Alloc
gzablf	rg_IMAGE_DIMENSIONS_Free
gzablg	rg_IMAGE_DIMENSIONS_Set
gzablh	rg_IMAGE_DIMENSIONS_Get
gzabli	rg_IMAGE_DIMENSIONS_Deflt
gzablj	rg_PEL_SPACING_WDbug
gzablk	rg_PEL_SPACING_Copy
gzabll	rg_PEL_SPACING_CkVal
gzablm	rg_PEL_SPACING_Alloc
gzabln	rg_PEL_SPACING_Free
gzablo	rg_PEL_SPACING_Set
gzablp	rg_PEL_SPACING_Get
gzablq	rg_PEL_SPACING_Deflt
gzablr	SEQUENCE_ASN1_OBJECT_ID_WDbug
gzabls	SEQUENCE_ASN1_OBJECT_ID_Copy
gzablt	SEQUENCE_ASN1_OBJECT_ID_CkVal
gzablu	SEQUENCE_ASN1_OBJECT_ID_Alloc
gzablv	SEQUENCE_ASN1_OBJECT_ID_Free
gzablw	SEQUENCE_ASN1_OBJECT_ID_Set
gzablx	SEQUENCE_ASN1_OBJECT_ID_Get
gzably	SEQUENCE_ASN1_OBJECT_ID_Deflt
gzablz	SEQUENCE_at_MEDIUM_TYPE_WDbug

Table I-58. Sorted by Real Name, XXIV

Real Name	Alias
gzabma	SEQUENCE_at_MEDIUM_TYPE_Copy
gzabmb	SEQUENCE_at_MEDIUM_TYPE_CkVal
gzabmc	SEQUENCE_at_MEDIUM_TYPE_Alloc
gzabmd	SEQUENCE_at_MEDIUM_TYPE_Free
gzabme	SEQUENCE_at_MEDIUM_TYPE_Set
gzabmf	SEQUENCE_at_MEDIUM_TYPE_Get
gzabmg	SEQUENCE_at_MEDIUM_TYPE_Deflt
gzabmh	SEQUENCE_BINDING_PAIR_WDbug
gzabmi	SEQUENCE_BINDING_PAIR_Copy
gzabmj	SEQUENCE_BINDING_PAIR_CkVal
gzabmk	SEQUENCE_BINDING_PAIR_Alloc
gzabml	SEQUENCE_BINDING_PAIR_Free
gzabmm	SEQUENCE_BINDING_PAIR_Set
gzabmn	SEQUENCE_BINDING_PAIR_Get
gzabmo	SEQUENCE_BINDING_PAIR_Deflt
gzabmp	SEQUENCE_BYTE_WDbug
gzabmq	SEQUENCE_BYTE_Copy
gzabmr	SEQUENCE_BYTE_CkVal
gzabms	SEQUENCE_BYTE_Alloc
gzabmt	SEQUENCE_BYTE_Free
gzabmu	SEQUENCE_BYTE_Set
gzabmv	SEQUENCE_BYTE_Get
gzabmw	SEQUENCE_BYTE_Deflt
gzabmx	SEQUENCE_CONSTITUENT_WDbug
gzabmy	SEQUENCE_CONSTITUENT_Copy
gzabmz	SEQUENCE_CONSTITUENT_CkVal
gzabna	SEQUENCE_CONSTITUENT_Alloc
gzabnb	SEQUENCE_CONSTITUENT_Free
gzabnc	SEQUENCE_CONSTITUENT_Set
gzabnd	SEQUENCE_CONSTITUENT_Get
gzabne	SEQUENCE_CONSTITUENT_Deflt
gzabnf	SEQUENCE_COPYRIGHT_INFO_WDbug
gzabng	SEQUENCE_COPYRIGHT_INFO_Copy
gzabnh	SEQUENCE_COPYRIGHT_INFO_CkVal
gzabni	SEQUENCE_COPYRIGHT_INFO_Alloc
gzabnj	SEQUENCE_COPYRIGHT_INFO_Free

Table I-59. Sorted by Real Name, XXV

Real Name	Alias
gzabnk	SEQUENCE_COPYRIGHT_INFO_Set
gzabnl	SEQUENCE_COPYRIGHT_INFO_Get
gzabnm	SEQUENCE_COPYRIGHT_INFO_Deflt
gzabnn	SEQUENCE_FONT_PAIR_WDbug
gzabno	SEQUENCE_FONT_PAIR_Copy
gzabnp	SEQUENCE_FONT_PAIR_CkVal
gzabnq	SEQUENCE_FONT_PAIR_Alloc
gzabnr	SEQUENCE_FONT_PAIR_Free
gzabns	SEQUENCE_FONT_PAIR_Set
gzabnt	SEQUENCE_FONT_PAIR_Get
gzabnu	SEQUENCE_FONT_PAIR_Deflt
gzabnv	SEQUENCE_INT_WDbug
gzabnw	SEQUENCE_INT_Copy
gzabnx	SEQUENCE_INT_CkVal
gzabny	SEQUENCE_INT_Alloc
gzabnz	SEQUENCE_INT_Free
gzaboa	SEQUENCE_INT_Set
gzabob	SEQUENCE_INT_Get
gzaboc	SEQUENCE_INT_Deflt
gzabod	SEQUENCE_INT_PAIR_WDbug
gzaboe	SEQUENCE_INT_PAIR_Copy
gzabof	SEQUENCE_INT_PAIR_CkVal
gzabog	SEQUENCE_INT_PAIR_Alloc
gzaboh	SEQUENCE_INT_PAIR_Free
gzaboi	SEQUENCE_INT_PAIR_Set
gzaboj	SEQUENCE_INT_PAIR_Get
gzabok	SEQUENCE_INT_PAIR_Deflt
gzabol	SEQUENCE_NAME_ORG_WDbug
gzabom	SEQUENCE_NAME_ORG_Copy
gzabon	SEQUENCE_NAME_ORG_CkVal
gzaboo	SEQUENCE_NAME_ORG_Alloc
gzabop	SEQUENCE_NAME_ORG_Free
gzaboq	SEQUENCE_NAME_ORG_Set
gzabor	SEQUENCE_NAME_ORG_Get
gzabos	SEQUENCE_NAME_ORG_Deflt
gzabot	SEQUENCE_RESOURCE_MAP_WDbug

Table I-60. Sorted by Real Name, XXVI

Real Name	Alias
gzabou	SEQUENCE_RESOURCE_MAP_Copy
gzabov	SEQUENCE_RESOURCE_MAP_CkVal
gzabow	SEQUENCE_RESOURCE_MAP_Alloc
gzabox	SEQUENCE_RESOURCE_MAP_Free
gzaboy	SEQUENCE_RESOURCE_MAP_Set
gzaboz	SEQUENCE_RESOURCE_MAP_Get
gzabpa	SEQUENCE_RESOURCE_MAP_Deflt
gzabpb	SEQUENCE_REVISION_HISTORY_WDbug
gzabpc	SEQUENCE_REVISION_HISTORY_Copy
gzabpd	SEQUENCE_REVISION_HISTORY_CkVal
gzabpe	SEQUENCE_REVISION_HISTORY_Alloc
gzabpf	SEQUENCE_REVISION_HISTORY_Free
gzabpg	SEQUENCE_REVISION_HISTORY_Set
gzabph	SEQUENCE_REVISION_HISTORY_Get
gzabpi	SEQUENCE_REVISION_HISTORY_Deflt
gzabpj	SEQUENCE_STRING_WDbug
gzabpk	SEQUENCE_STRING_Copy
gzabpl	SEQUENCE_STRING_CkVal
gzabpm	SEQUENCE_STRING_Alloc
gzabpn	SEQUENCE_STRING_Free
gzabpo	SEQUENCE_STRING_Set
gzabpp	SEQUENCE_STRING_Get
gzabpq	SEQUENCE_STRING_Deflt
gzabpr	SEQUENCE_STRING_OR_ID_WDbug
gzabps	SEQUENCE_STRING_OR_ID_Copy
gzabpt	SEQUENCE_STRING_OR_ID_CkVal
gzabpu	SEQUENCE_STRING_OR_ID_Alloc
gzabpv	SEQUENCE_STRING_OR_ID_Free
gzabpw	SEQUENCE_STRING_OR_ID_Set
gzabpx	SEQUENCE_STRING_OR_ID_Get
gzabpy	SEQUENCE_STRING_OR_ID_Deflt
gzabpz	SEQUENCE_TAB_STOP_WDbug
gzabqa	SEQUENCE_TAB_STOP_Copy
gzabqb	SEQUENCE_TAB_STOP_CkVal
gzabqc	SEQUENCE_TAB_STOP_Alloc
gzabqd	SEQUENCE_TAB_STOP_Free

Table I-61. Sorted by Real Name, XXVII

Real Name	Alias
gzabqe	SEQUENCE_TAB_STOP_Set
gzabqf	SEQUENCE_TAB_STOP_Get
gzabqg	SEQUENCE_TAB_STOP_Deflt
gzabqh	SEQUENCE_VALUE_WDbug
gzabqi	SEQUENCE_VALUE_Copy
gzabqj	SEQUENCE_VALUE_CkVal
gzabqk	SEQUENCE_VALUE_Alloc
gzabql	SEQUENCE_VALUE_Free
gzabqm	SEQUENCE_VALUE_Set
gzabqn	SEQUENCE_VALUE_Get
gzabqo	SEQUENCE_VALUE_Deflt
gzabqp	STRING_WDbug
gzabqq	STRING_Copy
gzabqr	STRING_CkVal
gzabqs	STRING_Alloc
gzabqt	STRING_Free
gzabqu	STRING_Set
gzabqv	STRING_Get
gzabqw	STRING_Deflt
gzabqx	STRING_EXPR_WDbug
gzabqy	STRING_EXPR_Copy
gzabqz	STRING_EXPR_CkVal
gzabra	STRING_EXPR_Alloc
gzabrb	STRING_EXPR_Free
gzabrc	STRING_EXPR_Set
gzabrd	STRING_EXPR_Get
gzabre	STRING_EXPR_Deflt
gzabrf	STRING_OR_ID_WDbug
gzabrg	STRING_OR_ID_Copy
gzabrh	STRING_OR_ID_CkVal
gzabri	STRING_OR_ID_Alloc
gzabrj	STRING_OR_ID_Free
gzabrk	STRING_OR_ID_Set
gzabrl	STRING_OR_ID_Get
gzabrm	STRING_OR_ID_Deflt
gzabrn	VALUE_WDbug

Table I-62. Sorted by Real Name, XXVIII

Real Name	Alias
gzabro	VALUE_Copy
gzabrp	VALUE_CkVal
gzabrq	VALUE_Alloc
gzabrr	VALUE_Free
gzabrs	VALUE_Set
gzabrt	VALUE_Get
gzabru	VALUE_Deflt
hasatt	HasAttr
hldatt	HoldAttr
hldcns	HoldConstituent
hldstr	HoldString
int2ds	Ints2Dsclass
intdvl	InitializeDefaultValueLists
iopbyt	IO_PrintByte
iopchr	IO_PrintChar
iopint	IO_PrintInt
iopstr	IO_PrintString
iorbts	IO_ReadBytes
iorbyt	IO_ReadByte
iorchr	IO_ReadChar
iorcrs	IO_ReadChars
iowbts	IO_WriteBytes
iowbyt	IO_WriteByte
iowchr	IO_WriteChar
iowcrs	IO_WriteChars
isprdf	isode_PrintODIF
isred	isode_ReadODIF
iswrt	isode_WriteODIF
lbldoc	LabelDocument
lblstr	LabelStructure
lblsty	LabelStyles
lkobid	LookUpObjectId
mkcomp	MakeComponent
mkcont	MakeContent
mkdoc	MakeDocument
mkdocp	MakeDocumentProfile

Table I-63. Sorted by Real Name, XXIX

Real Name	Alias
mksbgn	MakeSubgraphIterator
mkseq	MakeSequence
mkstgn	MakeStylesIterator
mkstr	MakeString
mkstyl	MakeStyle
nxtcns	NextConstituent
rdodif	ReadODIF
rematt	RemAttr
rganrd	ANDY_ReadRaster
rganwr	ANDY_WriteRaster
rgentr	Entering
rgetgn	ETC_GetNumber
rgetpn	ETC_PutNumber
rgetrd	ETC_ReadRaster
rgfaal	alloc_fbm
rgfafr	free_fbm
rgfbbo	machine_byte_order
rgfbgl	get_long
rgfbgs	get_short
rgfbpf	fbmprintf
rgfbpl	put_long
rgfbps	put_short
rgfbrd	FBM_ReadRaster
rgfbre	fbmread
rgfbwi	fbmwrite
rgfbwr	FBM_WriteRaster
rgfgcm	compress
rgfgrd	read_gif
rgfgwr	write_gif
rgfird	read_iff
rgfitk	FinalRGToolKit
rgfiwr	write_iff
rgfsrd	read_sun
rgfswr	write_sun
rgg3do	G3_DecodeOne
rgg3eo	G3_EncodeOne

Table I-64. Sorted by Real Name, XXX

Real Name	Alias
rgg3gb	G3_GetBit
rgg3nr	G3_NextRun
rgg3pb	G3_PutBit
rgg3pe	G3_PutEOL
rgg3pr	G3_PutRunLength
rgg3r2	G3_ReadBitmap2
rgg3rd	G3_ReadRaster
rgg3tr	G3_Traverse
rgg3w2	G3_WriteBitmap2
rgg3wr	G3_WriteRaster
rgg4gb	G4_GetBit
rgg4nr	G4_NextRun
rgg4rd	G4_ReadRaster
rgg4tr	G4_Traverse
rgg4wb	G4_WriteBitmap
rgg4wr	G4_WriteRaster
rggfen	GIFEncode
rggfnp	GIFNextPixel
rgintk	InitRGToolKit
rgleav	Leaving
rglsad	LST_AddElement
rglsde	LST_DeleteElement
rglsfl	LST_FreeList
rglsmh	LST_MoveHead
rglsml	LST_MakeList
rglsne	LST_NextElement
rgmcrd	MACP_ReadRaster
rgmcwr	MACP_WriteRaster
rgmmfr	MEM_Free
rgmmma	MEM_Malloc
rgmmre	MEM_Realloc
rgobrd	OBM_ReadRaster
rgobwr	OBM_WriteRaster
rgpbrd	PBM_ReadRaster
rgpbwr	PBM_WriteRaster
rgpeab	AssignBit

Table I-65. Sorted by Real Name, XXXI

Real Name	Alias
rgpeap	AssignPel
rgpegb	GetBit
rgpegp	GetPel
rgpswr	PS_WriteRaster
rgraal	RAST_AllocRaster
rgradr	DeleteRaster
rgraex	ExtractRaster
rgrafi	RAST_FinalRasterList
rgraho	HoldRaster
rgrain	RAST_InitRasterList
rgrais	InsertRaster
rgramr	MakeRaster
rgrapf	RasterPelFormat
rgrare	ReleaseRaster
rgrarh	RasterHeight
rgrarw	RasterWidth
rgrasa	RAST_SwapAlloc
rgrasi	RAST_SwapIn
rgrast	RasterStatus
rgrird	ReadRaster
rgrirm	ReadMemRaster
rgriwm	WriteMemRaster
rgriwr	WriteRaster
rgvbfl	PEL_BitOffL
rgvbfr	PEL_BitOffR
rgvblm	bl_make
rgvblt	bl_term
rgvbnl	PEL_BitOnL
rgvbnr	PEL_BitOnR
rgvbtt	BlackTreeTop
rgvemb	ETC_MagicBuffer
rgvems	ETC_MagicSize
rgverr	RGError
rgvfgb	fbm_GetByte
rgvfpa	fbm_Parameter
rgvfpb	fbm_PutByte

Table I-66. Sorted by Real Name, XXXII

Real Name	Alias
rgvfrm	RIO_FromMemory
rgvmal	MEM_Alloced
rgvrgp	INIT_RGParms
rgvwhm	wt_make
rgvwht	wt_term
rgvwtt	WhiteTreeTop
rgxrd	X_ReadRaster
rgxwr	X_WriteRaster
rlsatt	ReleaseAttr
rlscns	ReleaseConstituent
rlsstr	ReleaseString
setatt	SetAttr
setlat	SetLayoutAttr
setrat	SetPresAttr
sinatt	SetIntAttr
stffar	SetFontFamilyArray
stffst	SetFontFamily
str2sn	String2StrctName
ststat	SetStringAttr
stypri	StyleAttrPriority
svlatt	SetValueAttr
tkfinl	FinalToolKit
tkinit	InitToolKit
tocbit	at_TYPE_OF_CODING_bitmap
toccc	at_TYPE_OF_CODING_cc
tocgg	at_TYPE_OF_CODING_gg
toct41	at_TYPE_OF_CODING_T4_1
toct42	at_TYPE_OF_CODING_T4_2
toct6	at_TYPE_OF_CODING_T6
unsdvl	UnspecifyDefaultValueLists
unsfnt	UnspecifyFont
unsprm	UnspecParms
wrodif	WriteODIF
wsneoc	WASN1_EOC
wsnint	WASN1_Integer
wsnlen	WASN1_Length
wsnvlt	WASN1_ValueType

Table I-67. Sorted by Real Name, XXXIII

J
Source Code Organization

This appendix includes information on the internal structure of the CMU ODA Tool Kit. As such, it is not of interest to the ODA applications programmer. However, we include it for those interested in the internal structure of the tool kit. The first section discusses how the tool kit is modularized, and the second section details how include files are organized.

J.1. CMU ODA Tool Kit Modules

Following good software engineering practice, we modularized the implementation of the tool kit. We also expected several programmers to be working simultaneously, and needed to control interactions among them. In this section, we present information on the module structure of the tool kit.

The tool kit contains 25 modules, whose sizes vary greatly. The largest module is the attribute module, which is large because it must be capable of dealing with all the different attribute types found in ODA to perform the copying required by our semantics. In addition, it also includes support for the ODA defaulting semantics. The attribute module includes a total of 54 files and about 38,000 lines of code. An example of a small module is the string module, which comprises only three files and 423 lines of code. Table J-1 lists all of the tool kit modules and their functions.

J.2. The Structure of Include Files

We had several objectives in mind when we devised the elaborate structure of our include files. We were attempting to enforce inter-module hiding. Also, we needed to permit the source code of a module to span more than one source file. In addition, we wanted a programmer to have only to include a single include file to obtain all appropriate definitions. Lastly, given our goal of portability,

Module name	Function
attr	Attribute routines
atts	Attribute support routines
const	Constituent routines
debug	Debug routines
doc	Document routines
expr	Expression routines
fmt	Formatting support
font	Font support
init	Initialization and finalization
io	Input/Output support
iter	Iterator routines
label	Labeling routines
mem	Memory support
mm	Memory management
prior	Run time routines
rasn1	Manually-created Read ASN.1
rodif	Manually-created Read ODIF
seq	Sequence routines
str	String routines
stubs	Stubs for attribute types
tkerr	Error support
wasn1	Manually-created Write ASN.1
wodif	Manually-created Write ODIF

Table J-1. CMU ODA Tool Kit Modules

we needed to meet the requirements of systems with non-permissive linkers, which require that variables and procedures be declared exactly once, and that all other declarations use the C storage class specifier `extern`. Examples J-1 and J-2 show two include files, ***str.h*** and ***str1.h***. The reader is urged to refer to these throughout the discussion that follows.

We needed to deal with the problem of multiple inclusions of the same .h files. There are two possible solutions. The first prevents the problem by including the .h files in an order carefully chosen to prevent multiple inclusions. Given that we expected the tool kit to require hundreds of source files, we chose not to do this. The housekeeping required is tedious and unmanagable for a system

```
/*

(c) Copyright 1988 Carnegie Mellon University

See "The CMU ODA Tool Kit" (available as CMU/ITC tech
report CMU-ITC-071) for full text & permitted uses.

*/

#ifndef TK_INCLUDED_STR
#define TK_INCLUDED_STR

#ifdef TK_TKI_STR
#define TK_STR
#else
#ifdef TK_IMPL_STR
#define TK_STR
#else
#define AP_STR
#endif
#endif

#include <site.h>
#include <apityp.h>
#include <tags.h>
#include <seq.h>
#include <errors.h>
#include <apirt.h>
#include <tkerr.h>

/*
    Put the stuff seen across the user interface here
    as given in the API.
*/
#ifdef  TK_STR
#endif

/*
   put module internal #defines, macros, typedefs after the
   #ifdef TK_IMPL_<filename>.

*/
#ifdef TK_IMPL_STR
#endif

/*
 the STRING's defined in the API. Note that storage for the
 ERROR_STRING is allocated in STR_Init.
*/

/*
    Include(s) for the subordinate .h file(s)
*/
#include <str1.h>

#endif
```

Example J-1: The Include File ***str.h***

```
/*

(c) Copyright 1988 Carnegie Mellon University

See "The CMU ODA Tool Kit" (available as CMU/ITC tech
report CMU-ITC-071) for full text & permitted uses.
*/

#ifndef TK_INCLUDED_STR1
#define TK_INCLUDED_STR1

#undef EXTERN
#ifdef TK_GVARS_STR1
#define EXTERN
#else
#define EXTERN extern
#endif

/*
    Declarations seen by everyone go here.
*/
EXTERN STRING_type  Chars2String();
EXTERN STRING_type  ConcatString();
EXTERN INT_type     DeleteString();
EXTERN STRING_type  MakeString();

    /* Note that ERROR_STRING must be the same
        as ERROR_SEQUENCE */
#define ERROR_STRING
    ((STRING_type) ERROR_SEQUENCE)

    /* Note that NULL_STRING must be the same
        as NULL_SEQUENCE */
#define NULL_STRING
    ((STRING_type) NULL_SEQUENCE)

/*
   Declarations seen only inside the tool kit go here
*/
#ifdef TK_STR

#ifndef USE_LONG_NAMES
#define STR_Init        gstint
#define STR_Final       gstfnl
#endif

EXTERN INT_type STR_Final();
EXTERN INT_type STR_Init();
#endif

/*
    Declarations seen within the STR module go here.
*/
#ifdef TK_IMPL_STR
#endif
#endif
```

Example J-2: The Include File ***str1.h***

this large. The second possibility is to have the .h files themselves prevent multiple inclusions—we chose this approach. Therefore, all tool kit .h files start with a test to see if the file has been included previously. This test is made on a TK_INCLUDED_<name> macro where the <name> is that of the .h file (without the .h suffix). The body of the .h file always includes a definition of this macro to prevent subsequent inclusions.

Given our modularity goals for the tool kit, we wanted to provide a means to selectively hide the information in .h files. We have a model of storage classes that accomplishes this. In this model, the universe of data accessors is divided into three nested classes for each module. The outermost class includes application programs and all of the tool kit modules. Examples of information in this class include the `ERROR_CONSTITUENT` and `NULL_CONSTITUENT` identifiers. The middle ring in the storage hierarchy for a module is visible only to other tool kit modules. This ring is the interface layer that permits routines in one module to call internally defined routines in another (internally defined routines are not visible to the application programmer). The innermost ring is restricted to the implementer of the module. This level permits a module to hide its own data.

Corresponding to the inner two levels are switches that must be defined by the including .c file. These switches are given the names TK_TKI_<module_name> and TK_IMPL_<module_name>, where TK stands for Tool Kit, TKI stands for Tool Kit Interface, and IMPL stands for IMPLementor. Examples for the string module are TK_TKI_STR and TK_IMPL_STR. In example J-3, you can see how these swithces are used in the .h files to accomplish information hiding.

To allow modules to span more than one source file, we partition the module's include file information. There is always one master .h file, named <module_name>.h, which includes macro and type definitions. Corresponding to each source file there is a numbered .h file, where the numbered file name is the <module_name> with a number appended (such as ***str1.h*** whose source file is ***str1.c***). The slave .h files include declarations for variables and procedures only. Each slave .h file is associated with a given source file, and, thus contains the procedure and variable declarations associated with the source file. The slave .h files are included at the bottom of the master .h file. Therefore, the source file need only include the master .h file.

To accommodate non-permissive linkers, we needed to have one additional switch, the TK_GVARS_<numbered_file_name> switch, where the numbered file name is the <module_name> with an appended number. The GVARS switch is associated with individual slave .h files (and, hence, a .c source file). For example, the GVARS switch for ***str1.h*** is named TK_GVARS_STR1. The GVARS switch controls the value given to the EXTERN macro. The corresponding source file defines its GVARS switch before including the module's master .h file. Since the GVARS switch is defined only by the corresponding source file, EXTERN will be defined as blanks only once, and as `extern` everywhere else. To illustrate how a .c source file defines these switches, consider example J-3, which is a part of the source file, ***str1.c***. Because this file is part of the string module, it defines the string module switch, TK_IMPL_STR. It also defines its own GVARS switch, TK_GVARS_STR1. In addition, as it makes use of the interfaces of other modules, it defines TK_TKI switches for the memory management module (TK_TKI_MM), and the sequence module (TK_TKI_SEQ).

```
#define TK_IMPL_STR
#define TK_GVARS_STR1
#define TK_TKI_MM
#define TK_TKI_SEQ

#include <mm.h>
#include <seq.h>
#include <str.h>

#undef TK_IMPL_STR
#undef TK_GVARS_STR1
#undef TK_TKI_MM
#undef TK_TKI_SEQ
```

Example J-3: The Include Part of ***str1.c***

K
Ordering Information

The source code for the CMU ODA Tool Kit and the Raster Graphics Tool Kit, along with the validation suite, sample translators, programs and installation instructions may be obtained from two sources.

The software is available as part of the Andrew Toolkit contribution to the X Distribution. The X Distribution is a collection of software implementing the client server and applications of the X Window System for a variety of platforms. This distribution can be obtained for fee, from

MIT Software Distribution Center
Technology Licensing Office
Room E32-300
77 Massachusetts Avenue
Cambridge, MA 02139
USA
+1-617-258-8330

In addition, the software (as described above) and upgrades are available from:

Eureka Software
940 Greenfield Avenue
Pittsburgh, PA 15217
USA
+1-412-521-6303 (voice and FAX)
E-mail: Eureka.Software@cup.portal.com
...!sun!portal!cup.portal.com!Eureka.Software

Index